Pleistocene Environments in the British Isles

Pleistocene Environments in the British Isles

R. L. Jones
and
D. H. Keen
Senior Lecturers in Geography, Coventry University, Coventry, UK

CHAPMAN & HALL
London · Glasgow · New York · Tokyo · Melbourne · Madras

Published by Chapman & Hall, 2–6 Boundary Row, London SE1 8HN

Chapman & Hall, 2–6 Boundary Row, London SE1 8HN, UK

Blackie Academic & Professional, Wester Cleddens Road, Bishopbriggs, Glasgow G64 2NZ, UK

Chapman & Hall, 29 West 35th Street, New York NY10001, USA

Chapman & Hall Japan, Thomson Publishing Japan, Hirakawacho Nemoto Building, 6F, 1-7-11 Hirakawa-cho, Chiyoda-ku, Tokyo 102, Japan

Chapman & Hall Australia, Thomas Nelson Australia, 102 Dodds Street, South Melbourne, Victoria 3205, Australia

Chapman & Hall India, R. Seshadri, 32 Second Main Road, CIT East, Madras 600 035, India

---

First edition 1993

Typeset in 10/12 Times by Excel Typesetters Company, Hong Kong
Printed in Great Britain by Clays, Bungay

ISBN 0 412 44190 X

A catalogue record for this book is available from the British Library

Library of Congress Cataloging-in-Publication data available

Contents

List of Figures x
List of Tables xiii
Acknowledgements xiv
Preface xv

1 Background **1**
1.1 Basic concepts 1
1.2 The nature of the evidence 5
1.3 Types of evidence 5
1.3.1 Sediments 6
1.3.1.1 Sedimentation in cold climates 6
1.3.1.2 Sedimentation in temperate climates 10
1.3.2 Soils 12
1.3.3 Geomorphology 14
1.3.4 Biological evidence 15
1.3.4.1 Pollen and spores 16
1.3.4.2 Diatoms 18
1.3.4.3 Plant macrofossils 18
1.3.4.4 Vertebrates 18
1.3.4.5 Mollusca 19
1.3.4.6 Insects 21
1.3.4.7 Ostracoda 22
1.3.4.8 Foraminifera 22
1.4 Dating 23
1.4.1 Geochronometric age determinations 23
1.4.1.1 Laminations 23
1.4.1.2 Tree rings 23
1.4.1.3 Lichens 24
1.4.1.4 Radiocarbon 24
1.4.1.5 Thermoluminescence 24
1.4.1.6 Uranium series 24
1.4.1.7 Potassium–argon 25
1.4.2 Relative age determination 25
1.4.2.1 Amino-acid diagenesis 25
1.4.2.2 Tephrochronology 26
1.4.2.3 Palaeomagnetism 26
1.4.2.4 Flora and fauna 27

1.4.2.5	Artifacts	28
1.5	The nature of the evidence and the means of division of the Pleistocene of the British Isles	28
1.6	Conclusion	30
2	**Prelude to the Pleistocene**	**31**
2.1	Miocene environments	31
2.2	Pliocene environments	33
2.3	The Pliocene–Pleistocene boundary	38
2.4	Summary	39
3	**The Lower Pleistocene before the Pre-Pastonian**	**40**
3.1	Problems of evaluation and correlation of Lower Pleistocene deposits	40
3.2	The stages and their environments	42
3.2.1	The Pre-Ludhamian	43
3.2.2	The Ludhamian Temperate Stage	44
3.2.3	The Thurnian Cold Stage	46
3.2.4	The Antian Temperate Stage	46
3.2.5	The Baventian Cold Stage	47
3.2.6	The Bramertonian Temperate Stage	49
3.3	Gravels and Lower Pleistocene environments	52
3.4	The offshore sequence	54
3.5	Correlation and conclusion	54
4	**The Cromer Forest-bed Formation and associated deposits**	**55**
4.1	The stages and their environments	55
4.1.1	The Pre-Pastonian Cold Stage	55
4.1.2	The Pastonian Temperate Stage	58
4.1.3	The Beestonian Cold Stage	59
4.1.4	The Cromerian Temperate Stage	63
4.1.4.1	Flora and vegetation	64
4.1.4.2	Non-marine Mollusca	65
4.1.4.3	Marine Mollusca	66
4.1.4.4	Vertebrates	66
4.1.4.5	Other fossils	68
4.1.4.6	The presence of man	68
4.1.4.7	Pedogenesis	69
4.1.4.8	The offshore sequence and sea-levels	70
4.2	Summary and conclusions	70
5	**The Anglian Cold Stage**	**73**
5.1	East Anglia	73
5.2	The Vale of St. Albans and the Thames Valley	77
5.3	The Midlands of England	79
5.4	Wales, and England south of the Lower Severn and Thames valleys	80
5.5	Northern England and Scotland	81
5.6	Ireland	82

5.7	The offshore sequence	82
5.8	Correlation, age and archaeology	82
6	**The Hoxnian Temperate Stage**	**84**
6.1	Flora and vegetation	84
6.2	Non-marine Mollusca	89
6.3	Marine Mollusca	91
6.4	Vertebrates	91
6.5	Other fossils	92
6.6	The presence of man	93
6.7	Pedogenesis	95
6.8	The offshore sequence and sea-levels	95
6.9	Correlation, age and duration	96
6.10	Conclusion	97
7	**Post-Hoxnian and pre-Ipswichian events**	**98**
7.1	The Wolstonian Cold Stage	98
7.1.1	The Midlands of England	98
7.1.2	Lincolnshire, East Anglia and the Thames Valley	102
7.1.3	Northern England and Scotland	105
7.1.4	Wales, England south of the Lower Severn and Thames valleys, and Jersey	106
7.1.5	Ireland and the Isle of Man	107
7.1.6	The offshore sequence and sea-levels	108
7.2	Additional temperate episodes and intervening cold intervals	109
7.2.1	Re-evaluated sites	109
7.2.2	New sites	115
7.3	Towards a revised succession	119
7.4	Correlation with mainland Europe	121
7.5	Conclusion	122
8	**The Ipswichian Temperate Stage**	**123**
8.1	Flora and vegetation	123
8.2	Non-marine Mollusca	130
8.3	Marine Mollusca	132
8.4	Vertebrates	132
8.5	Insects	135
8.6	Other fossils	135
8.7	The presence of man	136
8.8	Pedogenesis	136
8.9	Rivers and sea-levels	137
8.10	The offshore sequence	140
8.11	Correlation, age and duration	140
8.12	Conclusion	142
9	**The Devensian Cold Stage**	**143**
9.1	The Early Devensian	145
9.1.1	The Cheshire–Lancashire lowlands	145

9.1.2	The Midlands of England	147
9.1.3	Yorkshire, Lincolnshire, the Peak District, northern Dukeries and East Anglia	148
9.1.4	The Thames Valley	152
9.1.5	Wales	152
9.1.6	England south of the Lower Severn and Thames valleys	153
9.1.7	Jersey	155
9.1.8	Scotland	156
9.1.9	Ireland and the Isle of Man	156
9.1.10	The offshore sequence and sea-levels	159
9.1.11	Age and correlation	159
9.2	The Middle Devensian	162
9.2.1	The Midlands of England	162
9.2.2	Yorkshire, Lincolnshire, the northern Dukeries and East Anglia	165
9.2.3	The Thames Valley	167
9.2.4	Wales, and England south of the Lower Severn and Thames valleys	168
9.2.5	Jersey	169
9.2.6	Scotland	169
9.2.7	Ireland	170
9.2.8	The offshore sequence and sea-levels	170
9.2.9	Age and correlation	171
9.3	The Late Devensian	171
9.3.1	The Midlands of England, Cheshire–Lancashire lowlands and northern Borders of Wales	174
9.3.2	North-eastern England and East Anglia	178
9.3.3	North-western England	186
9.3.4	The Thames Valley	189
9.3.5	Wales, and England south of the Lower Severn and Thames valleys	190
9.3.6	Scotland	195
9.3.7	Ireland and the Isle of Man	199
9.3.8	The offshore sequence and sea-levels	203
9.3.9	Age and correlation	206
9.4	Conclusion	207
10	**The Flandrian Temperate Stage**	**208**
10.1	The presence of man	210
10.1.1	Later Upper Palaeolithic and Mesolithic	210
10.1.2	Neolithic	213
10.1.3	The Bronze Age	214
10.1.4	The Iron Age	215
10.1.5	Historic time	217
10.2	Flora and vegetation	218
10.2.1	East Anglia	218
10.2.2	The Midlands of England	221
10.2.3	Wales	222
10.2.4	England south of the Lower Severn and Thames valleys	223
10.2.5	The Scilly Isles, Channel Islands and Isle of Wight	224

10.2.6	Northern England	225
10.2.7	Scotland	228
10.2.8	Ireland	229
10.2.9	Spatial and temporal variations in Flandrian vegetation	230
10.3	Non-marine Mollusca	231
10.3.1	Terrestrial woodland and grassland assemblages	231
10.3.2	Freshwater and marsh assemblages	233
10.3.3	Introduced species	235
10.4	Marine Mollusca	236
10.5	Vertebrates	236
10.6	Insects	240
10.7	Other fossils	245
10.7.1	Diatoms	245
10.7.2	Ostracoda	248
10.7.3	Foraminifera	248
10.7.4	Tree rings	248
10.8	Non-biological data	250
10.8.1	Magnetism	250
10.8.2	Chemical analysis	252
10.9	Soil and peat formation and erosion	253
10.9.1	Soil	253
10.9.2	Peat	255
10.10	Rivers	257
10.11	Sea-levels and crustal movements	258
10.11.1	North-eastern England, East Anglia and the Thames Estuary	260
10.11.2	The south coast of England, and Jersey	261
10.11.3	The Bristol Channel, western and northern Wales coasts	263
10.11.4	North-western England	263
10.11.5	Scotland	263
10.11.6	Ireland	268
10.11.7	Overview of crustal movements	269
10.12	The offshore sequence	270
10.13	Climate	270
11	**Epilogue**	**275**
References		**277**
Index		**333**

List of Figures

1.1	Chronostratigraphic ranges of the sedimentary facies represented in the Pleistocene deposits of the UK sector of the North Sea	3
1.2	Pleistocene climatic signals and chronologies	4
1.3	Till	6
1.4	Head	7
1.5	Fluvioglacial sediment	8
1.6	Loess	9
1.7	Coversands	10
1.8	Ice-wedge casts	11
1.9	Blanket peat	12
1.10	Speleothems	13
1.11	Raised-beach material	14
1.12	Raised shoreline	15
1.13	Organic mud	16
1.14	Palaeosols	17
1.15	Correlation of D-alloisoleucine/L-isoleucine non-marine events in the British Isles with odd-numbered oxygen isotope stages	26
2.1	Neogene materials and localities	32
2.2	Pliocene to early Middle Pleistocene sequences in the Netherlands, East Anglia and the southern North Sea	34
3.1	The Crag Basin and Lower Pleistocene sites	41
3.2	Pollen diagram from the Ludham borehole	44
3.3	Baventian ice margins and ice-movement directions	48
3.4	Pollen diagram from Bramerton	49
3.5	Pleistocene evolution of the Thames system prior to the Anglian Glaciation	50
3.6	The Thames terrace sequence	51
4.1	Cromerian and related sites	56
4.2	Pollen diagram of the Pastonian	57
4.3	Beestonian ice margins and ice-movement directions	60
4.4	Pollen diagram of the Cromerian	64
4.5	Climate curve and chronostratigraphy for the Middle Pleistocene of the Netherlands	65
5.1	Anglian localities and glacial limit	74
5.2	Palaeogeographical reconstruction of the Elsterian/Anglian Stage at the glacial maximum	75
5.3	The Anglian Glaciation and initial diverted course of the Thames	78
6.1	Hoxnian sites and coastal positions	85

6.2	Pollen diagram of the Hoxnian at Marks Tey	86
6.3	Section across Barnfield Pit, Swanscombe	90
6.4	Position of Lower and Middle Palaeolithic flint industries within Pleistocene stages of Britain	94
7.1	Wolstonian sites	99
7.2	Wolstonian proglacial lakes and their overflows	101
7.3	The Wolstonian and associated glacial episodes	104
7.4	Sites with evidence of additional post-Hoxnian/pre-Ipswichian temperate and cold episodes	110
7.5	Minchin Hole stratigraphy	112
7.6	Terrace sequence of the Midland Avon	114
7.7	Terrace sequence of the Upper Thames	117
7.8	Marsworth stratigraphy	118
7.9	Vegetational succession in the Early Saalian of the Netherlands	121
8.1	Ipswichian sites and coastal positions	124
8.2	Ipswichian pollen zone range diagram	125
8.3	Pollen diagram of the Ipswichian	130
9.1	Early and Middle Devensian/Midlandian localities	144
9.2	Deglacial retreat and readvance of the North Atlantic Polar Front	145
9.3	Average July temperature variation in the southern and central lowlands of the British Isles since the Ipswichian Interglacial	146
9.4	Four Ashes stratigraphy	148
9.5	Devensian ice advances and proglacial lakes in eastern England	148
9.6	Bacon Hole stratigraphy	153
9.7	The Midlandian Cold Stage	158
9.8	The Malin–Hebrides sea areas, stratigraphy	160
9.9	Devensian–Weichselian correlations	161
9.10	Upper Palaeolithic archaeology and timescale	166
9.11	Late Devensian/Midlandian localities and ice limits	172
9.12	Stratigraphic evidence for the Dimlington Stadial in eastern England	173
9.13	The Devensian Glaciation on the North Welsh border	176
9.14	The Pitstone Soil	177
9.15	Distribution of loess and coversands in England and Wales	179
9.16	Model of the Late Devensian ice sheet	180
9.17	Devensian glacial features of east Yorkshire and nearby areas	182
9.18	Periglacial features in Britain and Ireland	183
9.19	Pollen diagram from Low Wray Bay, Windermere	188
9.20	The extent of Devensian glaciation in Wales	191
9.21	The Aberdeen–Lammermuir and Perth readvances	196
9.22	The Wester Ross and Loch Lomond readvances. The distribution of Lateglacial marine deposits. Isobases for the Main Lateglacial Shoreline	197
9.23	Late Midlandian ice-sheet characteristics and Late Pleistocene landbridge routeways	201
10.1	The Postglacial archaeological and historical sequence and its timescale	211
10.2	Archaeological sites	212
10.3	Pollen and diatom localities	219
10.4	Pollen diagram of the Flandrian at Hockham	220

10.5 Woodland types in the British Isles 5000 years ago 232
10.6 South Street molluscan diagram 234
10.7 Localities referred to in sections 10.3–10.6, 10.7.2, 10.7.3, 10.8 and 10.9 241
10.8 Insect evidence of Flandrian climatic variation 243
10.9 Round Loch of Glenhead physical, chemical and biological data 247
10.10 Le Marais de St. Pierre palaeoenvironmental data 249
10.11 Loch Lomond palaeomagnetic and radiocarbon dating evidence 251
10.12 River, sea-level and crustal movement localities 259
10.13 Transgressive and regressive overlap tendencies in north-west England 264
10.14 Isobases for the Main Postglacial Shoreline 266
10.15 Estimated current rates of crustal movement in Great Britain 271
10.16 Temperatures in the British Isles during the past 22 000 years 273

List of Tables

1.1	Major divisions of and gross timescale for the later part of the Cenozoic	2
1.2	Zonal nomenclature system for the interglacial periods of the Late and Middle Pleistocene	27
3.1	Correlation of lithostratigraphic units in the early Thames Valley and their interpreted ages	53
4.1	Key elements of the Middle and Late Pleistocene mammal faunas of Britain	68
7.1	Nomenclature of Wolstonian lithostratigraphic units	100
7.2	Hoxnian to Ipswichian geological sequence in East Anglia	120
8.1	Age estimates for the (D/L) stages	141
10.1	Sub-divisions, terminology and chronology of the Flandrian/Littletonian	209
10.2	Red Moss pollen assemblage zones, radiocarbon chronology and chronozones	210
10.3	Zonation scheme for Lateglacial and Postglacial deposits using land Mollusca	235
10.4	Present-day native British terrestrial vertebrate fauna	237
10.5	The anthropogenic element of the Irish terrestrial mammal fauna	240

Acknowledgements

We are grateful to the following for permission to reproduce copyright material:

Academic Press and the authors for Figs 7.3 (part) and 9.23 (part), Tables 10.3 and 10.5; The American Association for the Advancement of Science and W. G. Mook for Fig. 1.2d; D. J. Briggs, G. R. Coope and D. D. Gilbertson for Fig. 7.7; The British Ecological Society for Figs 10.9b and 10.9c; Elsevier Science Publishers BV and the authors for Fig. 9.2; Geologisches Landesamt Nordrhein-Westfalen for Table 1.2; The Geological Society of London and the authors for Figs 1.1, 2.1 (part) and 7.6; The Geological Society of America and the authors for Fig. 1.2a (part); The Geologists' Association for Figs 3.5, 5.3 and 10.13; The Institute of British Geographers for Fig. 3.6; Leicester University Press for Figs 10.1 (part) and 10.6; The Linnean Society and the author for Table 4.1; Longman Group UK Ltd for Table 10.4; J. J. Lowe, J. M. Gray, J. E. Robinson and J. B. Sissons for Fig. 9.22 (part); Macmillan Magazines Ltd for Figs 1.2c, 1.3, 7.8, 10.9a, 10.11 and 10.16; The Natural History Museum for Fig. 9.6; *The New Phytologist* for Figs 3.2, 3.4, 4.2, 4.4, 6.2, 8.3 and 10.4; J. Neale, J. Flenley and the authors for Figs 3.3, 4.3, 9.13 and 9.22 (part); P. J. Osborne for Fig. 10.8; Oxford University Press for Figs 9.10 (part), 9.15 and 9.16; Pergamon Press Plc for Figs 1.2b, 6.4 and 7.5, Table 8.1, Figs 9.7, 9.22 (part), 9.23 (part) and 10.10; The Quaternary Research Association for Fig. 2.2b, Tables 3.1 and 7.1, Figs 9.17, 9.20, 9.22 (part) and 10.14; A. C. Renfrew for Figs 9.10 (part) and 10.1 (part); The Royal Irish Academy for Fig. 9.18 (part); The Royal Society of London and the authors for Figs 2.2a, 4.5, 5.2, 7.9, 8.2, 9.3, 9.4, 9.9 (part), 9.18 (part) and 9.19, Table 10.2; Scandinavian University Press for Figs 9.8 and 9.12; *Scottish Journal of Geology* for Figs 9.21 and 9.22 (part); A. Straw for Fig 9.5; The University of Washington (*Quaternary Research*) for Fig. 1.2a (part); John Wiley & Sons Ltd for Figs 10.5 and 10.15; J. J. Wymer for Fig. 6.3 and Table 7.2.

We are also indebted to Dr T. C. Atkinson, Dr P. C. Buckland, Dr P. R. Cundill, Dr C. P. Green, the Photographic Section of H. M. S. Osprey, Professor N. Stephens and Professor P. Worsley for providing photographs, whose sources are indicated in their captions.

Preface

The study of Pleistocene environmental changes has been an important facet of earth science in the British Isles for over 150 years. During the first half of the nineteenth century, observations of landforms and deposits that had resulted from processes operative during the most recent geological epoch, notably by Louis Agassiz, William (Dean) Buckland, Charles Darwin and Charles Lyell, laid the foundation for continued and increasingly complex investigations. Indeed, because of the amount of detail now available concerning Pleistocene events in the British Isles, enough published and unpublished material exists to fill more than one volume of this size. Hence the account which follows, whose aim is to provide as comprehensive a picture as space permits, must of necessity be selective in its subject matter. In recent years, and in no small measure due to the existence of the Quaternary Research Association, a multi-disciplinary approach to Pleistocene studies has become increasingly frequent in the British Isles. We have benefited enormously from attendance at field meetings of this association and found the accompanying handbooks produced by their leaders invaluable as sources of reference for this publication. Along with this welcome multi-disciplinary trend, there has been a burgeoning of scientific methods applicable to the tasks in hand. Thus, any overall account of the Pleistocene of the British Isles would become enormous if details of the many methods and techniques now involved in its study were expounded as a precursor to the presentation of the evidence accruing from them. Accordingly, this volume will concentrate on the environmental changes rather than the ways in which they have been elucidated. Those interested in more than the briefest outline of the latter, which is given where necessary as an aid to understanding their application, will be referred to specialist sources. As this book is aimed at undergraduates taking options in Pleistocene studies, and postgraduate and more experienced scientists, we have also assumed a working knowledge of fundamental geological, biological and archaeological principles, and have not attempted to define or provide an explanation of every term or phenomenon referred to in the text.

At the outset, it should be stated that the British Isles will be treated in both their geographical and national contexts. Consideration will thus be given to the Republic of Ireland, the Isle of Man and the Channel Islands. Our prime objective is to examine the terrestrial record. However, a recent substantial increase in knowledge of Pleistocene events on the continental shelf offshore of the British Isles (an area over twice that of the land and with thicker deposits), and of relationships between phenomena in the two regions, means that attention will be given to these topics.

After an introductory chapter dealing largely with concepts, methods and techniques, the remainder of the text is devoted to the description and discussion of successive environments. Its chapters deal with either formal or informal divisions of the timespan under review. Each of these chapters contains one or more maps showing the sites mentioned within them. Where ages are quoted, they are in years before present (BP). In archaeological contexts, reference is sometimes also made to years BC or AD. As geochronometric age determinations of several kinds are given, in order to avoid confusion, the archaeological convention of using bp (or bc/ad) for uncalibrated and BP (or BC/AD) for calibrated (to calendar years) radiocarbon dates has not been followed. For reference to work cited in figures and tables obtained from other authors, but not quoted in the text, the sources of the illustrations should be consulted.

Our reason for writing this book is that major advances have taken place in Pleistocene studies since we began our careers in the subject over 20 years ago. These advances have not all been routine and some have caused controversy. They have opened up exciting new avenues of investigation and lines of interpretation, as a result of which a more complete understanding of processes, patterns and changes within the Pleistocene is beginning to emerge. What follows is an attempt to view the Pleistocene of the British Isles in the light of these recent developments. We hope that our efforts will serve as a working basis for future syntheses.

The direct and indirect assistance of many fellow Pleistocene scientists has been noted above. We are especially grateful to Dr Chris Green, who read and constructively criticized the manuscript. We are also indebted to Caroline Mee, who drew the figures, and to Ruth Cripwell, Helen Heyes and Cam Appleby of Chapman and Hall for their editorial assistance. The publication would not have been generated without the backing of Roger Jones and Dr Clem Earle, formerly of Unwin-Hyman, to whom we offer our thanks.

1 Background

This chapter firstly introduces some basic concepts pertaining to the Pleistocene. It then provides a brief resumé of the nature of the evidence for environmental changes over this timespan and of the means available for dating them. Against this backdrop, it concludes with an assessment of the potentialities of and the problems related to the Pleistocene sequence of the British Isles.

1.1 Basic concepts

The time covered by this study is approximately the last 2 million years. It thus belongs to that part of the Cenozoic Era traditionally called the Quaternary Period, and consisting of the Pleistocene and Holocene epochs. West (1977a) has proposed the use of the term Pleistocene to include all of this duration of geological time. It is *sensu* West that Pleistocene is employed in this book, which means that the terms Quaternary and Holocene are excluded. In addition to division into formal stages, the Pleistocene has been separated into Lower (Early), Middle and Upper (Late) parts as a matter of expedience. Although the Pleistocene has been characterized by a series of distinctive environmental changes, as with the rest of geological time, it forms part of the continuum of earth history. In consequence, summary detail is also provided of environments during the Miocene and Pliocene, the epochs which preceded the Pleistocene in the later part of the Cenozoic (Table 1.1).

The most significant feature of the Pleistocene has been a sequence of marked climatic changes. Some of these have given conditions as warm, or warmer, than those of today; others have been cold enough to allow arctic environments to develop. There is a great variety of Pleistocene deposits representative of temperate and cold conditions in the British Isles, and these have formed the basis for the traditional division of the epoch. In the lower part of the land-based sequence the sediments are of mainly shallow-water marine origin, while those of its later stages are largely terrestrial. In sum, this sequence has been interpreted as indicating an initial preglacial phase, followed by one when glaciation and periglaciation alternated with temperate environments. Interglacials have been temperate intervals between cold stages, with interstadials representative of climatic ameliorations within the latter (sections 1.4.2.4 and 1.5). Seven cold, together with eight temperate stages have been recognized in the Pleistocene (Mitchell *et al.* 1973a; West 1977a; Funnell *et al.* 1979; Stuart 1982) (Table 1.1). Offshore, beneath the shallow water over the continental shelf, longer sequences of Pleistocene sediment have been found in superposition than on land. However, hiatuses and regional unconformities are present, and the number of stages represented appears analogous to that identified in the deposits of the terrestrial area (Davies *et al.* 1984; Pantin and Evans 1984; Cameron *et al.* 1987; Long *et al.* 1988) (Figure 1.1).

These sequences are at variance with that derived from oxygen isotope analyses and palaeomagnetic studies of deep-ocean sediments, in which up to ten glacial/interglacial cycles have been identified since about 800 000 years ago, and many more oscillations of climate are evident earlier in the Pleistocene (Shackleton and Opdyke 1973, 1976) (Figure 1.2(a)(b)). While it is clear

Table 1.1 Major divisions of and gross timescale for the later part of the Cenozoic. Pleistocene divisions in the British Isles: sources, Mitchell *et al.* (1973a), Funnell and West (1977), West (1977a), Funnell *et al.* (1979), Stuart (1982), Warren (1985). Additional sources: Bowen (1978), Curry *et al.* (1978), Zagwijn (1985)

Approximate age of commencement (Ma BP)	*Epochs*			*Stages*			
				Britain	*Ireland*	*North-western Europe*	
0.01				Flandrian (t)	Littletonian	Holocene	
0.115			Upper (Late)	Devensian (c)	Midlandian (Fenitian)	Weichselian	Late
				Ipswichian (t)	Last Interglacial	Eemian	
				Wolstonian (c)	Munsterian	Saalian	
0.3				Hoxnian (t)	Gortian	Holsteinian	Middle
0.5			Middle	Anglian (c)	Pre-Gortian	Elsterian	
				Cromerian (t)		Cromerian	
	Quaternary	Pleistocene		Beestonian (c)			
				Pastonian (t)			
				Pre-Pastonian (c)			
				Bramertonian (t)			
			Lower (Early)	Baventian (c)			
				Antian (t)			
				Thurnian (c)			
				Ludhamian (t)			
2				Pre-Ludhamian			
		Pliocene	Late				
			Early				
5	Tertiary (Neogene)						
		Miocene	Late				
			Middle				
			Early				
23							

t, temperate; c, cold

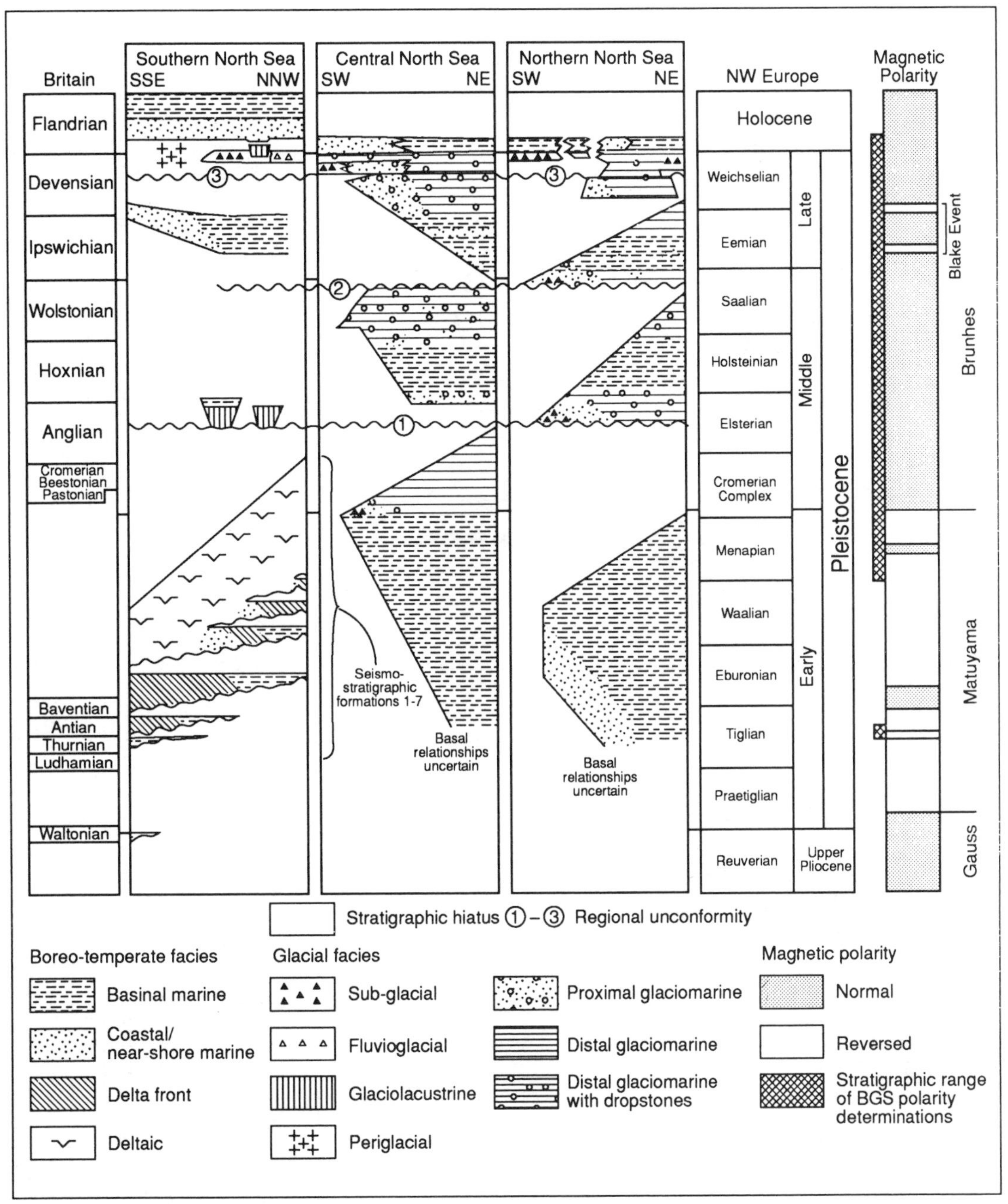

Figure 1.1 Schematic diagram summarizing the probable chronostratigraphic ranges of the sedimentary facies represented in the Pleistocene deposits of the UK sector of the North Sea. Units 1–7 were deposited during seven stages in the Lower and early Middle Pleistocene development of the southern North Sea. From Cameron *et al.* (1987). Reproduced by permission of the Geological Society of London and the authors.

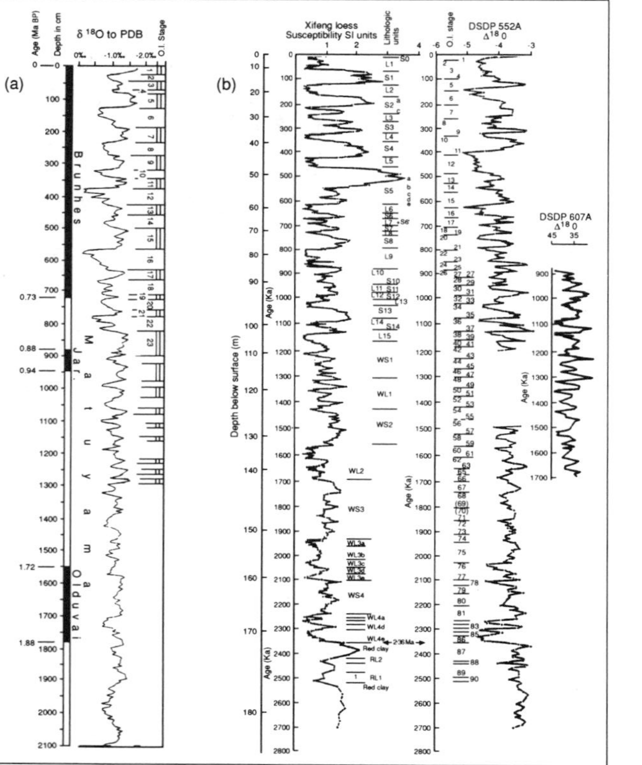

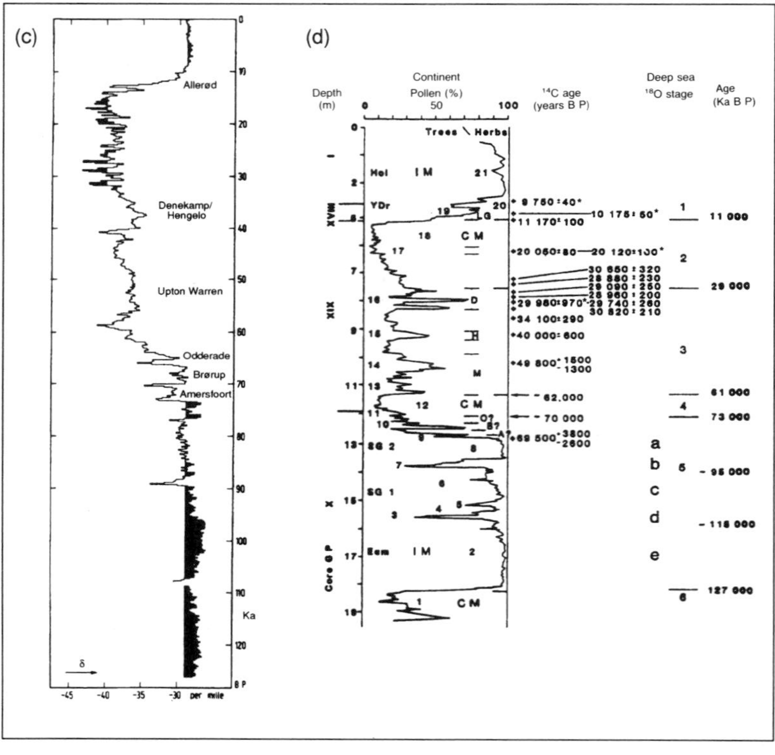

Figure 1.2 Pleistocene climatic signals and chronologies. (a) Oxygen isotope and palaeomagnetic record in core V28-239 (03°15′N 159°11′E; Western Equatorial Pacific). JAR, Jaramillo. From Shackleton and Opdyke (1976) and Berggren *et al.* (1980). After N. J. Shackleton and N. D. Opdyke in *Geological Society of America Memoir* **145**, 449–464, 1976. (b) Magnetic susceptibility record in the Xifeng loess (X. M. Liu *et al.* 1985) and proposed correlation of the Chinese loess and soil units with the oxygen isotope stages in the DSDP (Deep Sea Drilling Program) core 552A (Shackleton *et al.* 1984; Zimmerman *et al.* 1984). Also shown, correlation with the DSDP core 607 for the interval missing in DSDP 552A (Ruddiman *et al.* 1986). Depth in metres from the top of the plateau and from the top of the core. From Kukla (1987). Reprinted with permission from *Quaternary Science Reviews* **6**, G. J. Kukla, Loess stratigraphy in central China, © 1987 Pergamon Press Plc. (c) Oxygen isotope record for the Camp Century ice core, north-western Greenland. Tentative interpretations in European terminology are shown (Dansgaard *et al.* 1971). From Johnsen *et al.* (1972). Reprinted with permission from *Nature* **235**, 429–434; Copyright © 1972 Macmillan Magazines Ltd. (d) Correlation between the Grand Pile continental deposit, north-eastern France and the deep-sea record: time series of total tree and shrub pollen from Grand Pile (Woillard 1978). The depth scale is uniform except between GP XVIII and 1 (2.5–4.5 m), but verifications have shown that the pollen record has to be considered as perfectly continuous. Pollen zones are designated by the numbers 1–21 on this time series; local stratigraphic names are designated as follows: Eemian, Eem; St. Germain 1, SG1; St. Germain 2, SG2; Amersfoort, A; Brørup, B; Odderade, O; Moershoofd, M; Hengelo, H; Denekamp, D; Late Glacial, LG; Younger Dryas, Y Dr; Holocene, Hol; cold maximum, C M; interglacial maximum, I M. All the ^{14}C ages are expressed in conventional ^{14}C years BP. Asterisks indicate dates obtained on a parallel core. The ages ~62 000 and ~70 000 years BP are deduced. The ^{18}O stages are after Emiliani (1955) and Shackleton (1969). The ages of stage boundaries are after Hays *et al.* (1976) and Kominz *et al.* (1979). From Woillard and Mook (1982). Copyright 1982 by the American Association for the Advancement of Science.

that the record on land is not as complete as that of the oceans, most scientists, having accepted the discrepancy, are prepared to admit that closer correlation between the two sequences is possible. This conclusion has been based both on new data from terrestrial and shallow-marine environments and re-interpretation of some existing information from these areas. The advent of additional methods of age determination has also played an important part in our increased understanding of both the spatial pattern and temporal scale of Pleistocene events (Bowen 1978).

1.2 The nature of the evidence

Consideration of environmental changes during the Pleistocene has to take account of the nature of the evidence, both for the types of conditions and the placing of events in chronological order. In the British Isles, most is known about the terrestrial record. As this is erosion dominated, no long, land-based sequences comparable to those from loess (Kukla 1970, 1975, 1977, 1987) (Figure 1.2(b)), ice cores (Epstein *et al.* 1970; Dansgaard *et al.* 1971, 1985; Johnsen *et al.* 1972; Robin 1977; Broecker *et al.* 1988) (Figure 1.2(c)) and lakes and peat bogs (Wijmstra 1969; Florschütz *et al.* 1971; Woillard 1978; de Beaulieu and Reille 1984) (Figure 1.2(d)) in other areas have been discovered. It is rare to find even one complete cycle of glacial and interglacial sediments superimposed on land in the British Isles although, as noted above, temporally longer sequences occur offshore. The terrestrial Pleistocene succession, in particular, is punctuated by non-sequences or unconformities which mark unknown timespans. There are also seldom good grounds for long-distance correlation of the scattered deposits, as sedimentological, geomorphological or palaeontological parameters are often insufficient for this purpose, even within the restricted area of the British Isles. The generally accepted yardstick for environmental changes is the near-complete sequence of the deep-ocean basins noted above (Shackleton and Opdyke 1973, 1976). Therefore, the establishment of a succession of environmental changes for the onshore and offshore continental area of the British Isles must involve the reconciliation of these two lines of evidence.

1.3 Types of evidence

As for earlier geological time, fossils and the sedimentary rocks which contain them, are the principal sources of Pleistocene environmental evidence. Prior to the Pleistocene, the fossil record is generally poor in both spatial and temporal contexts. Moreover, the remains of former living organisms mainly occur in consolidated rocks, where they have usually either been replaced by more stable chemicals, or decayed to leave impressions. By way of contrast, Pleistocene rocks are normally unconsolidated, and frequently contain a considerable variety of organic remains. In addition, the majority of these remains are in a sub-fossil state, with some or all of their original parts extant, thereby making their recognition easier. The sediments containing the fossils may provide important information on environmental conditions. For example, the size and form of sediment particles can define the energy characteristics of fluvial environments, while glacial and periglacial deposits allow climatic regimes to be ascertained.

The reconstruction of Pleistocene environments is, therefore, multi-disciplinary, with the best results being obtained by synthesizing different lines of evidence (for example, Green *et al.* 1984) (Chapter 7). The virtue of this approach is that detailed theoretical representations of the past can be made. However, problems also occur. Sometimes there are conflicts between the evidence suggested by different disciplines. For instance, inferences concerning climate in the British Isles from pollen and insect data are substantially different for certain times during the Pleistocene (for example, West *et al.* 1974) (Chapter 9). It is important not to view such conflicts negatively, however. They are likely to be explicable with reference to the innate characteristics of the organisms concerned, especially dispersal ability and ecological tolerance.

Before presenting information about the environments of approximately the last 2 million years in the British Isles, a cursory look at the major types of evidence is necessary. Full discussion is precluded, and the reader is referred to West (1977a), Birks and Birks (1980) and Lowe and Walker (1984) for more detail concerning them.

1.3.1 Sediments

Sedimentation is a complex and variable process to which few general rules apply. However, prevailing physical, chemical and biological circumstances mean that the notion of deposition being inorganic in cold conditions and organic in temperate ones is not without foundation.

1.3.1.1 Sedimentation in cold climates

The most obvious expression of a cold climate in Pleistocene sediments is the occurrence of glacial till (Figure 1.3). This is characterized by a lack of sorting or bedding and by matrix-supported clasts, and occurs widely in the glaciated areas of the British Isles. The only real possibility of confusing till is with head, a cold-climate slope deposit (Figure 1.4) which may also lack sorting and bedding (Watson and Watson 1967). None the less, the fact that bodies of head are normally closely related to the local topography, and lack far-travelled erratic stones, usually serves to distinguish the two types of sediment. Glacial till can conveniently be divided into flow till, melt-out till and lodgement till (Embleton and King 1975), with the two former kinds developing during ice-

Figure 1.3 Till. Flow till and gravelly till. Late Devensian. Ellesmere Moraine, Ellesmere, Cheshire. (Photograph: D. H. Keen.)

Figure 1.4 Head. Devensian. Holdstone Down, near Combe Martin, north Devon. (Photograph: N. Stephens.)

sheet decay, and the latter being deposited by moving ice. Glacial tills are often highly complex (Boulton 1972; Eyles *et al.* 1983), but their indication of cold conditions is unambiguous. The direction of ice movement can be determined by examination of local and regional erratic-trains (Sissons 1967a; Shakesby 1978; Sutherland 1984), and till fabrics (West and Donner 1956; Andrews 1971). However, the use of the latter requires care and a regional approach, as local conditions, for example flow-till deposition down the maximum slope, may mask true ice-movement directions.

In contrast to till, fluvioglacial sediment (Figure 1.5) is stratified. It is deposited within and beneath ice, and beyond its margin by streams, and also accumulates in lakes and the sea. Such material can contribute to the identification of glacial limits. For example, outwash terraces can often be linked to terminal moraines, while stepped sequences of terraces may relate to successive periods of ice advance and retreat (Maizels 1985). Fluvioglacial deposits are particularly well developed in the lowlands fringing the mountainous parts of the British Isles (Sparks and West 1972).

Beyond glaciers and ice sheets, the periglacial area may also yield important data from its sediments. These have been summarized by French (1976) and Washburn (1979). Angular slope

Figure 1.5 Fluvioglacial sediment. Sand and gravel. Late Devensian. Wrexham Delta, near Wrexham, Clwyd. (Photograph: D. H. Keen.)

deposits have been noted above as indicators of a cold climate. Although landslides under temperate conditions may also produce head-like materials, the widespread nature and considerable thickness of angular slope deposits in cold conditions makes them readily recognizable. Periglacial aeolian deposits are primarily represented by loess (Figure 1.6) and coversands (Figure 1.7), whose grain-size parameters are usually sufficient for their recognition. Periglacial fluvial sediments are usually coarse grained and massively bedded, as befits their deposition under nival regimes. The recognition that most terrace sediments in south and east England are of cold-climate origin (Wymer 1968; Gibbard 1985) was an important step in the evaluation of palaeoenvironments in extraglacial areas. Associated with periglacial sediments are structures produced by ground ice and periglacial stresses. Foremost among these are ice wedges (Péwé 1966; Black 1976) (Figure 1.8). The occurrence of these, either in section as wedge pseudomorphs, or in plan as polygonal nets, is clear evidence of permafrost, with mean annual temperatures below −6°C. Pingos may indicate similar conditions. However, their surface expression as circular ramparts (Watson 1977) means that they are geomorphological rather than sedimentological features. Involutions (Williams 1975) are indications of lesser conditions of cold.

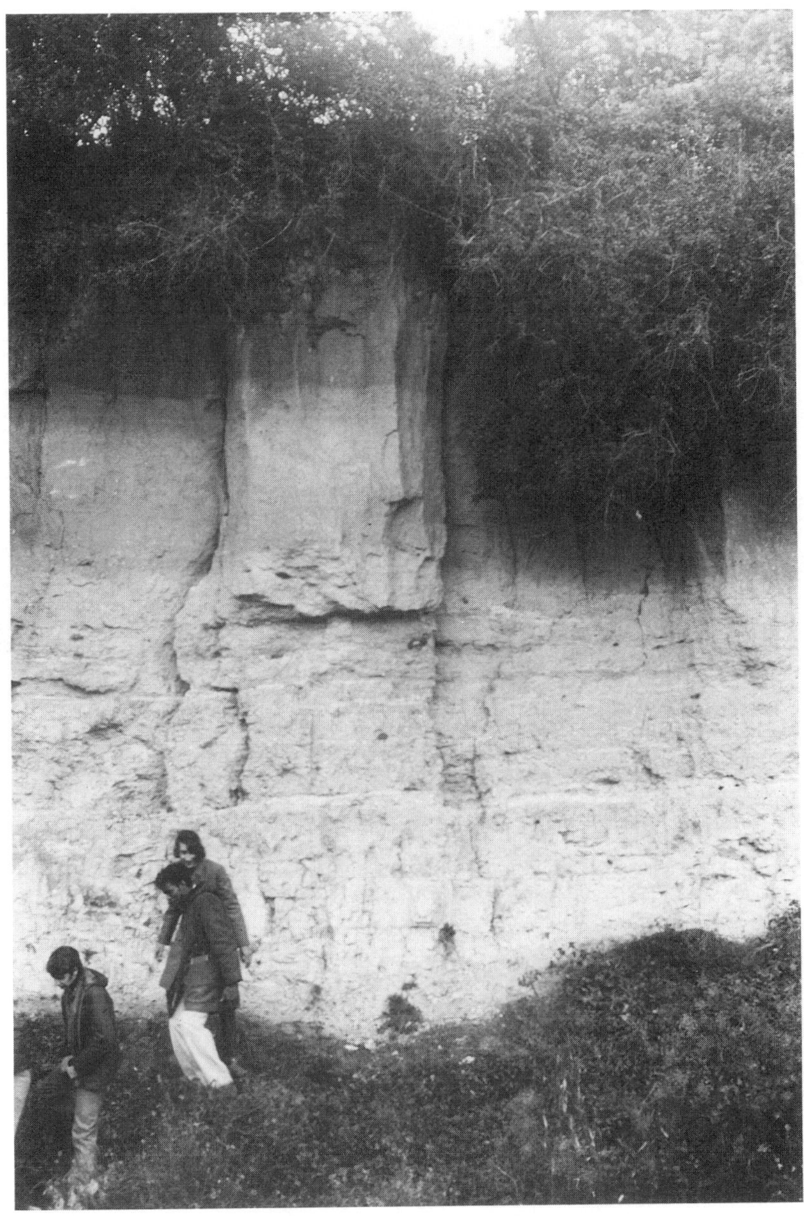

Figure 1.6 Loess. Silt occupies the top third of the section, above lighter coloured Thanet Beds (Palaeocene). The darker, upper part of the silt is decalcified. Late Devensian. Pegwell Bay, Kent. (Photograph: D. H. Keen.)

Figure 1.7 Coversands. Late Devensian. Messingham, South Humberside. (Photograph: P. C. Buckland.)

Most data on periglacial structures come from present-day arctic latitudes. However, this area may not provide a good analogue for Pleistocene periglacial palaeoenvironments in the British Isles. Then the climate here may have been wetter than in Siberia or arctic Canada now, and differences in slope aspect and day length would have caused significant variations in climatic regimes (French 1976). However, climatic conditions suggested from periglacial structures are similar to those indicated by insect fossils (Coope 1977a), so comparisons seem feasible. It is also worth observing that active periglacial phenomena exist in parts of the uplands of the British Isles today. These have been reviewed by Ballantyne (1987), who noted that the principal characteristics of the climate in the uplands of the British Isles are high levels of precipitation and powerful winds. A lack of extremely cold conditions means that frost penetration of the ground is restricted, so that processes and landforms characteristic of patterned ground are limited, and solifluction is shallow and occurs at a low rate. Wind action is involved in the development of deflation surfaces, sand deposits and terraces, while wetness leads to boulder ploughing and debris flows.

1.3.1.2 Sedimentation in temperate climates

Sediments are seldom useful as indicators of temperate climates. While organic productivity is greater in temperate environments than in cold environments, the latter preclude it only when

Figure 1.8 Ice-wedge casts. Developed in fluvial gravels. Devensian. Baston, Lincolnshire. (Photograph: P. Worsley.)

temperatures are below 0°C for considerable periods. Thus, while organic deposits such as blanket peat (Figure 1.9) may be characteristic of temperate stages, they are not necessarily confined to them, as neither are inorganic ones to cold stages. Of especial relevance in the latter context is soil and sediment erosion, which can lead to inorganic deposition under temperate conditions.

One of the most reliable indicators of a temperate climate is speleothem (Figure 1.10). The hydrological conditions necessary for these usually calcareous cave deposits to be precipitated effectively seem to occur almost exclusively in such climates. Uranium-series ages (section 1.4.1.6) of speleothems in the British Isles correlate with episodes which, from other data, are thought to have been temperate (for example, Atkinson *et al.* 1978) (Chapters 8 and 9).

Although there is little in the sedimentology of marine deposits to indicate the climate under which they accumulated, the majority of raised-beach material (Figure 1.11) seems to be indicative of temperate conditions. This is because the sea-levels which have produced beach sediments above modern high-water mark are most likely to have been those of interglacials, when ice melting was maximal. None the less, caution

Figure 1.9 Blanket peat. Flandrian. Collier Gill, Egton High Moor, north-east Yorkshire. (Photograph: P. R. Cundill.)

must temper the above statement when considering raised marine deposits. For example, the Devensian Lateglacial, Clyde Beds of western Scotland (Peacock 1981) (Chapter 9) possess a more cold-tolerant fauna than that characteristic of the sea around this area today, and appear to have been isostatically uplifted (their maximum elevation is now *circa* 35 m OD) since deglaciation. This isostatic rebound has also caused the Main Postglacial Shoreline in Scotland (Figure 1.12) to be raised up to 14 m above present mean sea-level (Sissons 1983) (Chapter 10). In non-glaciated areas, tectonic downwarping has been the overall trend over this timespan. Alterations in the standing of the sea surface compared with that of the land as a result of localized uplift or subsidence of the latter are termed relative sea-level changes. Oscillations in the quantity of water occupying the world's oceans, due to a combination of plate tectonic activity (which changes the dimensions of ocean basins) and ice-sheet growth and decay, constitute eustatic sea-level changes.

Theories of fluvial sedimentation (Leopold *et al.* 1964) indicate that stream sediments from former temperate stages should be similar to those of the present day, with for instance, sands and silts representative of meandering and equable flow regimes. In fact these are seldom encountered, for like many other temperate-stage deposits they have been in large part destroyed in subsequent cold stages. Sometimes, however, channel sediments have been buried beneath subsequent deposits. When such channel sediments have included organic-rich beds (Figure 1.13), they have been fruitful sources of palaeoenvironmental data (for example, Briggs *et al.* 1975; Green *et al.* 1984) (Chapters 4 and 7).

1.3.2 Soils

Soil-forming processes affected Pleistocene sediments during both cold and temperate climatic phases (Catt 1986). Vestiges of former soils may either be covered by subsequent deposits or occur on current land surfaces (Catt 1988). Soils formed

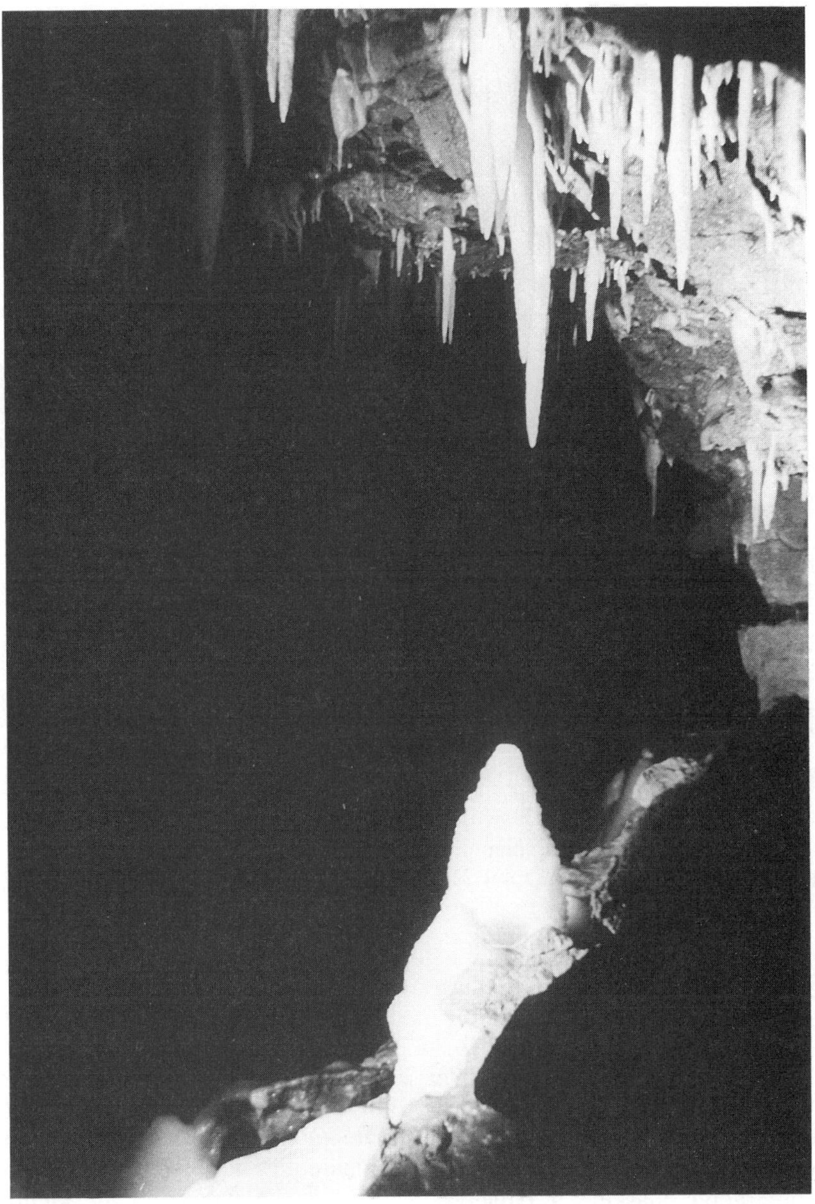

Figure 1.10 Speleothems. Stalagtites and stalagmites. Late Pleistocene. Bat Passage, G. B. Cave, Mendip Hills, Somerset. (Photograph: T. C. Atkinson.)

in past landscapes are termed palaeosols (Ruhe 1956) (Figure 1.14). If buried soils possess a complex history they are referred to as polycyclic or composite. Soil characteristics acquired from former time and existing on current land surfaces are termed relict. A compound soil or pedocomplex refers to two or more soils, possibly incomplete, which are isolated by thin bands of sediment that exhibit pedogenesis. The recognition and relative dating of palaeopedological

Figure 1.11 Raised-beach material. Post-Hoxnian/pre-Ipswichian. Portland West Raised Beach, Isle of Portland, Dorset. (Photograph: HMS Osprey, Photographic Section.)

characteristics has profited most from micromorphological investigations. For example, a sequence involving clay illuviation, iron mobilization and disruption as a result of frost action can be revealed by the thin-section technique (Catt 1988). Palaeosols can possess extensive distributions. As a stratigraphic marker, a palaeosol could provide unambiguous evidence of temperate conditions in a single section, such as any locality where it was exposed beneath later till. However, problems may arise when attempts are made to use the soil as a chronostratigraphic marker over a wider area. For example, the use of the Valley Farm Soil as an indicator of the Cromerian Interglacial over much of northern Essex and southern Suffolk (Rose *et al.* 1976) has had to be modified in the light of its polycyclic nature (Kemp 1985) (Chapter 4). Similar problems attend the use of soils as indicators of the type of temperate conditions prevailing. This has largely been done on mainland Europe (van Vliet-Lanoë 1988), but as a technique, it may be hampered by a strong element of subjectivity in the identification of the intensity of soil-forming processes.

1.3.3 Geomorphology

Apart from sediments and soils and their structures, the other non-biotic evidence which may assist in the interpretation of former environments comes from geomorphology. It is often difficult to disentangle morphological from sedi-

Figure 1.12 Raised shoreline. The Main Postglacial Shoreline. Flandrian. North of Drumadoon Point, Isle of Arran. (Photograph: D. H. Keen.)

mentary evidence, as consideration of the literature on river terraces shows (Gibbard 1985). Strictly morphological evidence is seldom valid as a guide to the palaeoenvironment. River terraces signify changed environmental conditions which may have involved movements in base level and modifications of discharge and sediment supply. In a glacial environment, geomorphology (meltwater channels, moraines and trimlines, for instance) is often the only evidence for defining ice limits. Indeed, the contrast between the fresh morainic topography of the 'Newer Drift' of the north of England and the subdued landscape composed of 'Older Drift' in the south is still regarded as a valid means of defining the limits of multiple glaciation. Aside from these particular examples, landforms alone provide, at best, supplementary evidence of particular palaeoenvironments.

1.3.4 Biological evidence

Whatever data sediments, soils and geomorphology can provide, the prime source of Pleistocene environmental reconstruction is the fossil record. Again, full consideration of the various techniques available will not be attempted here, and more detail on them can be found in West

Figure 1.13 Organic mud. The mud is the darkest layer in the foreground, where associated channel sediments can also be observed. Post-Hoxnian/pre-Ipswichian. The Lower Channel, Marsworth, Buckinghamshire. (Photograph: C. P. Green.)

(1977a), Birks and Birks (1980), Lowe and Walker (1984) and Berglund (1986), as well as the references cited below. The following summary examines the advantages and limitations of various fossil groups as environmental indicators in the context of the British Isles.

1.3.4.1 Pollen and spores

The pollen grains of higher plants and the spores of lower plants are a major source of information concerning Pleistocene environments. Pollen and spores are used in two main ways. Firstly, as a means of environmental reconstruction, whereby information on flora and vegetation allows inferences about other aspects of ecosystems, such as climate, soil and human impact. Secondly, via biostratigraphy as a method of dating, which will be dealt with later in this chapter. The history of the flora and vegetation of the British Isles (Godwin 1956, 1975; Pennington 1969, 1974) has been established mainly by palynology, with that of the current interglacial best known (Chapter 10). The reconstruction of past flora and vegetation using pollen and spores is far from straightforward. Although these microfossils are abundant, not all plants produce the same quantities of them, and further problems of representativity may occur due to their selective preservation or transport. In palynology, the use of absolute numbers, rather than relative

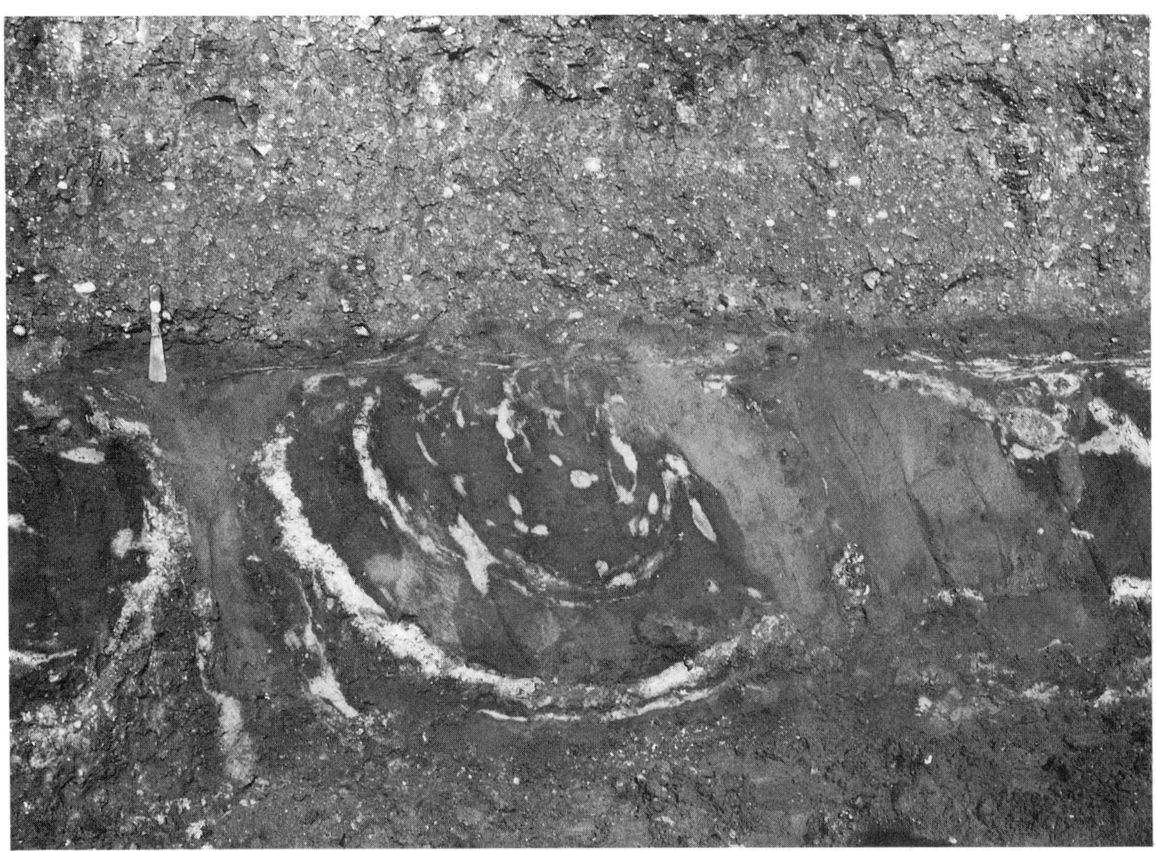

Figure 1.14 Palaeosols. The Valley Farm Soil (Cromerian and earlier Pleistocene), a rubified sol lessivé, and the Barham Soil (early Anglian), occupy the bottom two-thirds of the section. The whitish-grey (gleyed) layers of the Valley Farm Soil were deformed by cold-climate processes during the early Anglian, and arctic structures characterize the Barham Soil. The Barham Coversand (early Anglian) also became involuted, with one involution extending below the knife. Chalky (Lowestoft) till of Anglian age lies above the truncated surface of the Valley Farm Soil. Newney Green, near Chelmsford, Essex. (Photograph: C. P. Green.)

percentages, means that data from each pollen and spore producer can be assessed independently, and that inferences concerning vegetation should be sounder (Faegri and Iversen 1975; Faegri *et al.* 1989). However, as Faegri and Iversen (1975) and Faegri *et al.* (1989) have noted, in practice this is not altogether the situation. In addition to uncertainties as to the amounts of pollen and spores produced by plants in a particular locality, the dimensions of the catchment area from which the microfossils deposited at a site have come are difficult to quantify. Most pollen and spores seem to travel a relatively short distance from their source, probably no more than a few hundred metres, to their place of deposition (Turner 1964a; Tauber 1965; Birks and Birks 1980). Jacobson and Bradshaw (1981) have demonstrated how site dimensions influence the relative proportions of airborne pollen and spores from local and regional sources, with larger diameter sites having the greatest input from the latter. As Moore (1986) has observed, their model will differ according to the type of plant communities in an area, the relief of the latter and whether there is a stream entering the site.

1.3.4.2 Diatoms

These microscopic unicellular algae currently reside in a large number of aquatic habitats, whose sediments often contain substantial quantities of their siliceous frustules. Although their taxonomy is complex, diatoms are useful palaeoecological indicators. They are sensitive to variations in the nutrient status, salinity and pH of water bodies, and can provide information concerning changes in level of the latter. Perturbations in lake systems caused by anthropic impact can also be detected from diatom assemblages (Battarbee 1986a).

In the British Isles, the few analysts of fossil diatoms have concentrated on Devensian Lateglacial and on Postglacial (Flandrian) time (for example, Evans 1970; Haworth 1976). Within the last 10 years, diatoms have proved especially useful in the investigation of the relationship between recent acid deposition and lake acidification (Battarbee and Charles 1987) (Chapter 10).

1.3.4.3 Plant macrofossils

As Dickson (1970) has noted, the identification of plant macrofossils was among the earliest aspects of Pleistocene investigations. While numerous different parts of plants visible to the naked eye may be encountered in Pleistocene deposits, fruits and seeds are most abundant. Birks and Birks (1980) have enumerated the uses and limitations of plant macrofossils in environmental reconstruction. Of most value is that identification to species is often possible. This means that should either the species concerned or a close relative be extant, there are reasonable grounds for extrapolating their present and past ecologies. Also their dimensions normally ensure that they are not moved a long way from where they were formerly part of a living community. In consequence, a plant macrofossil assemblage is highly likely to have emanated mainly from proximal vegetation, especially that of the wetland or water body where it has been preserved. The usefulness of macrofossil plant parts is restricted because assemblages from the same site are often different, due to either *in situ* deposition or limited transport. This means that inferences from them should be about plant communities close at hand. Furthermore, local dispersal can be a variable phenomenon, which leads to marked variations in the quantity of macrofossils deposited at successive depths. When this is allied with the fact that only small quantities of macrofossil material are normally produced, thereby necessitating the procurement of a considerable quantity of sediment for analysis, the limitations are compounded.

In spite of its longevity in Pleistocene studies, it is fair to state that plant macrofossil analysis has been rather neglected. Compared with palynologists, there are currently relatively few scientists with expertise in plant macrofossils in the British Isles. However, their number is increasing, as is the application of the technique, most notably in archaeological contexts, and those in which firm indicators of particular environments, salt-marsh for instance (Jones *et al.* 1987) (Chapter 10), are sought (Mannion 1986a,b,c).

1.3.4.4 Vertebrates

Stuart (1974, 1982) has provided comprehensive accounts of Pleistocene vertebrates in the British Isles. The remains of large mammals are the most spectacular evidence. However, while the finds of either entire skeletons of, for example elephant or rhinoceros, or even their major bones are obvious, their interpretation may be problematical, especially if disarticulation has taken place. Difficulties also occur in the stratigraphic and environmental evaluation of such remains. Although evolution has meant that Lower and Upper Pleistocene faunas can be readily distinguished in the British Isles, its tempo and mode has been such that assemblages of animals from successive cold and temperate stages within these broad timespans are frequently so similar as to preclude their use as chronological markers. Also while assemblages with, for example, *Palaeoloxodon antiquus* (straight-tusked elephant), *Dicerorhinus* spp. (rhinoceroses), and *Hippopotamus* spp. (hippopotamuses) are indicators of optimal interglacial conditions, other faunas, mixtures of horse and deer together with mammoth for instance, may characterize either the end of a cold stage or an interglacial, or an interstadial. In the British Isles, the only large mammal which appears to be of use as a time-stratigraphic indicator is *Hippopotamus amphibius*. This is confined to Ipswichian deposits and appears particularly diagnostic of this stage (Stuart 1982) (Chapter 8).

By virtue of their size, bones of large mammals are relatively rare, except in stratified bone-beds, such as the Stoke Tunnel Bed, Ipswich (Wymer 1985) (Chapter 7), or in caves (Stuart 1982; Scott 1986). In contrast, bones of small mammals such as rodents and shrews are often abundant, especially when concentrated by birds of prey (Mayhew 1977). These small bones are also relatively fragile and thus are unlikely to be reworked, as may often occur with large bones and teeth. None the less, so far their use in either an environmental or a stratigraphic context has been limited. As with large mammals, there has been evolutionary progression through the Pleistocene (Sutcliffe and Kowalski 1976; Mayhew and Stuart 1986; Currant 1989). However, as general indicators of age, the use of small vertebrate bones is marred by the fact that they are very difficult to determine to specific level. In numerous cases, species identity is based on minute differences in tooth characteristics (for example, Mayhew and Stuart 1986). Also, not all small mammals are unequivocal indicators of particular environments. For instance, while the association of lemming with open terrain, and *Apodemus sylvaticus* (wood mouse) with woods is clear, there are other animals which are less habitat specific. For example, in the British Isles today the mole and common shrew occupy a variety of habitats including woods and grassland (Corbet and Southern 1977).

Other Pleistocene vertebrates are either seldom found as fossils, or may often be encountered but pose major difficulties of interpretation. Fossil bird bones are fairly frequently found in the British Isles (Bramwell 1960; Harrison 1979a,b), but problems of identification have limited their usefulness. Bones of fish and amphibians are common in aquatic and wetland deposits, although once again the presence of a few species in many types of deposit of different ages makes their use difficult. One suitable climatic indicator appears to be *Emys orbicularis* (pond tortoise). This chelonian, currently distributed on mainland Europe to *circa* 55°N (Arnold and Burton 1978), requires summer temperatures some 2°C higher than now pertain in the British Isles. Thus, its fossil presence seems to confirm a climate that was warmer and more continental than that of today (Stuart 1979) (Chapters 8 and 10).

Some Pleistocene vertebrate remains which have been little studied to date, may prove useful environmental or stratigraphic indicators when more is known about their distribution in both space and time. For example, Stinton (1985) has shown that in the British Isles, fish otoliths exist in Pleistocene marine deposits.

In an attempt to elucidate the number and sequence of Middle and Late Pleistocene temperate woodland episodes, Currant (1989) has presented a scheme of fossil mammal assemblages for Britain (Table 4.1) (Chapters 4, 6 and 8).

1.3.4.5 Mollusca

In calcareous deposits of Pleistocene age, snails are usually the most abundant fossil group. Marine, land and freshwater molluscs are known and these present different problems of evaluation. Marine Mollusca are especially abundant in the Lower Pleistocene Crags of East Anglia (Norton 1977) (Chapter 3). They are generally less common later in the Pleistocene. However, some Late Pleistocene raised beaches have substantial marine molluscan faunas which provide useful environmental information (for example, Headon Davies and Keen 1985) (Chapters 7 and 8). Marine Mollusca are not particularly sensitive to water temperature changes. Many species which now occur in waters around the British Isles have geographical ranges from Norway to West Africa (Tebble 1976). None the less, genuine arctic and southern species do occur in Pleistocene sediments. For example, *Portlandia arctica*, with a current distribution about eastern Greenland, has been recorded from the Devensian

Lateglacial, Errol Beds (Peacock 1975) (Chapter 9), while *Astralium rugosum*, having its present limit at the Ile de Ré on the French Atlantic Coast, occurs in the 8 m raised beach in Jersey (Keen *et al.* 1981) (Chapter 8). More usually, assemblages of species can be used to determine water depth, salinities and energy conditions (Norton 1967, 1977; Mottershead *et al.* 1987). Marine molluscs are of little use as indicators of age, as many taxa range through the Pleistocene. However, the appearance of certain species (for example, *Macoma balthica*) is generally believed to mark the onset of a cooler thermal regime in the North Sea in the Early Pleistocene (Norton 1977). Marine Mollusca are also a prime source of material for amino-acid diagenesis (section 1.4.2.1).

Non-marine Mollusca occur most frequently in the Middle and Upper Pleistocene deposits of the British Isles. They are especially common in calcareous sediments in southern and eastern England. The basis of their ecological evaluation was laid by Sparks (1957, 1961), although information concerning them had existed for some time previously; for example, Kennard and Woodward (1922). Non-marine Mollusca are most suitable as indicators of the local environment. For instance, freshwater molluscs can indicate the degree of movement and oxygenation of water bodies, while land taxa may denote the degree of shade or openness of the vegetation and also the nature of the substrate, whether dry or marshy (Sparks 1961). Non-marine Mollusca are less good at indicating thermal regimes, as they often have wide latitudinal ranges at present (Kerney and Cameron 1979). However, some species seem to provide a clear indication of habitat. The open grassland faunas of cold stages usually include *Columella columella*, currently restricted to such a habitat in arctic and alpine localities (Chapter 9). The past occurrence of the now southern European *Belgrandia marginata*, and continental *Clausilia pumila*, indicates warmer conditions than at present (Kerney 1977a). Assemblages of shade-loving species also indicate closed woods typical of interglacials (Kerney 1977b). Another crude indicator of climate is the diversity of molluscan species (Meijer 1985). In general, the numbers of land Mollusca in the Northern Hemisphere are reduced northward, so that the current British fauna has *circa* 135 species, while that at the Arctic Circle contains about 35, and some of these may be recent introductions (Kerney and Cameron 1979; Keen 1987). In British Pleistocene deposits, therefore, faunas with 10–15 species that possess a catholic habitat (*sensu* Sparks 1961) are generally of cold-climate origin; interstadials may yield *circa* 25 species (Holyoak 1982), and interglacials some 80–90 species. Taphonomic processes are complex and variable but some general characteristics exist. Thus, molluscan assemblages from lakes or wetlands are usually low in species diversity because there is little concentration of fossils from extraneous habitats by slope wash or stream action. However, larger numbers of species are normally found in rivers, where especially at times of flood, fluvial, marsh and land snails may become incorporated in the sediments.

Mollusca are less suitable as chronostratigraphic than palaeoenvironmental indicators, as they appear to have undergone little evolutionary change during the Pleistocene. Early investigators of Pleistocene molluscs, Kennard for example, used them in a similar way to those of earlier geological periods. However, current knowledge of Pleistocene stratigraphy appears to render such a methodology inappropriate. The assertions of Kerney (1977a), whereby the presence/absence of certain species (for example, *Gyraulus laevis* and *Gyraulus albus*) indicates a particular interglacial, appear to be less firm in this context. Nevertheless, some diagnostic features may exist. For instance, during the course of the Middle Pleistocene, species extinction seems to have occurred. *Tanousia* (*Nematurella*) *runtoniana* is thought not to have persisted later than the Cromerian Interglacial (Gilbertson 1980a,b; Preece 1989).

Temperate Mollusca seem to have been relatively rapid colonists of the British Isles following Devensian deglaciation (Keen *et al.* 1984, 1988) (Chapter 9), and remained in favourable habitats here at the end of the last interglacial despite a deteriorating climate (West *et al.* 1974). Thus, snails appear more reliable than plants as indicators of climatic amelioration at the commencement of a temperate episode but may be equivocal during the transition from the latter to a cold stage. Additionally, they are capable of providing information on climate in the absence of plant or other fossil data (Kerney *et al.* 1982).

1.3.4.6 Insects

Nowhere has more information accrued on Pleistocene insect faunas than in the British Isles, with assemblages of these animals especially well known from Late Pleistocene contexts (Chapters 9 and 10). The biological characteristics and applications to Pleistocene studies of the main group investigated, Coleoptera (beetles and weevils), have been reviewed by Coope (1970, 1979, 1986). In terms of environmental reconstruction, beetles seem to be ideal. Today they are abundant and occur in a wide variety of habitats, from below high-water mark to the summits of high mountains. Moreover, Coleoptera appear to have undergone relatively little evolution during the Pleistocene, with virtually all species occurring as fossils being still in existence. While they are thus of limited use as time-stratigraphic markers, inferences concerning their ecological tolerances in the past can be made with considerable confidence. A substantial number of beetles are stenotopic, with climatic parameters, substrates and food sources of particular relevance. In terms of climate, summer temperatures are critical for successful reproduction among insects, which because of hibernation are largely indifferent to those of winter.

Their ubiquitous nature has led to a profuse fossil record of Coleoptera, with well-preserved specimens often encountered. A fundamental property of insects is mobility, especially among those able to fly. Provided a suitable food supply is present, they can reach and colonize an area more quickly than almost any other animal, and considerably faster than plants, whose limited mobility is associated with their propagation mechanism. Thus, Coleoptera are arguably the most sensitive indicator of climatic changes and conditions during the Pleistocene. For instance, for the Devensian interstadials in progress *circa* 40 000 and 13 000 BP (Chapter 9), insect faunas have indicated temperatures in the British Isles that were warm enough for woodland growth, while the corresponding palaeobotanical records denote open vegetation traditionally indicative of colder climates (Coope 1977a). In addition to identifying changes in climate, assemblages of Coleoptera can also provide quite detailed evidence of climatic regimes. For example, in the Devensian a reasonably oceanic phase is indicated *circa* 43 000 BP (A. Morgan 1973; Girling 1974), while after about 41 000 BP, a more continental one replaced it, with insects which currently inhabit Tibet present in the British Isles (Coope 1962; Coope *et al.* 1971). Interglacial insect faunas appear less distinctive than those of stadials and interstadials, although they have not been as well studied. In general, interglacial faunas tend to be those of closed forest, and are relatively similar (Chapters 4, 6, 8 and 10).

Although, as noted above, a lack of evolution militates against using Coleoptera for chronostratigraphic purposes, their presence or absence may have some value in this context. The ability to identify species and the abundance of insect fossils means that the occurrence or non-existence of particular taxa may be a real feature of the succession. For example, large numbers of *Anotylus gibbulus* and *Stomodes gracilis* have only been recorded from deposits which may be of interglacial rank, and are aged to between the Hoxnian and Ipswichian stages (Green *et al.* 1984) (Chapter 7). However, the number of sites known for this temperate episode is small, so that such a faunal grouping must be tentative.

Other insect groups have not been so well studied as Coleoptera, but Hemiptera and Diptera (Buckland 1976; Crosskey and Taylor 1986) and Trichoptera (Moseley 1978) have been recovered from Pleistocene sediments in the British Isles. The majority of these are far more fragile and less distinctive in their species morphology than Coleoptera, so their use is limited. Most potential may exist in the robust cocoons of caddis-fly larvae (Trichoptera), the shed skins of juveniles being preserved in a cemented cover of leaf material or sand.

1.3.4.7 Ostracoda

As with Mollusca, ostracods (minute crustaceans) occur in both marine and freshwater environments. Like the former, Ostracoda underwent limited evolutionary change during the Pleistocene. Therefore, the present limiting factors of species should be either identical or similar to those found fossilized.

Marine ostracods have been found mainly in Pleistocene estuarine deposits, such as those in Somerset (Kidson *et al.* 1978) (Chapter 8) and Sussex (West *et al.* 1984) (Chapter 6). Higher energy sedimentary facies, such as raised beaches and the Crags (Chapters 2 and 3), are poor in Ostracoda due to the relative fragility of the carapaces of these organisms. The use of fossil ostracods from Pleistocene freshwater deposits in environmental reconstruction has been hindered by difficulties of species identification (Preece *et al.* 1984). However, fossil faunas identified in Britain have provided useful data on substrate conditions, amount of aquatic vegetation and water salinity (Keen *et al.* 1984, 1988; Preece *et al.* 1986). The latter condition is especially well shown by ostracods. Their sensitivity to small concentrations of salts in fresh water is evident at inland localities, such as Upton Warren in Worcestershire. Here brackish species, probably taking advantage of enhanced salinity in pools over permafrost (as occurs in the present tundra of arctic Canada), were living during the Devensian (Coope *et al.* 1961) (Chapter 9).

As in the cases of Mollusca and Coleoptera, Ostracoda appear to be rapid immigrants at the transition from cold to temperate conditions. For example, in Airedale, West Yorkshire, ostracods are present in the basal Devensian Lateglacial sediment of a kettle-hole lake infill, having arrived in this water body in advance of Mollusca (Keen *et al.* 1988).

Similarities between present and past species notwithstanding, the Pleistocene ostracod fauna of the British Isles is not well enough known to provide clear stratigraphic indicators. However, some taxa may be diagnostic of particular periods (Robinson 1980).

1.3.4.8 Foraminifera

These small, shelled protozoans, formally Foraminiferida (Brasier 1980), are marine, and like ostracods are found chiefly in fine-grained estuarine sediments. Certain taxa, such as *Elphidium* spp., are more robust than others (*Quinqueloculina* spp., for example) and than ostracods. Such robust forms are amenable to preservation in raised-beach sands or gravels (Headon Davies and Keen 1985; Mottershead *et al.* 1987). As with Ostracoda, foraminifera do not yet have a stable nomenclature, a situation which has hampered their study. Environmental reconstruction has been by comparison with factors influencing the modern range of species. In the British Isles, such a methodology has led, for example, to the identification of arctic faunas in Devensian Lateglacial deposits in western Scotland (Lord 1980), and also to a temperate assemblage including *Elphidium clavatum* (which today has an extreme northern limit in the English Channel) in the Ipswichian Interglacial raised beach at Portland (Haynes cited in Headon Davies and Keen 1985) (Chapter 8). Foraminifera were especially abundant in the Early Pleistocene marine and estuarine environments of the British Isles. Work by Funnell (1961a), West (1961a) and Funnell and West (1962, 1977) in East Anglia has enabled a series of pollen and foraminifera assemblage biozones to be set up (Chapter 3). The foraminifera data have supported the climatic oscillations inferred from palynological evidence, and have also afforded a check on the latter in shallow-marine sediments, where pollen taphonomy is complex (Traverse and Ginsburg 1966; Heusser and Balsam 1977; West 1980a). Thus, the exclusive use of palynological data from them may lead to erroneous environmental assumptions.

The other major use of foraminifera is in the delineation of cold and temperate climatic oscillations in deep-ocean sediments, a technique pioneered by Emiliani (1955). Although this succession (Figure 1.2(a)(b)) provides the background for global Pleistocene events and is the yardstick against which the terrestrial and offshore sequence of the British Isles must be assessed, the detail of it is outside the scope of this work. For further information, reference should be made to Emiliani (1955), Bowen (1966), Hecht (1976) and Shackleton (1977a,b).

While the majority of foraminiferid data have been used to infer climatic conditions, especially water temperature, in the British Isles (for example, Funnell 1961a), some may have chronological significance (Funnell 1987). The first appearance or extinction of species could delimit particular timespans. For instance, *Elphidiella hannai*, characteristic of Lower Pleistocene Crag sediments in East Anglia, seems to have been eliminated during the Anglian Cold Stage (Funnell 1988).

1.4 Dating

Two categories of age may be assigned to earth materials: relative, whereby a sequential order is established, and absolute, when an age in years before present (BP) is provided. Because of the large number of different Pleistocene deposits, a greater variety of dating techniques is applicable to them than to any other part of geological time. While relative methods can provide important information and are often the only ones applicable, it is absolute, or geochronometric (Bowen 1978), age determinations that are vital to Pleistocene chronology. For reliable absolute age estimates in the Pleistocene, determination by several methods is desirable. However, in most cases this is not possible and dating by one means has to suffice. This is because nearly all such methods are limited in applicability, either by virtue of their sensitivity to particular materials or as a consequence of a limited time range. For fuller discussions of dating methods and techniques, see for example, York and Farquhar (1972) and Lowe and Walker (1984).

1.4.1 Geochronometric age determinations

These are principally made in two ways. Firstly, by counting incremental units representative of a specific (often annual) time period; secondly, by radiometric means, using the mean decay rates of certain isotopes to approximate age in years. The most useful incremental accumulations are of laminated sediments, tree rings and lichens.

1.4.1.1 Laminations

Laminated sediments, found in lakes, are of two major types, proglacial rhythmites and organic varves due to algal blooms. Proglacial rhythmites are seldom encountered in the British Isles, and no chronologies comparable to those in Sweden (de Geer 1912) have been developed. However, short sequences of Devensian (de Geer 1935) and Wolstonian (Shotton 1953) (Chapter 7) rhythmites have been found. Varves formed by blooms of the diatom *Stephanodiscus astraea* have been described by C. Turner (1970) from Hoxnian lake deposits at Marks Tey, Essex (Chapter 6), while Simola *et al.* (1981) have reported complex algal laminations from the extant eutrophic freshwater lagoon of Loe Pool, Cornwall, which are of present interglacial age.

1.4.1.2 Tree rings

The equation of tree rings with calendar years is termed dendrochronology. In the southwestern United States of America, tree ring chronologies have been extended to in excess of 8000 BP, using *Pinus longaeva* (bristlecone pine) of the Sierra Nevada (Fritts 1976). In the British Isles a dendrochronology of a similar timespan, but containing hiatuses, has been established using *Quercus* (oak) (Pilcher *et al.* 1977) (Chapter 10). As elsewhere, an important aspect of dendrochronology in the British Isles has been its use, especially within archaeology, for the calibration of radiocarbon dates (Clark 1975; Suess and Clark 1976). Calibration is necessary because ages derived from dendrochronology and radiocarbon assay of wood diverge (Suess 1970; Renfrew 1973a). However, the dangers inherent in the acceptance of one calibration curve on a world-wide basis appear to be illustrated by dendrochronological and radiocarbon dating work in Ireland, where a different age relationship to that established in North America is evident (Pearson *et al.* 1977).

1.4.1.3 Lichens

A fundamental of lichenometry is that lichen growth rates can be established in a particular area by reference to their presence on surfaces of known age. Thereafter, size measurements of other lichens can be used to provide dates for their substrates. For fuller discussions of this technique, which is applicable to the last few hundred years of the Pleistocene, reference can be made to Beschel (1961), Locke *et al.* (1979), Innes (1985) and Worsley (1990a).

1.4.1.4 Radiocarbon

Details of the principles underlying the longest established, best-known and most widely used method of Pleistocene radiometric dating, and of the techniques used in age determinations, can be found, for example, in West (1977a), Lowe and Walker (1984) and Worsley (1990b). Its age limit appears to be *circa* 60 000 years, so that it is applicable only to Flandrian and some of Devensian time (Chapters 9 and 10). While this dating method has inherent problems, notably fluctuations in atmospheric ^{14}C levels and sample contamination by extraneous carbon, it remains the best available for establishing absolute ages in the Late Pleistocene. Indeed, the chronological framework for the latter period has mainly come from radiocarbon assays, with events during the 10 000 or so years of the present interglacial especially well dated.

1.4.1.5 Thermoluminescence

This fairly new method would appear optimal for dating Pleistocene sediments because it is applicable to quartz and feldspar silt. These mineral grains are abundant in a large number of Pleistocene contexts, so the potential for obtaining absolute ages by thermoluminescence (TL) is high (Wintle and Huntley 1982). The method has been mainly applied to loess. In the British Isles, considerable thicknesses of undisturbed aeolian silt (such as occur in mainland Europe and Asia), which would allow a timescale to be set up, are absent (Catt 1977a). A few ages have been obtained from Late Devensian loesses in England which are close to radiocarbon dates from organic material of this affinity (Wintle 1981; Wintle and Catt 1985). Attempts to date older fluvial sediments by thermoluminescence have met with mixed results. In one case, at Farnham, Surrey, silt has been dated to *circa* 106 000 and 107 000 BP by thermoluminescence, but organic matter from an analogous context has been dated to about 36 600 and 37 000 BP by ^{14}C (Bryant *et al.* 1983b; Gibbard *et al.* 1986a). However, silt from Kempton Park in the same county has given a mean thermoluminescence date of 36 600 BP (Southgate 1984), and included plant material with a radiocarbon age of *circa* 35 000 BP (Gibbard *et al.* 1982) (Chapter 9).

No demonstrably reliable thermoluminescence dates over about 125 000 BP have so far been obtained from Pleistocene sediments in the British Isles. However, thermoluminescence ages of around 228 000 and 202 000 BP for assumed Hoxnian fluvial deposits at Swanscombe, Kent (Bridgland *et al.* 1985), are not far from one estimate (*circa* 250 000 BP) of the age of this interglacial (West 1977a) (Chapter 6).

Thermoluminescence has also been used to date burnt flint from Palaeolithic archaeological sites at Pontnewydd, Clwyd (Huxtable 1984) and La Cotte de St. Brelade, Jersey (Huxtable 1986). In both instances, ages in the 200 000–250 000 year BP range have been obtained, which seem in keeping with the contexts of the sites (Chapter 7).

1.4.1.6 Uranium series

The potential age range of this disequilibrium dating method, using mainly ^{238}U and ^{235}U (Ford and Schwarcz 1990), makes it applicable as far back as the Middle Pleistocene (Ivanovich and Harmon 1982). If used on speleothems a fairly high level of accuracy can be obtained (Gascoyne *et al.* 1983; Atkinson *et al.* 1986). Results are less good for other precipitates such as tufa, although with corrections applied for contamination by detrital thorium, acceptable dates are possible (for example, Green *et al.* 1984) (Chapter 7). Perhaps the most difficult materials for U-series analysis are shell and bone, in respect of which *post mortem* contamination may necessitate careful calibration of any date obtained.

So far in the British Isles, U-series has been mainly used to date Ipswichian and Devensian cave deposits (for example, Atkinson *et al.* 1978; Gascoyne *et al.* 1981; Keen *et al.* 1981; Sutcliffe *et al.* 1984; Stringer *et al.* 1986) (Chapters 8 and 9). Dates of up to 285 000 BP have been obtained from bone (Szabo and Collins 1975) and tufa (Holyoak *et al.* 1983), but the problems noted above may have influenced these assays. In Jersey (Keen *et al.* 1981) and South Wales (Stringer *et al.* 1986), U-series dates have been used to calibrate the relative amino-acid timescale (section 1.4.2.1).

1.4.1.7 Potassium–argon

Although this method ($^{40}K/^{40}Ar$) has been widely used in the British Isles for earlier geological times, its use for the Pleistocene has been negligible. The problem is that to be effective, potassium–argon dating needs minerals with a high potassium content. This requires either volcanic material, or marine sediments rich in glauconite. Apart from tephra (section 1.4.2.2), the former is unrepresented, while the latter are scarce in the Pleistocene stratigraphy of the British Isles (Shotton 1977a). The potassium–argon method, however, is of considerable relevance to Pleistocene chronology in the British Isles. This is because it has been a major means of calibrating magnetic stratigraphy (section 1.4.2.3), which in turn has provided a measure for the environmental changes indicated in the deep-ocean cores.

1.4.2 Relative age determination

There are also two major ways in which relative ages can be ascertained. The first involves chemical analysis; the second, the identification of distinctive lithostratigraphic units, magnetostratigraphic signals, and biostratigraphic and artifact assemblages.

1.4.2.1 Amino-acid diagenesis

This chemical method of obtaining the age of shell and bone works by the racemization of amino-acids (AAR) of the L-form (L-isomers) to the D-form (D-alloisomers). It has been used mainly on Mollusca (Miller and Hare 1980; Bowen *et al.* 1985, 1989). Racemization is time and temperature dependent; the warmer the climate, the faster the diagenesis. It is also species dependent. For example, some gastropods racemize at rates up to three times those of certain bivalves. Thus, the method can only be applied over small areas where temperature regimes can be assumed to have been similar since the death of the analysed mollusc, and also it must be used on shells with analogous rates of diagenesis. The chemical measures made are amino-acid ratios, which are then used to impart relative ages to biostratigraphic and lithostratigraphic units.

In the Pleistocene of the British Isles, amino-acid ratios have been used to rank raised beaches of the English Channel and South Wales coasts in temporal order (Andrews *et al.* 1979, 1984; Keen *et al.* 1981; Headon Davies 1983; Headon Davies and Keen 1985; Bowen *et al.* 1985; Bowen and Sykes 1988) (Chapters 4, 7 and 9), and to determine the relative ages of estuarine, freshwater and terrestrial deposits in England (Miller *et al.* 1979; Bowen *et al.* 1989). The most reliable results appear to have been obtained from raised beaches, although even here the use of different preparation methods (Headon Davies and Keen 1985; Bowen *et al.* 1985) has caused problems with data comparability. Also the number of geochronometric dates available for calibration of the amino-acid information is small. None the less, a scheme of land–sea correlations for the Pleistocene of the British Isles has emerged (Bowen *et al.* 1989) (Figure 1.15).

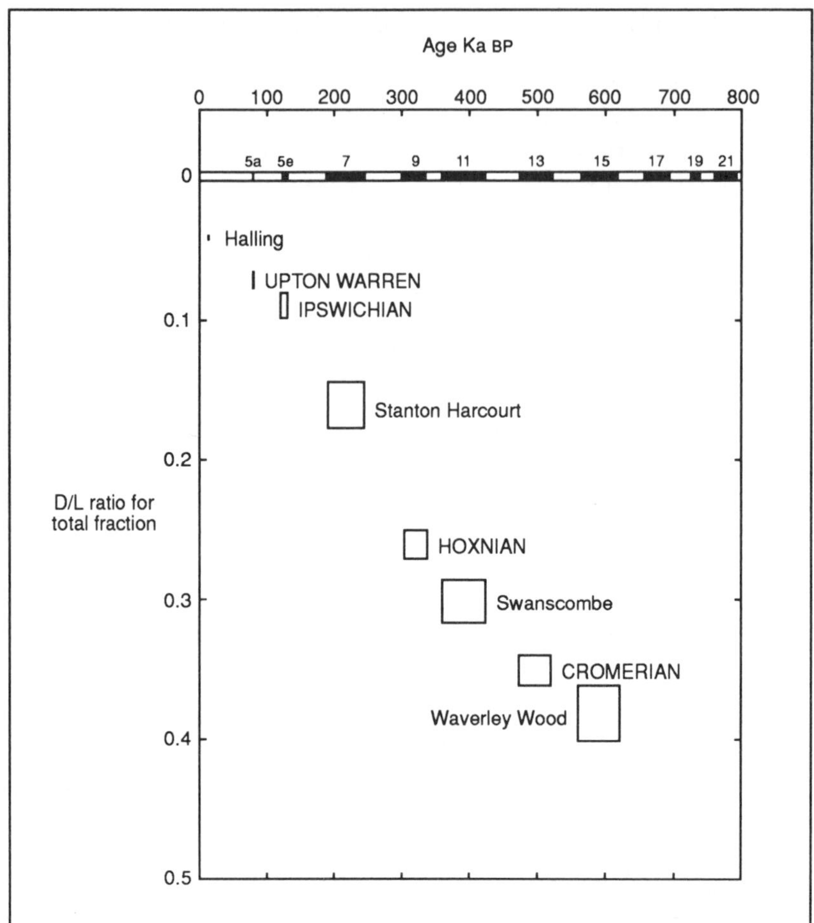

Figure 1.15 Correlation of D-alloisoleucine/L-isoleucine non-marine events in the British Isles with odd-numbered oxygen isotope stages as tuned to orbital timescales (Imbrie *et al.* 1984; Martinson *et al.* 1987). The vertical extent of the rectangles represents 1 standard deviation, and the horizontal extent is the duration of the oxygen isotope stage or substage with which correlation is made. From Bowen *et al.* (1989). Reprinted with permission from *Nature* **340**, 49–51; © 1989 Macmillan Magazines Ltd.

1.4.2.2 Tephrochronology

Layers of volcanic ash of known age and provenance can act as significant lithostratigraphic markers in the Pleistocene. To date, little tephrochronological work has been undertaken in the British Isles. However, Long and Morton (1987) have confirmed the existence on the United Kingdom continental shelf of a Devensian Lateglacial tephra layer of Icelandic origin (Chapter 9), identified by Mangerud *et al.* (1984) in western Norway. Ash sedimentation occurs in a wide range of environments over a considerable area. Thus, the search for such stratigraphic marker horizons may be among the more profitable of future Pleistocene chronological investigations.

1.4.2.3 Palaeomagnetism

The basis of this method, the changing polarity of the earth's magnetic field, has provided a general timescale for the Pleistocene when used in conjunction with $^{16}O/^{18}O$ ratios in deep-ocean cores (Berggren *et al.* 1980) (Figure 1.2(a)). The method involves identification of a relatively small number of normal and reversed polarity episodes. For these to be differentiated, long sedimentary sequences are necessary. As noted (section 1.2), these are not available on land in the British Isles. The Pleistocene deposits in the North Sea Basin have been related to the geomagnetic timescale by van Montfrans (1971). In the central and northern North Sea, the boundary between the Brunhes Normal and Matuyama Reversed polarity epochs has been recognized by Stoker *et al.* (1983) (Chapter 4). Skinner and Gregory (1983) have identified the Blake Event in the northern North Sea, and Stoker *et al.* (1985a,b) in the central North Sea (Chapter 8). On land, however, beyond recognition that some stages have normal polarity (for example, West 1980a; Zalasiewicz *et al.* 1988) (Chapters 3 and 4), little is known of the magnetostratigraphy.

There have, nevertheless, been a number of recent advances in the application of remanent magnetic measurements to Pleistocene deposits. For example, magnetic susceptibility provides an indication of magnetite concentration, while saturation remanence can indicate the types of magnetic minerals present. Comparative studies of sediments using the latter measure can point to sediment source areas, and assist in the study of patterns of erosion (Dearing 1986). Also, secular variation magnetostratigraphic dating has been successfully applied to Flandrian lake sediments in the British Isles (Thompson 1973) (Chapter 10). This technique is most secure for the past few hundred years, for which documentation of geomagnetic field fluctuations exists. These and other aspects of environmental magnetism have been examined in detail by Thompson and Oldfield (1986).

1.4.2.4 Flora and fauna

Variation in the make-up of plant and animal communities, as revealed by biostratigraphic investigations, has been a widely used means of relative dating in the Pleistocene of the British Isles. Sometimes, the presence or absence of a particular taxon (a vertebrate, insect or mollusc, for example), has been used to assign an age to a deposit. Of especial importance have been assemblages of pollen grains, from which inferences concerning flora and vegetation can be made (section 1.3.4.1). These pollen assemblages have been employed in the sub-division of interglacial stages in the British Isles (Turner and West 1968; West 1981a) (Table 1.2). The designation of site-specific (local) assemblage biozones can be followed by the delimitation of those with a wider (regional) significance (West 1970; Bennett 1988a). When biozone duration is determined by absolute age measurement, chronozones can be established. This has been accomplished for the Flandrian Interglacial (Hibbert *et al.* 1971;

Table 1.2 A zonal nomenclature system for the interglacial periods of the Late and Middle Pleistocene. From Turner and West (1968)

Zone names and numbers	*Stages of Middle and Late Pleistocene*				*Fossil pollen assemblage characters*
Early-glacial (e—)	e El	e Sa	e We		High NAP. Few or no thermophilous trees.
Post-temperate (IV)	Cr IV	Ho IV	Ee IV		*Pinus*, *Betula*.
Late-temperate (III)	Cr III	Ho III	Ee III	(Fl III)	*Carpinus*, *Abies* (*Fagus*), *Quercetum mixtum*.
Early-temperate (II)	Cr II	Ho II	Ee II	(Fl II)	*Quercetum mixtum*.
Pre-temperate (I)	Cr I	Ho I	Ee I	(Fl I)	*Betula*, *Pinus*.
Late-glacial (*l*—)	*l* Be	*l* El	*l* Sa	*l* We	High NAP, few or no thermophilous trees.

Stage name symbols
Interglacials: Cr: Cromerian, Ho: Holsteinian, Ee: Eemian (Fl: Flandrian, Post-glacial).
Glacials: Be: Beestonian, El: Elsterian, Sa: Saalian, We: Weichselian.

Hibbert and Switsur 1976; West 1980b) (Chapter 10).

The pollen-based scheme of Turner and West (1968) has reinforced a long-held view that there have been a relatively small number of Pleistocene interglacial stages in the British Isles (Godwin 1956). However, in an attempt to broaden the framework of Pleistocene events in the British Isles, Sutcliffe (1975) has suggested that some deposits assigned to the Ipswichian Interglacial on palynological criteria could be placed in separate temperate episodes on stratigraphic grounds and by using mammalian fossil evidence (Chapter 7).

The use of biostratigraphic information for dating purposes is thus controversial. As will now be apparent, plant and animal remains provide a clearer picture of past environmental factors than they do of age.

1.4.2.5 Artifacts

Variations in flint tool typology and in pottery characteristics have been used by archaeologists in order to effect intercultural and intracultural distinctions. However, such an approach has not been without problems. For example, Wymer (1977) has cautioned against the idea that technical refinement of stone tools was linked with temporal succession during the Lower Palaeolithic (Chapter 6), while Simpson (1979) has pointed out that the coincidence of beakers, food vessels and urns in Early Bronze Age contexts argues against their forming a chronological sequence in that order during the period (Chapter 10).

1.5 The nature of the evidence and the means of division of the Pleistocene of the British Isles

As noted (section 1.2), the most complete Pleistocene sedimentary sequences occur in the deep-ocean basins and in the loess deposits of continental interiors, where slow and reasonably continuous accumulation took place. These sequences have provided definitive information concerning both the number and character of climatic cycles (Figure 1.2(a)(b)). However, such successions are of limited value in the interpretation of Pleistocene environmental changes in the British Isles. Here, the legacy of rigorous processes is a suite of deposits whose temporal continuity on land rarely extends beyond one or two stages, so that close comparison is not possible. None the less, the British Isles will have been affected by the same gross climatic changes as the Atlantic deep-ocean basin and Eurasian continental interior. As a result, their terrestrial and offshore evidence must be considered in the context of information pertaining to these areas.

Although formidable exercises, some attempts have been made to link parts of the Pleistocene sequence of the British Isles with those of a more complete nature. For example, varve chronology has been used by Shackleton and Turner (1967) to correlate the duration of the Hoxnian Interglacial in the British Isles with that of Oxygen Isotope Stage 7 in the marine succession, as identified by Emiliani (1966). Bowen *et al.* (1989) have employed amino-acid data from Mollusca in a correlation of land-based evidence of Pleistocene events in the British Isles with those deduced from the oceanic record (Figure 1.15). In general, however, different fossil assemblages and dating methods make comparisons between terrestrial and oceanic sequences difficult. As evidence of the existence of many rather than few cold and temperate stages during the Pleistocene has emerged from the study of long records, the degree of confidence with which deposits from the fragmented sequence in the British Isles can be assigned to a particular climatic cycle has diminished. While the range of environmental evidence discussed earlier in this chapter can assist in ordering the succession, it is undeniable that scope now exists for the recognition of a considerably greater number of cold and temperate stages than those formally proposed for the British Isles by Mitchell *et al.* (1973a).

In the absence of any generally applicable method of dating, and of a universal means of correlation by either lithostratigraphy or biostratigraphy, Pleistocene successions have to be built in all possible ways. Traditionally, the definition of Pleistocene stages in the British Isles has been based on a limited number of lines of evidence, especially that of palynology. Interglacials and interstadials have been diagnosed by virtue of substantial arboreal pollen frequencies, with cold-stage floras identified on the basis of high values of non-arboreal pollen. Contrasting assemblages of tree pollen have been demonstrated as characterizing various parts of an interglacial, and have been considered sufficient evidence to distinguish between different episodes of this rank (Turner and West 1968; Turner 1975) (Table 1.2). However, when certain temperate-stage deposits, previously pigeon-holed on palynological criteria, have been reconsidered using other lines of evidence, for example, mammals, different ages have been suggested (Sutcliffe 1975, 1976; Currant 1989). Also, the use of a wide range of analytical techniques at new sites, whose investigators have been prepared to adopt a flexible approach to Pleistocene chronology, has led to the recognition of possible additional stages (for example, Briggs *et al.* 1985). Furthermore, some such studies have concluded that episodes during which the climate was temperate and of an interglacial or interstadial character may not have possessed forest vegetation (for example, Green *et al.* 1984).

Although cold-stage floras have been identified, the recognition of glacial and periglacial episodes has not relied mainly on plant or other biological evidence, but rather on that of stratigraphy and geomorphology (West 1988a). The major problems of the cold stages in the Pleistocene succession of the British Isles are, firstly, that not all produced glaciation as far south as the lowlands. Secondly, that only a small part of some of these stages was characterized by ice advances. Many of the even-numbered (cold) stages of the oceanic succession (Figure 1.2(a)(b)) may be represented in the lowlands of the British Isles by repeated deposition of fluvial gravels as river terraces and by periglacial landscapes, which are both very difficult to date.

In the highlands of the British Isles, glaciations have been more numerous, and the intensity of erosion has greatly reduced the stratigraphic evidence. In Scotland, only areas which may have been peripheral to the last major glaciation, such as north-east Aberdeenshire and the Western Isles, are likely to have any pre-Devensian cold-stage records (Sutherland 1984). Neither North Wales nor northern England have any unequivocal pre-Late Pleistocene deposits in their uplands.

Glaciation has also complicated the stratigraphic sequence in the lowlands. Here the maximum ice cover, perhaps in the Anglian Stage (Chapter 5), resulted in deposits that almost totally obliterated all but the major landscape features north of the current estuaries of the rivers Severn and Thames. Throughout this area, deposits of earlier stages are only visible as a result of subsequent landscape dissection. For example, over large areas of gentle relief in the English Midlands such exposures are few.

A further problem of the Pleistocene sedimentary record has been caused by recent (neotectonic) activity (Vita-Finzi 1986). Tectonic depression tends to preserve sediments from erosion. However, it also enhances their chances of burial beneath later deposits, thus making them difficult to examine. Conversely, tectonic uplift can accelerate erosion, thereby rendering deposits visible. The erosion though will lead to sediment removal. In the British Isles today, the major area of tectonic depression lies roughly between the rivers Humber and Thames and peripheral to a major tectonic basin in the North Sea (Smalley 1967; Hall and Smythe 1973; Sclater and Christie 1980). This tectonic basin has almost certainly persisted throughout the Pleistocene, during which there is strong evidence for subsidence in the southern part of the North Sea Basin (West 1972). Clarke (1973) has calculated a maximum subsidence rate in excess of 0.50 m per 1000 years for the North Sea Basin during the Pleistocene and has suggested that more than half of the sinking occurred over the last 15 million years. The effects of subsidence must be considered together with isostatic movements of the crust consequent upon glaciations. Adjustments following the last glaciation are still in progress in the northern and western sectors of the British Isles. Uplift is occurring in places where the land was depressed by the weight of ice. Elsewhere, stability has often been assumed, but recent local studies have indicated that this is questionable. For example, King (1977) has suggested that neotectonic uplift of south-east England is probably responsible for the current height above OD of the Red Crag in Surrey (Chapter 3), and late Flandrian deposits in Jersey may have been affected by block elevation between old faults (Jones *et al.* 1987) (Chapter 10).

To these tectonic matters must be added sea-level fluctuations consequent upon the glacial/interglacial cycles. Eustacy has had a profound effect on base level and hence erosion. Sea-level has risen during the current interglacial after its fall in the last cold stage. This rise has led to the submergence of sediments and landforms belonging to periods of lower sea-level. Data on some of these sediments and landforms are available from offshore boreholes and seismic surveys (McCave *et al.* 1977; Davies *et al.* 1984; Pantin and Evans 1984; Cameron *et al.* 1987). However, this information must be considered in the context of often widely separated boreholes and frequent lack of firm lithological or palaeontological control. Thicker deposits and more complete sequences notwithstanding, it is prudent to recall the problems involved in examining the terrestrial Pleistocene record, which is much more accessible, when considering the offshore evidence.

Over and above the uncertainties just discussed, there is the enigma of definition. It is noted above that on land, cold stages are most often defined by the occurrence of sediments and landforms, while temperate ones are usually identified on biological criteria. As Shackleton and Turner (1967) have stated, it is vital that the descriptions of such episodes enable their occurrence to be detected in ocean sediments. They have noted the difficulties inherent in this task, particularly with regard to interglacials. Cold stages (involving glaciation/periglaciation) can be recognized in the marine sequence from isotopic evidence of sea-level falls. Interglacials imply an ice cover analogous to that of today and a climate of sufficient warmth and longevity to enable particular modifications in biota, especially the plant cover, to take place. An interstadial is usually considered to be an episode of either insufficient length or warmth to permit the development of the climax interglacial plant cover of deciduous forest (West 1977a). Available data indicate that there are pitfalls in the rigid application of these terms. They point to the existence of a number of different events of variable duration along a continuum from full glacial to optimal temperate conditions. As West has observed, such terms have a rather local significance related to the type of climatic shift in a particular area, and care is needed in correlating the events which they represent.

1.6 Conclusion

It thus seems possible that both cold and temperate stages additional to those traditionally identified may be recognizable in the Pleistocene succession of the British Isles. In later chapters, this possibility is explored against the background of the conventional sequence.

2

Prelude to the Pleistocene

Miocene and Pliocene deposits are not extensive in the British Isles. However, Walsh *et al.* (1972) have demonstrated that most areas have minor occurrences of sediments and evidence of weathering related to this timespan (Figure 2.1). As a result, knowledge of environments during the approximately 21 million years of these epochs is limited. The paucity of Neogene (Table 1.1) rocks is largely attributable to tectonic activity associated with the convergence of the African and European plates. Miocene deformation was not uniform. Scotland, northern England, Wales and parts of Ireland were uplifted, intervening areas downwarped; a tectonic pattern that remained operative in the Pliocene. Erosion was dominant, with the sea seemingly playing a major rôle in landscape evolution (Owen 1976). Miocene and Pliocene environments were quite different from those of the Pleistocene. This difference was associated with a significant climatic change at the end of the Pliocene, information concerning which is essential for an understanding of subsequent events. Localities referred to within this chapter are shown in Figure 2.1.

2.1 Miocene environments

The plate movement that caused the Alpine Orogeny led to a marine regression from the British Isles. A North Sea whose margins were not far from those of today and which, as earlier in the Cenozoic, continued to be an area of subsidence, was thus established (Pomerol 1982). The erosional activity noted above meant that few Miocene deposits were preserved on land in the British Isles. Also, as we shall see shortly, the age of these and related deposits is uncertain, some perhaps forming earlier and others later in Cenozoic time. However, more extensive rocks referable to the Miocene have been recorded in the North Sea and Tremadoc Bay–Western Approaches basins offshore (Anderton *et al.* 1979). Palynological data from clayey lignites in the Mochras borehole, near to the landward margin of the Tremadoc Basin, suggest a Neogene (Herbert-Smith cited in Boulter 1971b) rather than Palaeogene (Wood and Woodland 1968) age for these deposits, some of which may belong to the Early Miocene (Rayner 1981). Away from the coast, most other potential Miocene deposits have been discovered in chalk and limestone solution hollows. The Lenham Beds have been primarily found in the former, at *circa* 190 m OD on the North Downs of Kent. They possess a marine molluscan fauna which has been referred to the Late Miocene by Curry *et al.* (1978).

The most informative sediments from an environmental viewpoint have been clays located beneath Pleistocene deposits in Carboniferous Limestone sink holes in the Peak District of Derbyshire (Boulter 1971a,b). These clays, belonging to the Brassington Formation (Boulter *et al.* 1971; Walsh *et al.* 1972), contain pollen and spores of woody and herbaceous angiosperms, conifers and pteridophytes. Macrofossils of

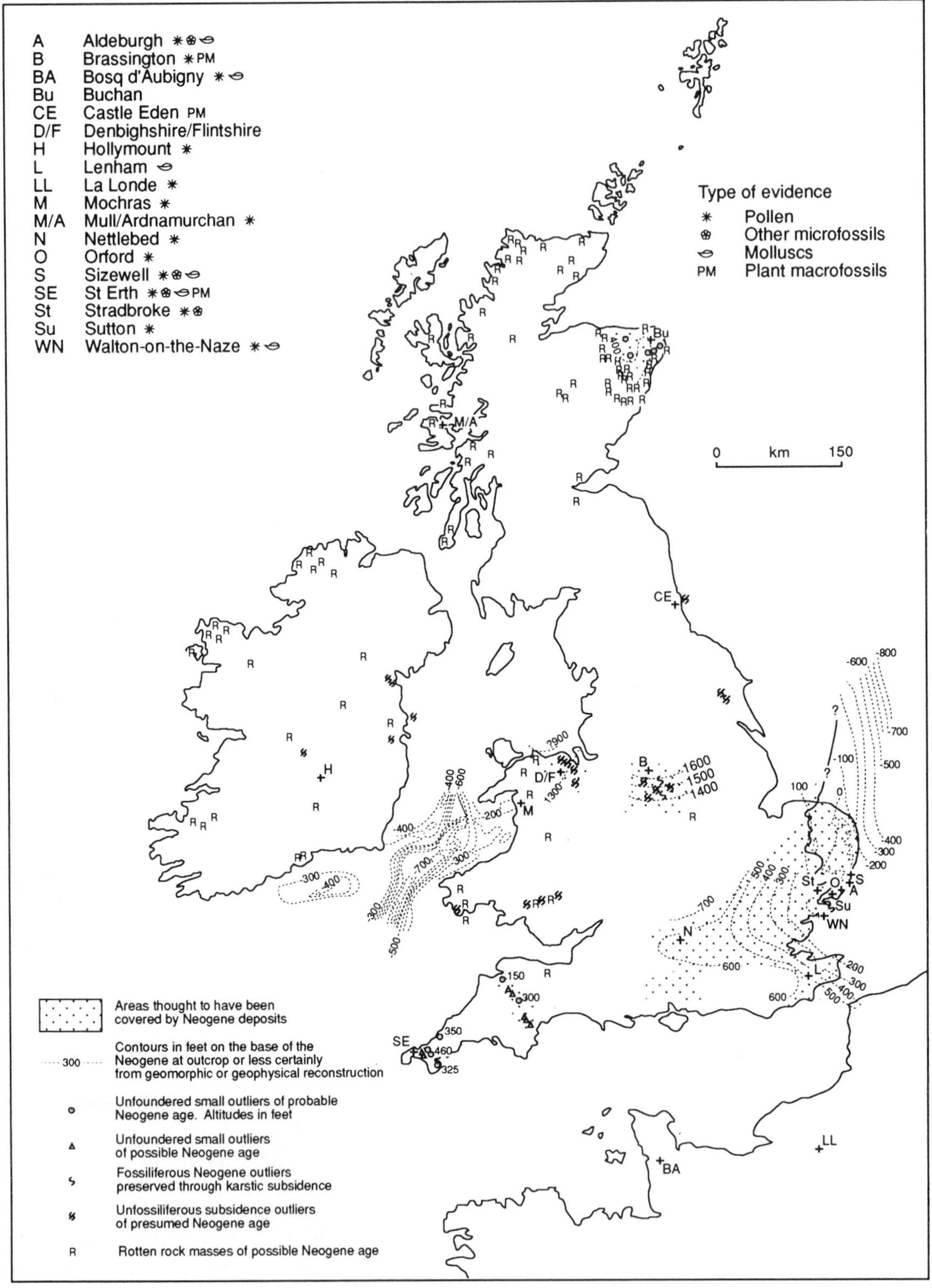

Figure 2.1 Location of some materials formed in the Neogene period in the British Isles and on the adjacent continental shelf. From Walsh *et al.* (1972). Reproduced by permission of the Geological Society of London and the authors. Localities referred to in Chapter 2 are also shown.

angiosperms, conifers, fungi and moss have also been recovered. The mosaic of vegetation probably included heath at higher elevations and woodland at lower levels in this part of the Pennines. Some 30% of genera in the flora are currently confined to tropical/sub-tropical localities, and a warm, oceanic climate can be inferred. A Late Miocene–Early Pliocene age has been assigned to these deposits, mainly on palynological evidence.

Palynology has also been used by Simpson (1961) to ascribe an Oligocene, Miocene or Pliocene age to sediments intercalated with basalts in Ardnamurchan and Mull. However, Mountford (1970) and Boulter (1971b) have disputed this, the latter stating that these Scottish pollen floras have little in common with those from the Derbyshire Neogene. Reid (1920) considered that a species-rich macroflora from an older fissure clay at Castle Eden, County Durham (Trechmann 1919) was of Middle Pliocene age. Boulter (1971b) has suggested that there could be links between this flora and those floras of Late Miocene–Early Pliocene age in Derbyshire. However, West (1977a) has preferred an Early Pleistocene age for the Castle Eden plant remains.

In Ireland, Neogene biogenic deposits discovered below till occupying a closed hollow in limestone at Hollymount, County Laois, have yielded a palynoflora including Palmae type (palms), *Symplocos* (sweetleaf), *Sciadopitys* (umbrella pine), *Liquidambar* (sweet gum), *Tsuga* (hemlock) and *Taxodium* type (swamp cypress). The dominant taxa in the assemblage are *Pinus*, *Quercus*, *Corylus* (hazel), *Myrica* (bog myrtle) and Ericaceae (heaths) (Hayes 1978; Watts 1985). Boulter (1980) has referred this assemblage to the Late Pliocene. However, Watts, noting a diverse and quite modern tree flora containing warm-temperate to sub-tropical swamp inhabitants, and comparing it with others from north-west Europe assigned by van der Hammen *et al.* (1971) to the Miocene or Early Pliocene, has suggested an equivalent age.

Walsh and Brown (1971) have used geomorphological criteria to assign pocket deposits in the Carboniferous Limestone of Denbighshire and Flintshire to the Neogene. Walsh *et al.* (1972) have considered the implications of the Brassington Formation for the development of the southern Pennines and other uplands in the British Isles. A land surface formed before either the Late Miocene or Early Pliocene, at an altitude of *circa* 450 m, has been demonstrated in the Peak District. Uplift of this and other upland areas, at the rate of about 1 m every 15 000 years since the Main Alpine Orogenesis in the mid-Tertiary, has been postulated. In north-east Scotland, the deep-weathered Buchan Gravels, which have been mapped up to *circa* 160 m OD, may be marine (McMillan and Merritt 1980; Merritt and McMillan 1982) or fluvial (Hall 1982, 1983, 1984a,b) Miocene or Pliocene deposits.

2.2 Pliocene environments

Although there are more Pliocene than Miocene deposits in the British Isles, they are also of restricted extent. East Anglia has the best-known sediments, and they also occur in Cornwall. Late Tertiary materials in Kent, Wales, Derbyshire and Scotland are equivocal, and may be either Miocene or Pliocene (section 2.1) (Figure 2.1). Also a Pliocene or Pleistocene age has been suggested for deposits in a number of localities in the British Isles, for example, west-central Ireland (Coxon and Flegg 1987) (Chapter 3).

In East Anglia the Pliocene deposits are part of the Crags (Figure 3.1). These are marine and estuarine sediments which accumulated close to the western limit of the southern North Sea. The oldest deposit, the Coralline Crag, has been identified only in south-east Suffolk and is probably of Early Pliocene age (Andrew and West 1977a; Hodgson and Funnell 1987). The Red Crag is next youngest and more extensive, having been located in Essex and Norfolk, as well as Suffolk. The Norwich Crag Formation has been found in Suffolk and Norfolk, above the Red Crag (Mathers and Zalasiewicz 1988). Part of these Crags is also probably of Pliocene age (Curry *et al.* 1978). However, the currently accepted view of the Pleistocene succession is that it begins at the base of the older Red Crag, whose type locality is at Walton-on-the-Naze, Essex (Mitchell *et al.* 1973a). Marine clays at St. Erth in Cornwall were probably deposited in the Late Pliocene (Mitchell *et al.* 1973b). Jenkins and Houghton (1987) have presented foraminiferal and calcareous nannofossil data which support a Late Pliocene age (2.1–1.9 million years BP) for the St. Erth Beds, and suggest that the climate was sub-tropical, with sea temperatures ranging between 18 and 24°C. The same fossil groups indicate that the Coralline Crag was deposited 3.6–2.3 million years ago, partly in the Early Pliocene and partly in the Late Pliocene. Like the St. Erth Beds, the Coralline Crag was laid down in fairly shallow water, but in temperatures ranging from 10 to 18°C. All of these sediments

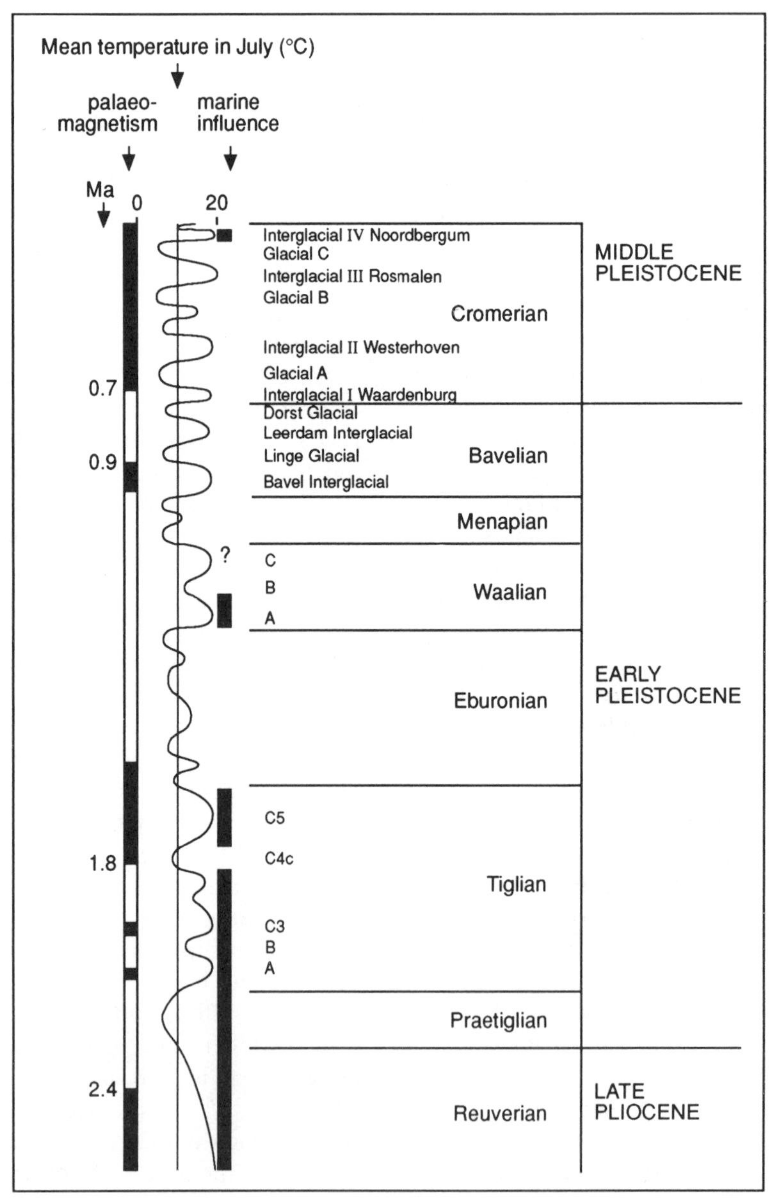

(b)

East Anglian stages	Age (Ma)	Magnetic polarity	East Anglian lithostratigraphic units			Netherlands stages possibly represented in East Anglia	Lithostratigraphic units in the southern North Sea	Netherlands stages ?unrepresented in East Anglia
CROMERIAN	<0.5		Bacton Member Mundesley Member West Runton Member	Cromer Forest-Bed	Kesgrave Formation	CROMERIAN IV	?	?CROMERIAN I–III BAVELIAN MENAPIAN (part)
BEESTONIAN PASTONIAN PRE-PASTONIAN B–D	<0.8 <1.3	■	Runton Member Paston Member Sheringham Member	Formation		?MENAPIAN WAALIAN	Yarmouth Roads Formation	
PRE-PASTONIAN A	<1.6	■	Sidestrand Member			?EBURONIAN	Winterton Shoal Formation	EBURONIAN (part)
BAVENTIAN BRAMERTONIAN ANTIAN	<2.0	▮	Westleton Beds Easton Bavents Clay Member Chillesford Clay Member Chillesford Sand Member	Norwich Crag Formation		?EARLY EBURONIAN TIGLIAN C	Smith's Knoll Formation	TIGLIAN B
THURNIAN LUDHAMIAN PRE-LUDHAMIAN	<2.4 <3.2	□ ▮	Thorpeness Member Sizewell Member	Red Crag Formation		PRAETIGLIAN REUVERIAN C	Westkapelle Ground Formation Red Crag Formation	TIGLIAN A REUVERIAN (part)
				Coralline Crag Formation		REUVERIAN BRUNSSUMIAN	Coralline Crag Formation	

Figure 2.2 Pliocene to early Middle Pleistocene sequences. (a) Climate curve and chronostratigraphy for the Late Pliocene to early Middle Pleistocene of the Netherlands. From de Jong (1988); source, mainly Zagwijn (1985). (b) Outline stratigraphy, and possible relationships of the East Anglian Pliocene to early Middle Pleistocene sequence to the sequences in the Netherlands and the southern North Sea. From Zalasiewicz and Gibbard (1988); sources, Funnell and West (1977), Cameron *et al.* (1984a), Mayhew and Stuart (1986), Zalasiewicz and Mathers (1985) and Zalasiewicz *et al.* (1988).

were either marine or estuarine in origin. Mollusca are well represented in them, while pollen and spores, foraminifera, ostracods, dinoflagellates (unicellular plankton) and bryozoans (colonial invertebrates) have also been recovered.

In the central and south-western Celtic Sea (south of the Scilly Isles), the upper member of the mainly sandy, Little Sole Formation is likely to be of Late Pliocene or Early Pleistocene age and of shallow-marine origin (Pantin and Evans 1984). Much of the Little Sole Formation was probably deposited by a meandering or braided-river system during an episode of low relative sea-levels. Near-coastal, perhaps inter-tidal or open-shelf marine deposition then occurred, indicating a transgression of the sea. Micro-palaeontological data from sediment belonging to the upper part of the Little Sole Formation indicate a marginally younger age than that of the St. Erth Beds (Evans and Hughes 1984). Thus, the marine episode close to the top of the upper Little Sole Formation, and the St. Erth Beds, are likely to represent the same transgression according to Pantin and Evans (1984).

The Coralline Crag has a long but patchy history of investigation, beginning with Charlesworth (1835) and including in its early phase, work by Prestwich (1871) and Harmer (1898). Following this there was no comprehensive study until that of Balson (1981). Much of the initial evidence for the environment of the Coralline Crag came from its Mollusca (Wood 1848–1882; Reid 1890); up to 20% of the genera having species with current Mediterranean distributions. A brief account of the foraminifera (Carter 1951) has suggested similar conditions, and the palaeoenvironmental conclusions of Jenkins and Houghton (1987) from foraminiferal evidence are noted above. Ostracod data (Wilkinson 1980) have also provided detailed ecological information. It has been suggested that the water was no more than 20 m deep, was homothermal with a temperature of *circa* 15°C, and had a salinity of 35 parts per thousand. An increase in brackish-water ostracods in the north of the outcrop implies the existence of estuarine conditions there. The ostracods are strongly indicative of currents from the north. Raffi *et al.* (1985) have studied the bivalve fauna of the Coralline Crag, concluding that it indicates temperatures of *circa* 20°C for several months annually.

A palynological investigation of silt from the base of the Coralline Crag at Orford, Suffolk, has led to the inference that a coastal marsh was present in the vicinity of the site. Pollen from now exotic trees, including *Sciadopitys*, *Taxodium*, *Sequoia* (redwood), *Pterocarya* (wingnut) and *Liquidambar*, have been recorded, together with that of the current European arboreal genera *Quercus*, *Pinus*, *Abies* (fir), *Picea* (spruce), *Betula* (birch), *Alnus* (alder) and *Ulmus* (elm) (Andrew and West 1977a). This assemblage is considered indicative of a diverse forest community beyond the local wetland vegetation. No modern analogue of this forest exists because of the varied distribution (in North America, eastern Asia and Europe) of its tree flora. Climatic inferences are thus also difficult, but the conditions suggested by the corresponding fauna are not incompatible with the palaeobotanical evidence. Gibbard and Peglar (1988) have reported the palynology of a sample from Coralline Crag deposits at Rockhall Wood, Sutton, Suffolk. This, dominated by *Pinus* and containing no *Sequoia*, probably belongs to the Reuverian C Stage of the Late Pliocene (Zagwijn 1960) (Figure 2.2). Mollusca, foraminifera and ostracods from the St. Erth Beds imply shallow coastal or estuarine waters and a warm-temperate or subtropical climate. There is a small pollen and plant macrofossil flora, which points to the proximity of salt-marsh and conifer-heath plant communities. The regional vegetation is not depicted, while typical Late Pliocene pollen taxa (*Liquidambar*, *Sciadopitys* and *Nyssa* (black gum), for example) are entirely lacking (Mitchell *et al.* 1973b). Correlation of the St. Erth Beds with the estuarine marl (marnes à *Nassa*) of the Bosq d'Aubigny in the south of the Cotentin Peninsula, Normandy (Buge 1957) has been suggested by Mitchell *et al.* (1973b) on the basis of comparable fossils (Margerel 1970). A study of their pollen record has shown that in addition to current western European genera such as *Quercus*, *Alnus*, *Ulmus* and *Corylus*, a number of exotics including *Tsuga*, *Pterocarya*, *Carya* (hickory), *Eucommia* (gutta-percha tree) and *Sciadopitys* are present (Clet 1982). The absence of *Taxodium* and *Liquidambar*, which are present in the Orford deposits, correlated with the Early Pliocene Brunssumian Stage of the Netherlands by Andrew and West (1977a) (Figure 2.2(b)), suggests that the marnes à *Nassa* are younger. Clet (1982) has proposed a Tiglian (Early Pleistocene) age for the French deposits and this may also be true of the St. Erth Beds, especially in view of their palaeobotanical record.

The maximum elevation at which the Coralline Crag has been discerned is *circa* 30 m OD (Balson 1981). The lagoonal St. Erth Beds have been recorded to 35 m OD, although their fauna implies a water depth of about 10 m, therefore indicating a sea-level of around 45 m OD (Mitchell *et al.* 1973b). The occurrence of these deposits at a similar height some 560 km apart may mean that the Pliocene sea-level was *circa* 30–50 m above that of the present. However, there are several complicating factors which make estimates of relative sea-level at this time particularly difficult. Firstly, sea-levels higher than those postulated above have been suggested for the Pliocene. The Lenham Beds of Kent, which may be of Miocene or Pliocene age (section 2.1), have been mapped at *circa* 200 m OD on the Chalk of the North Downs, as has the Nettlebed Gravel on the Chilterns near Henley-on-Thames (Horton 1977). Wooldridge (1927) thought that the latter sediment was also of Pliocene age and contemporaneous with the Lenham Beds. Wooldridge and Linton (1955) have indicated that these deposits both marked the height of the Pliocene sea and provided evidence for the amount of downwarping of the Coralline Crag in the North Sea Basin. As noted above (section 2.1), according to Curry *et al.* (1978), the Lenham Beds are of Miocene age, and thus relate to sea-level in this epoch. It is also worth observing at this point that Wooldridge (1960) has suggested an Early Pleistocene date for the Nettlebed Gravel, an interpretation currently favoured (Chapter 4).

None the less, some 300 km to the south of Suffolk, sequences in the Forêt de la Londe near Rouen, Normandy (Kuntz and Lautridou 1974), have revealed Pliocene estuarine deposits at heights of between 90 and 100 m above mean sea-level. Even if these sites in southern England and northern France are representative of varied environments and/or different episodes during the Pliocene, it seems improbable that sea-level can have fluctuated so much over a relatively short timespan, when the dominant climatic trend appears to have been one of gradual cooling. A possible explanation for the differing elevations of these sites may be forthcoming from the deposits around Bosq d'Aubigny in Normandy. These have been nearly split into two blocks by post-depositional faulting, with neotectonic movement reactivating Late Palaeozoic structural lines (Pareyn 1982). It is thus possible that the variation in altitude of the St. Erth, Coralline Crag and Normandy sequences is due to block movement along earlier faults. If so, reliable estimates of Pliocene sea-levels are not possible unless such movements can be quantified. However, the absence of numerous deposits of this age, except where they have been preserved by subsidence (Coralline Crag), down-faulting (marnes à *Nassa*), or in small graben structures (la Londe sediments), supports the contention that the general trend was of uplift, and that erosion has removed nearly all of the material, save where tectonic circumstances have been favourable. Pantin and Evans (1984) have concluded that the Celtic Sea shelf has been tilted towards the south-west because of the current level of the St. Erth Beds and upper Little Sole Formation, which are of nearly equivalent age. The latter now occur at 160 m depth and formed at about 50 m depth, the former are found at up to 35 m above sea-level. If, as Vail *et al.* (1977) have postulated, sea-level in the Late Pliocene–Early Pleistocene was similar to that of today, there has been uplift of the St. Erth Beds by some 45 m and depression of the upper Little Sole deposits by *circa* 110 m (Pantin and Evans 1984).

2.3 The Pliocene–Pleistocene boundary

The transition from the Pliocene to the Pleistocene in sedimentary sequences has been traditionally located where evidence of significant climatic deterioration is indicated by fossil records. In addition, radiometric, palaeomagnetic and oxygen-isotope analyses have been undertaken in order to date the event. At its original type locality at Le Castella, Calabria, Italy, and at Balcom Canyon, California, the Pliocene–Pleistocene boundary has been aged to *circa* 1.79 million years BP (Bandy and Wilcoxon 1970), while in the Netherlands a date of about 2.5 million years BP has been indicated by Zagwijn (1974a, 1975). Using biostratigraphic (calcareous nannoplankton and foraminiferal) correlations between deep-sea cores and the Le Castella section, Haq *et al.* (1977, 1978) have suggested a mean age of 1.6 million years BP. A new boundary stratotype has been designated at Vrica in Calabria (Bassett 1985), at the physical horizon immediately below the initial appearance of *Cytheropteron testudo* (Colalongo *et al.* 1981, 1982; Aguirre and Pasini 1985; Harland *et al.* 1990). Palaeomagnetic data from this locality indicate an age of *circa* 1.6 million years BP (Tauxe *et al.* 1983). According to Jenkins (1978), there are problems associated with the biostratigraphic correlations of Haq *et al.*, while Jenkins (1987) has suggested that Colalongo *et al.* (1981) have misidentified *Cytheropteron testudo*, and thereby misplaced the boundary.

In the British Isles, the usual indications of the progression from the Pliocene to the Pleistocene are a stratigraphic discontinuity, together with a significant increase in the incidence of marine Mollusca characteristic of northern seas and the first appearance of elephant and horse in the vertebrate fauna (Boswell 1952; West 1977a). The stratigraphic discontinuity occurs between the Coralline Crag and the Red Crag. The latter, whose fauna has demonstrated climatic cooling, has been referred to the Pleistocene (Mitchell *et al.* 1973a), with the former assigned to the Pliocene (Curry *et al.* 1978) (Figure 2.2(b)).

However, it has long been recognized that the Crag sequence is both complex and capable of alternative interpretation. For example, Reid (1890, 1899) has suggested that the Crags and the overlying Cromer Forest-bed Series (Chapter 4) are all of Pliocene age. Geikie (1894) has implied that the Pleistocene began with the Norwich Crag, while Zeuner (1937) has suggested that it commenced within the Red Crag. There have been a number of relatively recent studies which have a bearing on the ages of the Crag succession and the Pliocene–Pleistocene boundary. Stuart (1974, 1982) has examined the vertebrate fauna of the Red Crag, at the base of which is the Nodule Bed. The bulk of the vertebrate remains in the latter are referred to the Upper Pliocene, by comparison with Villafranchian faunas of this age on continental Europe. Most of the vertebrate fossils in the remainder of the Red Crag are considered to be reworked from the Nodule Bed. However, other vertebrate faunal evidence implies that the Red Crag could be assigned to the Lower Pleistocene (Stuart 1982). Funnell (1988) has presented data on Red Crag foraminifera. The occurrence of *Neogloboquadrina atlantica* in the Red Crag indicates that it was deposited prior to a cold episode which has been identified at *circa* 2.4 million years BP in the North Atlantic oceanic sequence. On these grounds the Red Crag is of Late Pliocene age.

In the North Sea offshore of the British Isles, the lower boundary of the Pleistocene has been placed at the top of the Red Crag, which has been referred to the Late Pliocene (Balson and Cameron 1985). The Pliocene–Pleistocene boundary in this area is adjacent to a polarity epoch change from Gauss Normal to Matuyama Reversed, dated to around 2.47 million years BP (Long *et al.* 1988) (Figure 1.1). In the southern North Sea, the Pliocene–Pleistocene boundary is proximal to a stratigraphic discontinuity which corresponds to a modification in the character of sedimentation (Cameron *et al.* 1984a). The magnetic polarity of the sediments above the unconformity is dominantly reversed (Cameron *et al.* 1984b).

For Suffolk, Zalasiewicz *et al.* (1988) have presented lithostratigraphic, biostratigraphic and chronostratigraphic data from the Red Crag and Norwich Crag between Aldeburgh and Sizewell. It has been concluded that the lowest part of the Red Crag probably was deposited in the latest Pliocene–earliest Pleistocene, within that time-span covered by the Reuverian C to Praetiglian stages of the Netherlands (Zagwijn 1960, 1985) (Figure 2.2(b)). Late Pliocene and Early Pleistocene deposits are particularly well developed in the Netherlands (Figure 2.2(a)). Zagwijn (1974a, 1975), assessing evidence from them and from the East Anglian Crags, has proposed that the transition from the Pliocene to the Pleistocene could be identified within the Red Crag sequence. The Pliocene – Pleistocene boundary is probably at the bottom of its Butley Crag Member, with the lower, Walton Crag being Pliocene. Such an age has been advocated by Hunt (1989) on the basis of palynological data from the Red (Walton) Crag at Walton-on-the-Naze. The pollen record, dominated by *Pinus* and containing Taxodiaceae together with taxa typical of mixed-oak forest, is comparable with that of Reuverian C (Late Pliocene) age. The assemblage differs from those in the Red Crag of the Stradbroke borehole (Beck *et al.* 1972) and at Sizewell (West and Norton 1974) (Chapter 3) which imply a more boreal plant cover. The Essex Crag has been referred to a Pre-Ludhamian stage, and contains an assemblage of foraminifera analogous to that of the Walton Crag (Beck *et al.* 1972), with which it has been correlated. The stage has been redesignated the Waltonian, the first of the Pleistocene (Mitchell *et al.* 1973a; Funnell and West 1977). Its palynology is similar to that of deposits assigned to the Praetiglian (earliest Pleistocene) in the Netherlands (Zagwijn 1974a).

It is noted above that Curry *et al.* (1978) have inferred that some of the Crag succession after the Coralline Crag could also be of Pliocene age. As Zalasiewicz and Gibbard (1988) have observed, revision of the date of the Pliocene–Pleistocene boundary to *circa* 1.6 million years BP could place a substantial part of the Crag succession within the Pliocene.

2.4 Summary

This chapter discusses work which has contributed to the identification, evaluation and correlation of what has proved to be a rather meagre quantity of Miocene and Pliocene deposits in the British Isles. It also addresses the complex and controversial issue of the position of the boundary between the Pliocene and Pleistocene, which is clearly some way from resolution.

3

The Lower Pleistocene before the Pre-Pastonian

The duration of the Early Pleistocene in the British Isles has not been quantified. However, reference to palaeomagnetic and radiometric data from other areas suggests that it was in progress for at least 800 000 years. In this account the stages referred to the Lower Pleistocene follow the scheme of Stuart (1982), who has justified the inclusion of the Pre-Pastonian and Pastonian in a Lower Pleistocene division (Table 1.1) on vertebrate faunal evidence. West (1980a) has noted that certain characteristics of Pastonian pollen assemblages link them to Early Pleistocene temperate-stage floras. However, in a scheme for East Anglia (West 1980b) the Pastonian and Pre-Pastonian stages have been designated as Middle Pleistocene. Sites referred to within this chapter are shown in Figure 3.1.

3.1 Problems of evaluation and correlation of Lower Pleistocene deposits

As noted in Chapter 2, the Lower Pleistocene Crag sediments of East Anglia (Figure 3.1) were deposited near to the edge of the southern North Sea Basin. Therefore, even minor movements of base level could have had a marked effect on deposition in this area. Small falls in sea-level would have been particularly important, leading to erosion and thus to local unconformities and hiatuses in the sediments. Gaps in a succession represent unknown spans of time. Although many erosion intervals are short, others can mark the passage of tens or hundreds of thousands of years, and thus may be responsible for either the removal or non-deposition of sediments representing a number of different climatic stages.

Other difficulties related to the Lower Pleistocene deposits of East Anglia are associated with their means of correlation from sections and boreholes (Mathers and Zalasiewicz 1985; Zalasiewicz and Mathers 1985). In general, as was recognized long ago (Harmer 1902), the Crag succession becomes progressively younger northwards. The Pliocene and earliest Pleistocene deposits are found in Essex and Suffolk, while the later preglacial Pleistocene (West 1980a) (Chapter 1) material occurs in Norfolk (Figure 3.1). Thus, firm lithological correlation by superposition has seldom been possible. Biostratigraphic criteria have enabled some association between units (West 1980a), but this approach too has been fraught with problems. The major means of association have been pollen and foraminifera assemblage biozones (Funnell and West 1977), and marine molluscan records (Norton 1977).

Taphonomic considerations hinder the interpretation of palynological data (West 1980a). The Crag sediments had shallow-marine and estuarine origins. Their pollen and spores may have been subjected to oxidation by groundwater and to

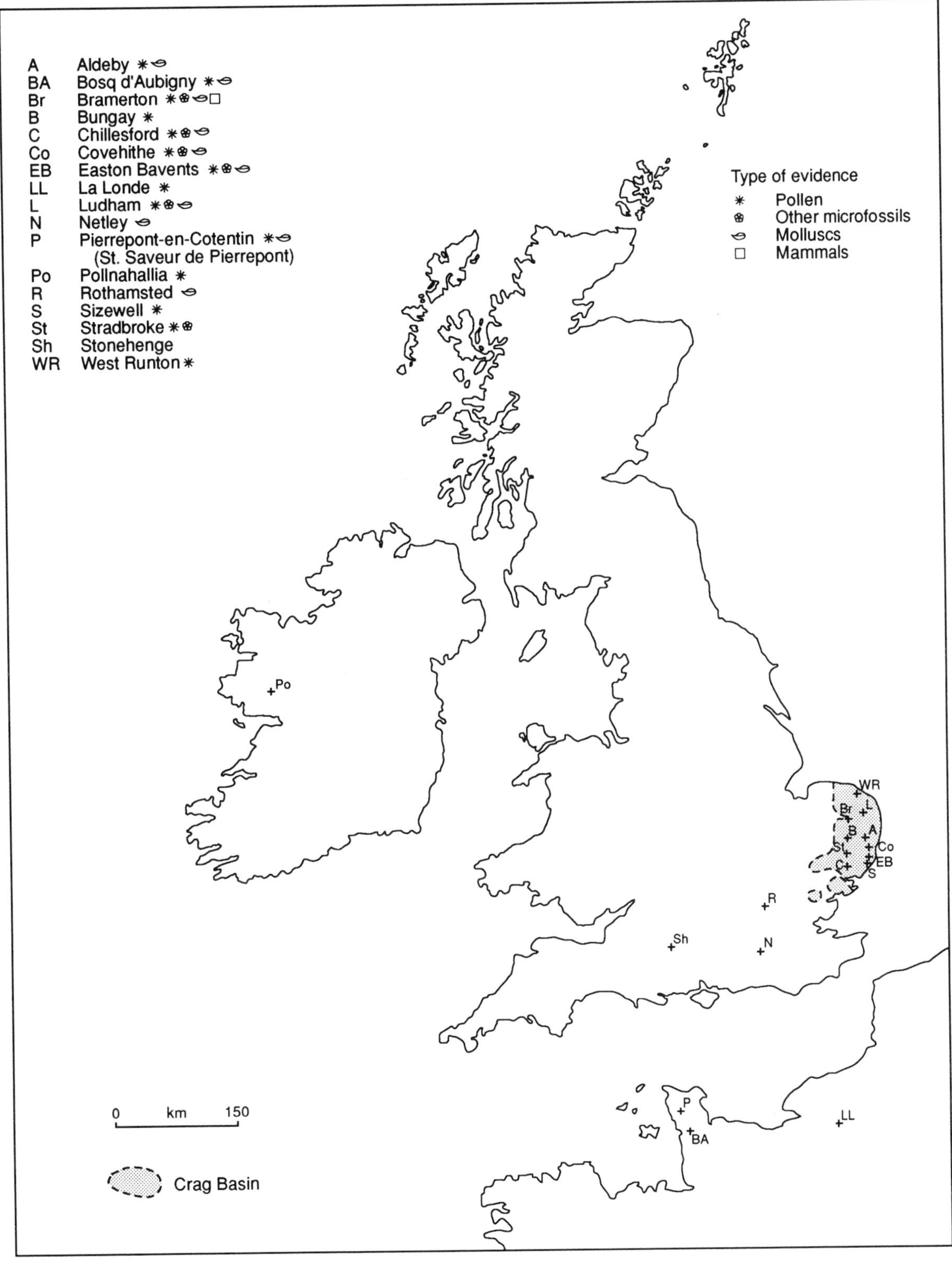

Figure 3.1 Location of the Crag Basin (source, West 1977a) and of sites referred to in Chapter 3 with types of evidence.

drying, which could destroy them (Gibbard 1988a). Pollen and spores also appear to be recruited differentially to estuarine and marine environments. This means that certain taxa are likely to be over-represented in palynological spectra from these environments. Over-representation is characteristic of conifer (especially *Pinus*) pollen. A winged form imparts bouyancy, which enables conifer pollen to float in water for considerable periods and be moved long distances in the atmosphere (Traverse and Ginsburg 1966; Faegri and Iversen 1975). In estuarine sequences, reworking of pollen and spores from pre-existing deposits may also be significant.

Different problems are apparent with Lower Pleistocene animal fossils. In the temperate stages, the abundance of fauna at some horizons has permitted reasonably definitive association between sites. However, correlation has been hindered by the similarity of faunas from rather uniform litho-facies at different stratigraphic positions. In the cold stages, low sea-levels and reduced biological productivity have combined to impoverish the record of estuarine and marine organisms. Reworking of foraminifera into the Red Crag and Norwich Crag from the Coralline Crag has probably occurred (Funnell 1988). The coarse sediments of the Red Crag were a poor medium for the preservation of diminutive and delicate Mollusca. In both the Red Crag and Norwich Crag, the marine molluscan faunas are indicative of having been transported (Long and Cambridge 1988). Also of particular importance for environmental reconstruction has been the recognition of significant gaps in the sequence of Lower Pleistocene deposits in the British Isles. Initial workers, such as Harmer (1902), believed that sedimentation had been continuous. However, detailed examination of the East Anglian succession (Funnell and West 1977) and comparison of it with that of the Netherlands (Zagwijn 1974a, 1975, 1985) have revealed this not to be the case, and has hampered correlation between them (West 1980a; Zalasiewicz and Gibbard 1988) (Figure 2.2(b)).

In Flanders and the Netherlands during the Early Pleistocene, similar shallow marine and estuarine conditions to those pertaining in East Anglia occurred beyond a shoreline cut in older formations (Zagwijn 1979, 1989; Long *et al.* 1988). The limit of this basin cannot be traced into either the Pas de Calais or Kent. However, Early Pleistocene deposits at la Londe (Kuntz and Lautridou 1974) and St. Sauveur de Pierrepont (Brebion *et al.* 1974) indicate the position of the coast in Normandy. It is thus unclear whether the Strait of Dover was in existence at this time. Faunal evidence suggests that for part of the Early Pleistocene, the southern North Sea was separated from the Atlantic (Funnell 1961a; Norton 1977). One view on sea-levels around this time was of heights in excess of 200 m OD, with shorelines on the Chilterns and North Downs (Wooldridge and Linton 1955) (Chapter 2). This interpretation would seem to be consonant with an enlarged Strait of Dover. However, Shotton (1962) thought that the latter was closed in the Early Pleistocene, and the notions of Wooldridge and Linton have been critically examined by Jones (1981) and Moffat *et al.* (1986). The most satisfactory reconciliation of the evidence may result from establishing that the Chalk, together with overlying sands and gravels of Red Crag age (possible remnants of which occur at Netley Heath, Surrey and at Rothamsted, Hertfordshire (Dines and Chatwin 1930; John and Fisher 1984)), have been uplifted, thereby forming a land bridge that extended to France (King 1977). Shotton (1962) considered that Ireland was not joined to the remainder of the British Isles during the Early Pleistocene. Water depth would then have been too great in the Irish Sea, and tectonic activity incapable of forging a land link. Mitchell (1977) has also stated that the Irish Sea was present at the start of the Pleistocene.

3.2 The stages and their environments

The stages described in this chapter account for most of Early Pleistocene time and occurred before the Pre-Pastonian (West 1980a) (Table 1.1). They have been proposed by Funnell and West (1977), and Funnell *et al.* (1979) on biostratigraphic criteria. The Pre-Pastonian and Pastonian deposits occur in association with Middle Pleistocene sediments on the Norfolk and Suffolk coasts, and are considered together with them in the next chapter. Funnell and West (1977) have also proposed a lithostratigraphic classification of the preglacial deposits of East Anglia. In this context, Zalasiewicz *et al.* (1988) have noted that lithostratigraphic studies have lagged behind those of biostratigraphy. They have stated that comprehensive lithostratigraphic information and its association with fossil evidence is confined to a sector of south-east Suffolk.

3.2.1 The Pre-Ludhamian

The deposits of this stage are sands and clays of tidal origin (Zalasiewicz and Mathers 1985). Palynological data are limited, but a *Pinus*-dominated assemblage, also containing *Picea*, *Betula* (birch), *Alnus*, Ericales (heaths) and Gramineae (grasses), and interpreted as representative of cool-temperate forest of this age, has been recorded from basal sediments in the Stradbroke borehole in Suffolk (Beck *et al.* 1972). A similar assemblage found in Crag sediments obtained from boreholes at Sizewell, Suffolk has also been referred to the Pre-Ludhamian (West and Norton 1974). Zalasiewicz *et al.* (1988) have delimited a *Pinus–Picea*–Ericales biozone, implying open boreal forest with associated heath and herbaceous vegetation, coincident with deposition of the lowest, Sizewell Member, of the Red Crag Formation. This pollen assemblage has been equated with that characteristic of the Praetiglian in the Netherlands (Zagwijn 1960). Funnell and West (1977) have stated that the Pre-Ludhamian was the first stage of the Pleistocene. However, using palaeobotanical criteria, West (1980b) considered the Pre-Ludhamian to be the probable equivalent of Substage C of the Reuverian (Late Pliocene) of the Netherlands (Figure 2.2(b)).

A possible Reuverian pollen assemblage has been reported from Pollnahallia, County Galway, Ireland by Coxon and Flegg (1987). Organic silt and clay overlain by sand and till, in the bed of a limestone gorge, contains *Carya*, *Pterocarya*, *Tsuga*, *Sequoia*, *Sciadopitys* and Taxodiaceae, as well as *Pinus*, *Picea*, *Abies*, *Betula*, *Ulmus* (elm), *Quercus*, *Alnus*, *Carpinus* (hornbeam), *Corylus*, *Salix* (willow), *Juniperus* (juniper), *Hedera* (ivy), *Ilex* (holly) and *Vitis* (vine). Coniferous trees and heath appear to have been of most importance in a diverse vegetation that also included shrub and herb communities. However, the authors have pointed out that these palynological data could alternatively be referred to the Tiglian Temperate Stage of the Early Pleistocene in the Netherlands (Zagwijn 1960).

In the British Isles, Stradbroke has been designated the type site for the Pre-Ludhamian (Funnell and West 1977). Its foraminifera have been described by Beck *et al.* (1972). On foraminiferal differences, the stage has been divided into a lower and an upper part. In the former, *Pararotalia serrata* is profuse and *Elphidium haagensis* occurs. These together with planktonic foraminifers, imply the presence of a warm-temperate environment, and of a link between the North Sea and Atlantic in the English Channel. In the upper part, *Pararotalia serrata* and *Elphidium haagensis* frequencies are considerably lower, with *Elphidium frigidum* occurring regularly. Cool-temperate conditions are suggested, with the planktonic foraminifera hinting that the North Sea and Atlantic had become separated in the English Channel area, with water from the latter reaching the former mainly via the north of Scotland. Funnell (1961a) has presented foraminiferal evidence from Red Crag deposits at the base of the Ludham borehole, Norfolk, which broadly agrees with that from Stradbroke, described above. Zalasiewicz *et al.* (1988) have assigned the lowest Foraminiferida assemblages in the Red Crag around Sizewell and Aldeburgh to the Pre-Ludhamian. They concluded that there are two parts to the stage; the early one equivalent to the Waltonian, the late one to the Butleyan division of the Crag. The fauna is dominated by forms characteristic of the open ocean, with the climate in the late Pre-Ludhamian analogous to that which currently pertains in the southern North Sea. Dinoflagellate fossils analysed by Zalasiewicz *et al.* (1988) have also suggested a Pre-Ludhamian age and a warm-temperate to sub-tropical depositional environment. The palaeomagnetic data of these authors support those of van Montfrans (1971) and Beck *et al.* (1972), revealing normal polarity in the Pre-Ludhamian sediments.

While some of the Red Crag molluscan faunas (Harmer 1902; Norton 1970; Cambridge 1977; Dixon 1977) are likely to be referable to the Pre-Ludhamian, little work has been accomplished on Mollusca from established deposits of this age (Funnell and West 1977; Long and Cambridge 1988).

The vertebrate remains in the Red Crag are discussed in the context of the age of the formation in Chapter 2. Here it is sufficient to note that Stuart (1982), referring much of the Red Crag to the Pre-Ludhamian on other faunal grounds, has suggested that vertebrate fossils not reworked into the Red Crag (including beaver, elephant, horse and deer) are probably of this age.

3.2.2 The Ludhamian Temperate Stage

This stage is represented by the lowest shelly marine sands and silty clays in the Ludham borehole, which extend to −49 m OD (West 1961a).

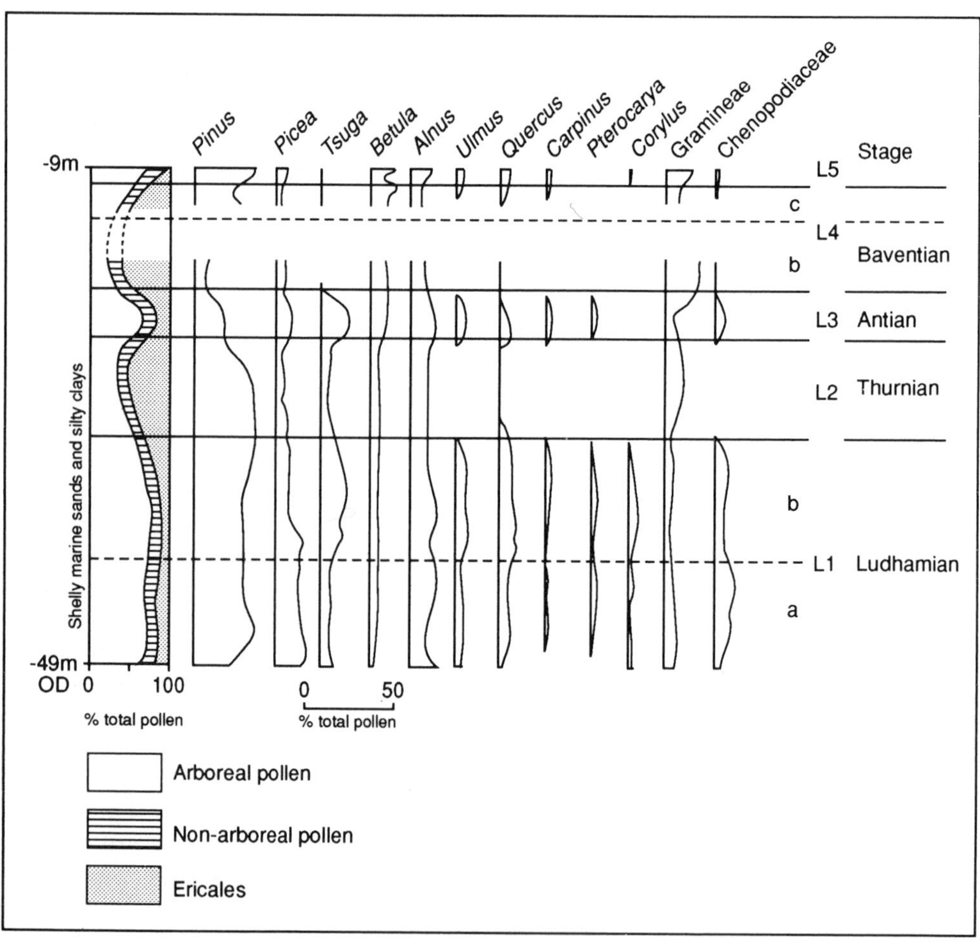

Figure 3.2 Pollen diagram from the Ludham borehole. From West (1980b); source, West (1961a).

The pollen flora of these sediments is dominated by the conifers *Pinus*, *Picea* and *Tsuga*. Of deciduous tree and tall-shrub pollen, *Quercus* and *Alnus* occur with greatest frequency; *Ulmus*, *Carpinus*, *Pterocarya* and *Corylus* having a subsidiary presence (Figure 3.2). This indicates a mainly mixed, temperate forest. However, there was also subsidiary grassland and heath, as Gramineae and Ericales pollen is present in reasonable amounts (West 1961a). Palynological data from Stradbroke indicate Ludhamian temperate forest in which *Pterocarya* and *Tsuga* were prominent (Beck *et al.* 1972).

The foraminiferal record from the type site for this stage, at Ludham, has been described by Funnell (1961a). The taxa in its lower part are not as numerous as, but are analogous to, those of the lower part of the Pre-Ludhamian. However, the upper part of the Ludhamian possesses a markedly contrasting assemblage of Foraminiferida. Elphidiidae dominates a fauna now characteristic of shallow water in estuaries and bays that possess a temperate climate. Part of the Stradbroke foraminiferal sequence is also found in the Ludhamian (Beck *et al.* 1972). The assemblages of foraminifera from the Thorpeness Member of the Red Crag Formation around Aldeburgh and Sizewell may permit its correlation with the Ludhamian Stage (Zalasiewicz *et al.* 1988).

Mollusca from the Ludham borehole have been investigated by Norton (1967, 1977). The overall picture is of a temperate Ludhamian fauna, containing species that today inhabit a variety of marine environments. The bulk of the Mollusca are currently of north-west European type. However, some of the taxa are now Lusitanian, and two species (*Macoma calcarea* and *Serripes groenlandicus*) are arctic. Such a mixed assemblage has made inferences concerning its environment difficult. For example, the presence of Mollusca characteristic of both colder and warmer conditions than now pertain in the southern North Sea could be either a pointer to the existence of traits (perhaps seasonal) of such climates, or signify that the taxa concerned have changed their ecological tolerance since the Early Pleistocene. On present day environmental criteria, water 40–15 m deep and getting shallower as the stage progressed has been postulated, with inter-tidal/brackish conditions suggested by the initial occurrence of *Hydrobia ulvae* in the late Ludhamian sediments.

Some members of the Ludhamian molluscan fauna may provide an insight into the position of the south coast of the North Sea at this time. Norton (1977) has noted the occurrence of aberrant large, thin-shelled specimens of *Spisula solida* and *Spisula subtruncata* within the Ludhamian. These deviant forms were considered to be possibly due to a lack of ecological competition which resulted from their residence in a limited marine basin. This limitation could have been heightened by the presence of an unbreached Chalk ridge from the Chilterns to the Paris Basin which enclosed the southern North Sea (section 3.1). Low salinity has also been cited by Norton as a potential cause of distortion in marine Mollusca. In the eastern Baltic today, there is more fresh than salt water because the sea can only penetrate via the narrow straits between the Danish islands. Aberrant marine Mollusca occur in this region. Separation of the southern North Sea from the Atlantic by a land bridge in the Early Pleistocene could have given rise to conditions similar to those which exist in the eastern Baltic at present. Large quantities of fresh water discharged by the Thames and Rhine river systems into an enclosed southern North Sea (Gibbard 1988b) may have reduced its salinity and thereby influenced molluscan morphology.

Outside the Crag Basin, sands and gravels at two localities have yielded a marine molluscan fauna of Red Crag type. They are both on Chalk, at Netley and Rothamsted in southern England. As already noted (section 3.1), these sands and gravels probably owe their current elevations to neotectonics. Their molluscan fauna, of Ludhamian affinity, has provided further evidence of the southern and western limits of the North Sea at this time. Mathers and Zalasiewicz (1988) have suggested that the altitudes of Red Crag in Essex (*circa* 90 m OD) and at Rothamsted (about 130 m OD) have been attained subsequent to its deposition, by the inclination of East Anglia in reaction to downwarping in the southern North Sea Basin.

The difficulty of identifying the Ludhamian in surface exposures of Crag has contributed to the absence of vertebrate remains assigned to this stage (Stuart 1982).

3.2.3 The Thurnian Cold Stage

Deposits ascribed to this stage occur in the Ludham and Stradbroke boreholes. It commenced when the forest vegetation depicted by the pollen record of the Ludhamian sediments changed in character to boreal, with *Pinus* dominant. A substantial increase in Ericales and Gramineae pollen implies that oceanic heath was widespread and the climate colder than that in the Ludhamian. A significant reduction in the frequency of Chenopodiaceae (goosefoot, orache, glasswort) pollen is probably the result of the less proximal presence of plant communities tolerant of saline conditions consequent upon a marine regression (West 1961a, 1980b) (Figure 3.2).

The Thurnian foraminiferal assemblage, with *Elphidiella hannai* important, is less diverse than that of the Ludhamian and is compatible with a lower water temperature (Funnell 1961a, 1988).

Mollusca are absent from the Thurnian deposits at Ludham (Norton 1967). This absence may have been associated with a eustatic fall in sea-level (Norton 1977). The possibility of an unconformity or hiatus in the Thurnian deposits of the Ludham borehole has been mentioned by Funnell (1961a).

Thurnian sediments have been difficult to identify in exposures and may belong to either the Red Crag or the Norwich Crag Formation. In consequence there is no knowledge of their vertebrate remains (Stuart 1982).

The Thurnian environment appears to have included climatic conditions that were colder than those which currently exist over the British Isles. Temperatures may have been low enough to permit the accumulation of at least localized ice masses in the northern and western highlands while this cold stage was in progress, but there is no evidence of these. In the Netherlands, part of the Tiglian marked by a significant deterioration in climate and when a mean summer temperature of a little over 10°C was likely, has been equated with the Thurnian. The palaeomagnetic timescale of Cox (1969) indicates that this episode occurred *circa* 1.90–1.85 million years BP (Zagwijn 1975).

3.2.4 The Antian Temperate Stage

The Thurnian deposits in the Ludham borehole are succeeded unconformably by marine sediments (West 1980b) which may have been laid down by a rising sea. The palynological record from the borehole supports this contention, having higher Chenopodiaceae frequencies than occur in the Thurnian. The regional vegetation seems to have been analogous to that of the Ludhamian. Temperate mixed forest, in which pine was of less, and birch greater, importance and where *Tsuga* and *Pterocarya* occurred, is indicated (West 1961a) (Figure 3.2). Evidence from foraminifera (Funnell 1961a) and Mollusca (Norton 1967) in the Ludham borehole is consonant with temperate conditions during the Antian, and with sea-levels akin to those of the Ludhamian.

The Antian is also represented in a surface exposure at Easton Bavents near Southwold in Suffolk, which has been designated its type locality. Pollen and spores, foraminifera and marine molluscs from this section have been studied (Funnell and West 1962; Norton and Beck 1972).

While fish remains are abundant in this part of the Norwich Crag those of other vertebrates are rare, although *Equus stenonis* (extinct horse) and *Archidiskodon meridionalis* (extinct elephant) fossils have been recovered, implying that open woodland was present (Stuart 1982).

Zagwijn (1975) has correlated the Antian with Tiglian Zone C5 of the Netherlands sequence (Figure 2.2). Mean summer temperature has been estimated to be just below 20°C, and palaeomagnetic dating placed this stage at *circa* 1.85–1.65 million years BP.

3.2.5 The Baventian Cold Stage

Coastal deposits at Easton Bavents (section 3.2.4) provide the type section for this stage. This locality has yielded pollen, spores, foraminifera (Funnell and West 1962) and Mollusca (Norton and Beck 1972), together with vertebrate remains (Stuart 1982). Sediments in the Ludham borehole have also provided fossil evidence for Baventian environments (Funnell 1961a; West 1961a; Norton 1967), as have deposits at Aldeby, Norfolk (Norton and Beck 1972), Sizewell (West and Norton 1974), West Runton, Norfolk (West and Wilson 1966) and Covehithe, Suffolk (West *et al.* 1980).

The palynological record from Ludham indicates that tree pollen frequencies were fairly low, and that the taxa present were of boreal forest affinity. High non-arboreal pollen values, to which Ericales made a substantial contribution, allow the inference that grass heath was widespread and the climate cooler than during the previous stage to be made. The curve for Chenopodiaceae pollen falls implying, as previously in this sequence, that a marine regression took place (West 1961a) (Figure 3.2).

Foraminiferal assemblages from Ludham and Covehithe contain boreo-arctic forms including *Elphidium subarcticum*, *Elphidium orbiculare*, *Buccella inusitata* and *Cibicides lobatulus* (Funnell 1961a; West *et al.* 1980). At Easton Bavents and Covehithe, both the pollen and foraminifera signal a climate which gradually became colder. The foraminifera denote a sub-littoral or low inter-tidal environment. Associated sediment could have been deposited during a marine regression (Funnell 1988).

Mollusca, although not common, point to similar conditions. At Ludham the early Baventian fauna resembles that of the first part of the Antian (Norton 1967). The Covehithe assemblages have been the most informative, with the cold-climate indicators *Astarte borealis*, *Macoma calcarea*, *Serripes groenlandicus* and *Yoldia myalis* recorded (Long 1974, 1988; West *et al.* 1980).

Vertebrate fossil data for the Baventian are equivocal. *Equus stenonis*, *Archidiskodon meridionalis*, *Eucladoceros falconeri* (extinct deer), *Anancus avernensis* (gomphothere mastodont), *Mimomys pliocaenicus* and *Mimomys blanci* (extinct voles) may have been members of the fauna early in the stage (Stuart 1974, 1982; Mayhew and Stuart 1986).

With the intensity of cold suggested from several lines of evidence, it is possible that glacial ice entered the North Sea Basin, perhaps for the first time, during the Baventian (Catt 1981) (Figure 3.3). Turner (1975) has stated that Baventian palynological data indicate an episode with very low temperatures. Hey (1976) has reported *Rhaxella* chert from Jurassic deposits in Yorkshire as a component of Baventian marine gravel at Easton Bavents. West (1980b) has cited the occurrence of metamorphic minerals, whose most likely provenance was Norway, in Baventian sediments of Suffolk. These materials may have been brought south to East Anglia either by icebergs or in glacial outwash. Scattered remnants of weathered material resembling till in Oxfordshire (part of the Northern or Plateau Drift) (Shotton *et al.* 1980) and Hertfordshire (Chiltern Drift) may be evidence for one or more Early Pleistocene glaciations (Evans 1971; Catt 1981; Jones 1981). It has been suggested by Kellaway (1971, 1977) that dolerite (bluestone) and ignimbrite from Wales could have reached the Marlborough Downs and Salisbury Plain by means of glacial activity in the Early Pleistocene, and that local Greywether sandstone boulders (sarsens), which like those of bluestone were used in the building of Stonehenge, were formerly incorporated in glacigenic sediments. Kellaway *et al.* (1975) have suggested that this glaciation was of Baventian age. It should be noted, however, that analysis of terrace gravels from rivers that crossed Salisbury Plain has refuted the notion of glaciation in this area (Green 1973). Further north in England, erratic content has been used by Catt (1982) to imply that residual glacial material on the Yorkshire Wolds could be of Baventian age.

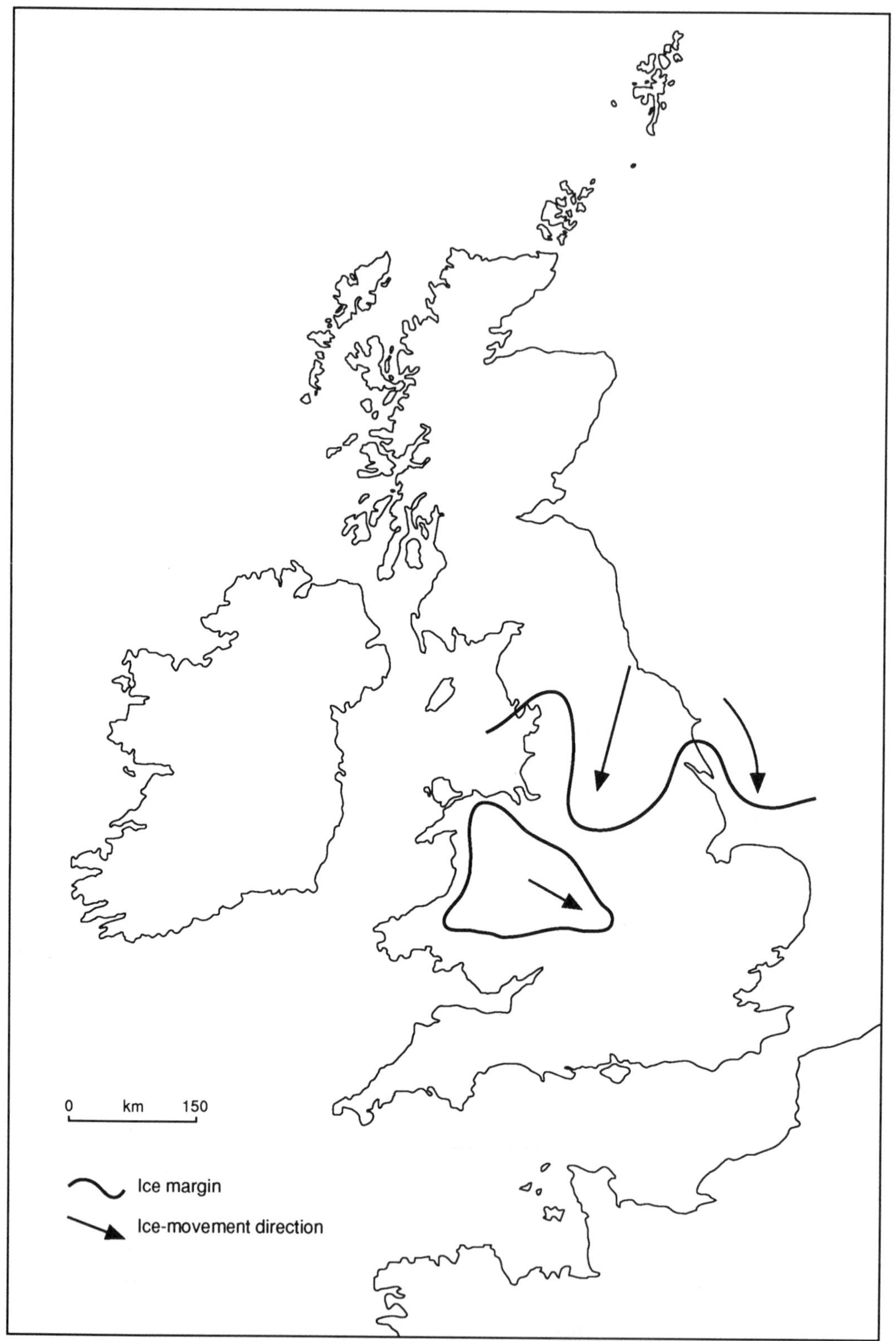

Figure 3.3 Possible ice margins and principal ice-movement directions during the Baventian. From Catt (1981).

3.2.6 The Bramertonian Temperate Stage

The timing of this stage is both problematical and controversial. Here it is assumed to be at the position in the Lower Pleistocene sequence suggested by Funnell *et al.* (1979) after examining assemblages of Foraminiferida, Mollusca and pollen and spores from sections in the Norwich Crag at its type locality of Bramerton near Norwich. The foraminifers are dominated by *Elphidiella hannai*, whose associated species at the base of a section at Bramerton Common indicate a shallow sub-littoral or inter-tidal temperate environment. The assemblages from Bramerton Common and Blake's Pit, Bramerton indicate that climatic conditions worstened as the stage progressed. Molluscan evidence shows the existence of numerous marine facies and supports the environmental interpretation based on foraminifera. The presence of *Macoma arctica* and *Yoldia* in the upper part of the sequence implies that the climate had deteriorated.

Palynological data from Blake's Pit reveal two pollen assemblage biozones. The first, with *circa* 70–90% tree pollen and *Quercus*, *Carpinus*, *Ulmus*, *Betula* and *Alnus* prominent, indicates regional temperate forest. The second, in which *Pinus*–Ericales–Gramineae dominates, has been referred to open forest, and to the rise of heathland and herbaceous vegetation. The temperate forest assemblage does not contain substantial values of *Tsuga* pollen and as a result has been assigned to a new stage. The open forest assemblage is considered to belong to the following Pre-Pastonian Stage (Figure 3.4) (Chapter 4). West (1980b) has noted the similarity of the *Quercus*–*Carpinus*–*Alnus* pollen assemblage with that from Norwich Crag-type sand and clay in Suffolk, termed Chillesford by West and Norton (1974). A pollen diagram from Norwich Crag sediments at Bungay, Suffolk has been reported by West (1988b). The lower of two assemblage biozones indicates the existence of boreal forest. The upper biozone (in which *Tsuga* is present) denotes the spread of mixed-oak forest. This assemblage has been equated with the temperate pollen spectra from Chillesford and with the palynological record from Bramerton.

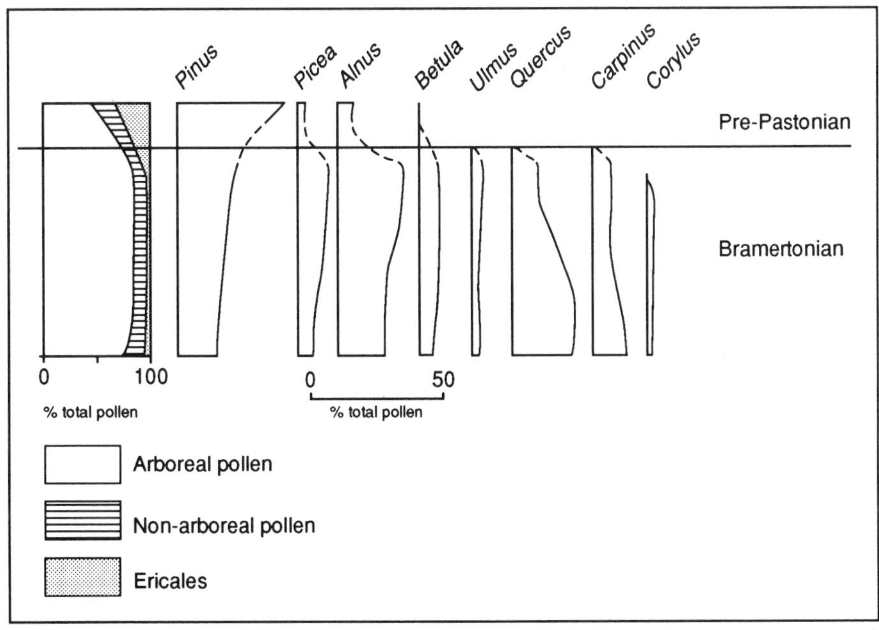

Figure 3.4 Pollen diagram from Bramerton. From West (1980b); source, Funnell *et al.* (1979).

Much of the current controversy over this stage stems from recent detailed investigations of the Norwich Crag Formation in Suffolk (Zalasiewicz and Mathers 1985; Zalazesiewicz *et al.* 1988). A single pollen spectrum from the Chillesford Sand Member in the Aldeburgh–Sizewell area has been identified as Bramertonian *sensu* Funnell *et al.* (1979). Lithostratigraphic and biostratigraphic

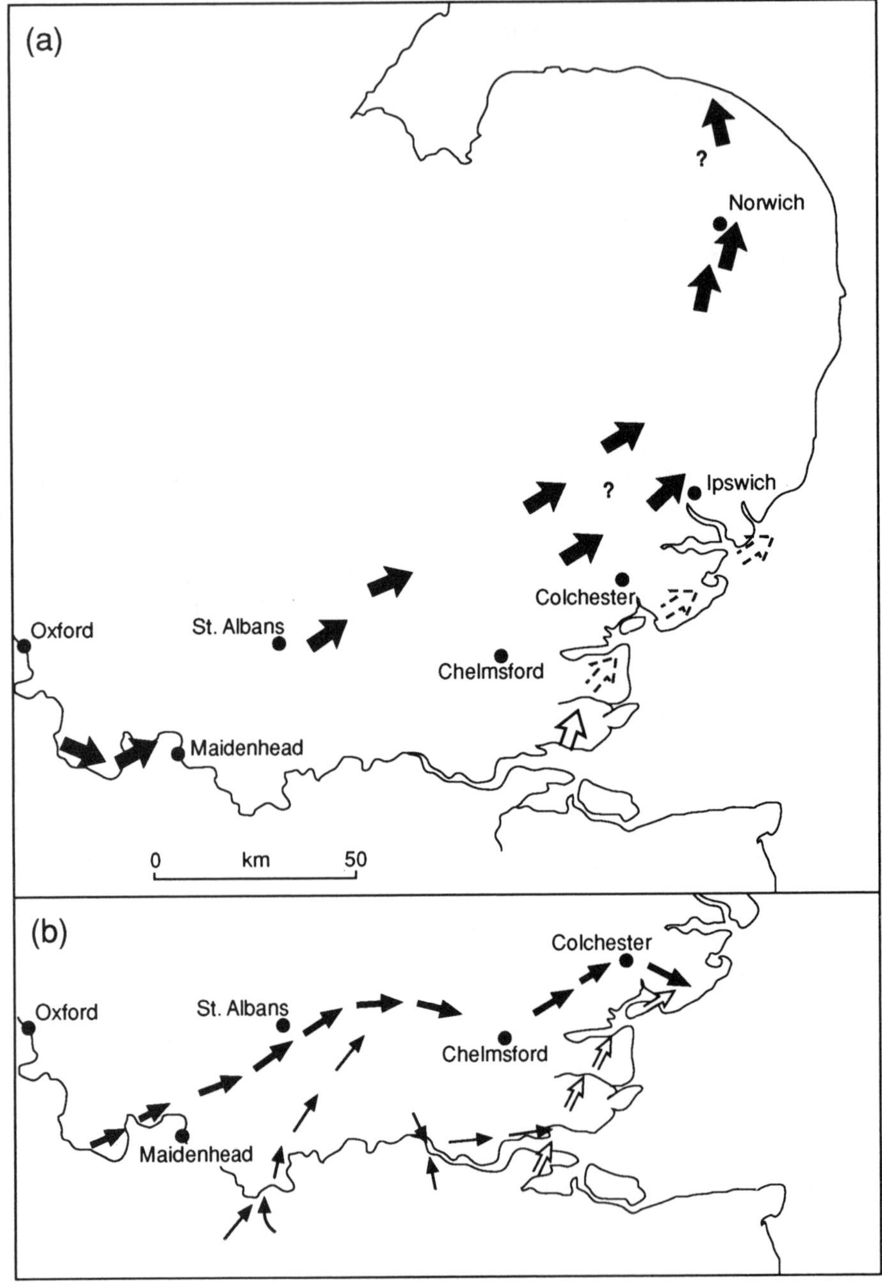

Figure 3.5 The Pleistocene evolution of the Thames system. (a) Course during deposition of the Westland Green Gravel. (b) Course immediately prior to the arrival of Anglian ice. From Bridgland (1988).

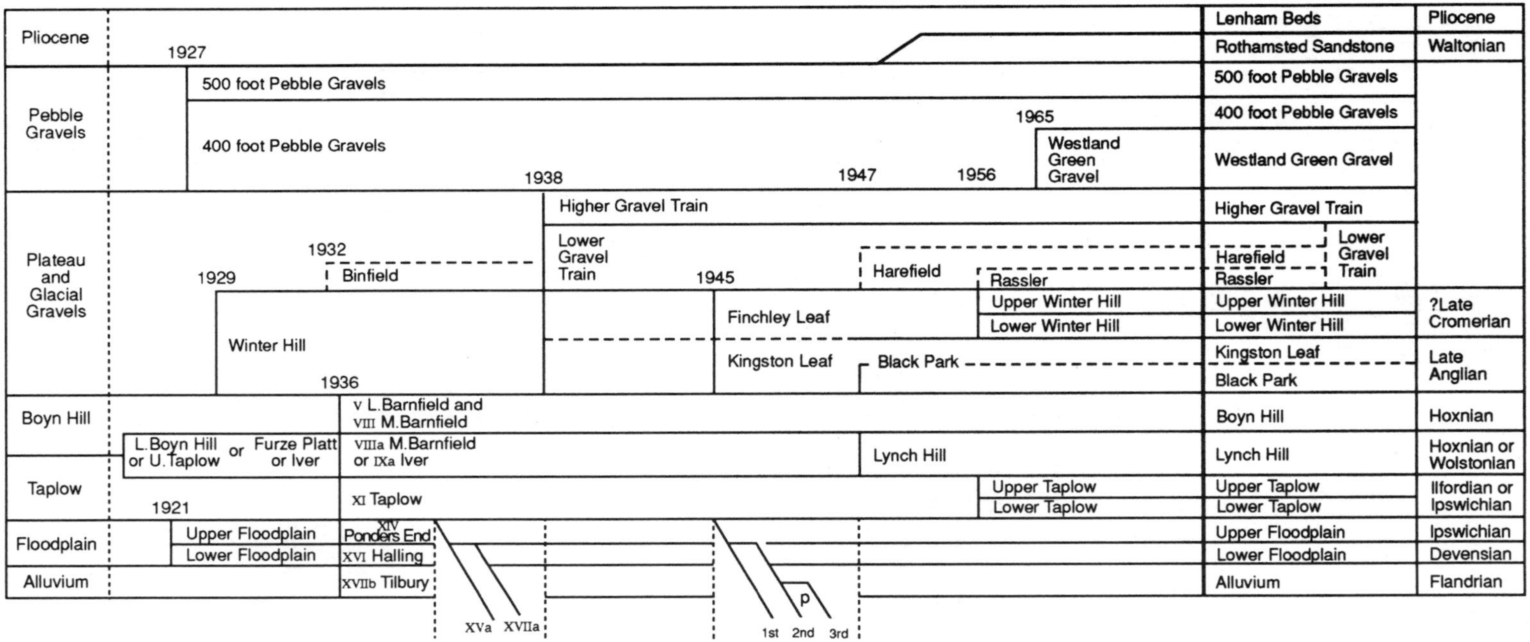

Figure 3.6 The Thames terrace sequence: the development of the nomenclature since 1912. Dates indicate the following sources: 1921, Dewey and Bromehead; 1927, Wooldridge; 1929, Saner and Wooldridge; 1932, Ross; 1936, King and Oakley; 1938, Wooldridge; 1945, Zeuner; 1947, Hare; 1956, Sealy and Sealy; 1965, Hey. Dashed lines indicate sub-divisions which either are of local significance only, or are disputed, or superseded. P: relationship of Ponders End sediments in Zeuner (1945); 1st: first sunk channel of Zeuner (1945); 2nd: second sunk channel; 3rd: third sunk channel. Roman numerals signify the stages of King and Oakley (1936). An attempt is made to show the position in the terrace sequence, as supposed by King and Oakley, of the sediments from which the stages are named. Pre-Boyn-Hill stages (i–iv), brickearths (vii, xii), periglacial deposits (x, xvb) and certain erosional stages (vi, ix, xiii) are omitted. The correlations in the final column reflect the authors' opinion in 1980. From Green and McGregor (1980).

data indicate that the Chillesford Sand, forming the bulk of the Norwich Crag, was deposited after the Ludhamian. No Thurnian, Antian or Baventian sediments have been encountered (Zalasiewicz *et al.* 1988).

Further uncertainty has been generated by Funnell (1987) who, in reconsidering the foraminiferal evidence, has suggested that the Antian sand beneath the Baventian clay at Easton Bavents (Funnell and West 1962) could be of Bramertonian age.

Vertebrate data may also assume significance in this controversy. Stuart (1982) has noted that *Aonyx reevi* (clawless otter), *Archidiskodon meridionalis*, *Eucladoceros falconeri* and *Gazella anglica* (gazelle) have been recorded from deposits of this stage. The most important remains, however, may be those of *Mimomys* spp. (Mayhew 1979, 1985; Mayhew and Stuart 1986). The vole remains from Bramerton represent the most primitive microtine assemblages identified in the British Isles and indicate that the Bramertonian Stage occurred prior to the Pre-Pastonian (Mayhew and Stuart 1986). The small mammal fauna in the locality may correspond to that designated as Tiglian C in the Netherlands (van der Meulen and Zagwijn 1974; Mayhew 1979; Mayhew and Stuart 1986; Stuart 1988a).

3.3 Gravels and Lower Pleistocene environments

The fluvial gravels laid down by the ancestral River Thames (Figure 3.5) have proved to be largely unfossiliferous, but their petrology and sequence of terrace landforms (Figure 3.6) have been intensively studied in Buckinghamshire, Hertfordshire, Essex, Suffolk and Norfolk (Hey 1976, 1980; Hey and Brenchley 1977; Green and MacGregor 1978, 1980; Gibbard 1979, 1985; Green *et al.* 1980, 1982; Bridgland 1988). The general conclusion reached by this work is of up to three phases of injection of exotic material, notably Welsh volcanic rocks and *Rhaxella* chert, during the Early Pleistocene. This far-travelled material most likely arrived in the Thames system as a result of glacial activity. Green *et al.* (1982) have suggested that the Westland Green Gravel and the Higher Gravel Train and Lower Gravel Train (the two last-named being approximately equivalent to the Beaconsfield and Gerrards Cross gravels) (Figure 3.6), are representative of between one and three glacial episodes. The correlation of these episodes with the Crag succession has been difficult because of the lack of palaeontological evidence in the gravels and of superposition of these fluvial deposits with the Crags. However, Hey (1980) has postulated that the Westland Green Gravel may have been deposited in either the Baventian or the Pre-Pastonian cold stage. This interpretation means that the first glaciation in the British Isles to be recognized in this part of southern England took place during one of these stages, with the Higher Gravel Train and Lower Gravel Train denoting subsequent glaciations. As all these gravel deposits may be represented in the Kesgrave Formation of East Anglia (Rose *et al.* 1976; Rose and Allen 1977), in which a Cromerian soil developed (Table 3.1), one of these glacial episodes may have taken place in the Beestonian (Chapter 4) (Table 1.1), and the other two earlier. At Easton Bavents and Covehithe the Baventian Clay is overlain by the Westleton Beds (Hey 1967), which may thus be either late Baventian or younger (Zalasiewicz and Gibbard 1988). These flint-rich gravels have been considered by Hey and Auton (1988) to be of uncertain age, but probably Bramertonian as has been suggested by Funnell *et al.* (1979). The Westleton Beds have a similar suite of far-travelled pebbles to those characteristic of the Baventian but of less overall importance in the assemblage (Hey 1976).

The Northern Drift of Oxfordshire (section 3.2.5) has been found to consist mainly of decalcified gravels (Shotton *et al.* 1980), which Hey (1986) has separated into a minimum of four units, which were deposited by a southward-flowing river in the Evenlode Valley. The oldest of these units is equated with the Westland Green Gravel and the two above it are linked to the Gerrards Cross Gravel (Green *et al.* 1980) and Satwell (Gibbard 1983, 1985) Gravel of

Table 3.1 The early Thames Valley: correlation of lithostratigraphic units and their interpreted ages. From Gibbard (1983)

Upper Thames	*Middle Thames*	*Vale of St. Albans*	*West Essex*	*East Essex*	*Stage*
Freeland Terrace Gravel	Black Park Gravel	Smug Oak Gravel	–	–	Anglian
Coombe Terrace Gravel	Winter Hill Gravel — Upper	Eastend Green Till Moor Mill Laminated Clay	Springfield Till	Lowestoft Till	
	Winter Hill Gravel — Lower	Westmill Gravel; Upper / Ware Till / Lower	Upper Chelmsford Gr. Maldon Till Lower Chelmsford Gr.	Barham Sands and Gravels Barham Arctic Structure Soil	
	Gerrards Cross Gravel	Leavesden Green Gravel	Widdington Sands		
				Valley Farm Temperate Soil	Cromerian
?Northern Drift	Beaconsfield Gravel Satwell Gravel			Kesgrave Sands and Gravels and Palaeosols	Pre-Cromerian
	Westland Green Gravel				
	Stoke Row Gravel				
	Nettlebed Interglacial Deposit				Nettlebedian
	Nettlebed Gravel	400 foot Pebble Gravel			Pre-Nettlebedian
		500 foot Pebble Gravel			

Beestonian age in the Middle Thames Valley (Table 3.1).

Away from the former course of the River Thames, the Buchan Gravels (Chapter 2), well rounded and rich in quartzite and flint (Hall 1984a,b), have been interpreted as Late Pliocene–Early Pleistocene glacial deposits by Jamieson (1906) and Kesel and Gemmell (1981).

3.4 The offshore sequence

A considerable thickness of Pleistocene sediments has been recorded below the North Sea (McCave *et al.* 1977). The Lower Pleistocene sediments are marine, deltaic and fluviatile (Cameron *et al.* 1987, 1989a). In the northern and central parts of the North Sea at this time, mainly basinal marine deposits, of the Shackleton Formation and Aberdeen Ground Formation, respectively, accumulated under overall warm-temperate conditions. In the southern North Sea the deposits are deltaic (Cameron *et al.* 1987; Long *et al.* 1988) (Figure 1.1). Although the corresponding marine fauna from the latter area indicates a climate at least as warm as that of the present day, palynological data hint at shifts from warm-temperate to cool-temperate conditions (Cameron *et al.* 1984b). The Lower Pleistocene deposits in the southern North Sea are thought to indicate a series of marine transgressions and regressions, with the latter perhaps a response to climatic deteriorations (Long *et al.* 1988). In the central and western Celtic Sea, the upper part of the Little Sole Formation may be of Early Pleistocene age (Pantin and Evans 1984). The clayey Skerryvore Formation on the inner continental shelf, west of Scotland, has yielded biostratigraphic and lithostratigraphic evidence of a climatic deterioration within a marine environment, which may have occurred during the Early Pleistocene (Davies *et al.* 1984) (Figure 9.8).

3.5 Correlation and conclusion

If an attempt is made to correlate the fragmentary evidence for the incidence of Lower Pleistocene cold stages and glaciations in the British Isles with the Netherlands sequence over this timespan, a number of possibilities arise. In the Netherlands there were four cold episodes in the Eburonian, one during the Waalian, two in the Menapian, two in the Bavelian, and at least three within the Cromerian Complex (Zagwijn 1975; de Jong 1988) (Figure 2.2(a)). In a correlation of the Netherlands and East Anglian sequences, Zagwijn (1975) has proposed a major hiatus (representing at least 1 million years) between the Baventian and Pastonian, within the deposits of the latter area. The glacial and fluvial sediments discussed above may at least in part be referable to that hiatus, and belong to cold stages over its duration, which is thought to have included much of the Eburonian, the Waalian, the Menapian, and most of the Cromerian Complex.

That part of the Lower Pleistocene succession examined in this chapter has been shown to possess numerous complex, and some contradictory, elements. Their resolution, and the correlation of the British Isles sequence with that of the neighbouring part of Europe, has been hindered by both spatial and temporal deficiencies in the lithostratigraphic and biostratigraphic records, and by a lack of geochronometric age determinations. In an attempt to accommodate the range of existing viewpoints, Zalasiewicz and Gibbard (1988) have presented an outline scheme for the Pliocene to Middle Pleistocene succession in East Anglia. This retains the preglacial Pleistocene stages of Funnell and West (1977), but places the Bramertonian between the Antian and Baventian, rather than after the latter, as has been proposed by Funnell *et al.* (1979) (Figure 2.2(b)).

4

The Cromer Forest-bed Formation and associated deposits

Most of the evidence concerning the remaining Lower Pleistocene stages (*sensu* Stuart 1982) and those of the subsequent early part of the Middle Pleistocene (Table 1.1) has come from the Norfolk and Suffolk coasts. Of especial relevance has been the north and north-east Norfolk coast between Weybourne and Happisburgh, while the north-east coast of Suffolk between Corton and Southwold has also yielded important information. In general, the older deposits in this sequence have been located either beneath or at the level of the current foreshore, with the younger ones occurring towards the base of the cliffs; the latter being composed mainly of Anglian glacial material (Chapter 5). Deposition of these late Lower Pleistocene and early Middle Pleistocene sediments included the final infill of the Crag Basin, which may have taken place in part behind a Wadden barrier-island complex (West 1980a).

It should be noted at the outset that while East Anglia remains the focus of attention regarding environments at this juncture in the Pleistocene, a number of sites of comparable age have been reported in other regions of the British Isles. These are also discussed below and are shown in Figure 4.1.

4.1 The stages and their environments

The first synthesis of the preglacial stratigraphy in coastal Norfolk and Suffolk was provided by Reid (1890) who recognized two main divisions: the Weybourne Crag and the (younger) Cromer Forest-bed Series, the latter sub-divided into three parts. The scheme followed here is based on lithostratigraphy and biostratigraphy, and was proposed by West (1980a). With the exception of the lowest, which has been referred to the Norwich Crag Formation (Funnell and West 1977), the lithostratigraphic members have been placed in the Cromer Forest-bed Formation. The biostratigraphic evidence has been derived mainly from palynology, and this has formed the basis for division into, and sub-division of, the stages, with the biozones being related to the lithostratigraphy. The most important biostratigraphical divisions are based on the evidence from sections at West Runton and Beeston, which thus have been designated type localities for them.

4.1.1 The Pre-Pastonian Cold Stage

Four substages have been recognized during this stage (West 1980a). In the first substage (a) shallow-water marine clays, sands and gravels were deposited in an embayment between Beeston and Covehithe. Analysis of the surface textures of sand grains belonging to units of the Sidestrand Member of the Norwich Crag Formation suggests that glaciation had taken place before their sedimentation (Krinsley and Funnell 1965). There was a marine transgression close to the

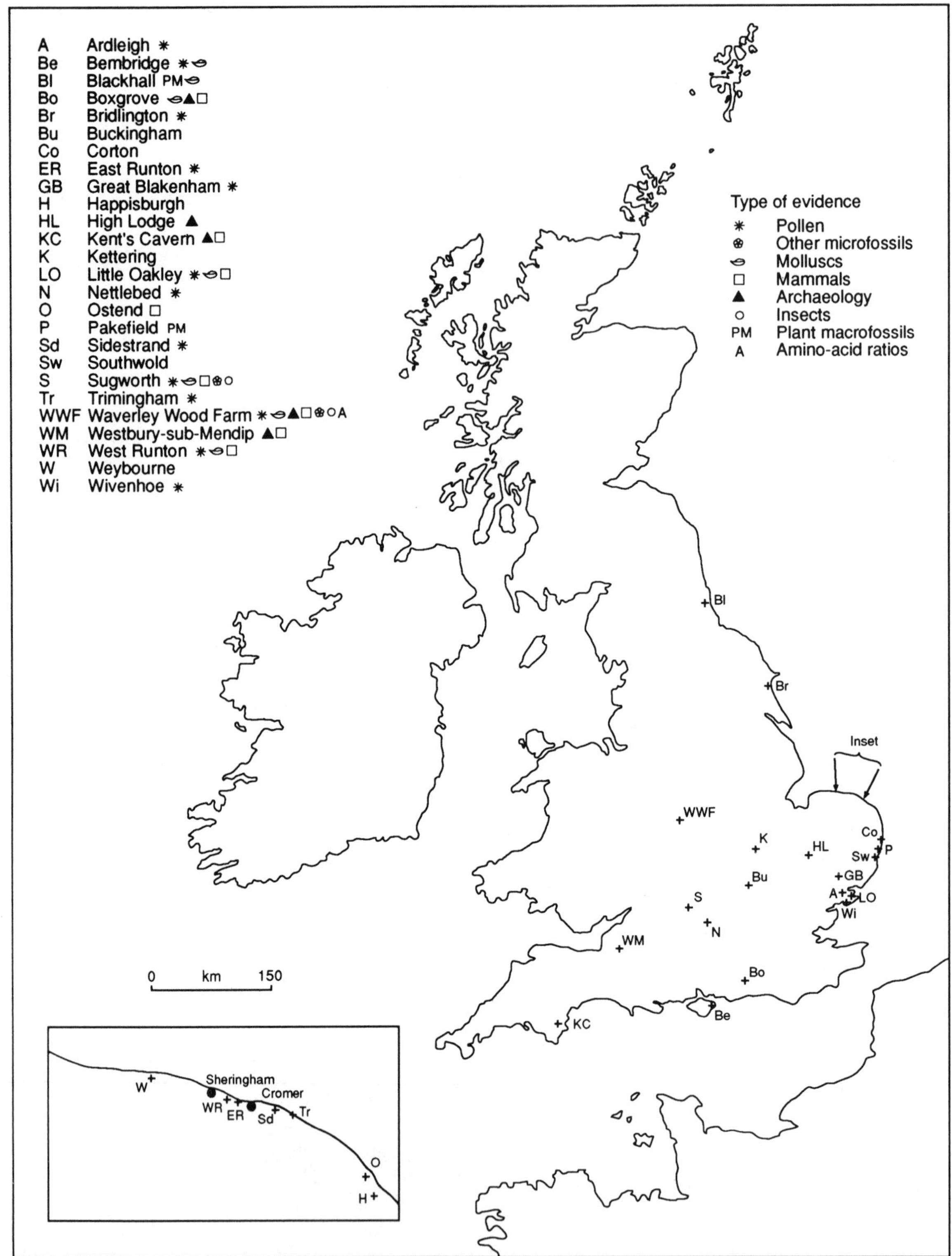

Figure 4.1 Sites referred to in Chapter 4 with types of evidence. Inset: the north and north-east coast of Norfolk.

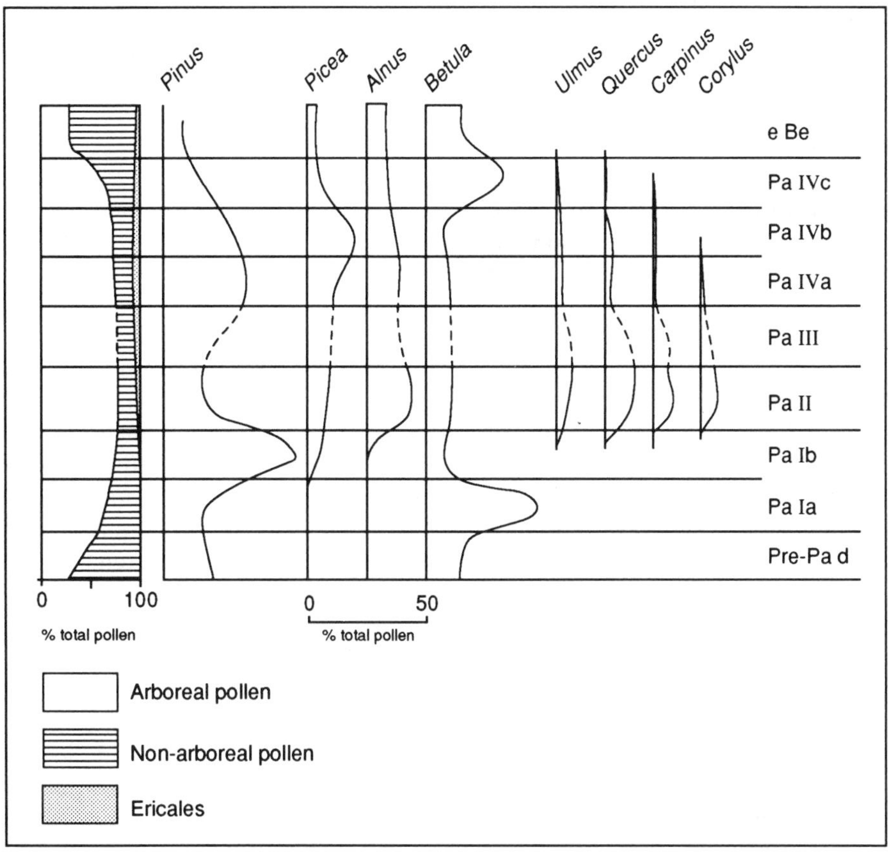

Figure 4.2 Pollen diagram of the Pastonian. Palynological data for substages Pre-Pastonian d and early Beestonian are also shown. From West (1980b).

start of the first substage of the Pre-Pastonian, as a result of which sea-level rose to about 5 m OD, falling later in this episode.

The pollen and spore assemblages of this substage are characterized by substantial frequencies of herbs, especially Gramineae and Compositae (composites), and of Ericales. The tree flora seems to have been mainly *Pinus*, *Picea*, *Betula* and *Alnus* (Figure 3.4). The palynological data have been interpreted by West (1980a) as indicative of a park–tundra landscape.

Foraminifera from the Sidestrand Member are diagnostic of a partly glacial environment (Funnell 1961a), while its meagre marine molluscan fauna, dominated by *Macoma balthica*, indicates boreal temperatures and water only a few metres deep (Norton 1967). Vertebrate remains include teeth of *Mimomys* spp. and *Archidiskodon meridionalis*, and an antler of *Alces gallicus* (extinct elk) (Stuart 1982).

Freshwater sediments belonging to the Sheringham Member were deposited during the remaining three substages of the Pre-Pastonian (b,c,d). This change in the sedimentation regime occurred as a result of the marine regression noted above. The palynological evidence from the first of these substages (b) is of dominantly herbaceous vegetation and a limited amount of boreal woodland. The harsh climate seems to have improved a little in the next substage (c). Tree pollen frequencies are higher and include *Ulmus*, but the substantial amounts of herbaceous pollen denoted the persistence of park–tundra. During the last substage (d) the vegetation was analogous to that of substage b of the Pre-Pastonian Stage (West 1980a) (Figure 4.2).

Vertebrate remains have been difficult to assign to these substages, although vole teeth from the Weybourne Crag at East Runton probably belong to one or more of them (Stuart 1982). This vole fauna suggests a correlation of the Pre-Pastonian with part of the Eburonian Stage of the Netherlands (Mayhew and Stuart 1986) (Figure 2.2(b)). In a chronological context, it is worth noting that West (1980a) has pointed out the possibility of a hiatus between the sediments of Pre-Pastonian a and b, which may mean that details of a further climatic episode have not been recorded.

Inland of the Norfolk and Suffolk coasts, a Pre-Pastonian date has been assigned to deposits overlying those of Bramertonian age at Bramerton (Funnell *et al.* 1979) (Figure 3.4). In addition, some units of the fluvial Kesgrave Formation (Rose and Allen 1977), which has been shown to underlie a considerable tract of East Anglia, may be of Pre-Pastonian age (Hey 1980; Hey and Auton 1988). A pollen flora possibly representative of an Early Pleistocene cool period has been reported by Holyoak (cited in Allen 1982, 1988) from the College Farm Clay, which occurs at the top of the Creeting Formation at Great Blakenham, in the Gipping Valley of Suffolk (Allen 1982, 1988). The clay has been thought to be equivalent to the Chillesford Clay Member of the Norwich Crag Formation (Zalasiewicz and Mathers 1985). The pollen spectrum is indicative of a vegetation mosaic including woodland and heath. The tree flora is dominated by *Pinus* and contains *Alnus*, *Picea*, *Betula*, *Quercus* and *Tsuga*. Holyoak has noted that the assemblage does not resemble those assigned to the Pastonian or Cromerian on the Norfolk coast. Lithostratigraphic and biostratigraphic data may mean that the College Farm Clay is a Pre-Pastonian deposit.

4.1.2 The Pastonian Temperate Stage

There may also be a non-sequence between Pre-Pastonian and overlying Pastonian sediments in a number of areas, due to the erosive effects of a marine transgression which occurred early in the latter stage. However, in some localities continuous sedimentation took place (West 1980a).

The Pastonian (West and Wilson 1966) is the oldest temperate episode in the Pleistocene of the British Isles with sediments containing sufficient biostratigraphic (palynological) data for the four substages typical of an interglacial stage (Turner and West 1968) to be recognized (Figure 4.2). The type site has been established at Paston in Norfolk, and the majority of sediments belonging to this stage are marine, although some freshwater facies occur (West 1980a).

Substage Pastonian I, during which both marine and freshwater sediments accumulated, possesses a pollen and spore record divisible into two biozones (West 1980a,b). The first (Pa Ia) demonstrates the existence of *Betula–Pinus* woodland, the second (Pa Ib) *Pinus–Picea* forest. Macroscopic remains of *Alnus*, *Betula* and *Salix* have been recovered from sediments of Pa Ib age. The environment was thus temperate, with boreal characteristics.

Substage Pastonian II has pollen assemblages dominated by *Pinus*, *Quercus*, *Carpinus*, *Ulmus* and *Corylus*, which indicate the development of temperate mixed-deciduous forest. Macroscopic tree remains include those of *Betula*, *Alnus* and *Picea* (Wilson 1973). Its mainly marine sediments illustrate that this was the time of maximum sea-level, *circa* 5 m OD.

Palynological data pertaining to Pastonian Substage III are sparse. However, they permit general vegetational traits to be ascertained. These include a change in the composition of the forest compared with that of Pa II. Some thermophilous taxa declined in importance and *Tsuga* entered the woodlands. Increased Ericales pollen frequencies indicate that heathland expanded its cover. Gravel in Pa III sediments may have been deposited during a marine regression.

Substage Pastonian IV pollen and spore assemblages are divisible into three biozones which demonstrate a boreal forest of changing composition. In Pa IVa *Pinus*, *Picea*, *Alnus* and *Betula* are the most important constituents; in Pa IVb *Pinus* and *Picea* assume greatest significance, while in Pa IVc *Betula* and *Pinus* are best represented. The proportion of non-arboreal pollen increases in Pa IVc, indicating that open woodland, together with herbaceous and heathland plant communities, were likely to have been in existence. The deteriorating climate was accompanied by a falling sea-level. The sediments of Pa IVa and Pa IVb are marine; those of Pa IVc accumulated in fresh water.

Animal fossils have been recorded from some Pastonian sediments. Norton (1967, 1980) has presented marine molluscan data from the Norfolk coast which include assemblages of probable Pa II age, when the marine transgression was at its peak. The most common species are *Astarte montagui*, *Macoma balthica* and *Yoldia*. Overall the fauna contains a mixture of temperate and boreal forms, but with a diminution in the Lusitanian element which was characteristic of the earlier part of the Crag succession. Vole remains from West and East Runton have been ascribed to the Pastonian Stage. In the former locality, their association with Pa II deposits implies that the animals were part of the regional temperate forest ecosystem (Stuart 1982). The vole fauna, which includes *Mimomys pliocaenicus*, is distinct from that in later stages present in the Cromer Forest-bed Formation (Mayhew and Stuart 1986). The earliest mammalian fauna from Westbury-sub-Mendip Cavern in Somerset (section 4.1.4) is of uncertain age. It is characteristic of open woodland and may belong to the Pastonian (Bishop 1982).

The extent of the marine transgression over East Anglia during the Pastonian is unknown, but lithostratigraphy may afford some insight. Hey (1976) has suggested that the Westleton Beds were beach shingle during this stage (Chapter 3), when they probably formed the equivalent of a modern Dungeness. The Westleton Beds overlie Baventian sediment at Easton Bavents, and they have been regarded as probably of Bramertonian age by Funnell *et al.* (1979). The age of the Westleton Beds has thus not been established beyond that of late Baventian or post-Baventian. Recent boreholes have shown flint-rich gravel of Westleton Bed-type to be present in the upper Waveney Valley to the west of its coastal outcrops (Hey and Auton 1988). To what extent upwarping and downwarping, which appear to have taken place in the coastal area since the deposition of the Pastonian sediments (West 1980a), were effective further inland is unknown. Hey and Auton (1988) have also drawn a parallel between the composition of the Kesgrave Gravels and those of Pastonian age on the East Anglian coast. As noted in Chapter 3, the Kesgrave Formation, buried by Anglian glacial deposits in most areas, has been shown to be characteristically capped by a palaeosol which Rose *et al.* (1976) and Rose and Allen (1977) have suggested developed during the Cromerian. However, Kemp (1985, 1987) has shown that while only one episode of temperate weathering took place on the youngest Kesgrave surfaces, on older ones the soil is polycyclic, having developed under both temperate and cold climates in the late Lower and early Middle Pleistocene, including during the Pastonian.

Beyond East Anglia, dinoflagellate and pollen and spore data from the Bridlington Crag in Yorkshire have led Reid and Downie (1973) to suggest that this may be of Pastonian age.

4.1.3 The Beestonian Cold Stage

The Beestonian (West and Wilson 1966) is represented by a complex of marine and freshwater lithological units and periglacial phenomena. The nature of the sedimentary evidence suggests that at least one major hiatus is present in the succession (West 1980a). The sediments indicate an environmental history that commenced with fluvial activity consisting of erosion and the deposition of sand, silt or clay. This was followed by a marine transgression, as a result of which sand and gravel with shells was laid down. A further fluvial episode ensued, with erosion and deposition. The final facies, a freshwater marl, was deposited in a fluvial channel cut during this episode.

The flora of the early Beestonian fluviatile sediments is indicative of cold conditions (Figure 4.2), while the late Beestonian marl has produced pollen assemblages indicative of a mainly herbaceous flora dominated by Gramineae, Cyperaceae (sedges) and Compositae (Figure 4.4), and

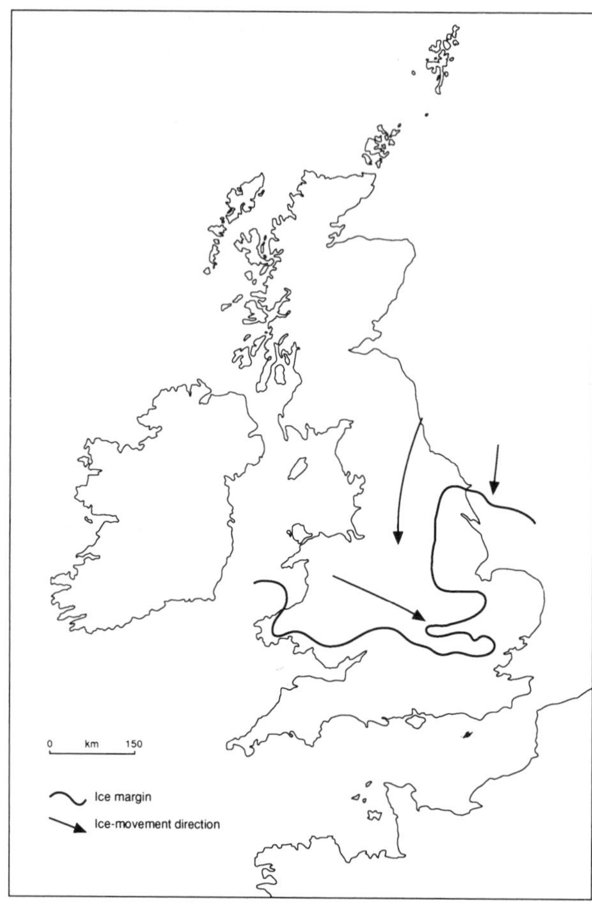

Figure 4.3 Possible ice margin and principal ice-movement directions during the Beestonian. From Catt (1981).

characteristic of cold climates (West 1980a,b). Mollusca recovered from the marine beds at West Runton imply the existence of a boreal climate but without indicators of severe cold (Norton 1980). A palaeopodsol has been identified in Beestonian sands and gravels at West Runton. The sands and gravels are overlain by Cromerian organic deposits, and the soil is likely to have formed in a cold climate near to the end of the Beestonian (Valentine and Dalrymple 1975).

Periglacial phenomena have confirmed the existence of a harsh environment in the Beestonian. Ice-wedge casts seem to have originated in the lower part of the upper fluvial material and penetrated the two units beneath (West 1980a). Such phenomena are associated with patterned ground in high latitudes today. Their presence in the Beestonian denotes at least discontinuous permafrost, and mean annual temperatures not exceeding 6°C (Chapter 1). No till referable to the Beestonian has been identified in East Anglia, so that the incidence of glacial ice there is unconfirmed. The presence of far-travelled glacial material in Thames river terraces which may have formed in the Beestonian has been reported by Green and MacGregor (1980). If such cold conditions existed in East Anglia during the Beestonian, glaciers in the northern and western highlands of the British Isles seem a distinct possibility. Indeed, Catt (1981) has suggested that the Northern Drift of Oxfordshire could be the result of Beestonian glaciation, and that the lower of two tills in the Kettering (Hollingworth and Taylor 1946) and Buckingham (Horton 1970) areas may be of equivalent age (Figure 4.3).

The occurrence of a marine unit in the Beestonian succession at heights of up to *circa* 7 m OD, is at variance with the pattern of sea-level movement which normally pertained during Pleistocene cold and temperate stages, whereby transgressions occurred in the latter and regressions characterized the former. The Beestonian transgression seems to have taken place during a cold stage, although as West (1980a) has pointed out, palynological data are lacking from the marine clays. If this biological sterility is not concealing a temperate episode within the Beestonian, when the sea-level could have risen (Zalasiewicz and Gibbard 1988), an alternative explanation is necessary. West (1980a) has suggested that initially there could have been a marine regression in the cold stage. As the stage progressed, ice built up in the North Sea, and thereby caused isostatic depression of the crust, which compensated for the regression. When the ice began to retreat, isostatic rebound led to a fall in sea-level. Glaciomarine sediments were deposited in the central North Sea at this time (Cameron *et al.* 1987; Long *et al.* 1988) (Figure 1.1).

The sands and gravels of the Kesgrave Formation of inland East Anglia were introduced in Chapter 3. Subsequent to their identification by Rose and Allen (1977), they have been recognized in numerous localities (Hey 1980). Many members of the Kesgrave Formation were probably deposited by a braided, ancestral River Thames under a periglacial climatic regime. The gravels have been shown to contain clasts derived from Triassic conglomerates from the western Midlands of England, and from North Wales volcanics. Assuming that the clasts were not the result of river capture in the Thames headwaters, then they almost certainly arose from glaciation in the western Midlands and North Wales (Bowen *et al.* 1986). If these gravels began to accumulate in the Pre-Pastonian (section 4.1.1) and continued during much of the late Lower and early Middle Pleistocene (Zalasiewicz and Gibbard 1988), their association with periglacial conditions suggests a Beestonian age for at least some of them. Quartz-rich gravels that comprise some of the Yarmouth Roads Formation of the southern North Sea (Balson and Cameron 1985; Cameron *et al.* 1987, 1989a; Long *et al.* 1988) have been related to the Kesgraves by Bowen *et al.* (1986). As Zalasiewicz and Gibbard (1988) have pointed out, this means that the Thames was now adding to the much enlarged deltas of the Rhine, Meuse and north German rivers, which were located in the southern North Sea (Zagwijn 1974a, 1989).

Work on the Early and Middle Pleistocene development of the Thames drainage system by Green *et al.* (1982) has equated the Westland Green Gravel (Chapter 3) with the higher members of the Kesgrave Formation and the Higher Gravel Train and Lower Gravel Train (Figure 3.6) with the lower Kesgraves (Table 3.1).

4.1.4 The Cromerian Temperate Stage

Environmental evidence concerning the Cromerian (West and Wilson 1966) has come mainly from marine and freshwater sediments in cliff sections on the Norfolk and Suffolk coasts. In north-east Essex non-marine deposits equated with the Cromerian Complex (*sensu* Zagwijn 1975) of the Netherlands (de Jong 1988; Zagwijn 1989) (Figure 4.5) have been reported. Inland East Anglia has yielded data on Cromerian pedogenesis. In other parts of the British Isles, localities in Berkshire, Somerset, the Isle of Wight, Sussex, Devon, County Durham and Warwickshire have produced fossils having affinities with the Cromerian (Figure 4.1).

Most Cromerian deposits identified along the Norfolk and Suffolk coasts have seemed to rest unconformably on those of the underlying Beestonian Stage. At the West Runton type site, sedimentation was at first in fresh water. A marine transgression ensued, then a regression, with freshwater deposits replacing those of marine origin at the top of the sequence. From this sequence, a full interglacial cycle has been identified from palynological evidence (Funnell and West 1977; West 1980a,b) (Figure 4.4). On the Tendring Plateau in north-east Essex, the Kesgrave Gravels of the Thames (Bridgland 1988) have associated pre-Anglian biogenic sediments of interglacial rank at Ardleigh, Little Oakley and Wivenhoe (Bridgland *et al.* 1988, 1990). At High Lodge, Mildenhall, Suffolk, there are pre-Anglian lake deposits from which floral and faunal evidence of temperate, possibly Cromerian, conditions has been obtained (Wymer 1988). At Sugworth, south-west of Oxford, old meander channels of the River Thames overlain by terrace deposits derived from Plateau Drift (Chapter 3) contain organic sands, silts and clays which provide fossil evidence that indicates a Cromerian IIIb age (Briggs *et al.* 1975; Shotton *et al.* 1980). The mammalian fauna of a cavern infill close to Westbury-sub-Mendip, between Cheddar and Wells in Somerset, is thought to represent two temperate episodes. The first may have been older than the Cromerian *sensu* West and Wilson (1966), perhaps Pastonian (section 4.1.2), the latter probably younger than this but pre-Anglian in age (Bishop 1975, 1982). Altitudinal (Bates 1986a; Bates and Roberts 1986) and faunal (Currant 1986) evidence from Boxgrove near Bognor Regis in Sussex, also suggests a pre-Hoxnian, Middle Pleistocene age that is not equivalent to the type Cromerian of the British Isles (Roberts 1986). Reference of this site to the Cromerian Complex seems likely. Marine interglacial deposits near Bembridge in the Isle of Wight also have been referred to this timespan (Preece *et al.* 1990). The lower deposits in Kent's Cavern, Torquay, may also have been laid down during the Cromerian according to the evidence of their animal remains (Campbell and Sampson 1971). A younger fissure clay at Blackhall, County Durham, investigated by Trechmann (1919) and Reid (1920) (Chapter 2), and containing plant, molluscan and vertebrate fossils, may be of Cromerian age (Francis 1970). At Waverley Wood Farm near Leamington Spa, Warwickshire, channel deposits below Wolstonian glacial sediments (Shotton 1989a) have produced a flora and fauna which would not be incompatible with those characteristic of the Cromerian Interglacial.

4.1.4.1 Flora and vegetation

The palynological data from West Runton (West 1980a,b) show that during the first substage of the Cromerian (Cr I), herbaceous plant communities of the late Beestonian were replaced by *Betula* and *Pinus* forest. In the first part of the substage (Cr Ia) this forest developed, while in its second part (Cr Ib), with pine the more important of the two trees, *Ulmus* became increasingly widespread and *Alnus* and *Picea* were present (Figure 4.4).

Cromerian II saw the development of mixed-deciduous forest consequent upon the entry of thermophilous tree taxa in Cr Ib. *Quercus* frequencies increased during Cr IIa and those of *Pinus* decreased. From this biozone at nearby Mundesley, *Alnus*, *Fraxinus* (ash) and *Quercus* macrofossils have been recovered (West 1980b). *Pinus* continued to be present in substantial amounts in both Cr IIa and Cr IIb. In the latter, *Alnus* became of considerable significance, probably locally, *Picea* was fairly well represented, and *Corylus* and *Tilia* (lime) consistently present.

Cromerian III pollen assemblages are characterized by *Carpinus* (Cr IIIa) and *Abies* (Cr IIIb). In Cr IIIa, pine, alder, oak, elm and hazel continued as important forest constituents. *Carpinus*, *Alnus*, *Betula*, *Pinus*, *Picea*, *Populus* (poplar/aspen) and *Acer* (sycamore/maple) have been recorded as macrofossils from Cr IIIa sediments at Pakefield in Suffolk (West 1980b). *Abies*, *Carpinus*, *Pinus* and *Picea* were of most importance in Cr IIIb; falls in the values of oak, elm, lime and hazel denoting the demise of deciduous forest.

Substage Cromerian IV contains three biozones associated with the development of boreal forest. Pine, spruce, alder and birch dominated the tree flora in Cr IVa, a time when the forest became more open and heathland cover expanded. *Pinus* and *Betula* were the principal trees in Cr IVb, with the latter genus gaining the ascendency during Cr IVc.

At Ardleigh, palaeobotanical evidence from

the organic sediments suggests that they formed during a temperate episode. The plant remains are not diagnostic of a known stage, but stratigraphic relationships point to an early Middle Pleistocene age, prior to the Cromerian *sensu* West and Wilson (1966). Bridgland *et al.* (1988) have postulated that this might represent a stage previously not delimited in the Pleistocene of the British Isles. Also of interest at Ardleigh is organic sediment in the Upper Gravel which has yielded cold-climate plant fossils. Overlying these deposits are the Valley Farm Soil (section 4.1.4.7) of pre-Anglian age, and the Barham Soil (Rose and Allen 1977; Rose *et al.* 1985a) which has been assigned to the early part of the Anglian Cold Stage (Chapter 5).

The organic silts and sands at Little Oakley, younger on lithostratigraphic grounds than the biogenic interglacial deposits at Ardleigh, have provided pollen records typical of Pre-temperate and Early-temperate interglacial substages. The principal palynological features are an absence of *Carpinus*, a rise to dominance of *Ulmus*, and the continued occurrence of *Picea* in spectra lacking relic taxa from the Tertiary. The assemblages are similar to those of the early substages at the Cromerian type site at West Runton (Gibbard and Peglar 1990). At High Lodge the pollen assemblages reported by Turner (1973a) are characterized by plentiful *Picea* and *Pinus*, together with some *Betula* and *Alnus*, and have been considered to portray interstadial vegetation of Chelford type (Chapter 9), rather than that characteristic of an interglacial.

In the organo-mineral sediments of one Sugworth channel (Figure 7.7), the pollen and spore record is diagnostic of the Late-temperate zone (Substage III) of an interglacial, but can not differentiate between that time in the Cromerian and Hoxnian (Gibbard and Pettit 1978) (Table 1.1). The macrofossil occurrence of *Potamogeton distinctus*, together with other environmental data (sections 4.1.4.2 and 4.1.4.5) points to a Cromerian age which would be Substage Cr IIIb on palynological criteria (Pettit and Gibbard 1980). However, it is worth noting at this point that these Sugworth deposits have been reinterpreted as Hoxnian on geomorphological evidence by Gibbard (1985).

Pollen from Boxgrove is mainly arboreal and is dominated by *Pinus*, *Picea* and *Abies*. *Carpinus* and *Betula* are absent but *Quercus* and *Fagus* (beech) occur in low frequencies. The high *Picea* values in the top sample may place it in a Post-temperate interglacial zone which is thought most likely to have been of the Hoxnian (Scaife 1986a). However, attention has been drawn to a similar pollen assemblage from Bembridge in the Isle of Wight (Holyoak and Preece 1983). Here an estuarine sediment, the Steyne Wood Clay, which occurs at *circa* 40 m OD, contains a tree flora dominated by *Pinus* and including *Picea* and *Abies*. The vegetation was probably that of the Post-temperate zone of a Middle Pleistocene interglacial and the clay is considered a correlative of the Slindon Sands at Boxgrove (Preece and Scourse 1987; Preece *et al.* 1990).

It may be pertinent here to provide detail of interglacial pollen spectra from Nettlebed near Henley-on-Thames. The Nettlebed Gravel has been mentioned as a possible Pliocene deposit (Chapter 2). However, Horton (1977) has considered this to be Pleistocene fluviatile material of the River Thames, an opinion endorsed by Turner (1983) who has presented the palynological evidence. This came from organic silts below gravel in an abandoned river channel. Three substages of a typical interglacial sequence have been recognized. In the opening, Pre-temperate zone, birch predominated, with lesser amounts of pine and a little spruce. The second, Early-temperate zone was characterized by mixed-deciduous forest in which oak was dominant and elm present. The Late-temperate zone (III) was marked by the appearance of *Carpinus* and increased frequencies of *Corylus*. Alder pollen occurred throughout the sequence which, however, lacks *Tilia*, *Hedera* and *Ilex*. The absence of these thermophilous genera and the behaviour of the *Picea* and *Corylus* pollen curves renders comparisons with known Lower and Middle Pleistocene interglacial vegetation successions difficult. Tertiary relic taxa have not been recorded. A deposit older than that at Sug-

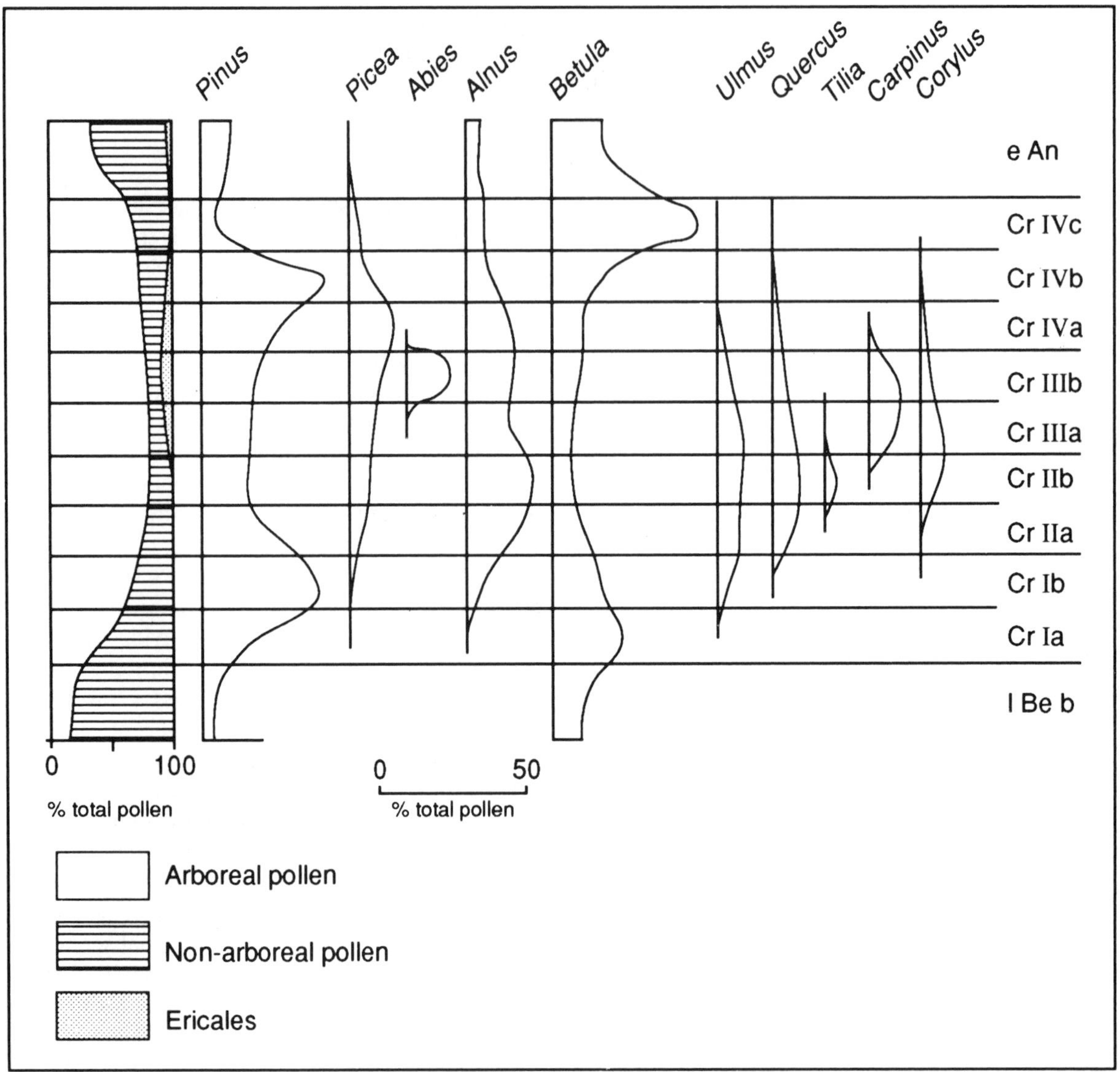

Figure 4.4 Pollen diagram of the Cromerian. Palynological data for the later division (b) of the late Beestonian Substage and for the early Anglian Substage are also shown. From West (1980b).

worth, and of early Middle or Lower Pleistocene age, has been suggested for what may have been a hitherto unrecognized interglacial stage. Its context invites reference to a Cromerian Complex, during which in the Netherlands at least four temperate episodes occurred (Zagwijn 1975, 1985; de Jong 1988) (section 4.2) (Figure 4.5).

A possible Tertiary fissure clay from County Durham is discussed in Chapter 2. Another fissure clay in the same locality near Blackhall Colliery, Castle Eden, and also lying below the earliest glacial deposit in the area (Trechmann

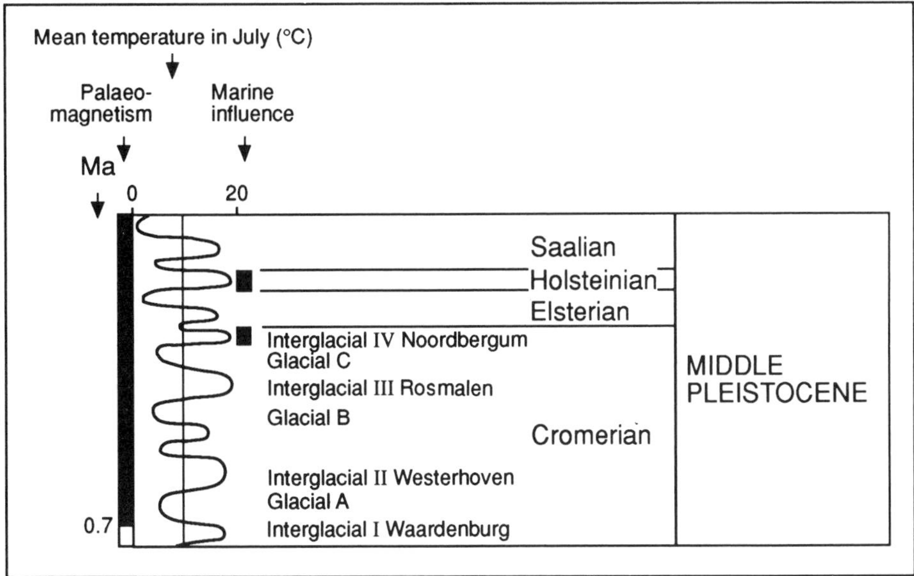

Figure 4.5 Climate curve and chronostratigraphy for the Middle Pleistocene of the Netherlands. From de Jong (1988); source, mainly Zagwijn (1985).

1919) (Chapter 7), has been found to contain plant remains. These, reported by Reid (1920), are macrofossils of species currently common in the British Isles, which have been referred to either Cromerian or slightly later time.

The tree pollen from Waverley Wood Farm is dominated by *Pinus* with *Picea* present but *Abies* absent. During the timespan covered by the record, tree pollen frequencies fell from *circa* 80 to 20% of the total sum (Peglar and Gibbard 1989), indicating forest reduction and the spread of dwarf shrub and herbaceous vegetation, perhaps in the later part of an interglacial.

4.1.4.2 Non-marine Mollusca

The majority of non-marine molluscan evidence for Cromerian environments has come from the Freshwater Bed at West Runton, whose sediments have been equated with substages Cr I and II. The Mollusca were dominantly freshwater in habitat, although land snails and slugs were present. The former, including numerous species of *Pisidium*, suggest the existence of moving, clean water, in either a stream or lake of fairly substantial dimensions that enlarged during Cr IIa when the climate became more temperate. The latter, with Succinidae, *Deroceras* and *Limax*, are mainly diagnostic of wetland marginal to water, although *Clausilia bidentata* and *Discus ruderatus* point to the presence of shaded habitats, perhaps in nearby woodland (Sparks 1980; Preece 1988). The occurrence of species such as *Valvata goldfüssiana*, *Tanousia* (*Nematurella*) *runtoniana* and *Bithynia troscheli* appears to typify the Cromerian (Sparks 1980).

At Little Oakley the molluscan fauna, including *Tanousia* and *Bithynia troscheli*, implies the existence of temperate conditions and points to a Cromerian age. The dominant, freshwater assemblage indicates a substantial body of well-oxygenated, moving water, considered to be the ancestral River Thames (Preece 1990). The molluscs from Sugworth, where *Tanousia runtoniana* and *Valvata goldfüssiana* have been recorded, are also indicative of a Cromerian age (Gilbertson 1980a,b). Preece (1989) has reported a re-examination of the molluscan fauna from

Sugworth. Several additional taxa have been recorded, including *Perforatella bidentata*. This currently inhabits woodland in eastern Europe, and had previously not been found in the Pleistocene deposits of the British Isles. Some of the Mollusca are indicative of a continental climatic regime, and the assemblage has been confirmed as interglacial in type and of pre-Anglian age. Similarity with the molluscan fauna of the type Cromerian has been noted, as have analogous amino-acid ratios in *Valvata piscinalis* from Sugworth and West Runton. The molluscan assemblage at Boxgrove has been interpreted as representative of temperate marsh/damp open woodland, with *Vitrinobrachium breve* suggesting a Middle Pleistocene age (Bates 1986b).

A non-marine molluscan fauna from Castle Eden has been reported by Kennard and Woodward (1919). Commonly encountered are *Lymnaea peregra*, *Planorbis* (*Gyraulus*) *laevis*, *Valvata piscinalis* and *Pisidium casertanum*, indicating a fluvial habitat but undiagnostic as to age. The Mollusca from Waverley Wood Farm are predominantly fluvial in character but contain none of the Cromerian indicator species noted above except *Bithynia troscheli*. The fauna of 56 species suggests an interglacial origin (Keen 1990).

4.1.4.3 Marine Mollusca

Norton (1967, 1980) has described the marine molluscan content of the *Leda myalis* Bed at West Runton (Reid 1882), which has been referred to Substage Cr IIIb by West (1980a). The characteristic molluscs, in addition to *Leda* (*Yoldia*) *myalis*, are *Mya truncata* and *Macoma balthica*. These species typify muddy or gravelly sublittoral, benthic habitats in boreal latitudes today, and support the incidence of mainly tidal sedimentation over this coastal area during the Cromerian transgression. As with other marine molluscan assemblages from the preglacial Pleistocene of East Anglia, however, the current wide latitudinal ranges of many of its constituents preclude detection of climatic shifts signified by other fossil evidence (Norton 1977). A small fauna from Boxgrove includes *Cardium*, *Mytilus* and *Neptunea contraria*, which would be consistent with a Cromerian age (Bates 1986b). The estuarine Steyne Wood Clay at Bembridge contains *Macoma balthica* which has its initial records in the Pre-Pastonian (Substage (a)) in East Anglia (Norton 1980). *Modiolus* or *Mytilus* sp. is also present in this deposit (Holyoak and Preece 1983).

4.1.4.4 Vertebrates

Evidence of these has been reviewed by Stuart (1975, 1982, 1988a). The West Runton type site has yielded a substantial fauna (Stuart 1988b) from sediments mainly equated with pollen assemblage biozones Cr Ib, IIa and IIb. Vertebrate remains from this locality have lent support to palynological and non-marine molluscan evidence of Cromerian environments. *Dama dama* (fallow deer), *Capreolus capreolus* (roe deer), *Sus scrofa* (wild boar) and *Apodemus sylvaticus* are temperate forest animals, although *Megaceros verticornis* (giant deer), *Alces latifrons* (extinct elk) and *Equus* fossils imply the existence of more open habitats, perhaps either within or marginal to the forest. Species of *Microtus* and *Pitymys* (voles), together with *Cricetus cricetus* (common hamster) would also have dwelt in open habitats, probably in floodplain grassland. *Castor fiber* (beaver), *Mimomys savini* (extinct water vole), and numerous birds, including *Phalacrocorax carbo* (cormorant), *Cygnus bewicki* (Bewick's swan) and several duck genera (Harrison 1979a), confirm the presence of aquatic conditions. Finds from nearby Ostend, assigned to Substage Cr IV, include *Sorex runtonensis* (extinct shrew) and *Arvicola cantiana* (extinct water vole) which lived in a boreal environment (Stuart and West 1976).

The Little Oakley fauna contains *Mimomys savini*, and although more limited than that from the Cr Ib–IIb sequence at West Runton, has a number of taxa in common with the latter (Lister *et al.* 1990). It differs from that at Ostend, however, in that *Arvicola cantiana* is absent. The evolutionary succession of *Mimomys savini* by *Arvicola cantiana* may have chronological significance. If faunas with the latter species are characteristic of late Cromerian (*sensu lato*) time (Stuart 1982, 1988a), then the Little Oakley vertebrates may equate with those of the type Cromerian (Bridgland *et al.* 1988, 1990) (see below).

Further insight into the possible chronological implications of Cromerian faunas *sensu lato* can be gained by examining those from the remaining localities listed at the outset of the discussion on this stage. The limited fauna recorded from the lake clays at High Lodge includes some taxa (*Dicerorhinus etruscus* (extinct rhinoceros) and voles) that have not been found in Hoxnian or subsequent stages (Wymer 1988). At Sugworth, *Mimomys savini*, *Sorex savini* (extinct shrew) and *Pliomys episcopalis* (extinct vole) are present (Stuart 1980), as at West Runton and Little Oakley, so that the Berkshire site is probably also of Cromerian age *sensu stricto*. As noted above, Preece (1989) has reported amino-acid ratios for Sugworth which are very similar to those for West Runton, also tending, therefore, to associate these two sites. However, both Sugworth and Little Oakley could belong to a prior interglacial about this time, which was not recorded in the Cromer Forest-bed Formation (Stuart 1988a). At Westbury-sub-Mendip, the Westbury 2 and 3 vertebrate faunas (the equivocal nature of Westbury 1 is mentioned in section 4.1.2) (Bishop 1982) have a number of similarities to that of West Runton (Stuart 1982). Perhaps of most significance at Westbury is the occurrence of *Arvicola cantiana*, which is also found at Ostend. Bishop (1982) has stated that the Westbury 2 and 3 faunas could belong to a new temperate stage in the Pleistocene of the British Isles, between the Cromerian and Hoxnian, and probably pre-Anglian; believing that the length of time available within one interglacial would have been insufficient for the emergence of *Arvicola cantiana*. As Stuart (1988a) has observed, this could refer the Ostend fauna to the same new temperate stage. At Westbury the fauna also contains a rich carnivore assemblage including *Panthera gombazogensis* (extinct leopard) and *Xenocyon lycaonoides* (extinct dhole), unknown from other Pleistocene stages in the British Isles (Stuart 1982). The lowest fauna in Kent's Cavern, dominated by *Ursus spelaeus* (extinct cave bear) (the less-evolved *Ursus deningeri*, also extinct, occurs at Westbury) and containing *Arvicola cantiana* and *Homotherium latidens* (extinct sabre tooth), may be of Cromerian age (Campbell and Sampson 1971). It has been correlated guardedly with Westbury by Bishop (1982). At Boxgrove, *Dicerorhinus etruscus*, *Pliomys episcopalis*, *Sorex savini* and *Arvicola cantiana* have been recorded (Currant 1986) indicating an age analogous to that of other Cromerian sites *sensu lato*, and perhaps equivalent to that of Ostend and Westbury (Stuart 1988a). Uncertainties regarding both the provenance and identification of the remains notwithstanding, it is worth noting that *Mimomys* (Hinton 1919) and *Archidiskodon meridionalis* (Andrews 1919) have been reported from the Castle Eden fissure deposits, one of which may be of Cromerian age. Both animals have been recorded in Cromerian contexts in the British Isles, with the latter becoming extinct in the early Middle Pleistocene (Stuart 1982).

At Waverley Wood Farm, *Arvicola cantiana* and *Sorex savini* have been found in early cold-stage deposits younger than those in the channel (Currant cited in Lister 1989a). *Arvicola cantiana* implies that the deposits are later than middle Cromerian and *Sorex savini* has not been found in deposits of after late Cromerian–early Anglian time (Lister 1989a).

The differential occurrence of *Mimomys savini* and *Arvicola cantiana* within deposits of early Middle Pleistocene age in continental Europe has been noted by Stuart (1988a), and may similarly reflect the existence of more than one temperate stage at this juncture. Currant (1989) suggested that British pre-Flandrian, Middle and Late Pleistocene mammalian faunas could be grouped into five assemblages (Table 4.1). The type locality of the Cromerian Interglacial has provided the Group 5 assemblage. Group 4 assem-

Table 4.1 Key elements of the Middle and Late Pleistocene mammal faunas of Britain arranged to show their use in deriving biostratigraphically significant groupings of temperate woodland assemblages. Group 1, youngest; Group 5, oldest; F, Flandrian. Asterisks denote occurrences. From Currant (1989)

		F	*1*	*2*	*3*	*4*	*5*
Mimomys savini	A water vole						*
Sorex savini	A giant shrew					*	*
Pliomys episcopalis	An extinct vole					*	*
Dicerorhinus etruscus	Etruscan rhinoceros					*	*
Trogontherium cuvieri	An extinct beaver				*	*	*
Pitymys subterraneus	Pine vole				*	*	*
Ursus spp. 'spelaeoid'	Cave bears				*	*	*
Macaca sylvanus	Macaque			*	*		*
Castor fiber	Beaver	*		*	*	*	*
Apodemus sylvaticus	Wood mouse	*	*	*	*	*	*
Clethrionomys glareolus	Bank vole	*	*	*	*	*	*
Equus ferus	Horse			*	*	*	*
*Arvicola cantiana	A water vole		*	*	*	*	
Dicerorhinus kirchbergensis	Merck's rhinoceros			*	*		
Dicerorhinus hemitoechus	Narrow-nosed rhinoceros		*	*	*		
Bos primigenius	Aurochs	*	*	*	*		
Crocidura spp.	White-toothed shrew			*			
Ursus arctos	Brown bear	*	*	*			
Hippopotamus amphibius	Hippopotamus		*				
*Arvicola terrestris	Water vole	*	*				

* The water vole represented in Group 1 is a transitional morphotype; some individuals have the molar characteristics of *Arvicola cantiana*, some have the characteristics of *Arvicola terrestris*, and others are true intermediates.

blages occur at Ostend, Westbury-sub-Mendip and Boxgrove, and have been considered to belong to a younger temperate stage.

4.1.4.5 Other fossils

Sugworth has produced Coleoptera. The fauna is of interglacial type and contains species that today occur in conjunction with deciduous trees (Osborne 1980). Of particular interest are *Oxytelus opacus*, *Pelochares versicolor* and *Valgus hemipterus*, whose current ranges are south of the British Isles. Their presence during the Cromerian is suggestive of summer temperatures in Berkshire that were substantially higher than those of today (Coope 1977b).

Ostracods are also found at Sugworth. The assemblage is typical of the middle of an interglacial, with abundant *Scottia browniana* indicating a Hoxnian or earlier age for the deposit. The uncommon members of the fauna are analogous to those present in late Beestonian sediments in north Norfolk (Robinson 1980). A temperate ostracod fauna, which does not militate against a Cromerian age for the channel deposits containing it, accrues from Little Oakley. The lack of a brackish-water element in the fauna denotes an absence of tidal influence in the river at this locality (Robinson 1990).

A diverse coccolith (protective scales of unicellular planktonic coccolithophores) assemblage is present in the Steyne Wood Clay (Preece *et al.* 1990) (section 4.2), indicating that the locality investigated was fully connected to the central English Channel.

4.1.4.6 The presence of man

Indications of the existence of Lower Palaeolithic humans in the British Isles during the timespan under consideration are slight (Mellars 1974; Roe 1981; Wymer 1977, 1981, 1988). As the assertions that artifacts (eoliths) have been recovered from the Lower Pleistocene Crag sediments of East Anglia are untenable, the initial evidence for the presence of man in the British Isles has come from deposits of Cromerian *sensu lato* affinity. Such deposits may be intra-Anglian, as well as pre-Anglian in age (Wymer 1988) (Chapter 5).

According to Wymer, there are four localities in the British Isles at which mammalian fossils unknown after the Anglian Stage have a certain or possible relationship with Palaeolithic artifacts. They are Kent's Cavern, Westbury-sub-Mendip, Boxgrove and High Lodge, which are discussed in other contexts earlier in this chapter.

In Kent's Cavern, Acheulian artifacts, including a number of hand axes (Figure 6.4), have been found. While their relationship to the sediment containing the fauna of Cromerian type is unclear due to nineteenth-century excavation methods (Campbell and Sampson 1971), coincidence seems likely (Wymer 1988). Westbury-sub-Mendip has produced flaked flint (but no hand axes) of Acheulian type in association with a cave fauna of Cromerian affinity (Bishop 1975, 1982). Boxgrove, where there is an *in situ* relationship of Acheulian flint tools and a pre-Hoxnian fauna (Roberts 1986), has provided the clearest indications of human presence and activity within this timespan. Palaeoenvironmental data indicate that a mosaic of environments existed around the site, including open habitats along a coastline and in its hinterland, and possibly forest on the Downs to the north. Stone implements were manufactured at the site, and the surrounding landscape exploited for a variety of terrestrial, freshwater and marine biotic resources (Roberts 1986). In the context of Boxgrove and its firm evidence of the earliest humans yet known in the British Isles, Stuart (1988a) has drawn attention to localities in Germany (Kahlke 1975) and Hungary (Janossy 1975, 1986) where early Middle Pleistocene faunas with similar compositions to those of the British Isles are associated with human remains and artifacts.

The lake clays at High Lodge have yielded refined flint scrapers analogous to those produced by later (Mousterian, for example) technologies, but which are probably pre-Anglian artifacts (Wymer 1988).

4.1.4.7 Pedogenesis

The data on Cromerian environments thus far advanced have come from a relatively small number of sites at which landforms, sediments, fossil biota and artifacts are present. Such a patchy record can be supplemented by evidence of soil formation, which will have affected all regolith to some degree, as it does today. Sometimes entire soils have been fossilized. However, former pedogenesis is often recognized from one or more soil horizons, or from relict phenomena associated with soil formation. The term palaeosol is usually applied to soil features from which information on past environments can be ascertained (Chapter 1).

The Valley Farm Soil of southern East Anglia (Figure 1.14) has already been introduced (Chapter 1 and section 4.1.4.1). The type site of this palaeosol is in south-east Suffolk (Rose and Allen 1977). As it formed in Kesgrave Sands and Gravels, considered to be of Beestonian age, and was covered by Anglian glacial deposits (Rose *et al.* 1976), this truncated soil of rubified lessivé type (therefore having developed in a temperate environment) has been assigned to the Cromerian Stage. Investigations by Kemp (1983, 1985, 1987) have demonstrated that the Valley Farm Soil, while probably of only Cromerian age on the lowest river-terrace surfaces in Suffolk, is also likely to have formed in older stages of the Pleistocene on higher riverine flats. This followed the suggestion that the sands and gravels of the latter accumulated in the Pre-Pastonian, with those in the lower terraces being of Beestonian age (Hey 1980). Kemp (1985) has stated that the rubification and clay illuviation in the Valley Farm Soil were produced by a climate either as warm as, or warmer, than that which has existed during the present interglacial.

4.1.4.8 The offshore sequence and sea-levels

In the southern North Sea, deltaic sedimentation continued into the Cromerian, while to the north, mainly marine deposits accumulated (Cameron *et al.* 1987; Long *et al.* 1988). In the Forth Approaches of the central North Sea, glacial and glaciomarine sediments have been located near to the top of the Aberdeen Ground Formation. As the Matuyama–Brunhes palaeomagnetic boundary (Figure 1.1) falls a short way below these sediments, a correlation of the glacial with Glacial A of the Cromerian in the Netherlands (Figure 4.5) has been suggested by Stoker and Bent (1985). As noted at the outset of this chapter, the early Middle Pleistocene witnessed the termination of deposition in the Crag Basin. After this time, marine and brackish-water sedimentation in the current onshore area of East Anglia was more limited.

A restricted amount of information has been obtained relating to sea-levels during the Cromerian. Along the coasts of Norfolk and Suffolk an incursion of the sea commenced in the latter part of Cr II and progressed during Cr III, leading to the deposition of marine sediments onshore, with regression commencing early in Cr IV. Height data suggest that, unlike those of the preceding temperate stage (Pastonian), the Cromerian sediments underwent little or no deformation after their deposition, and that sea-level attained an altitude of up to *circa* 8 m OD (West 1980a).

Around Bognor Regis on the Sussex coast, at least one episode of marine transgression and regression is evident in the sediments, probably related to the Cromerian timespan *sensu lato*. Here, however, the current height of the deposits indicates a sea-level of *circa* 40 m OD (Bates 1986a). Analogous circumstances have been observed in coastal locations both to the west and east of this area (Apsimon *et al.* 1977; Woodcock 1981). Additionally, in the Isle of Wight an estuarine deposit which may be Cromerian (section 4.2) has been recorded at *circa* 40 m OD (Holyoak and Preece 1983; Preece *et al.* 1990). Comparison of these data with those pertaining to sea-level in other parts of southern England and in eastern England at a similar time, has led to suggestions of a different palaeogeography and pattern of sedimentation from that which now exists (Bates 1986a), and to tectonic activity (Roberts 1986).

If the episodes along the coasts of East Anglia, Sussex and Hampshire were equivalent in time, then local circumstances must have played a rôle in adjusting the heights of the sediments that now signify marine conditions. An altitude of 40 m OD in Sussex and Hampshire at this time in the Pleistocene seems improbable, and subsequent uplift is perhaps the most likely mechanism to have been operating in this area. Equally, levels of only up to *circa* 8 m OD appear rather low in context. The location of East Anglia at the margin of the subsiding southern North Sea Basin (Chapter 1) may have led to downwarping since the Cromerian.

4.2 Summary and conclusions

Deposits on the Norfolk and Suffolk coasts have provided evidence of vegetational successions characteristic of all substages of two interglacials: the Pastonian Stage and the Cromerian Stage. Sequences of environments deduced from pollen and spore floras have been supported by molluscan and vertebrate data, and by sedimentary records. Similarities in biota have allowed deposits at Little Oakley, Sugworth, Waverley Wood Farm, Kent's Cavern, Westbury-sub-Mendip, Boxgrove, Bembridge and High Lodge to be equated with Cromerian time *sensu lato*, and there are accompanying artifacts from the last four localities. There are grounds for believing that the deposits at some of these sites do not represent the same episode as those of the Cromerian of coastal Norfolk and Suffolk. In the British Isles, potential Cromerian sites are few and their sequences fragmented, which has hindered correlation. The most spatially continuous phenomenon recognized has been the Valley Farm Soil of southern East Anglia. However, in addition to forming during the Cromerian and signifying the existence of temperate conditions during the stage, in some localities this palaeosol is polycyclic, having also developed in earlier Pleistocene time.

The situation is further complicated when reference is made to the Middle Pleistocene succession of the Netherlands, in which a Cromerian Complex including at least four temperate episodes has been designated (Zagwijn *et al.* 1971; Zagwijn 1975, 1985, 1989; de Jong 1988) (Figure 4.5). West (1980a) has suggested that the type Cromerian was the temperate stage that came directly before the Anglian Cold Stage, and thus may correlate with Cromerian Interglacial IV. With regard to the scheme of Zagwijn (1975), West (1980a) has postulated that the Pastonian might correlate with Interglacial III and that the Pre-Pastonian (a) Substage be referable to Glacial A or Glacial B of the Netherlands Cromerian Complex.

Zagwijn (1975) has indicated the possibility of a hiatus between the Baventian and Pastonian in the Pleistocene sequence of the British Isles, which may have lasted for about 1 million years. During this time, the Netherlands' climatic record shows that around nine temperate and the same number of cold episodes occurred (Figure 2.2(a)). The insertion of the Bramertonian and Pre-Pastonian into this gap may provide some of the missing evidence but the situation remains unresolved. The fact that some of the stages which have been recognized in the Lower and early Middle Pleistocene of the British Isles are themselves separated by gaps has not assisted in this task.

Palaeomagnetism has helped in ordering the sequence over this timespan. Normal polarity (the Brunhes Epoch) prevailed for the majority of the period occupied by the Cromerian Complex, and the next-earliest normal magnetism occurred during the Jaramilo Event within the Matuyama Reversed Epoch (Figure 1.2(a)). West (1980a) has suggested one possible correlation in which the Beestonian, Pre-Pastonian, Pastonian and Cromerian equated with the Brunhes Epoch (and Cromerian Complex), and another whereby the Pre-Pastonian and Pastonian fell within the Jaramilo Event, with the other stages during the Brunhes Epoch. At Little Oakley, palaeomagnetic measurements support other data which suggest that the deposit belonged to a temperate stage in the later part of the Cromerian Complex. This stage has been considered equivalent to the Cromerian *sensu stricto*, as represented at West Runton (Bridgland *et al.* 1990).

The Steyne Wood Clay has normal geomagnetic polarity. Amino-acid ratios from *Macoma balthica* in the clay are indicative of an early Middle Pleistocene age, while coccolith data, notably the dominance of *Gephyrocapsa oceania* and *Gephyrocapsa caribbeanica* (section 4.1.4.5), permit a timespan of 475 000–275 000 BP to be assigned by analogy with the biostratigraphy of cores from the north-east Atlantic Ocean (Pujos-Lamy 1977). Correlation of this sediment with the Slindon Sands (section 4.1.4.1) on altimetric and palaeontological criteria, and reference of these deposits to Oxygen Isotope Stage 9, 11 or 13 has been suggested (Preece and Scourse 1987; Preece *et al.* 1990).

In the aminostratigraphic scheme of Bowen *et al.* (1989), the Cromerian is correlated with Oxygen Isotope Stage 13. Measurements were made on shells from a number of sites assigned to the Cromerian. However, the ratios from Waverley Wood Farm allow the episode represented there to be placed in Oxygen Isotope Stage 15 (Figure 1.15).

If the type Cromerian of the British Isles is the equivalent of Interglacial IV of the Netherlands Cromerian Complex, and the Westbury-sub-Mendip faunas 2 and 3 do not represent this episode, but were part of a later temperate interval prior to the Anglian Cold Stage, a further interglacial may have occurred within the complex. As a result, perhaps the fauna of Westbury I is from the Cromerian *sensu stricto*, although it has most affinity with faunas of Pastonian age in the British Isles (Bishop 1982).

Bowen and Sykes (1988) have attempted to establish the chronology and spatial distribution of glaciations in the British Isles by using aminostratigraphy of molluscs from sediments associated with high stands of sea-level, in conjunction with geochronometric ages and oxygen-isotope ratios. One of their suggestions, following Stoker and Bent (1985), is that till over the Aberdeen Ground Formation (section 4.1.4.8) could be the product of a glaciation equivalent to a Cromerian Complex, Elster I event, postulated in East Germany by Cepek (1986). This glaciation may have occurred during Oxygen Isotope Stages 14, 16 or 18 (Chapter 5).

The Westbury events have alternative chronological interpretations as, will by now be apparent, have numerous others in the late Lower Pleistocene and early Middle Pleistocene. No good purpose would be served by presenting more arguments along the lines of those above as they are some way from resolution.

5

The Anglian Cold Stage

Geomorphological, stratigraphical, pedological and palaeontological evidence has demonstrated that following the temperate environments which pertained during what may have been the final interglacial of the Cromerian Complex, cold conditions of the Anglian Stage affected the British Isles. As will be revealed below, most information concerning this Middle Pleistocene stage has accrued from East Anglia, with successions elsewhere traditionally considered to contain little or no representation of it. In recent years, however, the latter view has changed. An important consequence of this has been a questioning of the status of the Wolstonian Glaciation, formally designated as later in time (Mitchell *et al.* 1973a) (Table 1.1). Localities with deposits and/or landforms referred to the Anglian are shown in Figure 5.1.

5.1 East Anglia

On the coasts of Norfolk and Suffolk, freshwater sediments above Cromerian deposits and below tills have yielded palaeobotanical data which demonstrate the development of cold environments (West 1980a; Zalasiewicz and Gibbard 1988). Pollen and spore records, whose general trend is from an assemblage with *Betula*, *Pinus* and substantial non-arboreal frequencies, to one in which the latter attained extremely large values, have indicated a regional vegetation including grassland, low-shrub communities and scattered woodland during an early Anglian (e An) substage (West 1977a) (Figure 4.4). The early Anglian sediments have ice-wedge phenomena developed within them which indicate the incidence of permafrost. They are succeeded by the earliest glacial deposits yet securely identified in the British Isles (West 1980a). Periglacial features in the freshwater sediments have been associated with the presence of the Barham Soil (Rose and Allen 1977; Rose *et al.* 1985a) (Figure 1.14). This soil has also been recognized on the surfaces of terraces built by the River Thames along its former course in Hertfordshire, Essex, Suffolk and Norfolk (Figure 3.5), which were buried by Anglian glacial and fluvioglacial sediments in many areas. The Barham Soil has related loess and coversand (Figure 1.14), mineralogical data on which has shown their provenance to be Anglian glacigenic, rather than earlier (Kesgrave) terrace material. Rose *et al.* (1985a) have suggested that the Barham Soil began to form first beneath the birch woodland that characterized the vegetation at the conclusion (Cr IVc) of the Cromerian. The tundra and later polar-desert landscape which emerged during the initial part of Anglian time, prior to glaciation, would have been conducive to the development of periglacial phenomena and aeolian deposits. There is little doubt that ice was nearby when the significant macromorphological and micromorphological features of the soil came into existence. Indeed, this time may have witnessed an environment whose frigidity and dryness have yet to be surpassed in the Pleistocene.

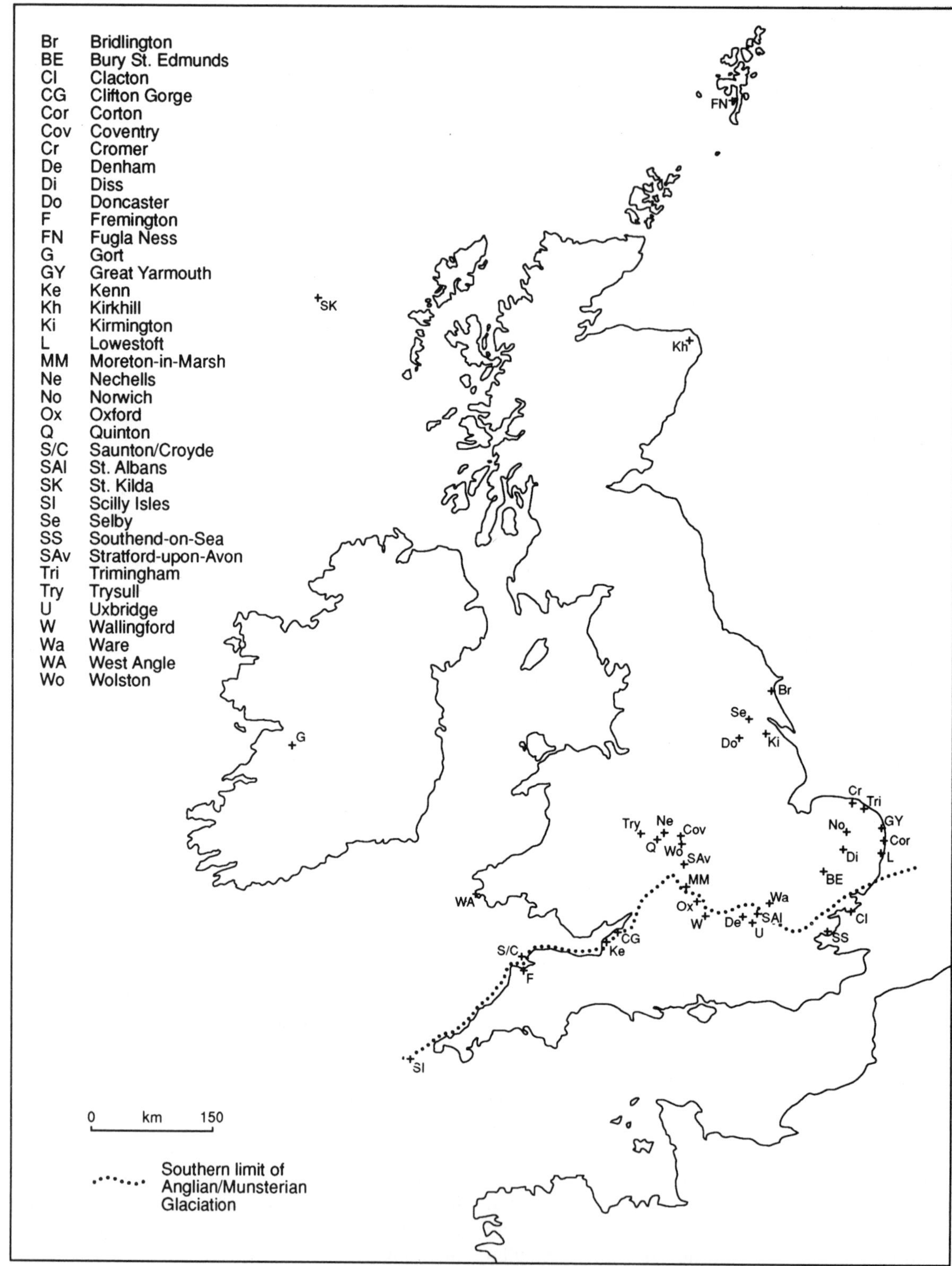

Figure 5.1 Localities referred to in Chapter 5. The southern limit of the Anglian/Munsterian Glaciation (source, Bowen *et al.* 1986) is also shown.

The length of the early Anglian periglacial episode is unknown but is likely to have been considerable, and sufficient for the production of a substantial regolith which subsequently contributed to the sediment load of the Anglian ice (Zalasiewicz and Gibbard 1988). The type locality for the Anglian Stage has been established at Corton in Suffolk (West 1961b; Banham 1971; Mitchell *et al.* 1973a; Shotton *et al.* 1977a; Rose 1989a) but deposits of this age are especially well developed in north-east Norfolk, where they have attracted the attention of numerous workers (for example, Banham 1968, 1975, 1977; Hart 1987, 1988; Eyles *et al.* 1989; Hart and Boulton 1991).

The earliest ice, of continental origin, came from the north/north-east (West and Donner 1956) and led to the deposition of what has been termed the North Sea Drift or Cromer Till Formation during the Gunton Stadial (West 1977a; Rose 1989a). Exceeding 100 m thickness in places, this has been shown to be a complicated sequence, but essentially consisting of three sandy tills with interposed outwash deposits. There is disagreement as to the depositional environment of the Cromer Till Formation. Banham (1977) believed that it was principally land based, with the tills being produced sub-glacially by basal melt-out. Hart (1987, 1988) has also inferred a mainly terrestrial environment, suggesting that the sands and gravels were deposited by a

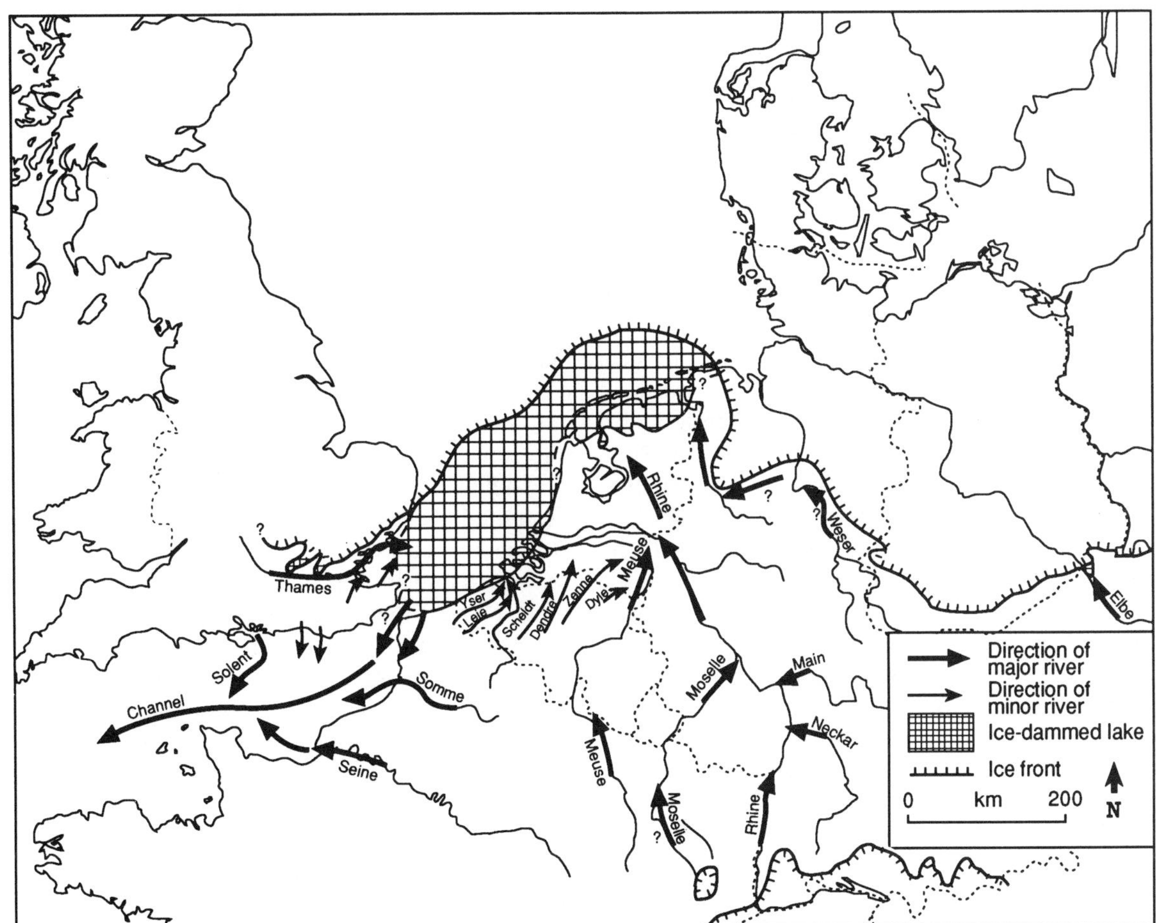

Figure 5.2 Schematic palaeogeographical reconstruction of the Elsterian–Anglian Stage at the glacial maximum. The ice-front positions and river courses are based on sources quoted in Gibbard (1988b). From Gibbard (1988b).

braided-river outwash regime. Zalasiewicz and Gibbard (1988) have postulated that the environment was glaciolacustrine. They pointed out that as Zagwijn (1974b, 1979) implied, during the Lower and early Middle Pleistocene, the southern North Sea was infilled with deltaic deposits from a number of northward-draining rivers including the Thames, Rhine and Meuse. With the progression of Anglian ice southwards, both from Scandinavia (the till contains erratics from Norway) and the British Isles, its ice sheets probably united, thereby closing the North Sea Basin. As a result, water from the latter was unable to enter the Atlantic and a large ice-dammed lake was formed in which the glacigenic sediments were laid down (Figure 5.2) (Gibbard 1988b). The Peelo Formation of the Netherlands consists of glaciolacustrine deposits referred to the Elsterian (Ter Wee 1983) (Table 1.1).

The Corton Beds, sandy deposits located in the East Anglian sequence between the Cromer Till Formation and the Lowestoft Till Formation (West 1961b), also imply sedimentation in an ice-dammed lake (Bridge and Hopson 1985; Zalasiewicz and Gibbard 1988). Water spilling from this lake is likely to have been responsible for breaching a Chalk ridge in the English Channel area, thereby opening the Strait of Dover, and probably caused the extension of a now vanished Channel River System (Gibbard 1988b). Eyles *et al.* (1989) have concluded that the North Sea Drift aggregated in a substantial body of water which was present about the isostatically depressed fringe of the Anglian ice sheet, and in an environment presumed to have been glaciomarine.

In the West Runton–Sidestrand area of Norfolk, examination of the North Sea Drift has shown it to be contorted and to include pieces of locally derived Chalk and Weybourne Crag of substantial size. Anticlinal folds of Contorted Drift extend upwards into the Gimingham Sands above the third till. Banham (1975) has invoked ice-loading and post-depositional compaction to explain this confused stratigraphy. None the less, it is generally accepted that the tills of the Cromer Formation were moved into East Anglia from the North Sea in three episodes. As noted above, the ice was not confined to the Cromer area. A North Sea Drift till (the Norwich Brickearth) was deposited between Great Yarmouth and Lowestoft, and this glacial deposit has been found inland around Norwich (Banham 1968). The outwash sediments intercalated with the Cromer tills are most likely to have been laid down in shallow water that was present subsequent to the withdrawals of the ice, with the margin of the latter responsible for the creation of a lacustrine environment (Zalasiewicz and Gibbard 1988).

At the Corton type-section, the North Sea Drift is overlain by the sandy Corton Beds (West 1961b). A flora from these has been reported by West and Wilson (1968). The vegetation was of tundra affinity and probably interstadial rank (the Corton Interstadial). Faunal evidence from the Corton Beds is equivocal (West 1977a). Marine Mollusca (Cambridge 1961) and foraminifera (Funnell 1961b) within them may be either *in situ* or derived.

A further till (the Lowestoft Till) was emplaced above the Corton Beds (West 1961b; Banham 1971) during the Lowestoft Stadial (West 1977a; Rose 1989a) by an ice sheet from the north-west (West and Donner 1956), which probably was initiated in Scotland (Bowen *et al.* 1986). Above the Lowestoft Till, lake clays (the Oulton Beds) and a flow till (the Pleasure Gardens Till), both of restricted distribution, were possibly laid down during ice decay (Bridge 1988). Lowestoft Till was not deposited in north-east Norfolk. Bowen *et al.* (1986a) have suggested that this was due to the continued residence there of the Scandinavian ice, which meant that the Scottish ice was diverted to the environs of Lowestoft. Cox and Nickless (1972) and Banham *et al.* (1975) have demonstrated that Cromer Formation and Lowestoft Formation deposits interpenetrate in Norfolk, thereby denoting contiguity of the ice masses during the Anglian. It should also be noted at this point that after analysing the particle size, heavy mineral and calcium carbonate content of the Cromer tills and the Lowestoft Till, Perrin *et al.* (1979) have been able to conclude that the former were distinct from the latter. However, stratigraphic, sedimentological and palynological (Peglar cited in Hart 1988) data from Trimingham have led Hart (1987, 1988) and Hart and Boulton (1991) to conclude that the last Anglian ice (either a readvance of a second sheet or the advance of a third) was present in north-east Norfolk late in this stage, and that the tills of the Lowestoft Formation and the North Sea Drift Formation were deposited at approximately the same time during the late Anglian.

Chalky till (Figure 1.14) comparable in mechanical composition and mineralogy with Lowestoft Till has been identified as a single sediment over a wide area of eastern England (Perrin *et al.* 1973, 1979). It has been suggested by Perrin *et al.* (1979) that this till was deposited during the Anglian. A conflict thus arose with the conventional opinions that a chalky Wolstonian till overlay one of Anglian age in East Anglia, and that the glacial sequence containing chalky till in the east Midlands was referable to the penultimate glaciation (see below and Chapter 7). This till has been shown to exhibit local modifications as a result of geological differences (Bowen *et al.* 1986). For example, a high Chalk content in north-west Norfolk has led to it being given the name Marly Drift (Straw 1965). In south-east Suffolk the Anglian sequence includes the Barham Soil, Barham Coversand and Barham Loess. These are succeeded by glacigenic gravels and tills belonging to the Lowestoft Till Formation (Rose *et al.* 1976). At Great Blakenham north of Ipswich, macrofabric analysis of the Lowestoft Till and the Gipping Till of Baden-Powell (1948), traditionally considered to be of Anglian and Wolstonian ages, respectively, has shown that they belong to different facies of the same Anglian till sheet (Allen 1982, 1988). In west Suffolk and Essex the Barham Coversand was deposited, then the outwashed Barham Sands and Gravels, and subsequently, Lowestoft Till (Rose *et al.* 1976). Typically, the Lowestoft Till Formation in these areas has been demonstrated to consist of a lower, thin till, such as the Maldon and Ware tills, followed by thicker, chalky till (the Springfield Till) (Clayton 1957a; Baker and Jones 1980). It should be noted at this point that the Springfield and Maldon tills have been assigned to the Wolstonian Glaciation by Clayton (1957a). A revised chronology (Clayton 1960) has equated the Maldon Till with the Lowestoft Till and the Springfield Till with the Gipping (Wolstonian) Till (Chapter 7).

The usual interpretation of the till sequence has not placed much emphasis on the lower member, which has been considered to reflect ice-sheet fluctuation prior to its main advance during the Lowestoft Stadial. However, Baker (1983) has suggested that this till may be indicative of a previous extensive Anglian ice advance across the region. A further controversial element in the Pleistocene chronology of this area is related to an alternative interpretation of the Kesgrave Sands and Gravels as fluvioglacial deposits (Cox and Nickless 1972; Bristow and Cox 1973). Baker (1977, 1983) has concurred with this, and assigned a possible early Anglian age to them. In consequence, the Valley Farm Soil developed in them might be referred to an Anglian interstadial. The loess, sands and gravels, and till above could also have been deposited during the Anglian Stage.

As Rose *et al.* (1985b) have emphasized, Anglian glaciation led to a major modification of the drainage pattern in East Anglia from a suite of river terraces inclined north-eastwards to a radial system developed largely on till.

5.2 The Vale of St. Albans and the Thames Valley

In Hertfordshire, where the Ware Till and related types have been identified (Cheshire 1983), most is known about Anglian events in the Vale of St. Albans and its environs (Gibbard 1977, 1979, 1983, 1985, 1988b) (Figure 5.3). The River Thames flowed through the Vale of St. Albans during the Anglian, depositing the West Mill Gravel (Table 3.1). A southerly ice advance via the Hitchin – Stevenage Valley progressed south-east and dammed the Thames, forming a lake north-west of Ware. This lake was eventually

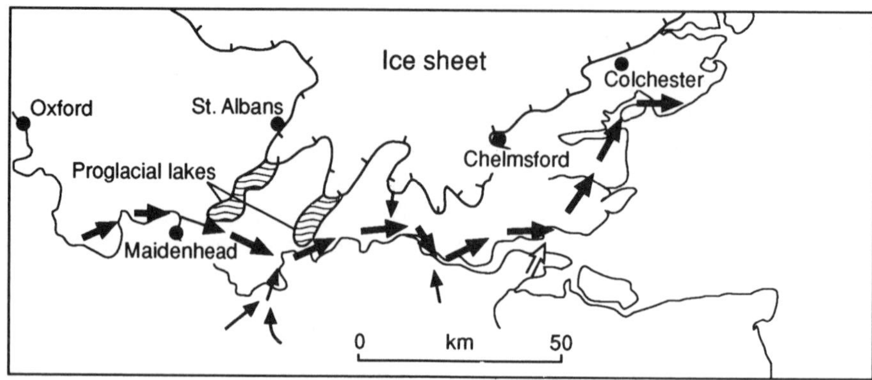

Figure 5.3 The Anglian Glaciation and initial diverted course of the Thames. From Bridgland (1988).

overridden by ice moving east and south-east which deposited the Ware Till around Ware and Much Hadham. Ice wastage saw fluvial conditions return to this area and renewed deposition of the Westmill Gravel. The Thames had continued to flow through the Vale of St. Albans during these events but now could not return to its eastward course around Ware which was blocked by till. It appears to have diverted via the Mole–Wey Valley at Hoddesdon before resuming its previous routeway to the east of Ware. A major ice advance then occurred from the north-east and moved south and south-west through the Vale of St. Albans. As a result of this advance, the Eastend Green Till (Lowestoft Till) was emplaced, the Thames impounded, and a large ice-dammed lake formed west of London Colney. A further south-westerly advance of the ice deposited more till. It also caused the water in the lake, and eventually that of the River Thames, to be deflected via spillways. An initial one near Uxbridge gave access to the Mole–Wey Valley. This was followed by others to this valley and probably that of the Wandle, with the Thames water finally entering the Medway Valley. Investigations of Pleistocene fluvial stratigraphy in Essex (Bridgland 1980, 1988) have revealed that the courses of the Thames and Medway joined in the Southend-on-Sea area. The combined rivers thereafter ran north-eastwards until around Clacton-on-Sea, where they reverted to the former route of the Thames (Figure 5.3).

In the Middle Thames region of Buckinghamshire, Berkshire and Oxfordshire, the Winter Hill Gravel (Table 3.1), the upstream equivalent of the Westmill Gravel (Gibbard 1977), has been widely recognized. Above Burnham this gravel was deposited by a braided river. Below this area, however, lower and upper sub-units have been discerned. The lower merges into the Westmill Gravel at Watford, while the upper, observed between Burnham and Denham, has a disposition indicative of deposition in an impounded valley. It is probably a deltaic deposit associated with valley blocking caused by Anglian ice in the Vale of St. Albans (Gibbard 1977, 1983).

The Wallingford Fan Gravel, located at the foot of the Chiltern escarpment about 45 m higher than the current channel of the River Thames (Whittow 1976), is largely a solifluction deposit, although also partly fluviatile. As it contains clasts derived from the Northern Drift of Oxfordshire (Chapter 3), it is younger than this deposit. Its relationship to the Thames terrace sequence indicates that it is older than the Chalky Till of Gloucestershire and Hertfordshire, and may mainly have accumulated during the early Anglian (Horton *et al.* 1981). In the Anglian gravels of the Thames, the amount of material having a northern and eastern origin built up while that of southern and western provenance declined as deposition progressed (Green and McGregor 1983).

In the Upper Thames region of Oxfordshire and Gloucestershire, glacial meltwater passed through the Evenlode Valley in the Cotswolds (Bishop 1958). The gravels of the Wolvercote Terrace (Figure 7.7) in the vicinity of Oxford include materials from this area (Briggs and Gilbertson 1973, 1980). These gravels, laid down after the assumed Cromerian deposits at Sugworth (Briggs *et al.* 1985), can be assigned to the Anglian according to Bowen *et al.* (1986) but conventionally have been thought of as Wolstonian in age (Chapter 7). Hey (1986) has suggested that the Combe and Freeland members of the Northern Drift gravels were deposited in the Anglian, and could be correlated with the Winter Hill and Black Park gravels in the Middle Thames Valley (Gibbard 1983, 1985) (Table 3.1).

5.3 The Midlands of England

It is stated in section 5.1 that Lowestoft Till has been recognized over a wide area in eastern England. According to Sumbler (1983a) it was deposited in Leicestershire, where it has been demonstrated interlocking with a till containing a high Triassic component, thereby indicating that it was formed at approximately the same time. Analogous till rich in Triassic material has been identified in both the eastern and western Midlands (Rice 1981), and thus also may be Anglian in age (Bowen *et al.* 1986). The chalky (Oadby) till of Leicestershire has been traced southwards into Northamptonshire, Bedfordshire and Buckinghamshire (Catt 1981).

However, the majority of these glacial deposits have been formally placed in the subsequent cold stage, the Wolstonian, whose type locality has been established near to Coventry (Shotton 1953, 1976, 1983a,b, 1989; Mitchell *et al.* 1973a) (Chapter 7). Rose (1987, 1989b) has reported that investigations into the distribution of the lower members of the Wolstonian sequence (the Baginton–Lillington Gravel and Baginton Sand) have established that they could be tracked from near Stratford-upon-Avon through Leicestershire and south Lincolnshire to East Anglia. They have been shown to occupy a buried valley of a river whose course was northeastward, and have been widely recognized (Shotton 1953; Rice 1968; Wyatt 1971; Wyatt *et al.* 1971; Douglas 1980; Hey 1980; Clark and Auton 1982a,b; Straw 1983). In the context of the status of the Anglian and Wolstonian glaciations, the most significant finding of Rose (1987, 1989b) is that around Diss in Norfolk, the equivalent of the gravel and sand in the Coventry area (the Ingham Sand and Gravel) occurs beneath Lowestoft Till, upon which were deposited sediments of the Hoxnian Temperate Stage that followed the Anglian (Chapter 6). The possibility of a hiatus between the gravel and sand and an overlying proglacial lake clay in the type area (indicating two glacial episodes) has been discounted by Rose, who has also noted that sand and gravel of this affinity near Bury St. Edmunds in Suffolk is intercalated with Kesgrave Sands and Gravels. These observations have led to the suggestion that the sands and gravels traditionally assigned to the Wolstonian Stage had begun to accumulate in the Early Pleistocene, contemporaneous with the Kesgrave Formation, and that this accumulation progressed up to the time of arrival of Anglian ice. It follows that the tills over much of eastern and midland England, which belong to the same lithostratigraphic unit, were deposited in the Anglian, and that their reference to the Wolstonian is erroneous. As Bowen *et al.* (1986) have stressed, such a revised glacial chronology does not affect the sequence of incidents represented by the morphological and stratigraphic evidence gathered by Shotton (1953), Bishop (1958), Rice (1968, 1981) and Douglas (1980). An advance of ice from the north-west early in the Anglian (probably at about the same time as the ice that laid down the North Sea Drift in East Anglia) deposited a till. On its retreat, a large proglacial lake (Lake Harrison) (Figure 7.2) impounded by ice to its north-west came into existence, where clay and sand were deposited. Subsequently there was an ice advance from the north-east which laid down a further till. This ice was also responsible for depositing the Lowestoft Till. As in East Anglia, confluence of the two ice sheets took place during this second episode, so that in a like manner, the tills of the eastern Midlands were placed over each other. In the context of the glacial succession in the Midlands, it is interesting to note that Shotton (1953) has described a till-like clay (the Bubbenhall Clay) in Warwickshire which is older than the basal Wolstonian sediments. Using the traditional chronological framework, this scattered glacial deposit could be of Anglian age, but if the recent views of Rose are correct, evidence of an earlier glaciation in this area may be present.

At Nechells (Kelly 1964), Quinton (Horton 1974, 1989a) and Trysull (A. V. Morgan 1973; Morgan and West 1988) in the western Midlands, biogenic deposits of Hoxnian age are underlain by glacial materials which *a priori* were deposited in the Anglian. As in East Anglia, this was a time of major landscape evolution in the Midlands (Bowen *et al.* 1986), with the excavation of the Severn Basin and heightened definition of the Cotswolds of particular significance. In the Lower Severn Valley, the Woolridge Terrace between 70 and 85 m OD (Wills 1938) has been equated with the Anglian Cold Stage; its gravel perhaps being outwash (Stephens 1970a). The current drainage network into the Wash (including the rivers Ouse and Nene) was initiated on deposits of the Anglian Glaciation (Bowen *et al.* 1986).

5.4 Wales, and England south of the Lower Severn and Thames valleys

As noted in Chapter 3, rocks from the mountains of North Wales, identified in the gravel deposits of the Thames Valley and East Anglia, have pointed to the incidence of glaciation in the former area during the Early Pleistocene. The so-called Irish Sea Glaciation of Wales (Pringle and George 1948; George 1970) may have taken place during the Anglian (Bowen *et al.* 1986), although a Wolstonian age has formerly been postulated (Chapter 7). In this episode, ice moved across Wales from the north-west (Bowen 1970, 1973a). At West Angle in Pembrokeshire, till of this affinity was succeeded by estuarine sediments. John (1968) has identified the latter and concluded that they were probably deposited during an interglacial whose age is unknown. Palaeobotanical data from the estuarine deposits, together with their stratigraphic context, have led John (1969) tentatively to assign an Ipswichian age to them (Chapter 8), whilst noting, however, that the pollen assemblage possesses Hoxnian affinities. Further palynological investigations of the material have revealed a flora indicative of a temperate episode, but one unlike that of any known interglacial in the British Isles (Stevenson and Moore 1982) (Chapter 6).

Fragmentary till deposits have been recorded in the Bristol district (Hawkins and Kellaway 1971; Gilbertson and Hawkins 1978). These have been considered probably to be of Wolstonian age, while the preceding Anglian Glaciation was responsible for cutting valleys such as the Clifton Gorge and Court Hill Channel, the latter being among those landforms subsequently infilled by glacial deposits (Hawkins 1977). However, amino-acid ratios from shells in fluvial sands above the till at Kenn near Clevedon, indicate that the glaciation occurred prior to Oxygen Isotope Stage 9, thereby probably referring it either to the Anglian or to an earlier cold stage (Andrews *et al.* 1984; Bowen *et al.* 1985; Bowen and Sykes 1988).

Elsewhere in south-west England, the Fremington Till, which has been observed in the Barnstaple area (Mitchell 1960; Stephens 1966, 1970b), may be of equivalent age to the tills around Bristol. It pre-dates raised-beach formation at nearby Saunton and Croyde (Bowen 1969a), and together with associated fluvioglacial sediments was deposited by Irish Sea ice (Wood 1974; Kidson and Wood 1974). Large erratic boulders which have been discovered on low wave-cut rock platforms in the Croyde – Saunton area and in Cornwall (Taylor 1956; Stephens 1970b) were also probably deposited during this episode. Wood (1974) has noted that giant erratics below raised-beach material at Saunton were of analogous composition to those in the Fremington Till.

On the northern Scilly Isles, patches of till and outwash gravel have provided evidence of the presence of Irish Sea ice. Mitchell and Orme (1967) thought these to have been of Wolstonian age because of their relationship to raised beaches. However, Bowen (1969a, 1973b) has maintained that the ice advance took place prior to raised-beach formation and thus could have been during the Anglian (Bowen *et al.* 1986).

Erratics also have been found on wave-cut platforms along the southern coast of England as far east as Sussex (West 1977a). They could have been the result of an ice advance through the English Channel in this or a subsequent glaciation (Kellaway *et al.* 1975). Such an idea is not supported by the valleys of the Channel River System, which are fluvial rather than glacial in origin (Gibbard 1988b).

5.5 Northern England and Scotland

As Catt (1981) has pointed out, there is likely to have been a substantial thickness of Anglian ice over much of the northern sector of the British Isles if there were glaciers in the Thames Valley (Figure 5.1). None the less, in Lincolnshire and Yorkshire, relatively close to the southern limits of the ice, such agents as relief, the impact of Scandinavian ice, and ice surges would have assumed importance, with localized glacial deposition taking place in response. According to Perrin *et al.* (1979) the main Anglian ice moved onshore in eastern England across the chalk escarpment and eroded the Wash–Fen Basin. One of the effects of the glaciation appears to have been the diversion of the River Trent northwards from an easterly course (Hey 1976; Catt 1981). Thus, the assumed Wolstonian tills of Lincolnshire (Straw 1979a,b, 1983) (Chapter 7) could have been deposited during the Anglian (Catt 1979, 1981). Watts (1959a) has ascribed a Hoxnian age to interglacial estuarine sediments at Kirmington near Brigg, Lincolnshire, after examining their pollen and spore records (Chapter 6). This suggestion has been supported by Boylan (1966) on archaeological grounds and the underlying glacial deposits have been assigned to the Anglian Stage. However, these glacial deposits include what appears to be the Wragby Till, designated as Wolstonian in age by Straw (1969, 1983). As Catt (1987a) has pointed out, this may be evidence of the Anglian age of at least the Wragby Till.

Further north in eastern England, remnants of possible Anglian glacial deposits have been observed on the summit areas of the Yorkshire Wolds (although these may be Baventian in age – Chapter 3) and the North Yorkshire Moors (Catt 1987a), also in the southern part of the Vale of York in the environs of Selby and Doncaster (Gaunt 1981). The Bridlington Crag, which has yielded a cool (possibly late Pastonian) biota (Reid and Downie 1973), alternatively might be of Anglian age, as may be the Sub-Basement Clay of Holderness (Shotton 1981). Scattered till deposits have been recorded in the Peak District (Stevenson and Gaunt 1971). Those that are deeply weathered and at the higher altitudes were probably deposited during the Anglian (Briggs and Burek 1985). In the north-east Lake District, pre-Devensian glacigenic sediments contain palaeosol features likely to have developed in three temperate episodes (Boardman 1985). While the age of the sediments is unknown, they could be Anglian.

In Scotland, as for earlier stages in the Pleistocene, the evidence of Anglian environments is tenuous (Bowen *et al.* 1986). At Kirkhill, north-west of Peterhead in Buchan, a complex sequence of Pleistocene sediments accumulated over a long time-period. The lowest deposits, head, and sand and gravel, are mainly periglacial, but erratics along the contact of the former with bedrock imply a prior glacial episode in which ice flowed from a westerly direction. Also the highest unit of the lower sequence may be a fluvioglacial gravel (Connell and Hall 1984a). The lower sequence was truncated by erosion and on the surface produced, a soil developed. The characteristics of this soil together with the pollen record from associated sediments have led to it being considered of Hoxnian Interglacial age, with the deposits below thus referable either to the Anglian or to a previous cold stage (Connell *et al.*

1982). However, Connell and Romans (1984) have reported micromorphological data which indicate that the soil was of a cold rather than temperate kind, so that an interglacial status has been precluded. The presence of plant remains, suggested to be characteristic of the Hoxnian Interglacial, in peat at Fugla Ness on the Mainland island of Shetland (Birks and Ransom 1969) (Chapter 6) means that an underlying till could be of Anglian age. Pre-Devensian (and possibly Anglian) glacial material may also have been identified on St. Kilda (Sutherland *et al.* 1984) and Lewis (von Weymarn 1979; Sutherland and Walker 1984). Elsewhere, the multiple till sequences of the lower Spey Valley (Sutherland 1984) could contain Anglian deposits (Bowen *et al.* 1986).

5.6 Ireland

The status of deposits formally referred to the Middle Pleistocene in Ireland has also been a matter of debate. Mitchell *et al.* (1973a) have assigned pre-Gortian sandy clay (below which is solifluction gravel) from near Gort in Galway, to the Anglian. The clay possesses a pollen flora indicative of cold conditions (Jessen *et al.* 1959) and the Gortian organic material above has been considered to equate with the Hoxnian on palaeobotanical grounds. While this correlation has been adhered to by Watts (1985), it has been suggested by Warren (1979, 1985) that the Gortian can be equated with the last (Ipswichian) interglacial (Chapters 6 and 8).

Pre-Gortian cold-stage deposits have been found in very few places. Coxon and Flegg (1987) have described sand (likely to be aeolian) and glacigenic sediments overlying a possible Early Pleistocene deposit in Galway (Chapter 3) which could be of this age, although till in the locality has been ascribed to the last glaciation by Mitchell (1980). The oldest widely distributed glacial deposits are of Munsterian age. These have been traditionally correlated with those of the Wolstonian Glaciation in other parts of the British Isles (Mitchell *et al.* 1973a). None the less, after studying till composition, Synge (1979, 1981) has suggested that an additional glaciation, the Connachtian, followed the Munsterian but preceded the last or Midlandian. Then the Munsterian Glaciation could have been equivalent to the Anglian and the Connachtian to the Wolstonian. However, as Bowen *et al.* (1986) and McCabe (1987) have pointed out, Munsterian deposits have been shown to possess contrasting lithologies as they were emplaced by ice within Ireland and from Scotland and the Irish Sea (Synge and Stephens 1960), and thus could belong to one glacial episode. Bowen *et al.* (1986) have equated the Munsterian with the Anglian on the basis of their interpretation of deposits assumed to be Wolstonian by others (Figure 5.1).

5.7 The offshore sequence

Offshore of Scotland, in the central and northern North Sea basins, respectively, Middle Pleistocene Aberdeen Ground Formation and Shackleton Formation sedimentation was truncated by erosion. In the central sector but not the northern one, till, possibly of Anglian age or older (Chapter 4), was deposited on the erosion surface (Stoker and Bent 1985; Cameron *et al.* 1987). Sediments on the continental shelf west of Scotland (Davies *et al.* 1984) (Figure 9.8) could include Anglian material (Bowen *et al.* 1986). In the southern North Sea, the Winterton Shoal Formation and Yarmouth Roads Formation are deltaic deposits that continued to accumulate in the early Middle Pleistocene (Balson and Cameron 1985; Cameron *et al.* 1987; Long *et al.* 1988) (Figure 1.1). The Swarte Bank Formation of the southern North Sea contains glaciolacustrine sediments of Elsterian (Anglian) age (Cameron *et al.* 1987; Long *et al.* 1988).

5.8 Correlation, age and archaeology

Bowen *et al.* (1986) and Bowen and Sykes (1988) have proposed that the Anglian Glaciation could be correlated with Oxygen Isotope Stage 12, whose age is around 450 000 years BP (Shackleton and Opdyke 1973; Johnson 1982). Supporting such a correlation are amino-acid ratios from non-marine Mollusca in gravels overlying Anglian till in Essex which can be referred to Oxygen Isotope Stage 11, and from Hoxnian Interglacial deposits that best correlate with Stage 9 (Bowen *et al.* 1985, 1989; Bowen and Sykes 1988) (Figure 1.15). The Anglian Glaciation usually has been regarded as equivalent to the Elster Glaciation of continental Europe (Mitchell *et al.* 1973a) (Table 1.1). Bowen and Sykes (1988) have suggested a correlation of the Anglian with an Elster II episode, and that this had been preceded by an Elster I glaciation. The latter, postulated for East Germany by Cepek (1986), may be represented offshore of the British Isles by till above the Aberdeen Ground Formation in the North Sea (section 5.7). The Elster I event could have formed part of the Cromerian Complex (Chapter 4) and may have occurred during Oxygen Isotope Stage 14, 16 or 18.

There is also evidence that during the timespan normally associated with the Anglian there may have been additional cold and temperate episodes in the British Isles. For example, Wymer (1985) has demonstrated typological variations in Lower Palaeolithic artifacts, mainly from localities in East Anglia and the Thames Valley, which suggest that at least one temperate (possibly interstadial) event occurred in Anglian time. There is little doubt that man existed in the British Isles during parts of the Anglian. Evidence is also present of a cold period late in this timespan, subsequent to one with a temperate environment, that witnessed a reduction in sea-level and the accumulation of gravel in the Thames Valley but which was not associated with the presence of a glacier. This period, when conditions were harsh, may be equivalent to Oxygen Isotope Stage 10 (Bowen *et al.* 1986). Finally, it will be recalled that the upper deposits at Westbury-sub-Mendip (Chapter 4) may belong to a temperate episode which occurred between the Cromerian and Hoxnian (Bishop 1982).

There are clearly a number of contentious issues associated with Anglian time and it is hoped that this chapter has explored them in such a manner that they may be evaluated.

6

The Hoxnian Temperate Stage

This interglacial has been defined by West (1956) following an investigation of the palaeobotany of lacustrine deposits overlying chalky (Lowestoft) till at Hoxne in Suffolk. Apart from the type site at Hoxne, a number of other localities of this age have been described in East Anglia, while analogous deposits have been recorded from the Thames Valley and its environs, the English Midlands, Sussex, Lincolnshire, Wales, Scotland and Ireland (Figure 6.1).

6.1 Flora and vegetation

The sediments at Hoxne are representative of the late Anglian, Hoxnian I, II, III (part) and early Wolstonian substages (West 1956, 1980b). In East Anglia, however, a site at Marks Tey near Colchester, Essex, has revealed the full vegetational sequence from late Anglian to early Wolstonian time (C. Turner 1970) (Figure 6.2). It should be noted here that the chalky boulder clay beneath the lacustrine deposits at Marks Tey corresponds to the Springfield Till, regarded by Clayton (1957a, 1960) as of Gipping (Wolstonian) age (Chapter 5). However, in view of the Hoxnian deposits above this till, it is probably the equivalent of the Lowestoft Till which occurs beneath the interglacial sediments at Hoxne (C. Turner 1970, 1973b).

The Hoxnian vegetational history of East Anglia has been synthesized by West (1980b). The Anglian lateglacial flora, of which *Hippophaë* (sea buckthorn) and Gramineae were prominent members, was replaced by *Betula* forest in Hoxnian I. This forest underwent compositional changes with *Pinus* achieving greater representation late in the substage. *Quercus*, *Ulmus*, *Acer* and *Tilia* were present in the vegetation and the substage concluded when forest dominated by oak succeeded that in which birch had been of most importance. Temperate forest including *Quercus*, *Ulmus*, *Tilia*, *Alnus* and *Corylus* was characteristic of Hoxnian II, during which three phases have been demarcated. In Ho IIa, *Quercus* was of most significance; in Ho IIb *Alnus* and *Corylus* assumed importance in the forest; in Ho IIc *Ulmus*, *Taxus* (yew) and *Corylus* appear to have been well represented. At Hoxne, Marks Tey, Barford near Norwich (Phillips 1976), St. Cross South Elmham near Bungay (West 1961b; Coxon 1979, 1982), Athelington near Eye (Coxon 1979, 1982, 1985) and Sicklesmere near Bury St. Edmunds (West 1981b), late Ho IIc pollen records reveal deforestation, with herbaceous vegetation expanding in place of trees and tall shrubs. C. Turner (1970) has ascribed this vegetation change to widespread burning of the forest, having noted a rise in charcoal values at the corresponding stratigraphic horizon at Marks Tey, where lamination counts in the lacustrine sediment indicate that the deforestation episode had a duration of *circa* 350 years. After this timespan, forest similar to that which existed previously in Ho IIc was able to regenerate. Turner has estimated from laminations that Ho IIc lasted some 2700 years and that the opening

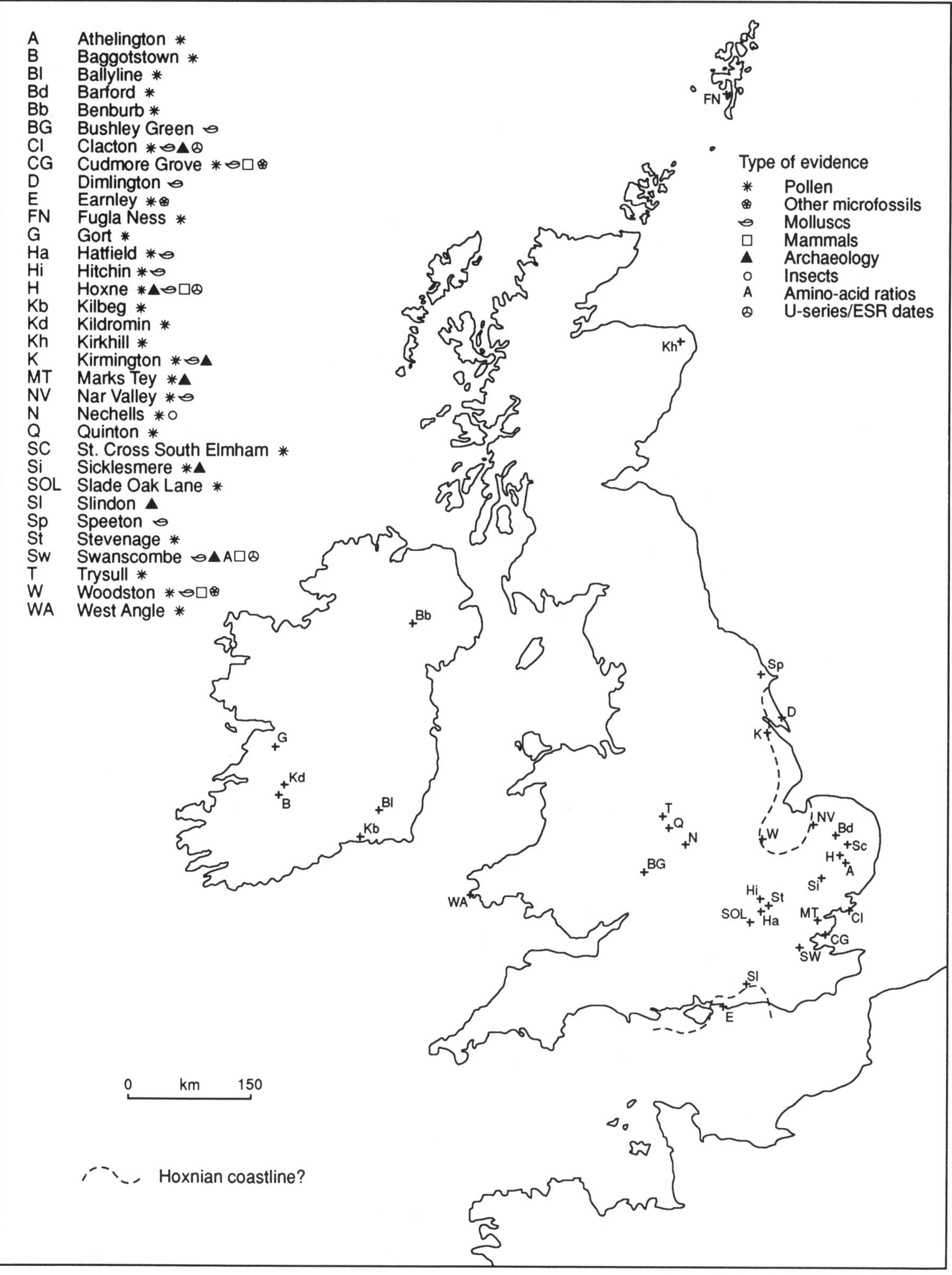

Figure 6.1 Sites referred to in Chapter 6 with types of evidence. Possible positions of the Hoxnian coastline are also shown.

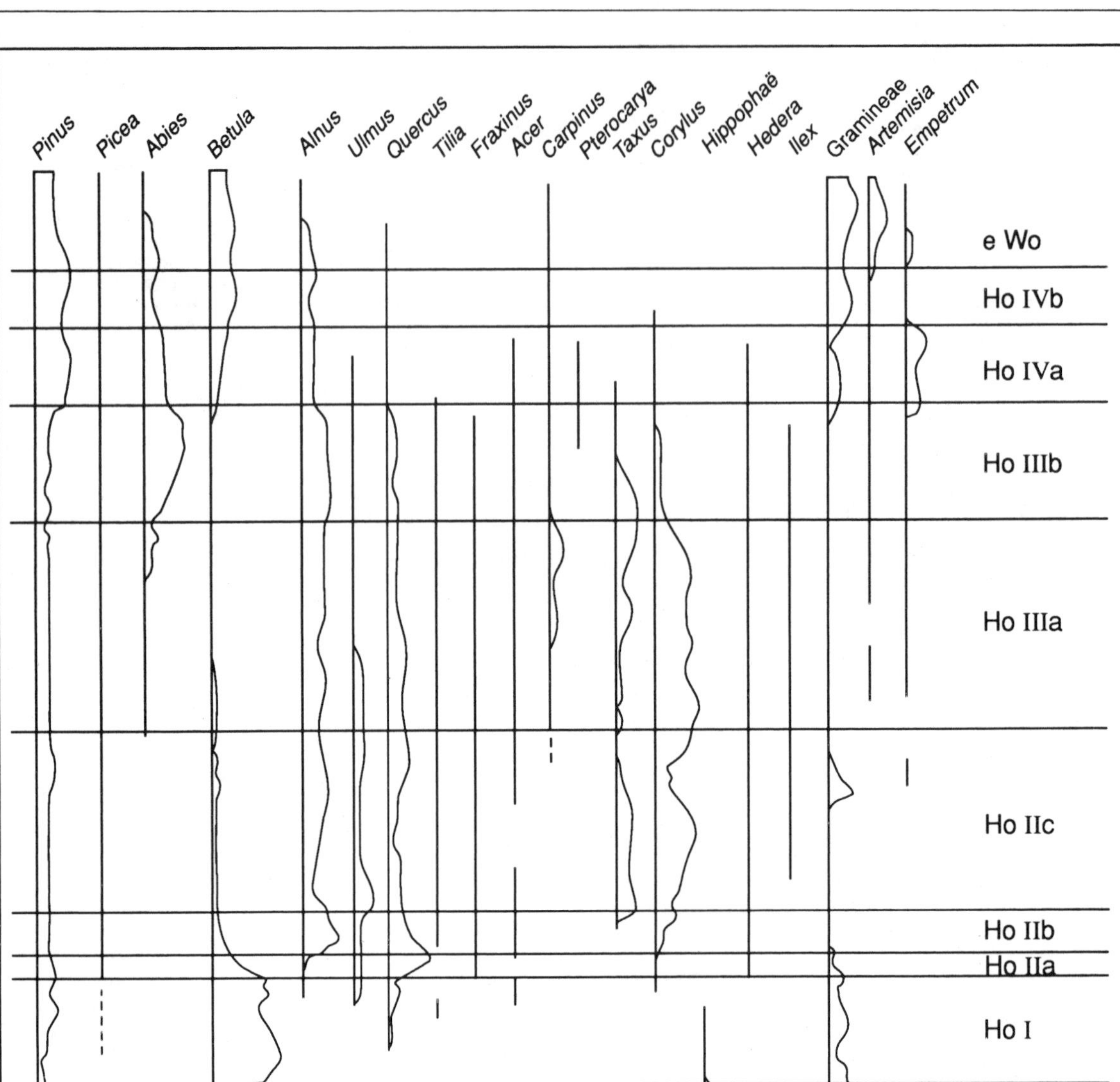

Figure 6.2 Pollen diagram of the Hoxnian at Marks Tey. Palynological data for the late Anglian and early Wolstonian substages are also shown. From West (1980b); source, C. Turner (1970).

phase of Hoxnian III (Ho IIIa) was in progress for around 2000 years. In Ho IIIa the arrival and spread of *Carpinus* occurred and the demise of *Quercus* and *Ulmus*, with *Corylus* remaining an important forest component. In Ho IIIb, hornbeam and hazel were of less significance, with *Abies* coming to dominance in a forest whose composition (including much *Alnus*) suggests an oceanic climate (West 1980b). Hoxnian IV witnessed a shift to boreal forest in which *Pinus* and *Betula* were important and where open habitats increased. In Ho IVa, *Abies* retained a substantial rôle and *Pinus* was significant, but *Empetrum* (crowberry) heath and grassland emerged. *Abies* and *Alnus* were reduced in Ho IVb and birch now assumed more importance in open woodland, heath became less, and grassland more, widespread. The vegetation of this last part of the interglacial has led to the suggestion that the climate was both colder and drier than during the previous phase of the substage. The onset of the subsequent cold stage, the Wolstonian, was characterized by the spread of grassland in place of boreal woodland during pollen zone e Wo (West 1977a).

While most of the sites in East Anglia from which Hoxnian palaeobotanical data have been obtained were inland lakes occupying hollows in Anglian till, some localities have yielded information on coastal plant communities and sea-levels (section 6.8) during this interglacial. Under the influence of a rising sea-level, the floor of the Nar Valley near Kings Lynn was occupied by a freshwater wetland vegetation early in the Hoxnian Interglacial when boreal forest was present in drier localities nearby (Stevens 1960; Ventris 1985, 1986). This forest, dominated by *Pinus* and *Betula*, was replaced by a temperate mixed-oak one. The eustatic rise progressed, with estuarine conditions being experienced in the valley during Ho IIc, and a peak of marine influence within Ho IIIb. The Woodston Beds, fluviatile freshwater and estuarine deposits, have been mapped at up to *circa* 16 m OD in the Peterborough area (Horton 1981, 1989b). Palynological data from fine-grained sediments of the Woodston Beds indicate the presence of mixed-deciduous forest vegetation of a type consonant with that of Ho IIc (Phillips cited in Horton 1981; Horton *et al.* 1992).

At Clacton-on-Sea in the Thames Estuary, freshwater sediments overlain by estuarine deposits have been referred to substages IIb and III of the Hoxnian Interglacial on palaeobotanical criteria (Pike and Godwin 1953; West 1956; Turner and Kerney 1971). An organic clay, above estuarine deposits in a channel at Cudmore Grove on Mersea Island at the mouth of the River Colne in north-east Essex (Bridgland *et al.* 1988), has a similar palynological record to that at Clacton some 10 km to the east and may also be of Hoxnian age (Roe cited in Bridgland *et al.* 1988).

In Hertfordshire and Buckinghamshire, Hoxnian (Ho I–IIIb) plant remains have been recorded from Hatfield (Sparks *et al.* 1969; Gibbard and Cheshire 1983), Stevenage (Ho I–IIIb) (Gibbard and Aalto 1977), Hitchin (Ho I–II) (Gibbard 1977) and Denham (Ho III–IV) (Gibbard *et al.* 1986b). At the latter locality the deposits, which accumulated in a doline, lend support to the notion of a rise in the regional water-table, probably as a result of higher rainfall during the second part of the interglacial.

In the western Midlands of England there are three localities of significance for Hoxnian environments. In the Nechells area of north Birmingham, Anglian fluvioglacial material is succeeded by lateglacial and interglacial lake sediments, the latter being replaced by fen and marsh deposits. A subsequent cold stage (the Wolstonian) is represented mainly by tills and sands and gravels (Chapter 7). The vegetational history around Nechells (Duigan 1956a; Kelly 1964) included a period with temperate deciduous forest in which *Alnus* and then *Taxus* were common in the environs of the lake. Later coniferous forest was initially dominated by *Picea*, then *Abies* and lastly *Pinus*. The pine forest was open, and together with heath and grassland present at this time, is considered indicative of the impending cold conditions at the start of the Wolstonian Stage (Kelly 1964). At Quinton in west Birmingham, organic sediments with a palynological record considered to extend from the late Anglian through the Hoxnian into the early Wolstonian (Herbert-Smith cited in Horton 1989a), are underlain by till and head and overlain by sand and till. The glacigenic sediments may be referable to the Anglian and Wolstonian stages, respectively (Horton 1989a). The palynological data from Nechells and Quinton are similar, and to these Hoxnian sites may be added Trysull, south-west of Wolverhampton (A. V. Morgan 1973; Morgan and West 1988). Here the pollen and spore flora from freshwater deposits overlying fluvioglacial outwash indicate late Anglian, Hoxnian I and Hoxnian II plant communities analogous to those around Nechells, some 25 km to the east. At first, Gramineae and

Cyperaceae dominated the herb component and *Hippophaë* and *Juniperus* were important shrubs in rather open vegetation. This was followed by forest in which *Betula*, then *Quercus*, *Corylus* and *Alnus* were of most significance.

Pollen analysis of marine sediments at Earnley in Bracklesham Bay, Sussex, has indicated a diverse mixed coniferous and deciduous forest in the region, possibly during Hoxnian III time (West *et al.* 1984). From Kirmington, close to the eastern foot of the Lincolnshire Wolds, peat associated with estuarine sediments which overlies glacial deposits (Chapter 5), has been referred to the Hoxnian on palynological criteria by Watts (1959a). At West Angle Bay in Dyfed, Wales, clays and muds above a raised beach and below head have yielded palynological data which are divisible into four assemblage zones (Stevenson and Moore 1982). Each of these zones is dominated by trees and shrubs characteristic of interglacial temperate forest. While firm correlation with other interglacials in the British Isles has not been possible, considerable amounts of *Abies* in the middle and upper part of the sequence, the presence of *Picea* throughout with a maximum in the upper part, the constant occurrence of *Tilia* in the middle two zones, and no peak of *Carpinus* have been cited in favour of a Hoxnian age for the sediments. Also of interest is the identification of forest destruction (as during Ho IIc at Hoxne and elsewhere in East Anglia – see above), considered most likely to have been caused by flooding of a coastal valley in the environs of the site.

Birks and Ransom (1969) have presented palaeobotanical data from peat which lies between tills at Fugla Ness on Mainland island of Shetland. Interglacial woodland (significant frequencies of *Abies* implying a Middle Pleistocene age) is indicated. This underwent changes, with deciduous taxa declining and *Pinus* expanding to dominate the tree component of open woodland. Heath and grassland were also present. A deterioration from a quite warm, oceanic climate occurred. There were considerable floristic similarities with the end of the Gortian Interglacial (see below), and the plant assemblages have been cautiously assigned to Zone IV of the Hoxnian or later. However, as Lowe (1984) has pointed out, the dominantly ericaceous pollen assemblages which characterize this and other purported interglacial intervals in Scotland may be representative of interstadial conditions. Also in Scotland, the pollen record from sediments above the truncated lower palaeosol at Kirkhill, Buchan, could be of Hoxnian age (Connell *et al.* 1982; Lowe 1984; Bowen *et al.* 1986).

The palaeobotany of interglacial lake deposits below tills considered to be of Munsterian and Midlandian age (Table 1.1) at Gort, County Galway in Ireland has been described by Jessen *et al.* (1959). The vegetational history comprised an initial stage with *Juniperus–Hippophaë* scrub, a second during which *Pinus–Betula* forest was dominant, and a third when *Taxus* thickets were of major importance. In a fourth stage, *Abies* and *Picea* became members of forest vegetation, and there were also juniper heath and *Rhododendron* thickets. *Taxus* experienced a resurgence in stage five, declining in stage six when *Buxus* (box) expanded. Watts (1959b) has presented floristic evidence of interglacial conditions from two sites in County Waterford. At Kilbeg, lacustrine sediments are underlain by solid rock and overlain by solifluction deposits and Munsterian till. At Newtown, peat occurs below tills and heads and above a rock platform forming a raised beach. The Kilbeg floristic sequence is very similar to that from Gort. The plant evidence from Newtown is not as diagnostic, showing dominance by *Pinus*. However, it probably belongs to the end of the Gortian Interglacial. Subsequent palaeobotanical data from two localities in County Limerick, Baggotstown (Watts 1964) and Kildromin (where six pollen zones have been designated) (Watts 1967), together with that from Benburb in County Tyrone (Boulter and Mitchell 1977; Gennard 1984) have also been assigned a Gortian age. A palynological investigation of lake sediments occupying a karst depression and overlain by glacial deposits at Ballyline, County Kilkenny, has been reported by Coxon and Flegg (1985).

Similarities to, and differences from, Gortian sites have been noted. While the authors concede that the vegetation may have been Gortian, they conclude that firm correlation with this episode is not possible and prefer to ascribe the deposit to a Middle Pleistocene interglacial hitherto unrecorded in Ireland.

The age of the Gortian deposits has caused controversy. On palaeobotanical grounds, Mitchell *et al.* (1973a) and Watts (1985) have favoured their correlation with those of the Hoxnian Interglacial of other parts of the British Isles (section 6.9). Watts has pointed to substantial *Hippophaë* pollen frequencies in the pioneer stage and considerable amounts of *Abies* later in the palynological sequence, and to the presence of *Azolla*, in support of this equivalence. It is also worth noting the presence at several Gortian sites (and at Ballyline) of an unknown pollen, designated Type X (C. Turner 1970; Phillips 1976), which has been recognized in numerous Hoxnian assemblages. However, Warren (1979, 1985) has cited stratigraphic evidence in support of a last interglacial (Ipswichian) age for the Gortian deposits. He has suggested that it was unusual for sediments of penultimate interglacial age to have been preserved in numerous localities when none belonging to the last episode of this rank have been identified, and has argued that the earlier ones were the more likely to have been destroyed by subsequent glacial activity. As McCabe (1987) has stated, the age of the Gortian sequences, the implications of their floristic composition and their relationships to drift stratigraphy are a major problem.

6.2 Non-marine Mollusca

A large fauna has accrued from Swanscombe in Kent (Kennard 1942; Kerney 1971). Here Thames terrace deposits (Figure 6.3), assigned to the Hoxnian by virtue of their animal remains and elevation above sea-level by Oakley (1957), contain freshwater and terrestrial Mollusca. The Lower Gravel and Lower Loam, perhaps of Ho II age, yield many temperate freshwater taxa (notably *Ancylus*, *Valvata piscinalis* and *Bithynia*), which suggest the existence of a large river with well-oxygenated calcareous water. Land snails occurred in numerous habitats nearby. They mainly reflect shifts in the river system but also suggest that open woodland existed. In the Lower Middle Gravel, temperate conditions and a fast-flowing river have been diagnosed. Terrestrial molluscs in this sediment imply that a reasonably closed forest was present in the area. The Upper Middle Gravel contains freshwater Mollusca characteristic of running, but quieter, water. Its terrestrial taxa, among which *Trichia hispida* is dominant, indicate open conditions and a colder climate, probably during late Ho III and subsequent interglacial and early glacial time (Kerney 1971). Variation in the composition of the freshwater faunas of the Lower Gravel and Middle Gravel has been noted by Kennard (1942). In the later deposits, the appearance of, for example, *Corbicula fluminalis*, *Theodoxus serratiliniformis* (*Theodoxus danubialis*), *Belgrandia marginata* and *Valvata naticina* has been thought to relate to the connection of the Thames and Rhine river systems at this juncture, as most of these species are now characteristic of central Europe. The Woodston Beds have also provided a substantial fauna (Kennard and Woodward 1922; Graham cited in Horton 1981; Horton *et al.* 1992). This mainly consists of taxa which indicate slow-moving, fresh water and adjacent wetland. Silty marls deposited during the late Anglian and Hoxnian at Hatfield have yielded abundant non-marine Mollusca. A small, shallow lake is indicated, with large numbers of *Valvata cristata* signifying clean water with a muddy substrate and abundant vegetation (Sparks *et al.* 1969). The lake deposits at Hoxne have produced a small, unremarkable molluscan fauna (Sparks 1956); those at Marks Tey also contain relatively few snails. In the latter locality, the existence of *Valvata piscinalis* and *Bithynia tentaculata* in marginal sediments of Ho III and IV age suggests slow-moving water (C. Turner 1970). Freshwater shells have been noted in the Kirmington interglacial deposit (Stather 1905; Penny *et al.* 1972; Catt 1977b), and an abundant molluscan fauna has been recorded from the Freshwater Member

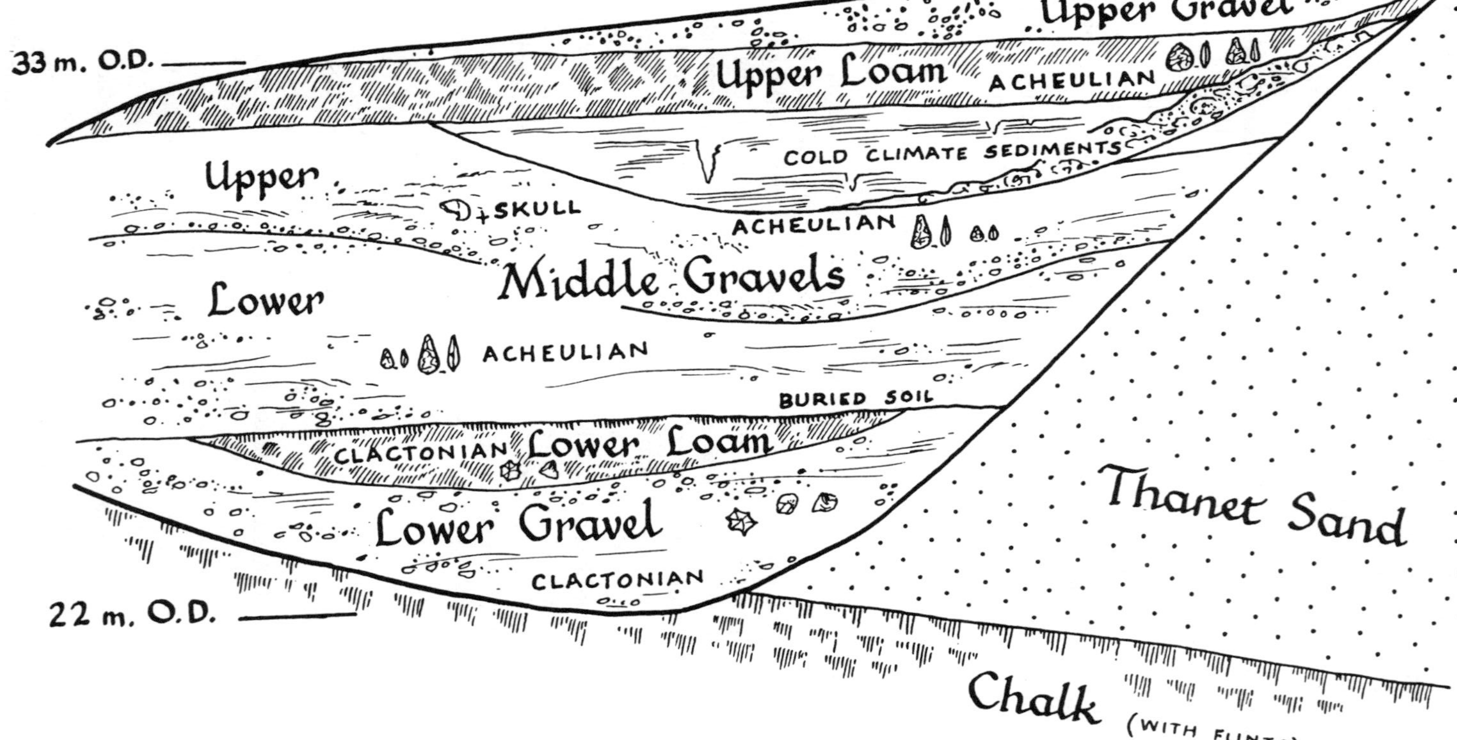

Figure 6.3 Section across Barnfield Pit, Swanscombe. Drawn by J. J. Wymer.

of the Nar Valley Formation (Ventris 1986). Freshwater beds of Ho IIb and Ho IIIa age at Clacton contain a suite of Mollusca characteristic of a river (Turner and Kerney 1971). At Cudmore Grove, a shelly detritus mud above estuarine silt has produced *Corbicula fluminalis*, *Bithynia*, *Valvata* and *Pisidium*, the assemblage being indicative of a brackish influence (Bridgland *et al.* 1988). Molluscs are infrequent at Nechells but the fauna, in which *Bithynia tentaculata* dominates and whose provenance is deposits equivalent to floral zones IN2, IN3 and IINa, is indicative of lacustrine conditions (Shotton and Osborne 1965). A molluscan fauna including *Bithynia tentaculata*, *Valvata piscinalis* and *Pupilla muscorum* from the Bushley Green Terrace of the River Severn, *circa* 30 m OD, has led Wills (1938) to suggest that some of its gravels were laid down during an interglacial (the Hoxnian) following the first glaciation (the Anglian) of this area (Stephens 1970a).

6.3 Marine Mollusca

Estuarine deposits of Hoxnian age from Clacton-on-Sea (Baden-Powell 1955), Cudmore Grove (Bridgland *et al.* 1988) and Kirmington (Stather 1905; Penny *et al.* 1972; Catt 1977b; Straw 1979a) have yielded faunas including *Cerastoderma edule*, *Mytilus edulis*, *Scrobicularia plana* and *Hydrobia ulvae*. These faunas are comparable to those present in the Humber and Essex estuaries today. The Nar Valley Clay has produced *Hydrobia*, *Ostrea* and *Cerastoderma*, indicating the existence of estuarine conditions in this part of Norfolk during the Hoxnian (Ventris 1986). The channel deposits at Earnley include *Ostrea edulis*, *Nucula*, *Sphenia binghami*, *Mysella bidentata*, *Chlamys varia*, *Abra*, *Cerastoderma*, *Gibbula*, *Odostomia unindentata*, *Turbonilla elegantissima*, *Bittium reticulatum* and *Akera bullata*, today characteristic of shallow, sub-tidal environments (West *et al.* 1984). *Hydrobia*, *Mytilus* and *Ostrea* in the upper part of the Woodston Beds marked the onset of brackish and more saline conditions associated with a marine transgression (Kennard and Woodward 1922; Graham cited in Horton 1981; Horton *et al.* 1992). Estuarine silt and sand, exposed at about 30 m OD in the cliffs near Speeton, south of Filey on the coast of Yorkshire, contain a molluscan fauna which includes *Scrobicularia*, *Macoma* and *Cardium*. This Speeton Shell Bed, whose fauna is temperate but which indicates water conditions colder than those of the current southern North Sea (Edwards 1987), is overlain by till of likely Wolstonian age (Chapter 7). Catt and Penny (1966) and Penny and Catt (1972) have suggested that the altitude and stratigraphic relationships of these estuarine deposits is consistent with a Hoxnian age, although palynological data reported by West (1969a) are more indicative of Ipswichian vegetation (Chapter 8). Rafts of marine clay within the same till near Dimlington in Holderness possess a marine molluscan fauna of boreal affinity, similar to that of the Bridlington Crag, and possibly of late Hoxnian date (Catt and Penny 1966; Catt 1987a,b).

6.4 Vertebrates

Stuart (1974, 1982) has synthesized data on Hoxnian vertebrates. The type site at Hoxne has yielded numerous vertebrate fossils. Lacustrine sediment of Ho IIc date has produced *Palaeoloxodon antiquus* and *Equus ferus* remains (Spencer 1956). It is interesting to note the presence of elephant and man (section 6.6) at the time during Ho IIc when palynological data indicate that forest receded and open-habitat vegetation developed in the environs of Hoxne (West 1956). Fluviatile, lacustrine and solifluction material above the lake muds, probably deposited during Ho IIIb and Ho IV, contain records of *Macaca* (macaque monkey), *Capreolus capreolus* (roe deer), *Dama dama* and *Castor fiber*, suggesting the existence of forest. The youngest beds in this sequence contain *Lemmus lemmus* (Norway lemming), today an animal of the tundra and its ecotone with boreal forest (Wymer 1974, 1988; Gladfelter 1975; Gladfelter and Singer 1975; Stuart 1982). Only fish remains (*Salmo* – salmon or trout) were recovered from Marks Tey, but other Hoxnian lake deposits at Copford, a short distance to the north-east (C. Turner 1970), have finds of *Trogontherium cuvieri* (extinct beaver) (also recorded in the organic muds at Hoxne) and *Ursus* (bear) (Stuart 1982). Fish remains from Nechells (including pike, perch and roach) imply the existence of a still water body (Shotton and Osborne 1965).

Swanscombe (Barnfield Pit) (Figure 6.3) has provided an important assemblage of vertebrate remains (Sutcliffe 1964; Conway and Waechter 1977; Stuart 1982). *Macaca* (macaque) occurs as do *Castor fiber* and *Palaeoloxodon antiquus* in a Lower Gravel and Lower Loam fauna which mainly reflects the existence of mixed-deciduous forest, probably of Ho II age. Some taxa (for example, *Equus ferus* and *Megaceros giganteus* (giant deer)) characteristic of open habitats were present at this time, and also later (perhaps during Ho IV) in the sequence, when they were accompanied by *Lemmus lemmus* and *Lepus timidus* (mountain hare), signifying a cooling of the climate. Dierden's Pit, Ingress Vale, a few hundred metres from Barnfield Pit, has a mammal fauna analogous to that of the Lower Gravel and Lower Loam in the latter location (Sutcliffe 1964). A notable occurrence here is *Emys orbicularis* (Stuart 1979). As observed in Chapter 1, the European pond tortoise now breeds in areas where summers are warmer, drier and sunnier than in the British Isles.

The Hoxnian deposits at Clacton (Pike and Godwin 1953) have been the source of many animal bones (Warren 1951, 1955; Sutcliffe 1964), most coming from freshwater deposits likely to be of Ho IIb age (Turner and Kerney 1971). Fluviatile gravels at Clacton Golf Course contain vertebrate remains assigned to the Anglian by Singer *et al.* (1973). However, Stuart (1982) has suggested that these sediments also belong to the Clacton Channel Complex and thus that their fauna is of similar age. The mammals, closely resembling those recorded in the Lower Gravel and Lower Loam at Swanscombe, are consonant with a mainly forested landscape. Also included are such animals as lion, horse and an extinct rhinoceros (*Dicerorhinus hemitoechus*) that are likely to have been inhabitants of a floodplain (Stuart 1982), the herbaceous vegetation of which is represented in the palaeobotanical record. At Cudmore Grove, the shelly detritus mud above estuarine silt has yielded numerous small vertebrates, including *Arvicola cantiana*, which signify an age younger than the type Cromerian, but not younger than the Ipswichian (Bridgland *et al.* 1988).

As Stuart (1982) has observed, the paucity of localities containing vertebrate remains militates against any firm conclusions concerning succession within, and distinctive characteristics of, the Hoxnian vertebrate fauna. As in the Cromerian and Ipswichian interglacials, lemming occurred towards the end of the stage, probably accompanying the shift to more open vegetation and a colder climate. In contrast to the Ipswichian, the straight-tusked elephant and fallow deer continued to be present into the later part of the Hoxnian. The absence of *Hippopotamus*, recorded from both the Cromerian and Ipswichian, may also assist in the diagnosis of Hoxnian vertebrate faunas.

The scheme of Currant (1989) (Table 4.1) has a Group 3 faunal assemblage that can be distinguished from Group 4 and Group 5 assemblages, and whose characteristics include a diversity decrease in small mammals and changed composition of the rhinoceros fauna. It has been concluded that Swanscombe, Clacton and the type site at Hoxne have faunas referable to this group.

6.5 Other fossils

Coleopteran assemblages from Nechells, described by Shotton and Osborne (1965), are very similar to those of the British Isles today; only three exotic species being recovered. The insects reflect the environments of a lake, its immediate surroundings and tributary streams, together with more distant, drier habitats. Of especial interest is the correlation between the time of first appearance of *Rhynchaenus* spp., phytophages which depend either entirely or largely on oak for food, and of *Quercus* pollen. The insect evidence points to an optimum climate during the Hoxnian which is analogous to that of today but with slightly higher temperatures. The beetle fauna from Hoxne suggests that the climate was similar to that in which the Nechells' insects were living (Coope 1977a).

Ostracoda contained in the Earnley deposits have assisted in the interpretation of coastal environments and sea-level changes (West *et al.* 1984). Some (*Carinocytheresis*, for instance) are very like taxa in the Hoxnian Nar-Valley Clay (Lord and Robinson 1978). The ostracod assemblage from Cudmore Grove, dominated by the euryhaline *Cyprideis torosa*, is diagnostic of brackish or inter-tidal environments (Bridgland *et al.* 1988). Ostracod remains from Nechells, with *Cytherissa lacustris* most prominent, indicate the presence of a large, deep, cold water-body early in the sequence (Sylvester-Bradley 1965).

Analysis of Cladocera from Fishers Green has provided information on climate and water chemistry (Beales 1977). For example, *Chydorus sphaericus*, a dominant above the Arctic Circle at present, occurs in sediment whose palynology indicates lateglacial vegetation, while the bulk of the assemblage suggests water of moderate or higher alkalinity and conductivity. Chironomids from Nechells, where *Microtendipes* and *Tantytarsus* were the initial colonizers, are suggestive of still water which had resulted from a recent deglaciation (Shotton and Osborne 1965).

6.6 The presence of man

Possibly the oldest human remains in the British Isles have come from Barnfield Pit, Swanscombe, where pieces of a skull have been discovered in the lower part of the Upper Middle Gravel (Oakley 1952; Molleson 1977) (Figure 6.3). Although le Gros Clark (1938) has considered the skull to be related to modern *Homo sapiens*, Stringer (1974) has demonstrated its affinity to early Neanderthal man. Remains of other mammals in the same part of the Upper Middle Gravel suggest that more grassland was present than previously, with a Ho IV age likely (Kerney 1971). The only other skeletal material of broadly comparable antiquity has been recorded from North Wales (Green *et al.* 1981; Green 1984) (Chapter 7).

Wymer (1968, 1974, 1977) and Roe (1981) have described and discussed the industries of the Lower Palaeolithic in the British Isles. The Clactonian, with chopper cores (Figure 6.4) and flakes most prominent, and in which unspecialized flake tools occur, has been considered by Wymer to be indicative of a low level of technology. The Acheulian was characterized by a variety of hand axes and both specialized and non-specialized flake tools occurred. Although it is tempting to place inelegant industries earlier in time than more sophisticated ones, Wymer (1977) has cautioned against such a practice until the artifacts can be dated. Wymer (1988) has synthesized information concerning palaeoliths and their relationship to Hoxnian deposits. Of sites ascribed to the Hoxnian on palynological criteria, only Hoxne and Clacton have produced artifacts and mammalian fossils in a primary context that are amenable to correlation with the interglacial stages based on vegetation succession. At Hoxne, the lake muds of Ho IIc contain a few hand axes and flakes, a number from the levels in the sediment that have provided palynological evidence of forest opening (West 1956). However, Wymer (1988) has suggested the likelihood that these had moved downwards, as they are of analogous type to the Lower Acheulian Industry present in higher, reworked sediments. These also contain an Upper Acheulian Industry, and both must have been operational after Ho III according to the pollen record from the deposits. Wymer has suggested that they spanned Ho IV and the Early-glacial zone of the next cold stage. At Clacton, palaeoecological data from deposits with artifacts have established a probable Ho IIb age for Palaeolithic activity (Turner and Kerney 1971). An early Hoxnian date for a Clactonian Industry at the Clacton Golf Course site (Singer *et al.* 1973) is also likely (Wymer 1988).

Other sites from which palaeobotanical data have been obtained and hand axes discovered are Marks Tey (C. Turner 1970) and Sicklesmere (West 1981b), but the sources of the artifacts are uncertain (Wymer 1988). Palaeolithic flint flakes

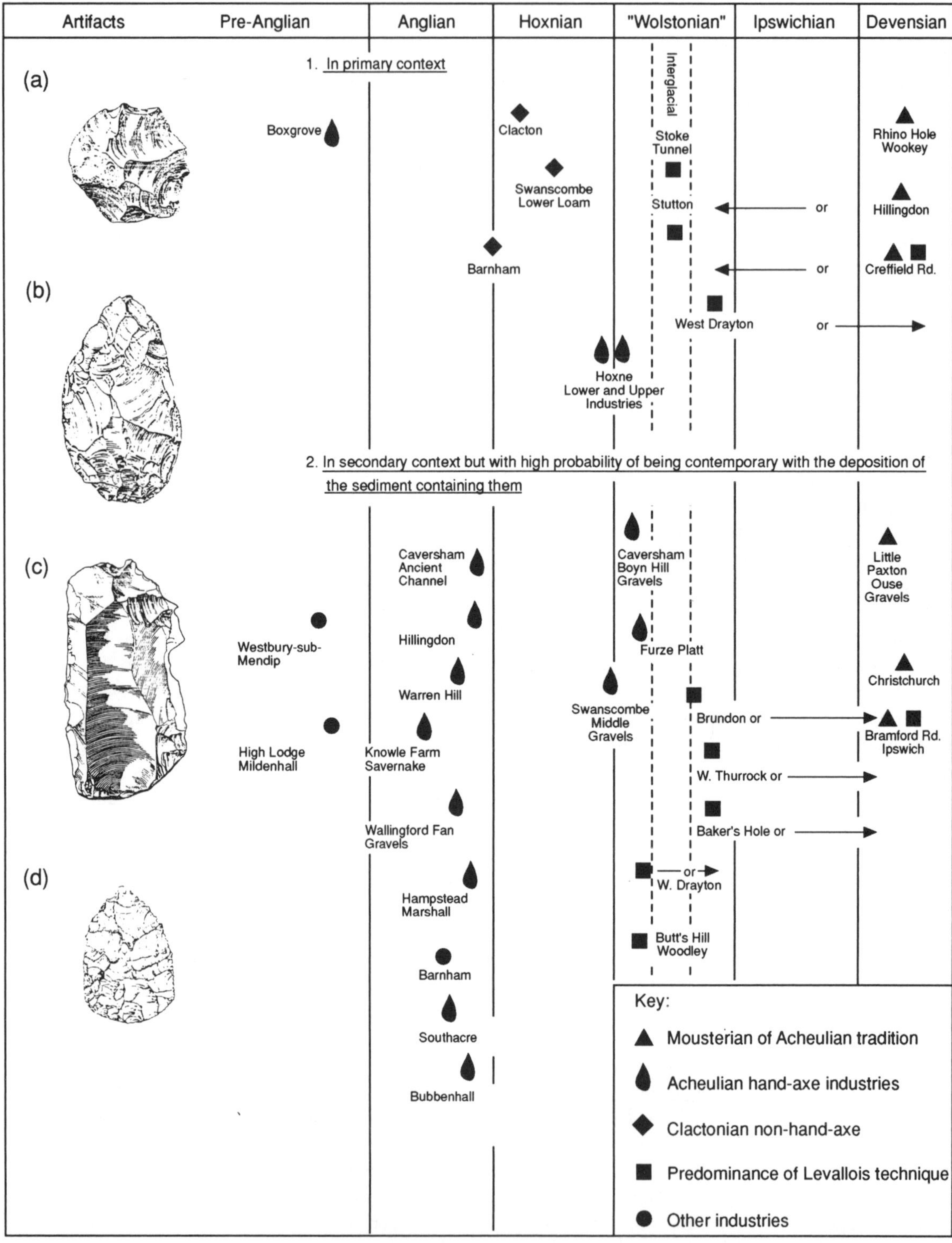

Figure 6.4 The position of Lower and Middle Palaeolithic flint industries within stages of the Pleistocene of Britain. From Wymer (1988). Artifacts: (a) Clactonian chopper-core, Clacton-on-Sea (source, Oakley and Leakey 1937); (b) Acheulian hand-axe, Swanscombe (source, Waechter 1973); (c) Levallois flake, Brundon (source, Wymer 1985); (d) Mousterian of Acheulian Tradition hand-axe, Little Paxton (source, Paterson and Tebbutt 1947). From Wymer (1988). Reprinted with permission from *Quaternary Science Reviews* **7**, J. Wymer, Palaeolithic archaeology and the British Quaternary sequence, © 1988 Pergamon Press Plc.

from marine shingle above the Hoxnian estuarine silts at Kirmington are consistent with this age (Boylan 1966). At Swanscombe, Clactonian artifacts occur at the base of the Lower Loam in a primary context. There are no reliable palynological data from the site, but Kerney (1971), extrapolating from palaeoecological evidence at Clacton, has suggested the loam to be of Early-temperate (Ho II) age.

Acheulian artifacts, mostly derived, have been found in abundance in gravels deposited subsequent to Anglian till, and which occur beyond the limit of the main Wolstonian ice advance in south-east England. Such artifacts have been infrequently recorded from post-Wolstonian gravels. In such a context, assignation of either a Hoxnian or a Wolstonian age to Acheulian sites has been problematical (Wymer 1968, 1977).

The archaeological and environmental data demonstrate that Lower Palaeolithic man favoured water-side habitations. Swanscombe and Clacton were riverine locations, while at Hoxne the margin of a lake was occupied (Wymer 1981).

6.7 Pedogenesis

Chartres (1980) has ascribed a possible Hoxnian age to illuviated, rubified clay in soil profiles in presumed Anglian terraces of the River Kennet in Berkshire. The lower palaeosol at Kirkhill in Buchan may have formed during the Hoxnian (Connell *et al.* 1982; Hall 1984a; Connell and Romans 1984). It is worth remembering that soil material may be polycyclic (Chapter 1) and contain evidence of more than one climatic episode. Thus, for example, palaeosol features observed on the North Yorkshire Moors (Bullock *et al.* 1973) and in the north-eastern Lake District (Boardman 1985) might have developed partly during the Hoxnian Interglacial.

6.8 The offshore sequence and sea-levels

The Hoxnian marine transgression covered the majority of the North Sea region (Cameron *et al.* 1987) and was responsible for the localized smoothing of its pre-existing relief (Long *et al.* 1988). Hoxnian marine sediments have been preserved in valleys cut during the Anglian (Cameron *et al.* 1987). A freshwater deposit containing plant macrofossils, including *Azolla filiculoides* (Griffin 1984), from the lower part of the Ling Bank Formation of the central North Sea has been referred to the Hoxnian (Stoker *et al.* 1985a; Cameron *et al.* 1987).

The Earnley inter-tidal channel sediments may be of late Hoxnian age. They were laid down when sea-level was falling and it has been estimated that mean tidal level was about 1 m lower than today when deposition ceased (West *et al.* 1984).

The Swanscombe terrace deposits at 23–30 m OD (Waechter 1971), which belong to the Late-temperate stage of the Hoxnian, contain the brackish-water mollusc *Hydrobia* (Kerney 1971) and remains of dolphin (Sutcliffe 1964). The inference from this evidence is of a higher sea-level than today, probably to at least 25 m OD (Kerney 1971). At Clacton, estuarine deposits of Ho IIIb age replaced freshwater sediments at about 3 m OD (Pike and Godwin 1953; Warren 1955; Turner and Kerney 1971). There is thus little doubt, as Gibbard (1988b) has observed, that sea-level rise led to the flooding of the Lower Thames Valley during the Hoxnian Interglacial.

In the Fenland, the Hoxnian marine transgression reached at least 23 m OD in the Nar Valley (Stevens 1960; Ventris 1986), while in the Peterborough area, there is evidence of a Hoxnian estuarine environment at about 12 m OD (Horton 1981, 1989b; Horton *et al.* 1992). At Kirmington, north of the Fenland, Hoxnian estuarine silts occur at 27 m OD (Watts 1959a). Shingle above the silts was probably deposited as a beach during the highest stand of the sea in the later part of the interglacial (Catt 1977a). Mean sea-level may have been about 22 m OD at its Hoxnian maximum in this area (Penny and Catt 1972). In eastern Yorkshire, the estuarine Speeton Shell Bed, at *circa* 30 m OD, could reflect a Hoxnian sea-level (Catt and Penny 1966).

Other statements concerning Hoxnian sea-levels have been made on the basis of raised beaches. Of especial relevance is a gravel beach which has been noted at numerous places along the south coast of England (Mitchell 1977). At Slindon in Sussex, this has an elevation of *circa* 30 m OD (Dalrymple 1957) and contains a temperate fauna together with Acheulian artifacts, which point to a Hoxnian age (Mitchell 1977). At Boxgrove in West Sussex, the Slindon Formation indicates one or more episodes of marine transgression and regression (Bates 1986a; Bates and Roberts 1986). A pre-Hoxnian Middle Pleistocene context is probable (Roberts 1986). Correlation of the Slindon Sands (which occur above the raised-beach gravel) at Boxgrove, with the Steyne Wood Clay in the Isle of Wight, and their reference to Oxygen Isotope Stage 9, 11 or 13 have been suggested by Preece *et al.* (1990) (Chapter 4). It should be noted that Kellaway *et al.* (1975) have concluded that gravels in these locations along the south coast of England are of fluvioglacial origin and have been raised by isostatic uplift after glaciation of the English Channel, rather than being emplaced by a higher sea-level. Support for a high, possibly Hoxnian sea-level, may be forthcoming from beaches recognized at *circa* 20 m OD in Cornwall and *circa* 25 m OD in the Isles of Scilly (Mitchell 1977), while in the Channel Islands a beach has been identified 30–40 m above sea-level (Keen 1978, 1982). Mitchell (1977) has hypothesized that as sea-level fell in the late Hoxnian, a sequence of beaches formed at lower altitudes around the southern coasts of the British Isles. Mitchell (1960) and Stephens (1970b) have suggested that the most widely developed raised beach in southwestern England, South Wales and southern Ireland (here termed the Courtmacsherry Beach), and whose elevation has been determined as below 10 m OD in most localities, was formed in the Hoxnian. This reasoning, based on the assumption that head and till above the beach are of Wolstonian age, has been disputed by, for example, Kidson (1971, 1977), Bowen (1973a) and Synge (1985) (Chapter 8).

6.9 Correlation, age and duration

The interglacial termed the Hoxnian in England, Wales and Scotland and the Gortian in Ireland, has been considered a correlative of the Holsteinian of continental Europe (Mitchell *et al.* 1973a) (Table 1.1). Palynological data have revealed that vegetational successions referred to these episodes possessed both similarities and differences. Regarding the former, the importance of *Abies* in Substage III forest of all areas is notable. Of the latter, the significant rôle of *Picea*, in eastern parts of Europe, and *Pinus*, in Ireland, in Substage II forests while those in the remainder of the British Isles contained a mixture of deciduous trees, is of likely significance (Turner 1975). Although faunal data from the Holsteinian are sparse, similarities exist between vertebrate assemblages of this age in Germany (Adam 1954, 1975) and those from Hoxnian deposits at Swanscombe and Clacton (Sutcliffe 1964; Singer *et al.* 1973; Stuart 1982).

Estimates of the age of the Hoxnian have reflected the dates and correlations available at particular times. Page (1972) has postulated its incidence 25 000–21 000 BP on the basis of a series of radiocarbon dates consistent with the stratigraphic sequence in East Anglia. This chronology also equates the Cromerian with an interstadial in the middle of the last cold stage and refers the glacial deposits of East Anglia to the latter. Sparks and West (1972) and West (1977a) have indicated an age around 150 000 BP. Szabo and Collins (1975) have obtained dates of around 245 000 BP, and in excess of 272 000 BP as a result of U-series determinations on bones from Hoxnian contexts at Clacton and Swanscombe, respectively (Chapter 1). Shackleton (1975) has suggested that this temperate episode could be accommodated within Oxygen Isotope Stage 11, which lasted from *circa* 428 000 to 352 000 BP (Figure 1.2(a)(b)). Bowen *et al.* (1986) have preferred to correlate the Hoxnian with Oxygen Isotope Stage 9, about 338 000–302 000 BP. This view is based on the equivalence of the later, Ipswichian Interglacial with part of Oxygen Isotope Stage 5 (Chapter 8), and the existence of a further temperate episode of interglacial rank between this and the Hoxnian (Chapter 7). Grun *et al.* (1988) have established a date of 319 000 ± 38 000 BP by using the electron-spin-resonance

(ESR) technique on tooth enamel collected from the sediments at Hoxne. Bowen *et al.* (1989) have employed this date as one of several geochronometric controls for amino-acid ratios in non-marine Mollusca from sites representative of different Pleistocene events. An Oxygen Isotope Stage 9 age has been assigned to the Hoxnian on these criteria. It has been concluded, however, that while the degree of isoleucine epimerization has enabled discrete recognition of the Hoxnian, the Swanscombe sediments hitherto assigned to this temperate episode are not part of it, having been deposited during Oxygen Isotope Stage 11 (Figure 1.15).

On continental Europe, Zagwijn (1975, 1989) has estimated that the Holsteinian occurred in the Netherlands around 220 000 BP (Figure 4.5), while Bowen (1978) has indicated an age range from *circa* 500 000 to 400 000 BP for this stage in northern Europe.

The duration of the Hoxnian Stage has been calculated at 20 000–25 000 years (Turner 1975) on the basis of counts of laminated sediments from Marks Tey (C. Turner 1970) (Chapter 1). Such a timespan is comparable with the 16 000–17 000 years indicated for the Holsteinian in Germany, using the same method (Müller 1974).

6.10 Conclusion

Although beset by a number of problems relating to correlation and dating, the Hoxnian Interglacial is less controversial than some stages in the Pleistocene of the British Isles. Its termination marked the passage from the Middle to the Upper Pleistocene (Table 1.1).

7 Post-Hoxnian and pre-Ipswichian events

The scheme of Pleistocene stages in the British Isles of Mitchell *et al.* (1973a) (Table 1.1), assigned time between the Hoxnian and Ipswichian interglacials to the Wolstonian Cold Stage. It is noted in Chapter 5 that a dispute exists regarding the status of the Wolstonian Glaciation; deposits conventionally referred to it being considered of Anglian age by some investigators. It is also stated in Chapter 6 that evidence is now available for an interglacial stage between those of the Hoxnian and Ipswichian. Against this backdrop, the account which follows firstly presents details of the orthodox Wolstonian sequence, and secondly, examines data which point to the existence of at least one substantial additional temperate episode during the timespan under consideration. Sites relevant to these issues are shown in Figures 7.1 and 7.4.

7.1 The Wolstonian Cold Stage

This stage witnessed the penultimate major glaciation of the British Isles according to Shotton (1983a). It affected the Midlands of England, where the stratotype for the stage has been designated (Mitchell *et al.* 1973a), but its southerly extent has not been well established (Shotton 1986). In Ireland, ice of the equivalent glaciation, the Munsterian (Mitchell *et al.* 1973a), covered most of the island (McCabe 1985, 1987).

7.1.1 The Midlands of England

Glacial, fluvioglacial and glaciolacustrine deposits have been recorded in the Coventry–Rugby–Leamington Spa area by Shotton (1953). The succession (Table 7.1) is most completely observed at Wolston, which has been chosen as the type locality (Mitchell *et al.* 1973a). The significance of these deposits has been reiterated by Shotton (1983a,b). The lowest members of the Wolstonian sequence, the Baginton–Lillington Gravel and the Baginton Sand, were deposited under cold conditions in a broad valley (termed the Proto-Soar) which ran from the environs of Evesham to those of Leicester. A till (making up the Lower Wolston Clay) containing abundant clasts of Triassic material and deposited above the Baginton Sand was the product of a glacier moving from north or north-west of the area. This glacier held up a major water body (Lake Harrison) (Figure 7.2) before it, during a northerly withdrawal of ice. Sand, silt and clay above the till were the result of sedimentation in this proglacial lake. The Upper Wolston Clay was also a till, indicative of a second, more vigorous glacial advance that came from a more north-easterly direction than the first, and extended to the Moreton-in-Marsh area. The Dunsmore Gravel at the top of the sequence was outwashed as this ice retreated. Following this glacial episode,

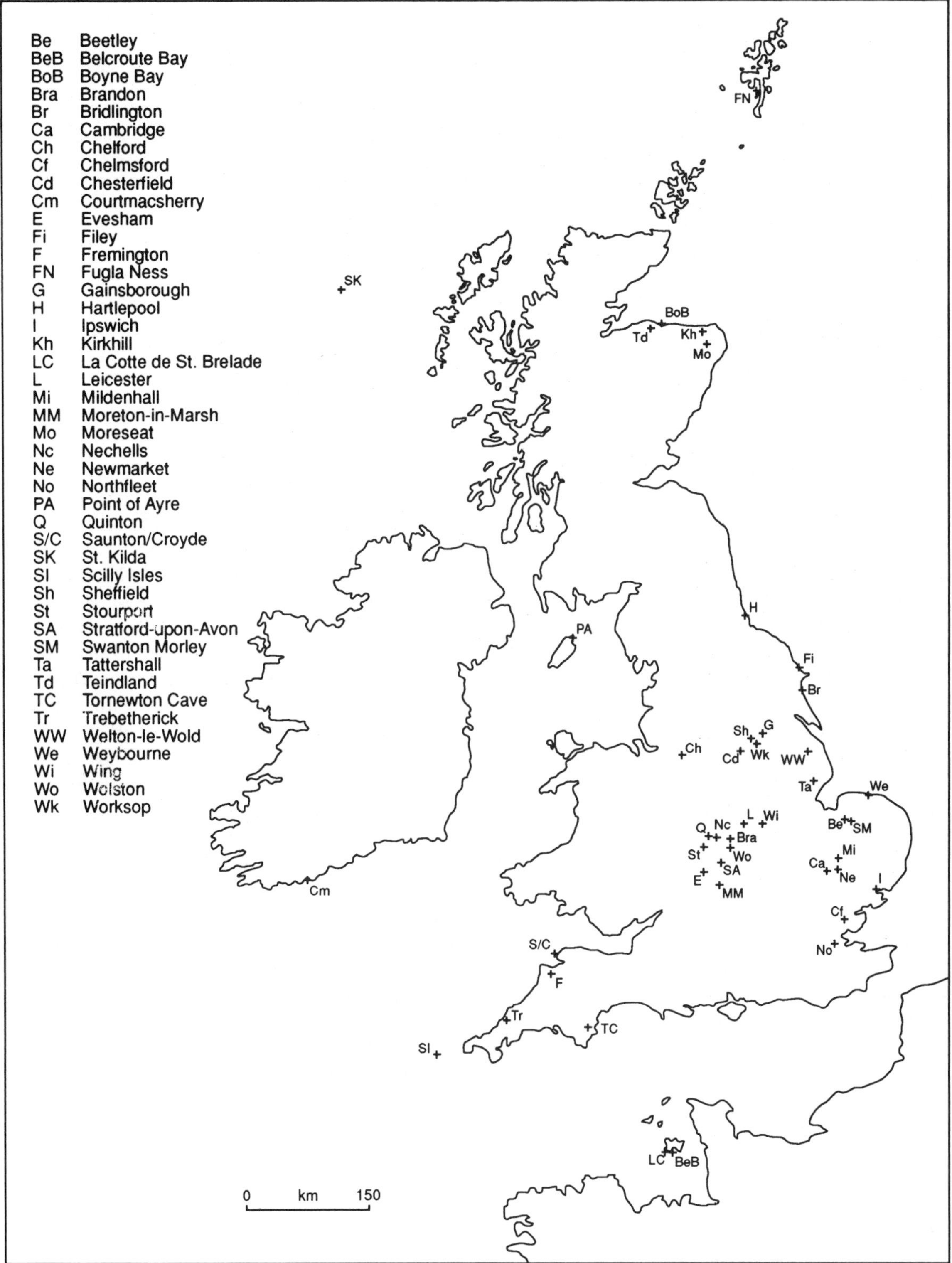
Be Beetley
BeB Belcroute Bay
BoB Boyne Bay
Bra Brandon
Br Bridlington
Ca Cambridge
Ch Chelford
Cf Chelmsford
Cd Chesterfield
Cm Courtmacsherry
E Evesham
Fi Filey
F Fremington
FN Fugla Ness
G Gainsborough
H Hartlepool
I Ipswich
Kh Kirkhill
LC La Cotte de St. Brelade
L Leicester
Mi Mildenhall
MM Moreton-in-Marsh
Mo Moreseat
Nc Nechells
Ne Newmarket
No Northfleet
PA Point of Ayre
Q Quinton
S/C Saunton/Croyde
SK St. Kilda
SI Scilly Isles
Sh Sheffield
St Stourport
SA Stratford-upon-Avon
SM Swanton Morley
Ta Tattershall
Td Teindland
TC Tornewton Cave
Tr Trebetherick
WW Welton-le-Wold
We Weybourne
Wi Wing
Wo Wolston
Wk Worksop
0 km 150

Figure 7.1 Sites with evidence of Wolstonian environments.

Table 7.1 Nomenclature applied to the lithostratigraphic units of the Wolstonian Stage. From Shotton (1983); sources, Shotton (1953, 1976), Rice (1981), Sumbler (1983a)

Shotton (1953)	*Shotton (1976), Rice (1981)*	*Sumbler (1983a)*
Dunsmore Gravel	Dunsmore Gravel	Dunsmore Gravel
Upper Wolston Clay	Upper Oadby Till Lower Oadby Till	Upper Wolston Clay
Wolston Sand	Wigston Sand and Gravel	Wolston Sand and Gravel
Lower Wolston Clay	Bosworth Clays and Silts	Lower Wolston Clay
	Thrussington Till	Thrussington Till
Baginton Sand	Baginton Sand	
Baginton–Lillington Gravel	Baginton–Lillington Gravel	Baginton Sand and Gravel

erosion of the Lake Harrison deposits took place and the drainage basin of the Warwickshire Avon developed. This river system, draining south-westwards, was reversed in direction to that of the Proto-Soar in the area. Shotton (1953) has reported that no interglacial deposits could be seen beneath the glacial sequence described but that patches of an earlier till-like clay (the Bubbenhall Clay) (Chapter 5) are present. Support for a penultimate glacial age (it lacked a stage name at this time) has come from vertebrate remains recovered from the Baginton–Lillington Gravel. These, including *Palaeoloxodon antiquus*, *Mammuthus primigenius* (woolly mammoth), *Rangifer tarandus* (reindeer) and *Equus caballus*, were first described by Buckland (1823) and have been equated by Shotton with the vertebrate fauna of the penultimate (Saalian) Cold Stage in continental Europe.

The highest member (Number 5, at *circa* 45 m OD) of the terrace sequence of the Warwickshire Avon (Tomlinson 1925) (Figure 7.6), and part of the Bushley Green Terrace of the River Severn (Wills 1938), have been considered products of the Wolstonian Cold Stage (Tomlinson 1935, 1963; Wills 1938; Shotton 1983b). However, at this point it should be noted that the sequence, correlation and age of the Avon–Severn terraces is controversial (section 7.2.1).

Subsequent research in various parts of the Midlands has extended knowledge of the spatial and temporal relationships of Wolstonian sediments and of their environments of deposition. Bishop (1958) has mapped the Pleistocene deposits and landforms in the area to the south of that described by Shotton (1953), concurring with the scheme of the latter and investigating possible overflows of Lake Harrison through gaps in the Jurassic escarpment into the Thames drainage system. Lacustrine sediments were deposited as Wolstonian ice advanced to the Moreton-in-Marsh area, and the episode was probably linked to the formation of the Wolvercote Terrace of the Upper Thames (Figure 7.7) (section 7.2). Lithostratigraphic and biostratigraphic data from Brandon, some 2 km from Wolston, have provided additional insight into Wolstonian events. Cryoturbation features have been identified within the Baginton–Lillington Gravel and between the Baginton Sand and Lower Wolston Clay (Shotton 1968). A silt-filled channel within the Baginton–Lillington Gravel contains plant fossils that suggest a birch–willow–juniper scrub with herb-rich grassland beneath. The characteristics of the flora suggest that it developed after a cold episode, and as it was followed by an even colder one, is thought to be that of an interstadial (Kelly 1968). This channel has also yielded fish and insects characteristic of a lake or gently moving river. There is a group of insects with

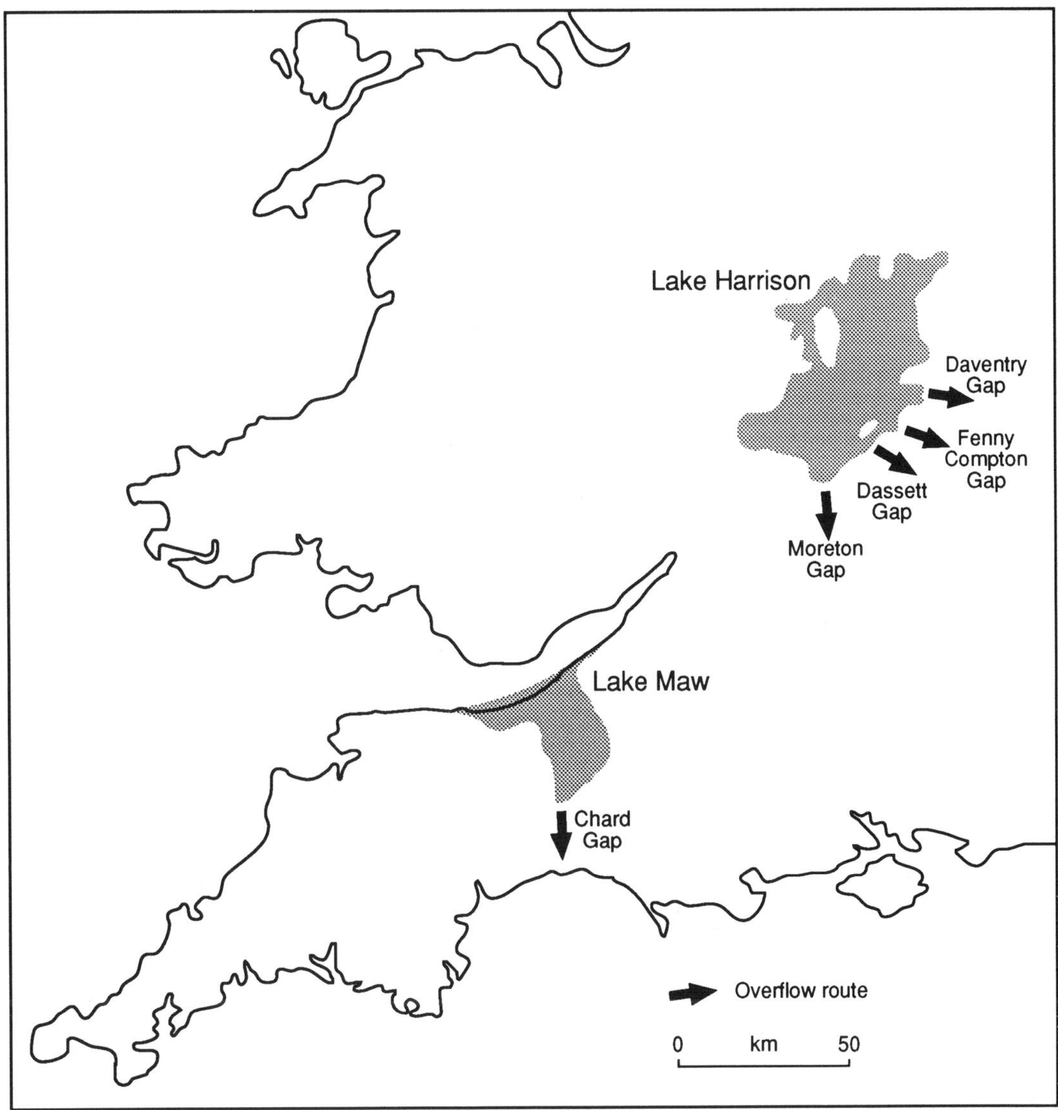

Figure 7.2 Proglacial lakes associated with the Wolstonian Glaciation and their main overflow routes. Lake Harrison; sources, Shotton (1953), Bishop (1958). Lake Maw; sources, Maw (1864), Mitchell (1960), Stephens (1970a,b).

a current northern distribution (including *Pycnoglypta lurida*, *Arpedium brachypterum*, *Otiorrhynchus* spp. and *Helophorus* spp.), which implies a cooler climate than that of today but not characteristic of the Arctic. Early-glacial, not necessarily interstadial, conditions are probably represented (Osborne and Shotton 1968). More recent exposures of the Baginton–Lillington Gravel in the same area have revealed additional channel deposits (Maddy 1989). The pollen flora of a lower channel indicates regional boreal forest, that of an upper one herb-dominated vegetation, very similar to that described by Kelly, and typical of a colder climate (Gibbard and Peglar 1989). However, Coleoptera from the lower channel indicate a temperate climate of interglacial type (Coope 1989).

In central and southern Leicestershire, Rice

(1968, 1981), and in the western part of the county, Douglas (1980), have established a Pleistocene succession closely resembling that of Shotton (1953). Elements of this succession also appear to be present in Charnwood Forest, although the sequence in the north of this area is uncertain (Bridger 1975, 1981). At Wing, Rutland, a chalky Jurassic till below Ipswichian interglacial deposits has been regarded as of Wolstonian age by Hall (1981).

The erratic content of the Wombourn Gravels of Staffordshire, south of, and older than, Late Devensian glacigenic deposits in the same area (Chapter 9), have led A. V. Morgan (1973) to postulate their being reworked Wolstonian material. Diamict deposits thought to be of glacigenic origin, which occur in a pre-Late Devensian context (Chapter 9) in the Severn Valley at Stourport, Worcestershire (Dawson 1988), may relate to the Wolstonian, or to an earlier cold stage.

On the basis of evidence from the Coventry and Leicester areas, Shotton (1976) has revised the nomenclature of the Wolstonian Stage (Table 7.1). A lower, lodgement till, rich in Permian and Triassic erratics (the Thrussington Till) was laid down by a glacier which went up the Proto-Soar Valley to the neighbourhood of Stratford-upon-Avon. The Bosworth Clays and Silts were deposited in Lake Harrison. The upper chalky till (the Oadby Till) was produced by a glacier fed by two ice streams. Its erratic content indicates one from the north-west and one from the north or north-east (Shotton 1983b). It terminated near Moreton-in-Marsh, where a moraine exists (Bishop 1958).

In the northernmost Midlands, the Oakwood Till at Chelford, Cheshire (Worsley 1978), if not Early Devensian (Worsley *et al.* 1983) (Chapter 9), must belong to a previous cold stage, perhaps the Wolstonian, or the Anglian (Chapter 5). At the southern extremity of this region, chalky till on the sides of the Great Ouse Valley in Buckinghamshire may be of Wolstonian age (Horton *et al.* 1974). Upper terrace gravels in this locality indicate the proximity of glacial ice during their deposition. They appear to be related to a channel, the deposits of which in the Stoke Goldington area have produced a flora and a fauna characteristic of a cold environment, probably during the Wolstonian (Green *et al.*, in preparation) (section 7.2).

The possibility that the majority of sediments traditionally assigned to the Wolstonian belong to the Anglian is explored in Chapter 5. While the evidence for this need not be re-stated, some points relevant to the area of the type Wolstonian, cited by Rose (1987, 1989b) in support of this hypothesis, need elaboration. Sumbler (1983a) has described the stratigraphy and lithology of these deposits, proposing some new formations and adjustments to the nomenclature of the succession (Table 7.1). Of especial significance is the conclusion that there is insufficient lithostratigraphic and biostratigraphic information for the type Wolstonian to be regarded as post-Hoxnian in age (Sumbler 1983b). Equivalence between the Oadby Till and the Lowestoft Till of East Anglia has been suggested and the likelihood of an Anglian age mentioned. Shotton (1983a,b), in a critique of Sumbler's views, has stated that the Birmingham district possesses evidence of Hoxnian sediments underlain by Anglian and overlain by Wolstonian glacigenic deposits (Kelly 1964; Horton 1974, 1989a) (Chapters 5 and 6). However, it has been observed by Shotton that it has not been possible to link deposits which correlate with the Oadby Till and occur to the south and east of Solihull, to the Nechells and Quinton areas of Birmingham less than 20 km away.

7.1.2 Lincolnshire, East Anglia and the Thames Valley

Geological and geomorphological evidence has led Straw (1966, 1979a,b, 1983) to postulate that the older, chalky tills of Lincolnshire were deposited by ice moving in a south-south-easterly direction during a Wolstonian glaciation (Figure 7.3). Most of these tills have been shown to rest on bedrock, and the distinctions between them have been regarded as spatial rather than temporal. Their different lithologies can be explained in terms of adjoining ice streams moving across particular rock types during a single glacial episode. Continuity of tills in this sequence with those of similar type in Leicestershire (Rice 1968), Cambridgeshire and Northamptonshire (Horton 1970) is thought likely (Straw 1983). Catt (1981) has suggested that central Lincolnshire was unaffected by the passage of Wolstonian ice which, however, impinged on the coast of the county and terminated in north Norfolk (Figure 7.3). Thus, the central Lincolnshire tills are probably of Anglian age, as suggested by Perrin *et al.* (1979) (Chapter 5). Indeed, these deposits have stimulated considerable discussion. For example, Catt (1977a, 1987a) has drawn a parallel between the glacigenic sediment beneath Hoxnian interglacial deposits at Kirmington (Chapter 6) and the Wragby Till designated by Straw (1969), thereby implying that the latter is Anglian rather than Wolstonian in age. Madgett and Catt (1978) have provided mineralogical evidence to assist correlation of the Welton Till of Straw with others of Wolstonian age in Yorkshire and Durham (section 7.1.3).

The evidence marshalled by Straw (1983) in support of a Wolstonian age for the older tills of Lincolnshire includes a number of lithostratigraphic, biostratigraphic and archaeological details. At Welton-le-Wold near Louth, one of the few localities where other deposits have been discovered stratified between solid rock and the tills in question, a mammalian fauna including elephant and horse, together with Acheulian hand axes, occurs in sub-till valley gravel, thereby introducing a Hoxnian and post-Hoxnian context to the succession (Alabaster and Straw 1976; Wymer and Straw 1977). At Tattershall, deposits assigned to the Ipswichian Interglacial (Holyoak and Preece 1985) (Chapter 8) occur immediately above the Wragby Till. Also, a connection between the glacigenic material below the Hoxnian sediments at Kirmington and the Wolstonian tills has not been considered possible because of the spatial isolation of the former. The significant erosion accomplished by the Wolstonian ice sheet (Straw 1979a,b; Rice 1981) could have been responsible for the removal of biogenic deposits representative of previous interglacials. Furthermore, as Straw has pointed out, it is puzzling that if the Lincolnshire tills were deposited in the Anglian, no Hoxnian sediments have been encountered above them, as in East Anglia. In the Witham–Trent river-terrace sequence in Lincolnshire, sands and gravels older than those bearing a characteristic Ipswichian fauna were probably deposited during the Wolstonian (Brandon and Sumbler 1988).

On the basis of till stratigraphy in East Anglia, Baden-Powell (1948) has proposed the incidence of a glaciation subsequent to the Lowestoftian (Chapter 5). This glaciation, termed the Gipping, reached a limit on the plateau of central Norfolk and Suffolk (Figure 7.3). The interglacial deposits at Hoxne have been considered of intermediate age by Baden-Powell (1951). West and Donner (1956) have supported the notion of a Gipping Glaciation, and West (1955, 1963) has correlated the Lowestoft with the Elster, and the Gipping with the Saale glaciations of continental Europe (Table 1.1). The Gipping Till, like the Lowestoft Till, is chalky. In Essex, Clayton (1960) has considered the Maldon Till to be equivalent to the Lowestoft Till and the Springfield Till to equate with the Gipping Till. West (1968) has proposed that the Gipping Glaciation terminated in the area south of Ipswich, Chelmsford and Cambridge. West (1977a) has adjusted the limit northward and westward across East Anglia (Figure 7.3). The Marly Drift (Chapter 5), which Straw (1965, 1967) has maintained to be closely related to the Calcethorpe Till of Lincolnshire, is inter-mixed in west Norfolk (Straw 1982) with a till thought equivalent to the Wragby Till (Gallois 1978). There was no Wolstonian ice limit within the Fen Basin according to Straw (1982), the terminus of this glaciation in East Anglia being about a curved line running from Weybourne to Newmarket (Straw 1983) (Figure 7.3). Outwash gravels associated with the retreat of Wolstonian ice in central Norfolk are overlain by Ipswichian interglacial deposits at Beetley north of East Dereham. At nearby Swanton Morley, palaeobotanical investigations have identified late Wolstonian organic deposits which accumulated

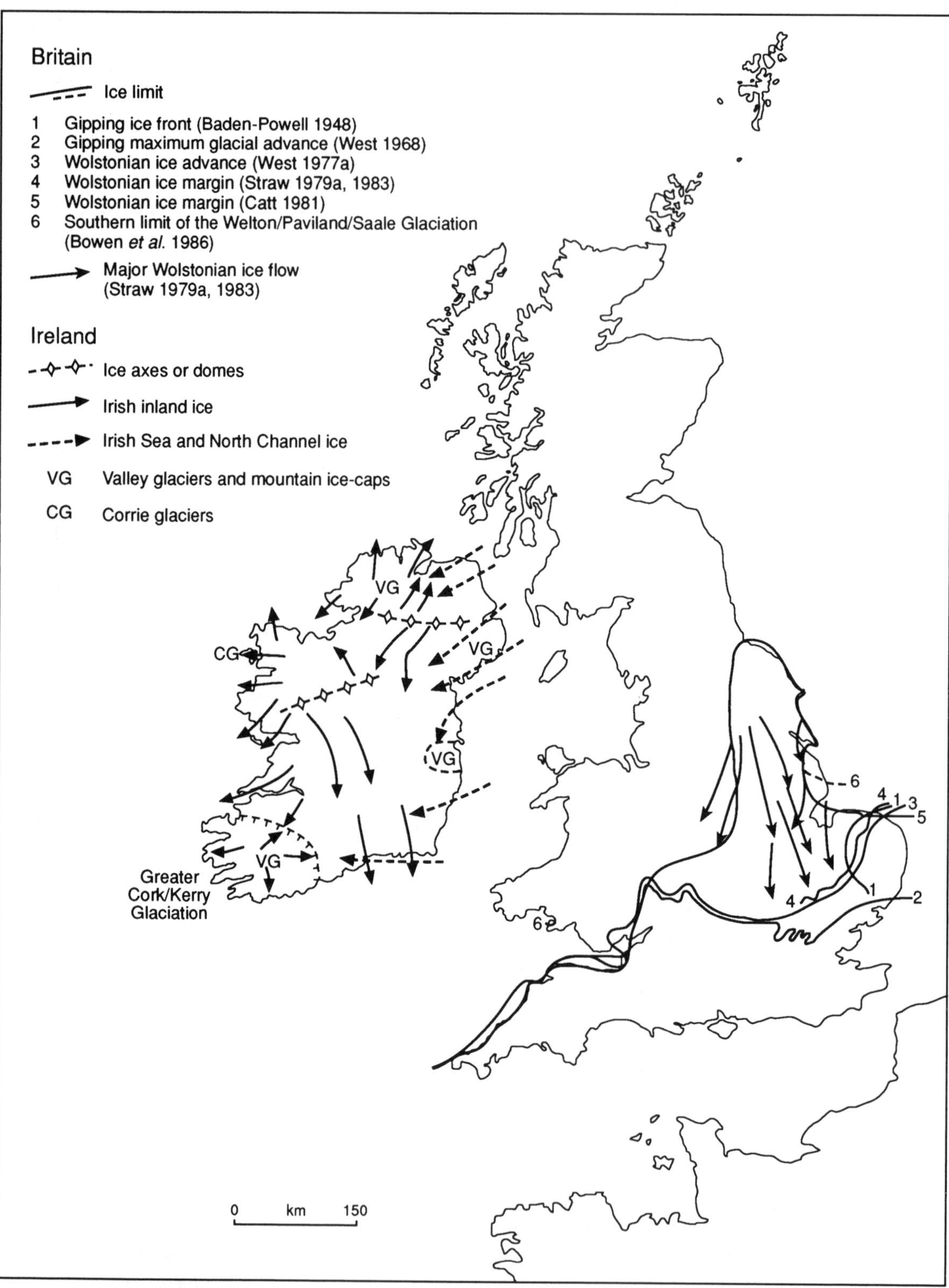

Figure 7.3 The Wolstonian and associated glacial episodes. Britain; postulated ice limits, and major Wolstonian ice-flow directions in eastern England. Sources as in key. Ireland; general directions of ice-sheet movement during the Munsterian. From McCabe (1985); sources, the published works of Charlesworth, Creighton, Colhoun, Farrington, McCabe, Mitchell, Stephens and Synge.

in association with herb-rich vegetation probably including scattered willow shrubs and birch trees (Phillips 1976).

The Pleistocene deposits at High Lodge, Mildenhall, may be Wolstonian not Anglian (Chapter 5) and *in situ* (Archaeology Correspondent 1968; Holmes 1971) rather than partly or wholly transported (Turner 1973a). The *in situ* hypothesis of Archaeology Correspondent (1968) envisages two Wolstonian ice advances that deposited chalky tills. Thus, intervening lake muds, whose palaeobotany is indicative of boreal forest and herb-rich grassland (Turner 1973a), would have been laid down during a Wolstonian interstadial (West 1980b). High Lodge has produced flint artifacts, which, while resembling those of a later industry, are probably pre-Anglian according to Wymer (1988) (Chapter 4).

The possibility that the Chalky Till (Boulder Clay) of East Anglia is Anglian in age is discussed in Chapter 5. However, at this point, it is necessary to introduce a further theory regarding the glacial deposits in this region. This has been propounded by Cox and Nickless (1972) and Bristow and Cox (1973), while Cox (1981) has reviewed it in the light of evidence accumulated during the intervening decade. The essence of their argument is that only one chalky till exists in East Anglia, and that this was deposited by Wolstonian ice. Thus, the chalky Lowestoft Till, conventionally assigned to the Anglian Glaciation, has been reclassified as Wolstonian, with no distinction made between this and the Gipping Till. Lack of superposition and lithological similarities within the body of till over a wide area has been cited in support of a single glacial episode. Reasons for a Wolstonian age include the lack of dissection of the drift-covered plateau, unlikely if two interglacial and two cold stages had followed the emplacement of the till; evidence of only one further cold (periglacial) period following subsequent interglacial conditions; and a relatively uncomplicated river-terrace sequence since the latter. This scheme also challenges the conventionally separate chronological status afforded to the Hoxnian and Ipswichian interglacial stages, as deposits of both would now overlie a chalky till of Wolstonian age. The biotic differences of the Hoxnian and Ipswichian are considered to be probably related to two episodes within the Ipswichian (Last) Interglacial, between which a marked cold phase occurred. Evidence of such environmental fluctuations has been forthcoming from deposits of assumed equivalent ages (Holsteinian and Eemian) on continental Europe (Jessen and Milthers 1928; Woldstedt 1966; Menke 1968; Ducker 1969). After reviewing the evidence available, Cox (1981) has concluded that the above interpretation of the glacial stratigraphy of East Anglia is correct, a single till having been produced by a glaciation prior to the Hoxnian, with the latter unable to be correlated with the Holsteinian Interglacial of continental Europe.

In the Thames Valley, sands and gravels were deposited in a periglacial environment during the Wolstonian (Gibbard 1988b). The Boyn Hill, Lynch Hill and Taplow terrace gravels of the Lower and Middle Thames were laid down (Green and McGregor 1978, 1983; Gibbard 1985, 1988b; Bridgland 1988) (Figure 3.6). The equivalent gravels of the Upper Thames area (Gibbard 1988b), in the Hanborough and Wolvercote terraces, have been attributed to the Wolstonian, with the latter aggradation yielding flint from till of this age in the Moreton-in-Marsh district (Briggs and Gilbertson 1973, 1980; Briggs *et al.* 1985). Molluscan evidence has prompted Briggs and Gilbertson (1980) to postulate that some of the Wolvercote terrace material accumulated during a Wolstonian interstadial.

7.1.3 Northern England and Scotland

North of the Humber, equivalents of the Wragby, Belmont and Calcethorpe tills of Lincolnshire have not been recognized (Catt 1987a). In eastern Yorkshire, till (Basement Till), observed at several localities along the coast from Filey Brigg to Spurn Point, and which occurs below an Ipswichian raised beach at Sewerby (Catt and Penny 1966) and above the Speeton Shell Bed in Filey Bay (Edwards 1981, 1987), has been referred to the Wolstonian Stage. Catt (1977a,

1987a) has considered the Basement Till to be the equivalent of the Welton Till of Lincolnshire. Isolated patches of deeply weathered glacigenic material on the Yorkshire Wolds and North Yorkshire Moors may have been deposited during the Wolstonian (Catt 1977a, 1981). In the southern part of the Vale of York, weathered and dissected till, deposited by ice which came from a northerly and westerly direction, together with channels deeply incised into bedrock by subglacial drainage and outwash from meltwater that entered the area from the south and west, could be features of the penultimate glaciation (Gaunt 1981). Some of the glacial deposits surveyed in the country around East Retford, Worksop and Gainsborough (Smith *et al.* 1973), and in the Sheffield (Eden *et al.* 1957) and Chesterfield (Smith *et al.* 1967) areas, could date from this time. Tills in the Wye, Derwent and Manifold valleys of the Peak District, deposited by ice from a north-north-westerly direction, seem most likely to be of Wolstonian age (Stevenson and Gaunt 1971; Briggs and Burek 1985).

Interglacial deposits, probably Ipswichian (Carter *et al.* 1978), are underlain by a till of presumed Wolstonian age (Letzer 1981) in the upper Eden Valley of Cumbria. The basal till in the Low Furness region of north-west England may also belong to this stage (Tooley 1977; Huddart *et al.* 1977). Till (Scandinavian Drift) occupying a buried valley on the Durham coast north of Hartlepool has been reported by Trechmann (1915). Loess overlying the till has been identified by Trechmann (1919, 1931). Although the loess was considered to be interglacial by Trechmann, both it and the till, the basal one in the sequence of this district, and renamed the Warren House Till by Smith and Francis (1967), have subsequently been assigned to the Wolstonian Cold Stage (Francis 1970; Shotton 1981). However, no comparable deposits have been recognized inland in this part of north-east England (D. B. Smith 1981).

Scotland has yielded similarly tenuous evidence of Wolstonian glaciation. The Pleistocene sediments of the Midland Valley may include a till of this age (Price *et al.* 1980), as could the multiple till sequence of the lower Spey Valley (Sutherland 1984). The Pleistocene deposits of north-east Scotland have led Bremner (1931) to propose that three glacial episodes took place. Support for one of these being Wolstonian has come from Buchan, where tills occur below presumed Ipswichian palaeosols at Kirkhill (Connell *et al.* 1982; Connell 1984; Hall 1984a), Boyne Bay (Peacock 1966, 1971a; Connell and Hall 1984b) and Moreseat (Hall and Connell 1982; Hall 1984a,b). At Teindland in Morayshire, a palaeosol (FitzPatrick 1965) referred to the Ipswichian, developed in outwash of assumed Wolstonian age (Edwards *et al.* 1976). On Shetland, the lower of two tills (Chapelhow 1965) above the Hoxnian deposit at Fugla Ness (Birks and Ransom 1969) might be Wolstonian (Sutherland 1984), and erratics on St. Kilda may date from this glaciation (Sutherland *et al.* 1984).

7.1.4 Wales, England south of the Lower Severn and Thames valleys, and Jersey

According to Bowen (1970), the central and south-eastern parts of South Wales were engulfed by a combination of Welsh and Irish Sea ice during the Wolstonian. Synge (1970) has concluded that Wolstonian ice spread across the majority of Wales, save for a few areas in the south of the country. However, Bowen *et al.* (1986) proposing this as a complete glaciation of Wales, have assigned it to the Anglian Stage (Chapter 5). Bowen (1973a) has postulated an advance of Welsh ice subsequent to that of the Irish Sea Glaciation and prior to that of the Devensian Glaciation. This separate event, when ice advanced north–south, is represented by glacial deposits in the Paviland Moraine on the Gower Peninsula (Bowen *et al.* 1985) (Figure 7.3). Bowen *et al.* (1986) have suggested that the glaciation involved (the Paviland Glaciation) was probably that which deposited the Welton, Basement and Warren House tills, the Lower Till at Kirkhill and the Ridgacre Till above the Hoxnian deposits at Quinton. They have pointed out that scant evidence of this glaciation in the British Isles is likely to be the result of a combination of the erosive power of later (Devensian/ Midlandian) ice and the burial capacity of its associated deposits.

Mitchell (1968) has suggested that Wolstonian ice occupied the Bristol Channel. The glacial deposits of the Avon and Somerset (Hawkins and Kellaway 1971; Gilbertson and Hawkins 1977) and north Devon (Mitchell 1960; Stephens 1961, 1966, 1970b; Kidson and Wood 1974) coastlands, and of the Scilly Isles (Mitchell and Orme 1967) are described in Chapter 5 in the context of a possible Anglian age. However, these deposits have been traditionally considered as Wolstonian (Mitchell *et al.* 1973a), although a Devensian origin has also been postulated for some of them (Synge 1985) (Chapter 9). Of the periglacial material identified in Devon and Cornwall, a lower, Main Head may be Wolstonian (Waters 1961, 1964, 1974; Stephens 1961, 1970b, 1974a, 1980). On Dartmoor, Wolstonian solifluction has been proposed as a possible mechanism for exposure of tors (Waters 1974), probably features of considerably greater antiquity formed below ground (Linton 1955; Eden and Green 1971).

Faunal remains including reindeer, *Ursus arctos* (brown bear), *Coelodonta antiquitatis* (woolly rhinoceros), *Gulo gulo* (wolverine), *Lagurus lagurus* (steppe lemming) and *Microtus nivalis* (snow vole) have been recorded from a Wolstonian context in Tornewton Cave at Torbryan, Devon (Sutcliffe and Zeuner 1962; Sutcliffe 1974a). According to Stuart (1982), a component of this fauna may be indicative of interstadial conditions.

The fauna of the Burtle Beds, in the Brue and Parrett valleys of Somerset, has indicated to Kidson and Haynes (1972) and Kidson *et al.* (1974) that these were marine and interglacial, and probably dated from the Ipswichian (Chapter 8). However, Kellaway (1971) and Hawkins and Kellaway (1973) have suggested that these sands and gravels were glacial in origin and of likely Wolstonian age.

Maw (1864), Mitchell (1960) and Stephens (1970a) have considered that the ice which occupied the Bristol Channel during the Wolstonian led to the formation of a proglacial lake (Lake Maw), whose outlet was via the Chard Gap to the English Channel (Figure 7.2). In support of this hypothesis, terrace gravels in the Axe Valley have been ascribed a fluvioglacial origin, with their aggradation thought to have taken place during the Wolstonian because of the typology of Acheulian hand axes contained within them (Stephens 1970a, 1974b). However, data presented by Green (1974) have led to the conclusion that these deposits were derived from Tertiary Plateau Gravel nearby and were periglacial.

Head found below a raised beach of assumed Ipswichian age at Belcroute Bay in Jersey may be Wolstonian (Keen 1986). Palaeoenvironmental evidence (Campbell and Pohl 1971; Jones 1986; Lautridou *et al.* 1986; van Vliet Lanoë 1986) and an average thermoluminescence date of 238 000 ± 35 000 BP from burnt flints (Huxtable 1986) of early Middle Palaeolithic type (Callow 1986a), have led to a correlation of some of the sediment and soil in La Cotte de St. Brelade on the island, with Saalian (Wolstonian) periglacial and interstadial environments. Human occupations of the site may have taken place during the interstadials (Callow 1986a,d).

In the Weald, deposits datable to the Wolstonian have been infrequently encountered (Shephard-Thorn and Wymer 1977). However, some of the periglacial Coombe Rock of the Downs and adjacent areas may have formed then (Zeuner 1959), while the older of two loess deposits at Northfleet in Kent (Burchell 1933, 1954), which is overlain by mud with a characteristic Ipswichian molluscan fauna (Catt 1977c), could have been deposited during the Wolstonian. At Baker's Hole, Northfleet, Wolstonian periglacial deposits have yielded *Mammuthus primigenius*, *Equus ferus* and *Coelondonta antiquitatis*, and human artifacts have occurred (Smith 1911; King and Oakley 1936; Stuart 1982). The Later Acheulian Industry in the Upper Loam at Swanscombe (Chapter 6) may be Wolstonian (Stuart 1982; Wymer 1988).

7.1.5 Ireland and the Isle of Man

Deposits assigned to the Munsterian Stage in Ireland have been correlated with those of the

Wolstonian in other parts of the British Isles by Mitchell *et al.* (1973a) but referred to the Anglian by Bowen *et al.* (1986). These sediments have been established as most extensive over the lowlands of southern Ireland to the south of the South Irish End Moraine (considered to represent the limit of the last glaciation in this area) (Chapter 9), and in the west of County Mayo (Synge 1969; Mitchell 1976). Munsterian deposits investigated have been characteristically deep weathered and fashioned into gentle relief features (Finch and Synge 1966). In south-east Ireland, they have been found above the Courtmacsherry Raised Beach of presumed Gortian age (Mitchell 1976). However, stratigraphic criteria have prompted Warren (1979, 1985) to propose that this beach and its associated biogenic deposits are of last interglacial date, a view endorsed by Synge (1981, 1985). It has been suggested by Warren (1985) that the Munsterian deposits may belong to the last glacial. Munsterian sediments, within which varied lithologies have been recognized, were probably deposited by two main ice sheets, one from within Ireland and the other from western Scotland (Synge and Stephens 1960) (Figure 7.3). As noted in Chapter 5, Synge (1981) has postulated two major episodes to account for the older drift sequence of southern Ireland. The first, termed the Munsterian, is equated with the Anglian, the second, the Connachtian, with the Wolstonian. This scheme has been rejected by Bowen *et al.* (1986) and McCabe (1987) who have argued for a single glaciation on the basis of the established lithological differences in the deposits.

In a review of the glacial stratigraphy of Ireland, McCabe (1987) has stated that this is poorly understood compared with that of England, and has concluded that while some basal members of multiple sequences of glacigenic sediments and certain erratic suites of inland Ireland may be related to glacial activity prior to the Midlandian, most of the deposits (including those of the south coast) are likely to be of this age.

In the Isle of Man, Mitchell (1965a, 1971a) has suggested that the lower of two tills (the Mooar Till) found on the north-west coast is attributable to local Wolstonian ice, a view which has been disputed by Thomas (1971, 1977, 1985a). In the south of the island, high-level granite erratics of local provenance have been thought by Thomas (1977) to be indicative of a glaciation, possibly during the Wolstonian. Boreholes in the Point of Ayre district (Lamplugh 1903; Smith 1927) have revealed what may be Wolstonian glacigenic deposits underlain by marine sediments (Thomas 1985b).

7.1.6 The offshore sequence and sea-levels

In the United Kingdom sector of the southern North Sea, no deposits which could be assigned with certainty to a Wolstonian glaciation have been recorded (Figure 1.1), so that the area below 54°N appears not to have been ice covered during this cold stage (Cameron *et al.* 1987). Saalian till has been mapped up to 40 km from the coast in the Dutch sector of the southern North Sea, with proglacial and periglacial sediments identified beyond (Cameron *et al.* 1986; Jeffery *et al.* 1988). The Dutch offshore evidence points to the existence of a periglacial environment across the majority of the southern North Sea during the Wolstonian, according to Long *et al.* (1988).

The central and northern North Sea adjacent to northern England and eastern and northern Scotland has produced evidence of Wolstonian glacial activity (Cameron *et al.* 1987) (Figure 1.1). Valleys cut during the Anglian continued to infill, now mostly with glaciomarine sediments. In the late Wolstonian there was a major erosional episode, with a new valley system mainly running north–south, incised, probably by glacial meltwater. Some sediments deposited after this downcutting are also of Wolstonian age. The offshore sequence has suggested to Cameron *et al.* (1987) and Long *et al.* (1988) that terrestrial ice did not move far into the North Sea and was not in contact with that from Scandinavia. *Circa* 55–57°N on the inner continental shelf west of Scotland, the widespread glaciomarine sediment of the Canna Formation, laid down on an erosion surface formed after deposition of the Malin

Formation, was followed by an episode of downcutting (Figure 9.8). The Canna Formation is probably representative of pre-Devensian stadial conditions (Davies *et al.* 1984), which may have been Wolstonian.

Erratics on the floor of the English Channel and in present and former (raised) beaches around its coasts have been used by Kellaway *et al.* (1975) to argue for Wolstonian glaciation of this area. Investigations of the sediments in the southern Irish Sea (Garrard and Dobson 1974; Garrard 1977; Delantey and Whittington 1977) have established the existence of two tills separated by marine interglacial deposits, the latter probably of Ipswichian age. Pantin and Evans (1984) have shown that there was negligible deposition in the central and south-western Celtic Sea between the Early and Late Pleistocene and have suggested that this was probably due either to sediment bypassing the area, or to it having been eroded after deposition.

7.2 Additional temperate episodes and intervening cold intervals

Evidence for interstadial environments within the orthodox Wolstonian Stage is presented above. The Pleistocene succession for the British Isles proposed by Mitchell *et al.* (1973a) envisages that such environments were the only temperate episodes to occur between those of the Hoxnian and Ipswichian interglacials. Over the last two decades, however, a school of thought has emerged which has allocated at least one additional temperate episode of interglacial rank, preceded, and followed, by cold conditions, to this timespan. The information due to which this modification has been made can be placed in two categories. Firstly, that which has accumulated as a result of re-evaluation of deposits traditionally placed in either an Ipswichian or a Hoxnian context. In a number of instances this has resulted from additional investigations at the sites concerned. Secondly, that provided by new sites whose geomorphology, lithostratigraphy and biostratigraphy have not been amenable to incorporation within the orthodox succession over this timespan (Figure 7.4). In both categories, some sites have yielded geochronometric data inappropriate to either the Hoxnian or Ipswichian stages.

7.2.1 Re-evaluated sites

In the first category, a significant contribution has been made by Sutcliffe (1975, 1976), who on the basis of variations in mammalian faunal assemblages, has suggested that two separate temperate episodes occurred between the Hoxnian Interglacial and Devensian Cold Stage. Both of these temperate intervals hitherto have been referred to the Ipswichian Interglacial on palynological criteria (Chapter 1). Sutcliffe has pointed to the conclusions of Evans (1971) that there were post-Hoxnian and pre-Ipswichian mild intervals *circa* 180 000 and 130 000 BP, and has argued that if two temperate episodes had occurred within a relatively short timespan, their botanical differences may have been slight, leading to their inclusion in one stage. The especial difficulty of correlating partial palynological sequences against such a backdrop was stressed. The mammalian evidence relates to the terrace sequence of the River Thames at Trafalgar Square in central London, and at Ilford and Aveley in Essex (Sutcliffe and Kowalski 1976). Palaeobotanical data from these sites (Franks 1960; West *et al.* 1964; West 1969b) have been considered indicative of the Ipswichian Interglacial; Substage Ip IIb, associated with its climatic optimum (Chapter 8), being recognized at each. Sutcliffe (1976) has stated that Trafalgar Square seems to be situated on a lower terrace than Ilford and Aveley (Sutcliffe and Bowen 1973), and that the mammalian faunas of the latter two localities are similar, and different from that of the former site. Of diagnostic importance in Ip IIb deposits at Trafalgar Square is the existence of *Hippopotamus*, *Palaeoloxodon antiquus*, *Dicerorhinus hemitoechus* and *Dama dama*, components of a typical Ipswichian fauna (Stuart 1976, 1982). At Ilford the bulk of the elephant fossils appear to be from a species resembling *Mammuthus trogontherii* (extinct elephant), which was a forerunner of *Mammuthus*

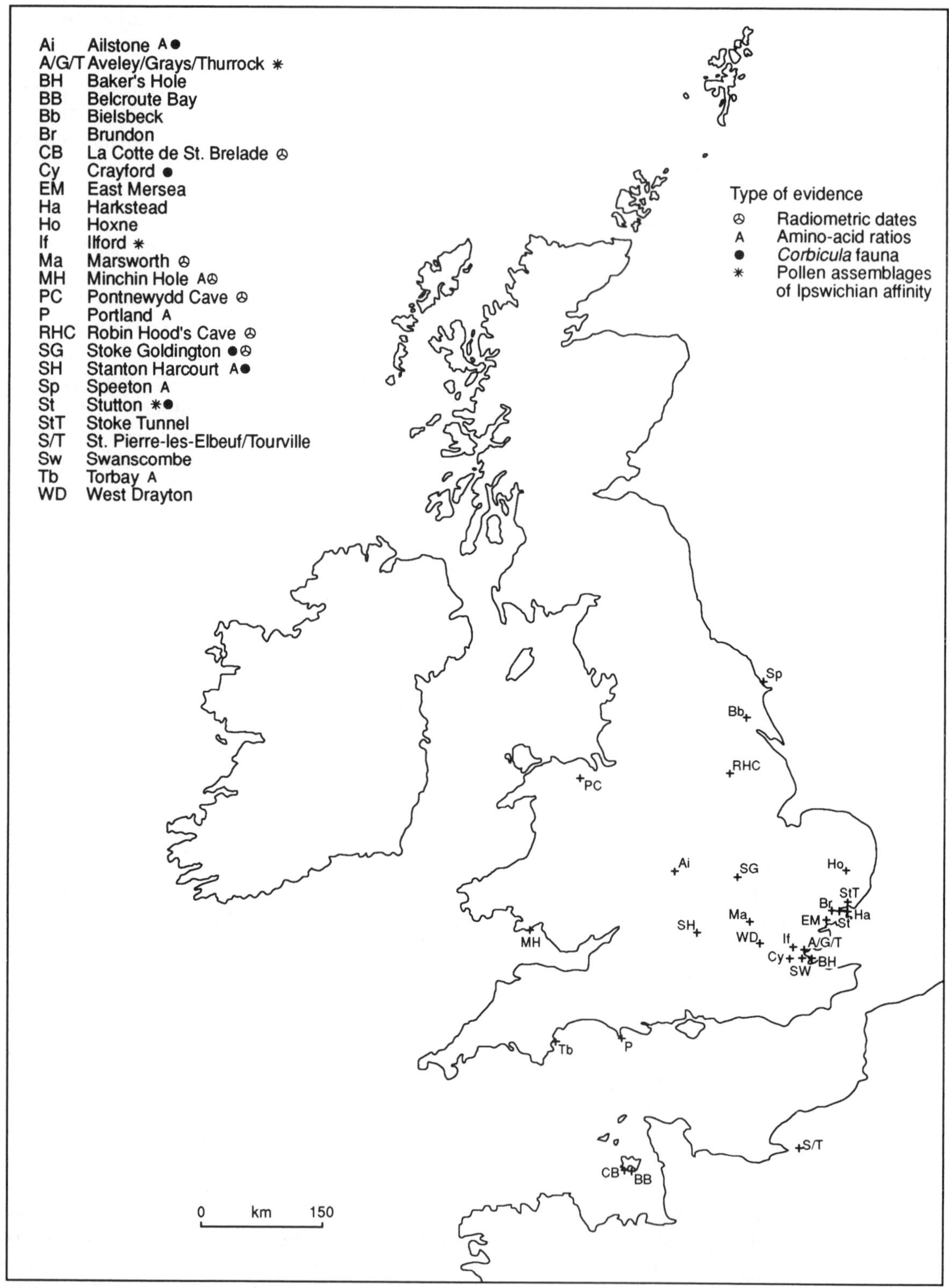

Figure 7.4 Sites with evidence of additional post-Hoxnian and pre-Ipswichian temperate episodes and intervening cold intervals.

primigenius, remains of which also occur. There are two extinct rhinoceros species recovered from Ilford, *Dicerorhinus hemitoechus* and *Dicerorhinus kirchbergensis*, the latter a frequent component of Hoxnian faunal assemblages. However, Stuart (1976) has disputed both the resemblance to *Mammuthus trogontherii* and the identification of *Dicerorhinus kirchbergensis* when assigning the Ilford fauna to the Ipswichian Interglacial (Chapter 8). *Equus* is abundant at Ilford but *Hippopotamus* and fallow deer have not been recorded, with horse and mammoth absent from Trafalgar Square. The Ilford fauna (Rolfe *et al.* 1958; Sutcliffe 1964) has been equated with Substage Ip II (West *et al.* 1964) but Sutcliffe (1976) has suggested that it might be younger because its provenance is brickearth (fine-grained inorganic sediment), much of which is above the deposits analysed for pollen and spores. However, remains of an evolutionarily earlier mammoth at Ilford imply a pre-Ip II age for those deposits. The Aveley deposits cover substages IIb and III of the Ipswichian according to West (1969b). Straight-tusked elephant in the former substage is succeeded by mammoth in the latter (Sutcliffe 1976). While it is possible that the Trafalgar Square, and the Ilford and Aveley faunas may belong to different segments of one interglacial, it is perhaps more likely that they represent two interglacials (Sutcliffe 1976). The latter suggestion has been discounted by Stuart (1976, 1982). It seems relevant to note here the observation of Shotton *et al.* (1983) that the insect fauna from Aveley bears little resemblance to those of Ipswichian Substage II age from Bobbitshole and Trafalgar Square (Chapter 8).

Sutcliffe (1976) has considered that there is sufficient evidence to identify a characteristic Ipswichian mammalian fauna, to which there were temporal modifications and, which demonstrates areal variation over the British Isles. However, Ilford–Aveley type faunas, including those of Grays (Hinton and Kennard 1901, 1907; West 1969b) and Thurrock (Carreck 1976), Essex; Crayford, Kent (Kennard 1944); Stutton (Sparks and West 1963), Stoke Tunnel (Layard 1920; Wymer 1985), Brundon (Moir and Hopwood 1939) and Harkstead (Wymer 1985), Suffolk, may belong to a separate temperate stage according to Wymer (1985, 1988) and Shotton (1986). Uranium-series dating of bones from Brundon has resulted in ages of 230 000 ± 30 000 and 174 000 ± 30 000 BP (Szabo and Collins 1975), both older than those generally accepted as delimiting Ipswichian time. In support of a pre-Ipswichian age for the additional temperate stage, Sutcliffe (1976) has pointed out that the Ipswichian layer in Tornewton Cave, Devon, (Chapter 8) is not succeeded by an Ilford-type fauna prior to the Devensian stratum. The absence of Ilford-type faunas in caves has been discussed by A. Turner (1981) who has suggested that a lack of *Crocuta crocuta* (spotted hyaena), thought to be the main bone-accumlating agent in such situations during Ip II, when most remains were deposited, may have been the reason. Noting that hyaena was not characteristic of Ip III and IV, Turner has cautioned against drawing chronological conclusions on the basis of absence of the Ilford faunal assemblage.

A scheme of grouping Middle and Late Pleistocene faunas of the British Isles (Currant 1989) is mentioned in Chapter 4 (Table 4.1). A Group 2 assemblage, with characteristic temperate woodland taxa and unique in the possession of one or more species of *Crocidura* (white-toothed shrew), occurs at Grays; in the lower part of the Aveley sequence; and in one of two deposits at East Mersea, Essex (Bridgland *et al.* 1988). A post-2 but pre-1 assemblage, characterized by a distinctive form of *Mammuthus primigenius* has been recovered from Ilford and from above the Group 2 assemblage at Aveley.

Minchin Hole on the Gower Peninsula, has also provided evidence of more than one temperate episode during this time interval. George (1932) has identified a *Patella* raised beach, above which is a *Neritoides* Beach, in this cave. However, subsequent stratigraphic studies (Sutcliffe and Bowen 1973; Sutcliffe 1981; Sutcliffe and Currant 1984; Sutcliffe *et al.* 1987) have established that the *Patella* Beach and *Neritoides* Beach are not discrete and comprise one feature separated from an older, sandy Inner Beach at about the same

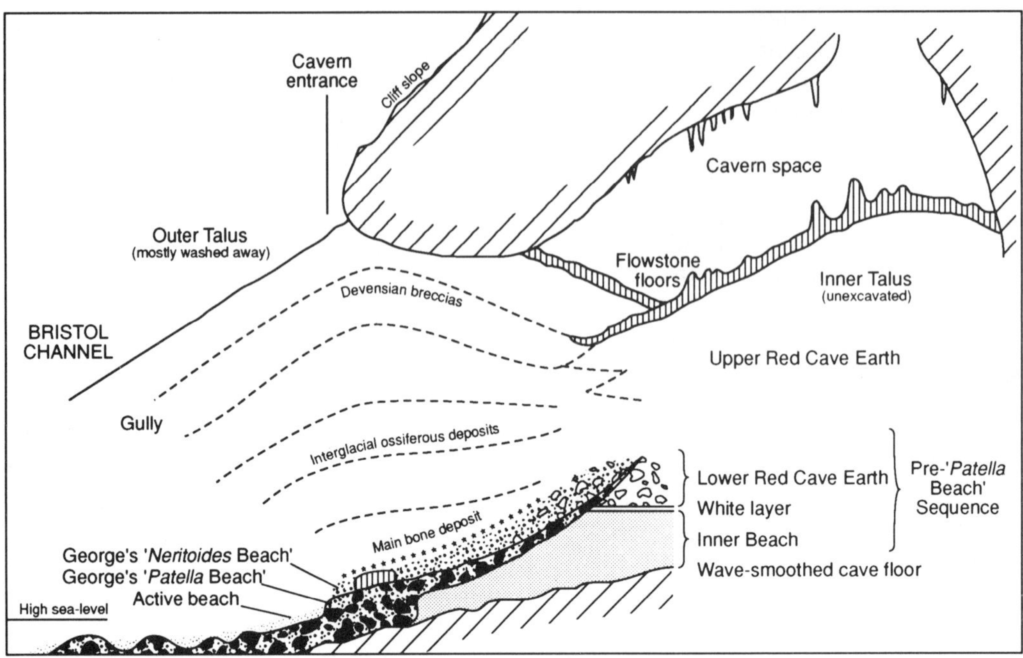

Figure 7.5 Schematic section of Minchin Hole, not to scale. The actual dimensions are: length of gully 50 m; length of cave 42 m; height of summit of inner talus above level of present beach 22 m. From Sutcliffe *et al.* (1987). Reprinted with permission from *Progress in Oceanography* **18**, A. J. Sutcliffe *et al.*, Evidence of sea level change from coastal caves with raised beach deposits, terrestrial faunas and dated stalagmites, © 1987, Pergamon Press Plc.

elevation (*circa* 12 m OD) by a terrestrial cave earth (Figure 7.5). Faunal investigations have revealed remains of *Microtus oeconomus* (northern vole) in the cave earth, whose lithological characteristics resemble those of head. Interglacial deposits overlying the *Patella* Beach contain *Dicerorhinus hemitoechus* and *Palaeoloxodon antiquus*, and although lacking *Hippopotamus* the fauna is most akin to that of the Ipswichian. At nearby Bacon Hole Cave, a Sandy Breccio-Conglomerate, Sandy Cave Earth and Shelly Sand (Figure 9.6), considered to be the equivalent of the *Patella* Beach, has yielded an analogous fauna. Beneath these deposits are sands, from which nothern vole (the same large ecotype as at Minchin Hole) and horse remains have been recovered. These sands have also yielded terrestrial Mollusca, amongst which *Lauria cylindracea*, *Pyramidula rupestris* and *Pupilla muscorum* are most common. A cool, dry climatic episode, during which the plant cover was fairly open, has been inferred (Stringer 1975, 1977; Currant *et al.* 1984; Stringer *et al.* 1986; Sutcliffe *et al.* 1987).

It appears that two separate marine transgressions are represented at Minchin Hole, with cold conditions having developed in the intervening period (Sutcliffe and Currant 1984; Sutcliffe *et al.* 1987). In the context of the oxygen-isotope record, Bowen (1973a,c) has postulated that the Inner Beach could belong to either Stage 7 or Stage 9 and the *Patella* Beach to Substage 5e (Figure 1.2(a)(d)), with the former denoting the existence of a separate interglacial before the Ipswichian. Headon Davies (1983) has produced amino-acid ratios that indicate different ages for the two beaches at Minchin Hole, the Inner Beach being the older. Schwarcz (1984a) has presented U-series dates of *circa* 120 000 BP from a fallen block of flowstone (speleothem) that was resting on (and hence no younger than) the *Patella* Beach, which support a correlation of the latter with Oxygen Isotope Substage 5e (Chapter

8), with the Inner Beach indicative of a pre-Ipswichian high sea-level (Sutcliffe and Currant 1984; Sutcliffe *et al.* 1987). This beach forms the stratotype for the Minchin Hole (D/L) Stage (Bowen *et al.* 1985) (Table 8.1). Thermoluminescence dates, the mean of which is around 190 000 BP, from the Inner Beach (Southgate 1985), have given support to this contention and point to a Stage 7 age for the deposit. Analogous amino-acid ratios from shells of the Inner Beach at Minchin Hole and of the Burtle Beds in Somerset (Andrews *et al.* 1979) have led Sutcliffe (1981) to hypothesize that these two deposits had resulted from the same pre-Ipswichian marine transgression.

At Bacon Hole, amino-acid data (Headon Davies 1983; Bowen *et al.* 1985) implied that the Sandy Breccio-Conglomerate, Sandy Cave Earth and Shelly Sand are of equivalent age to the *Patella* Beach, *Neritoides* Beach and part of the Earthy Breccia Series (Sutcliffe 1981), so that the sands below represent an earlier cold episode (Currant *et al.* 1984; Sutcliffe *et al.* 1987). Stalagmites from within, and upon, the Shelly Sand have given U-series ages ranging from 129 000 to 116 000 BP (Schwarcz 1984b; Stringer *et al.* 1986).

A re-investigation of the altimetry, lithostratigraphy and biostratigraphy of raised marine deposits at Portland, Dorset, and amino-acid ratios from their Mollusca, have enabled Headon Davies and Keen (1985) to identify two separate beaches, each representing a different temperate episode. The Portland deposits previously had been considered to represent one high sea-level event, probably of Ipswichian age (Mottershead 1977). That at Portland East has been assigned to Oxygen Isotope Substage 5e, and that at Portland West (Figure 1.11) to Stage 7. On the basis of linear first-order reversible kinetics (Miller *et al.* 1979), an age of 200 000 ± 30 000 BP has been tentatively suggested for the Portland West deposit. Using similar techniques, and employing the amino-acid data of Headon Davies (1983), Mottershead *et al.* (1987) have presented a re-evaluation of the raised beaches and shore platforms (*circa* 8–12 m OD) in Torbay, Devon. Two raised beaches, deposited in environments similar to those of the area today, have been identified. The beach at Thatcher Rock probably formed in Oxygen Isotope Substage 5e, with that at Hope's Nose accumulating in Stage 7. However, Bowen *et al.* (1985) have described amino-acid ratios which assigned the Hope's Nose deposit to a Pennard (D/L) Stage (stratotype the Outer Beach, *Neritoides* Beach and part of the Earthy Breccia Series at Minchin Hole) of Oxygen Isotope Substage 5e, 5c or 5a age, and that at Thatcher Rock to an Unnamed (D/L) Stage of possible Oxygen Isotope Substage 5e or Stage 7 age (Table 8.1).

At Belcroute Bay in Jersey, evidence of two periods of sea-level above that of today has been found in superposition (Keen 1978). The lower of two raised beaches may have been produced during a post-Hoxnian and pre-Ipswichian temperate episode (Keen 1986).

Bowen (1978) and Shotton (1986) have discussed the possibility of post-Hoxnian and pre-Ipswichian temperate episodes. Shotton has stated that in the valley of the Warwickshire Avon, Terrace 5 gravels, considered by Tomlinson (1925) to have been outwashed during the Wolstonian, have deposits beneath which contain a temperate fauna (Whitehead 1989). The latter may be equated with Oxygen Isotope Stage 7, according to Shotton. It has been demonstrated by Bridgland *et al.* (1986) that part of the deposits in the assumed correlative terrace of the Severn Valley (the Bushley Green Terrace) were laid down under temperate conditions, which they have considered incompatible with those represented by Avon Terrace 5. Evans (1971) has recognized a lower (later) terrace than the Bushley Green in the Severn Valley, and correlates this with Avon Terrace 5. Assuming that the interpretation of Tomlinson regarding Avon Terrace 5 is correct, the Bushley Green temperate episode occurred prior to the Wolstonian Cold Stage. Mainly on lithostratigraphic and biostratigraphic evidence, Bridgland *et al.* (1989) and Maddy *et al.* (1991) refuted part of the established model of Avon terrace development proposed by Tomlinson (1925), which allowed the inference to

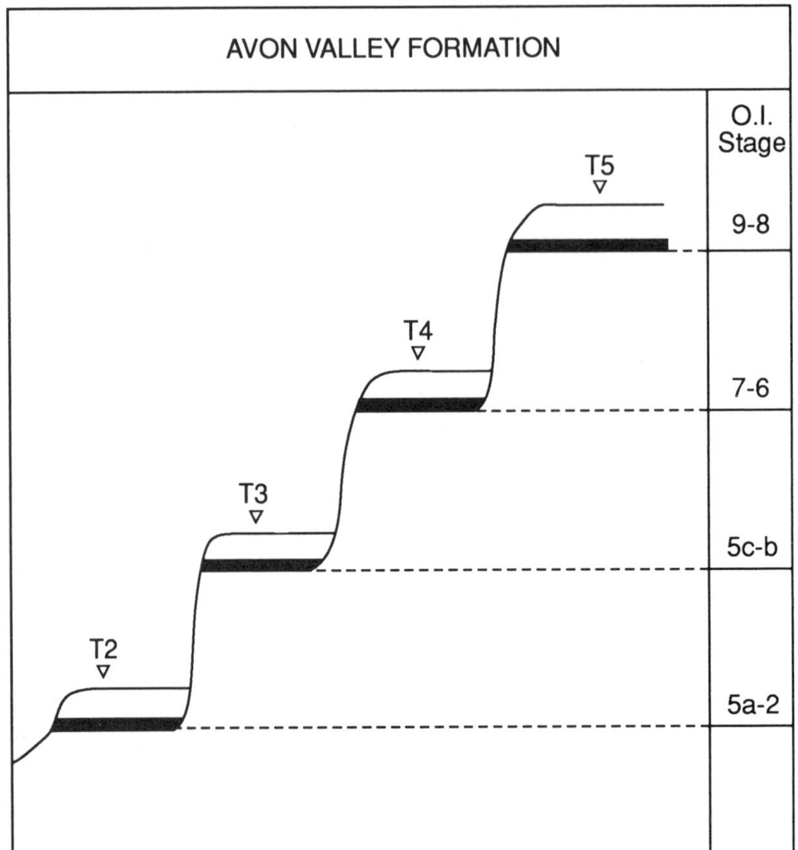

Figure 7.6 The Midland Avon terrace sequence and its correlation with oxygen isotope stages. Fossiliferous beds shown in black. From Maddy *et al.* (1991). Reproduced by permission of the Geological Society of London and the authors.

be made that the sediments below Terrace 4 are younger than those below Terrace 3, the latter being Ipswichian. On the basis of investigations at a number of localities including Ailstone near Stratford-upon-Avon, they have concluded that the sub-Terrace 4 deposits are older than those under Terrace 3, and that two separate interglacials are represented in the sediments of these two terraces. Terrace 3 has been confirmed as Ipswichian, with Terrace 4 probably of Oxygen Isotope Stage 7 age (post-Hoxnian and pre-Ipswichian) according to amino-acid ratios. Such a scheme could mean that the faunal remains of Avon Terrace 5 belong to an even earlier (perhaps Oxygen Isotope Stage 9) temperate episode, and that the main glaciation of the Midlands took place in Stage 10 or earlier (Figure 7.6). The relationship of the Bushley Green temperate episode to those established is problematical. Bridgland *et al.* (1986) have found no evidence for a glaciation subsequent to sand and gravel accumulation at Bushley Green. Hey (1958) has thought the high-level Woolridge Terrace (Chapter 5) to be composed of Wolstonian outwash gravel, which would mean that the Bushley Green temperate interval occurred subsequent to the main glaciation of the Midlands.

According to Wymer (1988), a clear indication of climatic environments between the Hoxnian and Ipswichian has been forthcoming from the terraces of the Middle Thames Valley. The work of Gibbard (1985) has demonstrated that the

Boyn Hill, Lynch Hill and Taplow terrace gravels, forming a progressively lower (younger) sequence (Figure 3.6), were deposited over this timespan. They were periglacial but could have been peripheral to glacial activity to the north and east. Evidence for temperate intervals is lacking in these gravels which, however, have yielded derived Acheulian palaeoliths, notably those of the Lynch Hill Terrace (Wymer 1968). The condition of some of the artifacts in the latter implies their contemporaneity with human activity in the vicinity of the Thames Valley. Levallois implements (Figure 6.4) have been found in the Taplow Gravel. The Levallois Industry discovered at the top of this deposit at West Drayton, Middlesex, is analogous to that at Baker's Hole, Northfleet (section 7.1.4) (Wymer 1988). In the Lower Thames Valley between Southend and the Blackwater, temperate deposits of the Rochford Channel (intermediate in age between the Boyn Hill and Lynch Hill gravels) could be placed in a post-Hoxnian and pre-Ipswichian context according to Bridgland (1988).

Possible intra-Wolstonian interstadial episodes at La Cotte de St. Brelade, Jersey, are mentioned above (section 7.1.4). Similar detailed archaeological and palaeoenvironmental investigations at Pontnewydd Cave in North Wales have revealed hominid, possibly early *Homo sapiens neanderthalensis*, fossils (teeth and bones) and other faunal remains including rhinoceros, lion, horse, vole and lemming, together with Upper Acheulian tools including hand axes, and Levallois implements (Green *et al.* 1981; Green 1982). Stalagmites above the hominid remains have been aged by the U-series method to 224 000 $^{+43\,000}_{-31\,000}$ BP (Ivanovich *et al.* 1984), which equates approximately with Oxygen Isotope Stage 7. Thermoluminescence dating of burnt flint found near to a human molar has given an age of 200 000 ± 25 000 BP (Huxtable 1984). Occupation of the cave *circa* 250 000–230 000 BP (during Oxygen Isotope Substage 7c) has been suggested by Green (1984).

From the Speeton Shell Bed on the coast of eastern Yorkshire, Wilson (1991) has identified a comparable pollen assemblage to that obtained by West (1969a) (Chapter 8). However, amino-acid ratios of *Macoma balthica* shells from the deposit suggest that it accumulated during Oxygen Isotope Stage 7.

7.2.2 New sites

The most significant sites in the second category have been confined to a fairly restricted area of southern England (Figure 7.4). At Stanton Harcourt (Linch Hill) west of Oxford, temperate organic channel deposits have been discovered beneath the widely developed Summertown–Radley Terrace of the Upper Thames Valley (Briggs 1976; Briggs and Gilbertson 1980; Bryant 1983; Briggs *et al.* 1985) (Figure 7.7). Two sedimentary units have been defined in this terrace. A lower, Stanton Harcourt Gravel possesses a fauna incorporating *Mammuthus primigenius*, *Coelodonta antiquitatis* (Sandford 1924; Currant 1985a) and *Pupilla muscorum*, and contains structures which are indicative of cold conditions including permafrost (Briggs *et al.* 1985). An upper unit, the Eynsham Gravel (not present at Stanton Harcourt), has yielded a fauna of which *Hippopotamus amphibius*, *Panthera leo* (lion) and *Bos primigenius* (aurochs) are vertebrate members and *Corbicula fluminalis*, *Unio littoralis* and *Helix nemoralis* are represented among non-marine Mollusca (Kennard and Woodward 1924; Briggs *et al.* 1985), which indicates an interglacial environment. At Magdalen College, Oxford, the Eynsham Gravel contains silt whose sparse pollen content came mainly from trees and herbs. *Pinus* and *Corylus* dominate the arboreal and shrub taxa, but *Carpinus*, *Quercus*, *Picea*, *Betula*, *Acer* and *Taxus* also occur, and the inference is of open interglacial woodland (Hunt 1985). The faunal and floral evidence points to an Ipswichian age for this deposit (Briggs *et al.* 1985). Thus, the Stanton Harcourt Channel, cut in Oxford Clay underlying the Stanton Harcourt Gravel, has been interpreted as representative of a pre-Ipswichian temperate episode. The age of the Stanton Harcourt Gravel is uncertain. However, the incidence of fresh flint in a correlative gravel in the Oxford area implies that its aggradation post-dated the advance of Wolstonian ice to Moreton-in-Marsh (section 7.1.1). The biogenic material in the channel contains *Picea*, *Quercus*, *Alnus* and some herb pollen, and wood of oak, hazel and *Cornus sanguinea* (dogwood) has been recovered (Beck 1985). Vertebrates found include *Panthera leo*, *Mammuthus primigenius*, *Equus* and *Ursus* (Currant 1985b), and among the molluscan fauna *Corbicula fluminalis*, which is widely distributed in Asia at present (Ellis 1978). Insect remains suggest an oceanic climate analogous to that of southern England today, and the overall impression from the flora and fauna is of a temperate episode of interglacial rank (Briggs *et al.* 1985). Locally developed sands and silts in slight hollows at the bottom of the Summertown–Radley Terrace contain plant and molluscan fossils which are diagnostic of the same episode but of a later, cooler environment. Plant (Bell 1904; Duigan 1956b), insect (Blair 1923) and vertebrate (Sandford 1924) data from the Wolvercote Channel, which was probably incised into the Wolvercote Terrace of this area, are thought to relate to either the same interglacial as that demonstrated in the Stanton Harcourt Channel or to a separate event of the same type (Briggs *et al.* 1985). The age of the Wolvercote Terrace is unclear but its stratigraphic position indicates accumulation prior to the Summertown–Radley Terrace (Sandford 1924), and its gravels have been designated as outwash from the Wolstonian Glaciation of the English Midlands (Bishop 1958) (section 7.1.1). Thus, the cool-temperate environment inferred from the channel biota has been considered to belong either to the early part of the interglacial recorded in the Stanton Harcourt Channel or to a previous temperate episode. Palynological information has confirmed a cool-temperate environment but indicates a vegetational succession characteristic of late rather than early interglacial time. These data thus appear to refer the Wolvercote Channel deposits to the Late-temperate substage of an earlier interglacial (Briggs *et al.* 1985). The Stanton Harcourt Gravel, which occurs above the interglacial sediments in the Stanton Harcourt Channel (Figure 7.7), has thus been interpreted as representative of a cold (periglacial) environment that developed subsequent to that during which the Wolstonian Glaciation took place.

At Marsworth in Buckinghamshire (Green *et al.* 1984), biogenic deposits filling part of the lower of two channels (the Lower Channel cut in Chalk) (Figure 7.8; Figure 1.13) have produced a pollen and plant macrofossil assemblage dominated by herbaceous taxa indicative of both dry and damp herb-rich grassland. The small amounts of tree and shrub pollen (mainly *Pinus*, *Picea*, *Abies*, *Betula*, *Quercus*, *Ulmus*, *Alnus* and *Salix*) imply very limited woodland in the region. Coleopteran remains signify local marshy grassland dominated by *Glyceria maxima* (reed-grass), the food plant of the commonly encountered *Donacea semicuprea* and *Notaris acridulus*, and the numerous dung beetles suggest that grazing animals were present. The mammal fauna supports the latter supposition. It is dominated by *Mammuthus* and *Equus*, and contains *Panthera leo*, *Canis lupus* (wolf) and *Microtus oeconomus*. Currant (1989) has assigned this mammalian assemblage to a post-2, pre-1 position (Table 4.1). The corresponding molluscan assemblage contains mainly grassland taxa including *Pupilla muscorum*.

At the bottom of the Lower Channel, travertine bearing the impressions of leaves of deciduous trees (including *Acer* sp.) and containing Mollusca (*Azeca goodalli*, *Discus rotundatus* and *Clausilia bidentata*, for example) characteristic of temperate woodland habitats, has been discovered. The travertine has been dated by U-series to 170 000–140 000 BP. Coombe Rock overlies the Lower Channel deposits and incised into it is an Upper Channel containing finely stratified sandy gravel and marl. These channel deposits have yielded only a mammalian fauna, which with *Hippopotamus amphibius* and *Dicerorhinus hemitoechus* is of typical Ipswichian temperate-forest type, and of the Group 1 assemblage of Currant (1989). Thus, at this locality there is evidence of two temperate episodes, the earlier pre-Ipswichian, separated by an interval during which intense periglaciation took place. While there is a possibility that the Lower

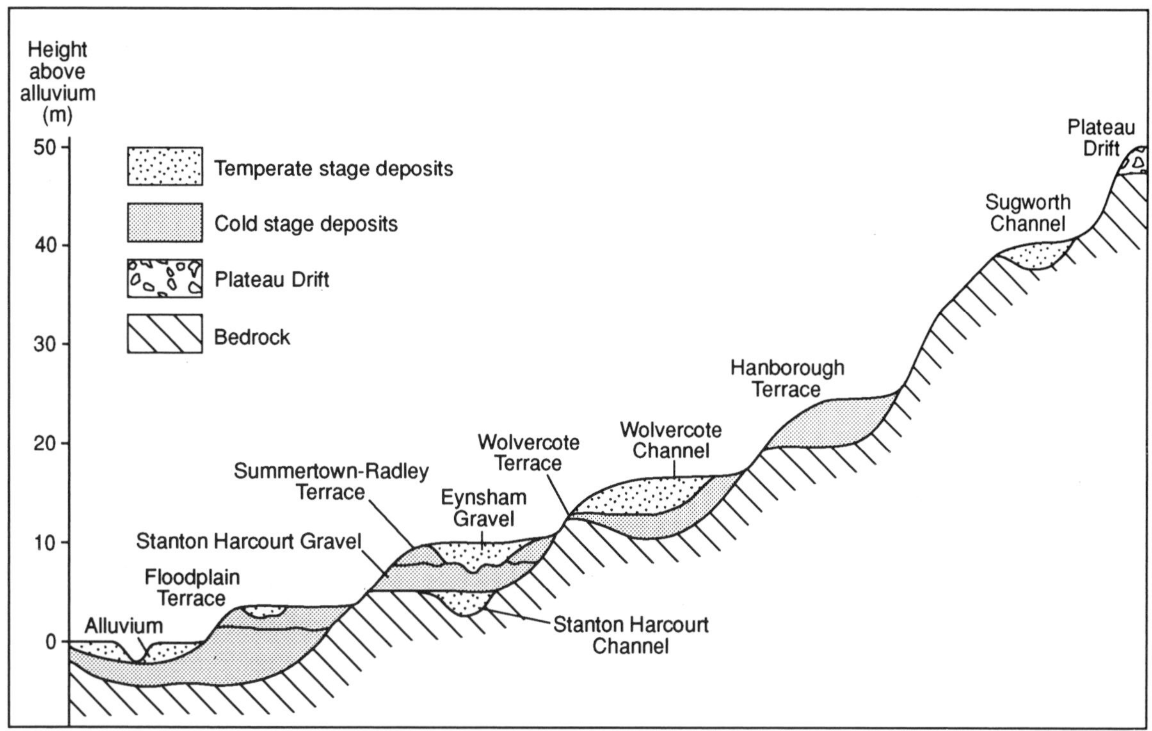

Figure 7.7 The terrace sequence of the Upper Thames. From Briggs *et al.* (1985).

Channel deposits are of interstadial rank, the biota within them tend to support the incidence of interglacial conditions. The biotic assemblages, rather atypical of those of established interglacials (and interstadials) in the British Isles, probably represent the later part of a post-Hoxnian and pre-Ipswichian interglacial stage.

In the valley of the Great Ouse near to the border of Bedfordshire and Buckinghamshire, chalky till has been observed mantling the sides and cold-stage terrace deposits occur on the floor. The latter include organic sediments with a temperate biota, which occupy a channel (the Lower Channel) close to the village of Stoke Goldington (Green *et al.*, in preparation). Plant remains in the organic deposits exhibit little temporal taxal variation and denote a predominantly herbaceous plant cover. Floodplain pools were present, as was adjacent wetland, and there may have been occasional waterside trees. There seems to have been open woodland on interfluves, with *Pinus*, *Picea*, *Carpinus*, *Abies*, *Corylus* and *Ilex* among its constituents, but the drier parts of the floodplain and the valley sides were covered by herb-rich grassland. The molluscan fauna is substantial, including representatives of river, marsh and floodplain habitats and notable for the absence of woodland taxa. The presence of *Corbicula fluminalis* indicates an interglacial environment, a conclusion which could also be drawn from plant and insect evidence. The coleopteran assemblage, lacking indicators of trees, is typical of that found in southern England today. However, six species of Coleoptera are not currently members of the fauna of the British Isles. Of these *Anotylus gibbulus*, *Stomodes gyrosicollis* and *Heterhelus scutellaris* have been considered of possible special significance. The former has also been recorded from Stanton Harcourt and Marsworth but not from any deposits of established Hoxnian

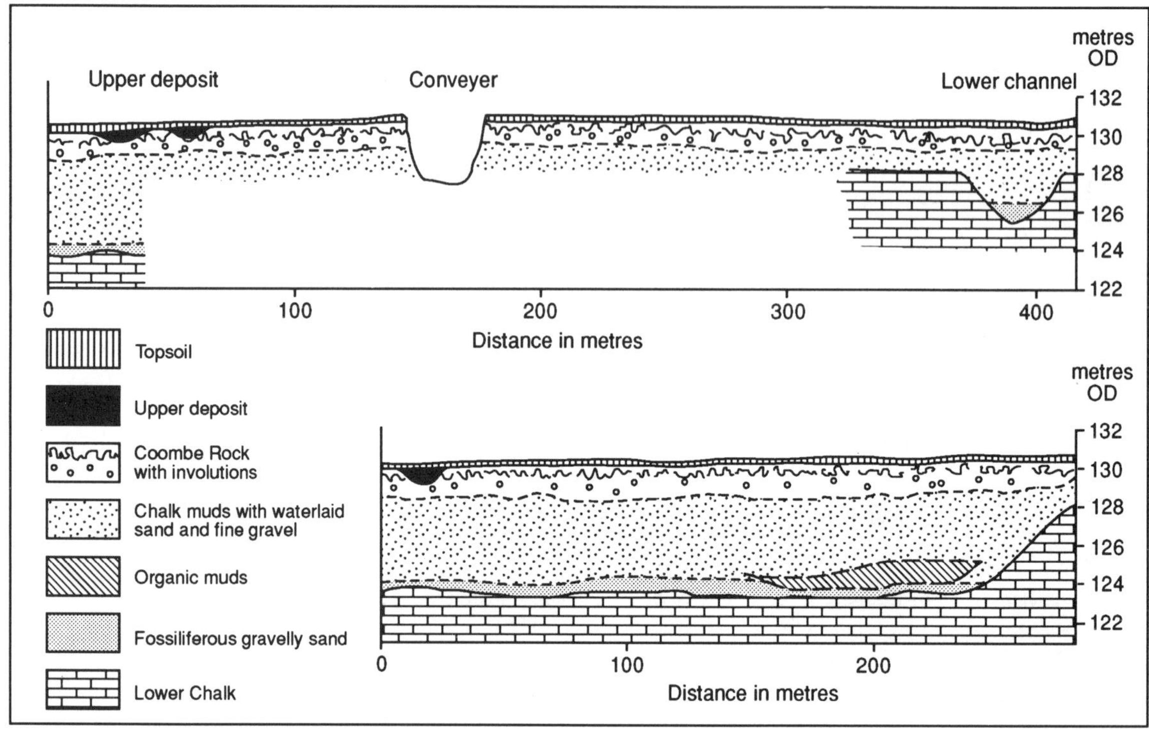

Figure 7.8 Sections in Pitstone No. 3 quarry, Marsworth. From Green *et al.* (1984). Reprinted with permission from *Nature* **309**, 778–781, © 1984 Macmillan Magazines Ltd.

or Ipswichian age. The others have also been found at Marsworth but nowhere else in the British Isles. The climatic implications of the insect fauna include temperatures slightly above those of the present. The presence of *Clausilia pumila* and a European ecotype of *Azeca goodalli* in the molluscan record at Stoke Goldington adds a continental dimension to the climate. The find of *Pelocypris atabulbosa* among ostracods from Stoke Goldington (and Stanton Harcourt) also supports a continental environment. This sub-tundra inhabitant of the Canadian interior today (Delorme 1970) is commensurate with the palaeobotanical inference of the Late-temperate substage of an interglacial, or of an interstadial. Uranium-series determinations on shells of *Valvata piscinalis* (the unusual morphology of which is matched by specimens from Stanton Harcourt) has given ages of *circa* 208 000 and 168 000 BP, with the latter considered the more reliable. The Lower Channel was cut in cold-stage fluvial gravel and covered first by a similar deposit, then by a chalky gravel resembling outwash. As this area lies to the south of that where outwash of Devensian age would be expected (Figure 9.11), the glaciation which produced this fluvial material must have been earlier, and in Wolstonian or possibly Anglian time. A cold-stage biota is found in a further, later channel (the Upper Channel) incised into the Middle Gravel (section 7.1.1). Whatever the age of the glaciation, its incidence and proximity (the chalky till on the adjacent valley sides may also have been produced by this event) has established that the organic deposits of the Lower Channel are pre-Ipswichian. Moreover, there are strong geomorphological grounds for regarding all the Pleistocene terrace gravels around Stoke Goldington as post-Anglian. Thus, a Wolstonian context for the succession in this area seems most likely, thereby supporting the incidence of a glaciation at this time in this part of southern England.

It has been satisfactorily demonstrated that the temperate deposits at Stanton Harcourt, Marsworth and Stoke Goldington are pre-Ipswichian, and it is very likely that they are post-Hoxnian in age. While the lithostratigraphic and biostratigraphic information from the sites makes it tempting to refer them all to the same episode (although not to an equivalent part of it) and to assign this the rank of interglacial, such correlations seem unwise for a number of reasons. For example, stratigraphic successions have indicated that glacial outwash of presumed Wolstonian age lies below the temperate deposits at Stanton Harcourt but above them at Stoke Goldington. Also with regard to the biological evidence, studies of Devensian insect faunas (Coope 1977a) (Chapter 9) have shown that interstadial climates are not necessarily cool. It has been pointed out by Woillard (1978) that a thermophilous plant assemblage, traditionally considered typical of an interglacial environment, could be present during an interstadial if the climate was suitable and refugia for such taxa were close enough to the area. The shorter timespan of an interstadial would mean that thermophiles would be unable to immigrate in successive well-ordered waves, as during an interglacial. The sudden and disorderly appearance of thermophilous plants in an interstadial is likely to be related to reduced competition compared with that during an interglacial. There is thus a possibility that some or all of the Stanton Harcourt, Marsworth and Stoke Goldington temperate deposits accumulated during one or more interstadial intervals.

In situ flowstone obtained from within sand and silts at Robin Hood's Cave, Creswell, north Derbyshire, has been dated to *circa* 165 000 BP by U-series analysis (Rowe and Atkinson 1985). The palynology of the sand, silts and stalagmites (Coles *et al.* 1985) suggests a vegetation that changed from grassland to open deciduous woodland, subsequently reverting to steppe then tundra. Equivalence of the temperate episode represented here with that identified at Marsworth is considered likely (Rowe and Atkinson 1985).

A Pleistocene faunal assemblage from Bielsbeck near Market Weighton in eastern Yorkshire, which contains *Palaeoloxodon antiquus*, *Dicerorhinus hemitoechus* and *Equus* but lacks hippopotamus (Stather 1910; de Boer *et al.* 1958), may be of an equivalent age to that from the Lower Channel at Marsworth (Catt 1987a).

7.3 Towards a revised succession

In a discussion of the Palaeolithic occupation of East Anglia against the backdrop of the evolution of the landscape of this region, Wymer (1985), drawing upon geological, biological and archaeological data, and adopting the temperature fluctuations and timescale provided by the oxygen isotope record, has presented a revised scheme of Pleistocene stages which may have applicability over a wider area of the British Isles (Table 7.2). This incorporates two temperate episodes (Wolstonian 1/2 and Ilfordian) separated by three colder intervals (Wolstonian 1, 2 and 3) in the time between the Hoxnian and Ipswichian interglacials. Palynological data from the Upper Sequence at Hoxne (Gladfelter 1975) indicate an initial episode (Wolstonian 1) in which the vegetation was open but conditions were not severe enough for periglaciation. This was followed by one in which the tree cover increased (Wolstonian 1/2), then a further interval (Wolstonian 2) with substantial frequencies of non-arboreal pollen, considered to reflect temperate and cold (periglacial) environments, respectively (Mullenders and Dirickx cited in Wymer 1985). A similar sequence has been postulated in the Middle Gravel and Upper Loam at Swanscombe (Wymer 1985). Wolstonian 2 is regarded as the time of a glaciation, whose limit in East Anglia accords with that proposed by Straw (1979a,b) (section 7.1.2). Probable correlatives with this activity have been considered to be the Lynch Hill (also formed in Wolstonian 1) and Taplow Terrace 1 materials of the Middle Thames, the Wolvercote Terrace of the Upper

Table 7.2 Summary of the probable geological sequence in East Anglia from the Hoxnian to the Ipswichian. From Wymer (1985)

Geological stage	*Years* BP	*Sea-level*
	75 000	
Ipswichian		High 6–8 m OD
	128 000	
Wolstonian 3		Low
	195 000	
Ilfordian		High 15 m OD
	240 000	
Wolstonian 2		Low
	297 000	
Wolstonian 1/2		
	330 000	
Wolstonian 1		
	367 000	
Hoxnian		High 20 m OD
	400 000	

Thames, and Lake Harrison in the English Midlands. The Ilfordian Temperate Stage, interglacial on biological criteria, followed Wolstonian 2 and has been thought to be represented at sites including Ilford, Aveley, Stoke Tunnel, Stutton, Brundon, Harkstead and Marsworth. Coombe Rock above Ilfordian deposits at Marsworth and Stoke Tunnel and overlying the Levallois Industry at Baker's Hole might have formed during Wolstonian 3, when intense periglaciation took place. At this juncture, Taplow Terrace 2 of the Middle Thames and the lower part of the Summertown–Radley Terrace (Stanton Harcourt Gravel) also could have accumulated. The time-span for all these events may have been *circa* 367 000–128 000 BP, with the temperate episodes occurring about 330 000–297 000 BP (Wolstonian 1/2), and 240 000–195 000 BP (Ilfordian), during Oxygen Isotope Stage 9 and Oxygen Isotope Stage 7, respectively.

Uranium-series ages defining episodes of speleothem growth in caves of the Craven District of north-west England (Gascoyne *et al.* 1983) have assisted in refining such a chronology. Mineral precipitation was profuse during 90 000–135 000 and 170 000–350 000 BP, but ceased during 140 000–160 000 BP and was reduced *circa* 260 000 BP. The temperate conditions required for speleothem growth from 170 000 to 350 000 BP have been equated with those of Oxygen Isotope Stage 7 and Stage 9. The period of zero speleothem assimilation has been assigned to Stage 6 (the Wolstonian Cold Stage) and that of diminished accumulation probably to Stage 8. Gordon *et al.* (1989), after examining the palaeoclimatic implications of speleothem growth in the British Isles since *circa* 220 000 BP, have concluded that a broad peak occurred around 180 000 BP in a pre-Ipswichian context.

Bowen *et al.* (1986) have suggested that a post-Hoxnian and pre-Ipswichian interglacial (identified by amino-acid ratios and exemplified at Stanton Harcourt) could be correlated with Oxygen Isotope Stage 7 by extrapolation with the stages assigned to the Hoxnian and Ipswichian (Figure 1.15). Bowen *et al.* (1985) and Bowen and Sykes (1988) have demonstrated that the Paviland Glaciation of the Gower Peninsula (section 7.1.4) occurred prior to Stage 7 on the basis of the relationships of its deposits with raised-beach material. Amino-acid ratios from the latter suggest its reference to Stage 7. Bowen *et al.* (1986) have equated the Paviland glacial deposits and their presumed equivalents in eastern England and north-eastern Scotland with Oxygen Isotope Stage 8. This glaciation probably did not move south to the Coventry area, where the glacigenic deposits are of Anglian age, but may be represented by till above the Hoxnian interglacial sediments at Quinton in Birmingham (Chapter 6). As deposits of Paviland age are not present at Wolston, the designation of glacigenic sediments there as Wolstonian has been considered inappropriate.

Bowen *et al.* (1989) have presented isoleucine

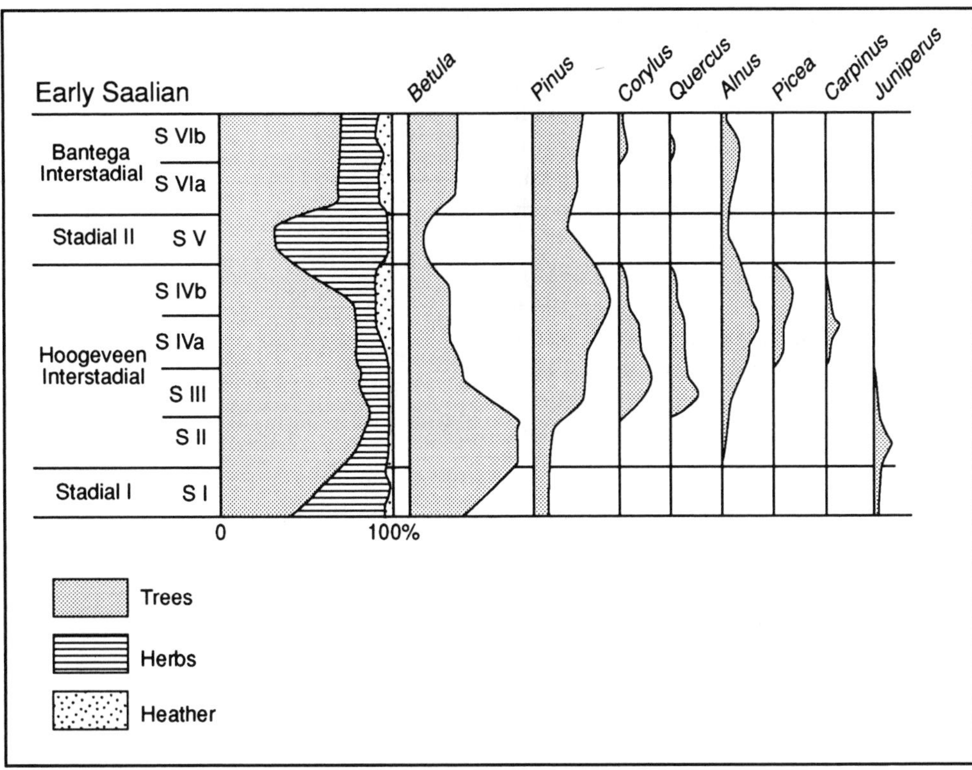

Figure 7.9 Schematic representation of vegetational succession in the Early Saalian of the Netherlands, based on data obtained in three localities. From de Jong (1988).

epimerization data from non-marine Mollusca which enabled a number of sites including Aveley, Crayford, Stutton and Ailstone, to be grouped with Stanton Harcourt. The Stanton Harcourt amino-acid signal has been shown to be distinctive and to define a major geological event that has been correlated with Oxygen Isotope Stage 7, a temperate episode which was in progress *circa* 200 000 BP (Figure 1.15).

7.4 Correlation with mainland Europe

Bowen *et al.* (1986) have suggested that the glaciation, of which there is limited evidence in the British Isles, represented in England and Wales by the Welton Till, Basement Till, Warren House Till, Ridgacre Till and Paviland glacial deposits, could be correlated with the Saalian Glaciation of the Netherlands. In France, evidence of two intra-Saalian temperate episodes (probably interglacials) has been forthcoming from two localities in the Seine Valley. At St. Pierre-les-Elbeuf, pedogenetic features characteristic of temperate environments occur at two levels between those representing the Holsteinian and Eemian interglacials. The intra-Saalian palaeosols are separated by loesses which were deposited in cold climates (Lautridou 1982). At Tourville, these temperate episodes are represented by estuarine and marine deposits, respectively, the latter attaining 14 m NGF (Nivellement General Français, 0.075 m above OD) (Carpentier and Lautridou 1982). Early Saalian interstadials have been recognized in the northern Netherlands (de Jong 1988) (Figure 7.9). Sediments of the Bantega Interstadial have Holsteinian deposits below them and Saalian till above (Zagwijn 1973). Palynological data suggest a quite temperate climate during the Hoogeveen Interstadial, whose flora possessed certain interglacial characteristics (de Jong 1988). Both episodes have been provisionally correlated with Oxygen Isotope Stage 7 by de Jong and by Zagwijn (1989). An intra-Saalian temperate interval has been dated by thermoluminescence to 270 000 ± 11 000, ± 22 000 BP at Maastricht – Belvédère in the Netherlands (Huxtable and Aitken 1985). In both France and the Netherlands, Levallois implements have been associated with intra-Saalian deposits (Fosse and Carpentier 1982; Roebroeks 1985). Skeletal remains in these contexts have been nearly all of a somewhat archaic form of *Homo sapiens neanderthalensis* (Vandermeersh 1978; Tuffreau 1982). In eastern Germany, the Lower Travertine at the Middle Palaeolithic site of Ehringsdorf has produced an average U-series age of 225 000 ± 26 000 BP (Schwarcz 1980; Jäger and Heinrich 1982; Cook *et al.* 1982), thereby relating it to Oxygen Isotope Stage 7. Parallels with the situation in the British Isles at this time, as represented at Pontnewydd, La Cotte de St. Brelade, Baker's Hole and West Drayton, may thus be drawn.

7.5 Conclusion

Although evidence has been forthcoming of additional climatic episodes between the Hoxnian and Ipswichian interglacials in the British Isles, and parallels exist with successions in continental Europe, both the number and status of such intervals remain uncertain. The picture is further complicated by a dispute concerning the status of deposits traditionally assigned to a glaciation within this timespan. Data indicative of these episodes are limited and problems exist with their correlation. The orthodox Pleistocene succession in the British Isles (Mitchell *et al.* 1973a) is not, as Shotton *et al.* (1983) have observed, sacrosanct and immutable. During the time allocated to the Wolstonian Stage in this succession, it appears that an interglacial occurred. An earlier temperate episode, many of whose biotic characteristics were unlike those of a typical interstadial, may also have taken place. A temperate stage followed the Wolstonian.

8

The Ipswichian Temperate Stage

The controversy concerning certain deposits which may or may not belong to this interglacial is presented in Chapter 7. The account which follows adopts the orthodox view that there is sufficient evidence, notably of a biological nature, to assign most if not all of these deposits to the Ipswichian. While the majority of sites referred to this temperate stage on the basis of their fossil flora and fauna have been in East Anglia, a scattering has been found in other parts of the British Isles (Figure 8.1).

8.1 Flora and vegetation

The stratotype for this interglacial has been established as Bobbitshole, a site in the Belstead Brook Valley, which joins the estuary of the River Orwell just south of Ipswich (Mitchell *et al.* 1973a). Here, West (1957) has described the stratigraphy and palaeobotany of freshwater lacustrine deposits which are underlain by a raft of chalky till of presumed Wolstonian age. The first part of an interglacial (the Pre-temperate and Early-temperate zones) has been recognized. Vegetation of park–tundra type was replaced by birch-dominated forest. Subsequently, pine then oak assumed prime importance in an arboreal context. The plant evidence suggests a quick improvement in climate at the beginning of the interglacial. From its latest sediments, there are signs (for example, *Najas minor* and *Salvinia natans* in the lacustrine flora, which today are distributed only as far north as the Netherlands) that summer temperatures at that time exceeded those current in East Anglia. Some of these interglacial deposits occur below present sea-level and West has concluded that the absence of an upper part to the interglacial sequence was probably due to a lowering of base level (linked to a falling sea-level), so that erosion or non-deposition of sediment took place. Sandy gravel with both fluvial and solifluction characteristics, which occurs above the interglacial deposits, has been considered to be the product of a succeeding cold stage.

A fundamental problem in the demonstration of an interglacial cycle *sensu* Turner and West (1968) for the Ipswichian has been that no site, in East Anglia or elsewhere in the British Isles, has extended throughout all of its substages (Figure 8.2). Interglacial muds in the Middle (Barnwell) Terrace of the River Cam at Histon Road on the northern fringe of Cambridge, first reported by Hollingworth *et al.* (1949), have been further investigated by Walker (1953) and Sparks and West (1959). Walker has demonstrated the existence of two pollen zones, the lower dominated by *Carpinus* and the upper by *Pinus*, a biozonation confirmed by Sparks and West. The high *Carpinus* frequencies have led to the correlation of these deposits with a later part of an interglacial than those at Bobbitshole. At Stutton in the Stour Estuary, Suffolk, 8 km south of Bobbitshole, Sparks and West (1963) have described a pollen record from brickearth. High *Carpinus* values have led to its referral to the second part of

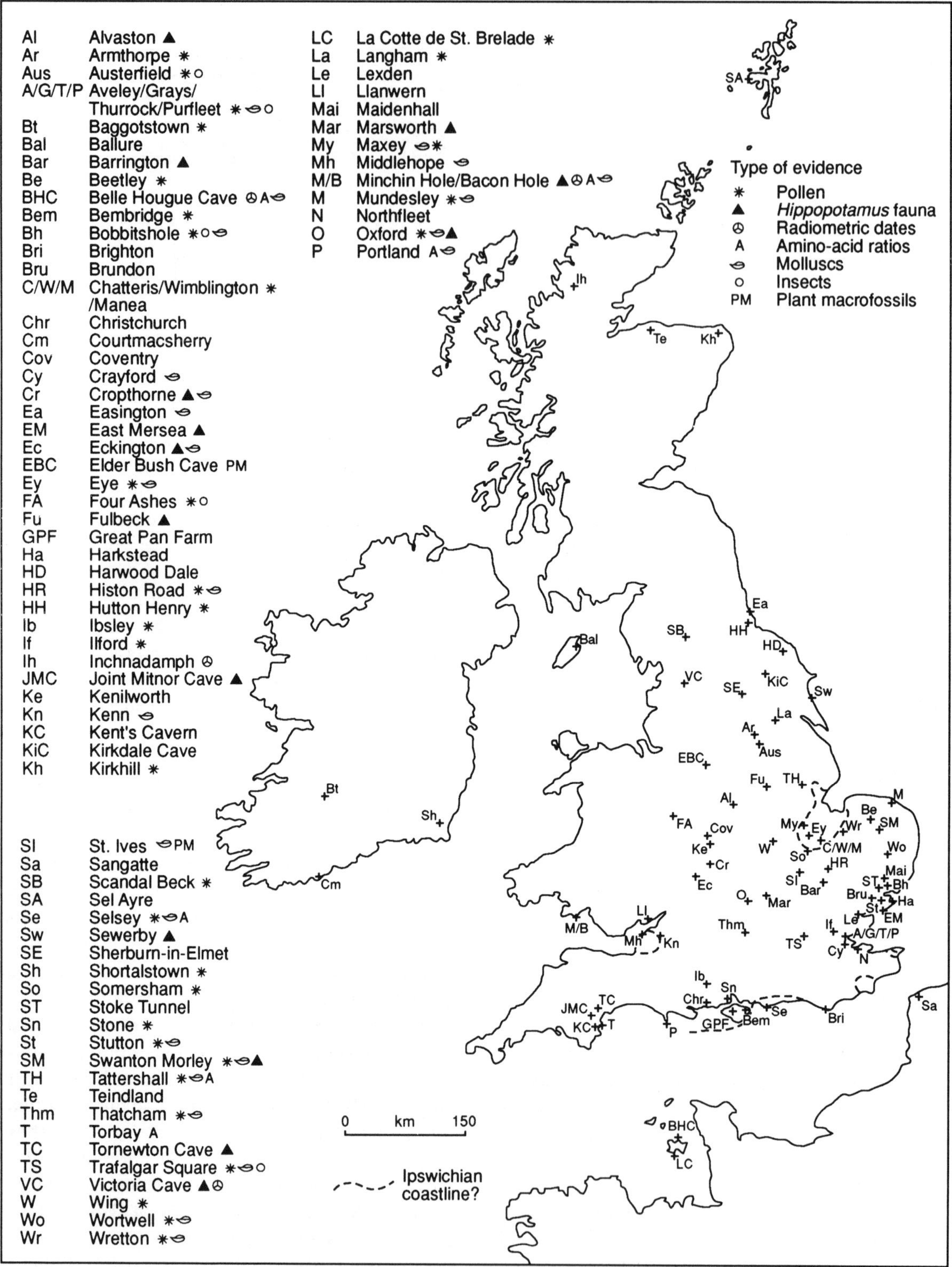

Figure 8.1 Sites referred to during Chapter 8 with types of evidence. Possible positions of the Ipswichian coastline are also shown.

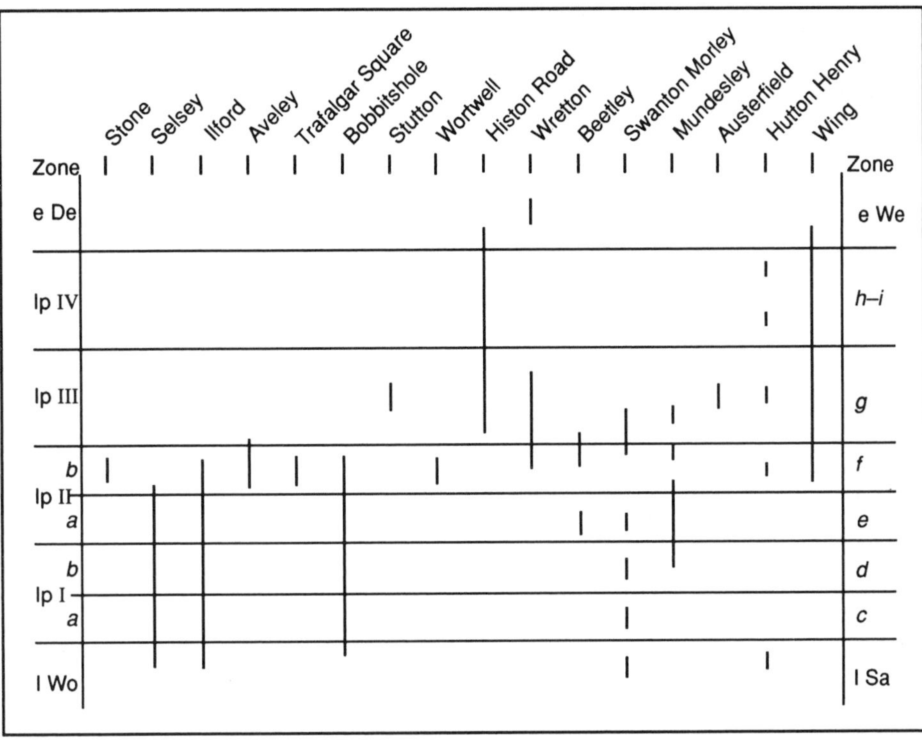

Figure 8.2 Pollen zone range diagram of Ipswichian interglacial sites in Britain. From Hall (1980).

the Ipswichian Interglacial. At Wortwell near Harleston in the Waveney Valley, Norfolk, Sparks and West (1968) have reported organic deposits in a low terrace (at *circa* 15 m OD), whose plant fossils allow their correlation with early and middle Ipswichian time. The Wortwell area has been reinvestigated by Coxon (1982), who has shown the existence of a floodplain backwater deposit with a pollen stratigraphy extending from the late Wolstonian to the last part of the Early-temperate substage of the Ipswichian. At Wretton in west Norfolk, the palaeobotany of fluviatile deposits occupying meander channels of an ancestral River Wissey, and overlain by Devensian terrace sediments (Chapter 9), has been presented by Sparks and West (1970). It has been concluded that pollen assemblage biozones dominated by *Quercus*, *Pinus*, *Corylus* and *Alnus*, and by *Carpinus*, *Quercus*, *Pinus* and *Corylus* could be ascribed to Early-temperate and Late-temperate Ipswichian substages, respectively. The Barrington Beds have been shown to comprise terrace remnants of a tributary of the River Cam south-west of Cambridge. Palynological data from sediment stuck to characteristic Ipswichian mammal bones (section 8.4) imply that the floodplain was dominated by herbaceous vegetation, amongst which some plants (certain composites and plantains, for example) could have thrived because of grazing and trampling by animals. Arboreal pollen from the same context suggests the existence of regional mixed-oak forest (Gibbard and Stuart 1975). At Maxey near Peterborough, organic channel deposits have yielded pollen indicative of Ipswichian Pre-temperate and Early-temperate vegetation (French 1982). In the southern Fenland of Cambridgeshire, brackish/marine deposits at Somersham have a pollen flora indicative of an Early-temperate Ipswichian age.

At Chatteris, both Early-temperate and Late-temperate Ipswichian substages are represented (West 1987). Muddy sands within the nearby March Gravels have produced pollen spectra referable to the Ipswichian Temperate Stage (West 1987), with silty clay in the same context at Eye providing palynological data suggestive of a late Ipswichian age (Keen *et al.* 1990).

At Mundesley in north-east Norfolk, a channel incised through Anglian glacigenic deposits into the Cromer Forest-bed Formation, probably during the Wolstonian, contains organic deposits of Ipswichian Pre-temperate, Early-temperate and Late-temperate age on palynological criteria (Phillips 1976). Wolstonian outwash gravels at Roosting Hill, Beetley, in the Wensum Valley of Norfolk, contained a channel with organic mud, from which Early-temperate Ipswichian plant remains have been recovered. At nearby Swanton Morley, the pollen record from organic deposits intercalated with fluviatile sands and gravels covers the period from the late Wolstonian to the onset of the Late-temperate Ipswichian zone (Phillips 1976; Coxon *et al.* 1980). Of particular interest at these sites are reductions in the ratios of arboreal to non-arboreal pollen in the Early-temperate and Late-temperate zone sediments, interpreted as possibly due to the grazing and trampling effects of animals, especially on the valley floors. The remains of large mammals have been found in association with these sediments (section 8.4). This topic has also been considered by Phillips (1974) in a discussion of the vegetational history of the Ipswichian Interglacial in the British Isles and the Eemian Interglacial in continental Europe. In common with many other localities of this age, evidence is present at Swanton Morley of either the cessation of deposition, or of its interruption and resurgence in the later part of the interglacial. In the Nar Valley of Norfolk, channel muds at Pentney have yielded boreal then mixed-oak forest pollen spectra referable to the Ipswichian (Ventris 1986). At Galley Hill near St. Ives, Cambridgeshire, an organic channel-fill beneath terrace gravels of the River Ouse possesses plant fossils that allow its correlation with the Early-temperate part of the Ipswichian (Preece and Ventris 1983).

Terrace deposits beneath Trafalgar Square, London, contain organic material whose plant remains denote deciduous forest in which *Quercus* and *Acer* were important and *Taxus* and *Carpinus* present. A substantial frequency of non-arboreal pollen has been recorded and an Early-temperate Ipswichian age suggested (Franks 1960). Further down the Thames system, West *et al.* (1964) have described organic deposits above sand and below brickearth, occupying a channel or depression in a tributary valley near Seven King's Station, Ilford. The succession of plant communities revealed from pollen and plant macrofossil analyses has been considered characteristic of the first half of the Ipswichian Interglacial. Substantial amounts of herbaceous pollen, especially Gramineae and Cyperaceae, characterize each of the biozones delimited. There are indications from the flora of a continental-type climate and the rise of summer temperatures to values above those of today in this area. For example, *Berula erecta* (lesser water-parsnip) and *Groenlandia densa* (opposite-leaved pondweed) have current distributions of a southern affinity in Europe; while seeds of *Hydrocharis morsus-ranae* (frogbit), which now fruits infrequently in the British Isles, are indicative of increased summer warmth. At Aveley in Essex, freshwater organic sediments overlain by brickearth have been found occupying a channel cut in London Clay. Pollen analysis has suggested late Early-temperate and early Late-temperate ages for the organic sediments (West 1969b). A rise in non-arboreal pollen in the upper part of the sequence is attributed to the likely spread of open-ground vegetation following the increase in alluviation in the valley represented by the deposition of the brickearth. At Little Thurrock, Grays, Essex, sandy silts have been observed above gravel, in a similar context (up to 15 m OD) to the localities at Ilford and Aveley (West 1969b). The pollen data from these silts have been thought most probably to reflect the growth of Ipswichian Early-temperate woodland including pine and oak. Hollin (1977) has reported the results of pollen analysis of analogous sediments at Little Thurrock, West Thurrock and Purfleet, suggesting their formation during a similar span of Ipswichian time. In the Middle

Thames area, a detritus mud from near the bottom of floodplain gravel of the River Kennet at Thatcham has yielded a pollen record which depicts mixed-oak forest including *Acer* and *Taxus*, which Holyoak (1983a) has referred to the Early-temperate substage of the Ipswichian. In the Upper Thames Valley, silts within the Eynsham Gravel (Figure 7.7), which make up the upper part of the Summertown–Radley terrace deposits observed at Magdalen College, Oxford, have produced pollen (including *Acer*, *Taxus* and *Carpinus*) from deciduous woodland, together with a substantial frequency of that of herbs implying the existence of grassland, probably of Late-temperate Ipswichian age (Hunt 1985).

Palynological information has allowed Pre-temperate and Early-temperate Ipswichian contexts to be assigned to deposits in a channel cut in Bracklesham Beds at Selsey, Sussex (West and Sparks 1960). These freshwater and estuarine sediments are older than an associated raised beach and brickearth (section 8.9). Between Stone Point and the mouth of the Beaulieu River on the Hampshire coast, estuarine deposits and intercalated freshwater mud have revealed a pollen assemblage comparable with that of the Early-temperate Ipswichian zone (West and Sparks 1960). Further investigation of the Stone sequence (Brown *et al.* 1975) has established organic material at a higher elevation – just above OD. Pollen analysis of this material has shown *Carpinus* to be present, and has allowed the inference of a later phase of Early-temperate Ipswichian time than that recorded by West and Sparks. At Ibsley, peat located beneath terrace gravels of the Hampshire Avon (Clarke and Green 1987) has produced a palynological spectrum with high herbaceous values, large amounts of *Ilex* and substantial frequencies of *Hedera*, thought likely to be representative of the vegetation during the Ipswichian climatic optimum by Barber and Brown (1987). Organic mud above a raised beach at Bembridge Foreland on the Isle of Wight has produced a pollen assemblage with *Quercus*, *Corylus*, *Alnus*, *Acer* and considerable amounts of *Carpinus*, estimated to have been deposited in the last part of the Early-temperate and the first part of the Late-temperate Ipswichian substages (Ip IIb and III). At nearby Lane End, peat containing mainly Gramineae and Cyperaceae pollen and overlain by gravel (probably equivalent to that above the organic deposit on Bembridge Foreland) probably belongs to the Post-temperate substage of the Ipswichian (Preece and Scourse 1987; Preece *et al.* 1990). Pollen from sediments associated with a raised beach at La Cotte de St. Brelade, Jersey (section 8.9) include taxa consistent with a wooded Ipswichian environment (Leroi-Gourhan 1961; Jones 1986).

The most complete record of Ipswichian vegetational history has come from Wing in Leicestershire, about 15 km north of Corby (Hall 1980). Here a small, deep, closed basin in Jurassic strata has been found to be lined with chalky till of presumed Wolstonian age, over which have accumulated clays, peats and peaty silts. Palaeobotanical information suggests a continuous vegetational succession extending from the latter part of the Early-temperate Ipswichian substage, through its later temperate and Post-temperate substages and into the Early Devensian (Figure 8.2). A pollen assemblage dominated by *Carpinus*, replaced by one in which trees and shrubs are of reduced importance and herbaceous taxa occur in greater frequency, has been interpreted as marking the change from a Late-temperate to a Post-temperate environment.

At the Devensian stratotype locality, Four Ashes near Wolverhampton (A. V. Morgan 1973) (Chapter 9), basal organic material in a bedrock hollow (Figure 9.4) has yielded macrofossils of *Taxus*, *Ilex* and *Alnus* (Loader, Conolly and Wilson cited in A. Morgan 1973) and a pollen assemblage in which *Alnus* and *Quercus* dominate (Zagwijn cited in A. Morgan 1973). The Ipswichian age postulated has been confirmed on palynological criteria by Andrew and West (1977b), who have suggested reference to Zone IIb of this interglacial.

Data on the Ipswichian vegetational history of Lincolnshire have come from two sites at Tattershall in the lower part of the Bain Valley (Holyoak and Preece 1985). At Tattershall Castle, a depression in Wragby Till (Chapter 7) contains interglacial deposits which are overlain by gravels belonging to the succeeding cold stage, the Devensian. A pollen assemblage in which *Quercus*, *Alnus*, *Corylus*, *Ulmus*, *Taxus*, *Acer* and *Tilia* are present has been referred to the Early-temperate substage of this interglacial. At Tattershall Thorpe, a similar lithostratigraphy and biostratigraphy have been encountered but with *Taxus* and *Tilia* absent from the forest pollen components. However, these absences are probably a local phenomenon. From nearby sediments, Phillips (cited in Holyoak and Preece 1985) has obtained later Early-temperate Ipswichian pollen spectra containing *Carpinus* and Post-temperate ones with *Pinus*, *Betula* and Gramineae.

At Austerfield in southern Yorkshire, south-east of Doncaster and north-east of Bawtry, sand and gravel deposited by an ancestral River Idle contain a bed of organic silt (Gaunt *et al.* 1972). Palaeobotanical investigation has demonstrated a single biozone representing mixed-deciduous forest including *Acer* and *Carpinus*, probably of Late-temperate Ipswichian age. At Langham near Goole, finer grained sediment within the lower of two gravel units has produced pollen and plant macrofossils suggestive of an estuarine environment during the middle of the Ipswichian Interglacial (section 8.9). At Armthorpe near Doncaster, a correlative gravel includes sediments, which on yielding substantial amounts of *Pinus* and *Carpinus*, have been equated with either the Late-temperate, or the early part of the Post-temperate Ipswichian zone (Gaunt *et al.* 1974). West (1969a) has suggested that a mixed-oak forest pollen assemblage obtained from the Speeton Shell Bed in Filey Bay (Chapters 6 and 7) is analogous to that characteristic of Zone II (f) of the Ipswichian.

In the Peak District, travertine from Elder Bush Cave in the Manifold Valley contains leaf impressions of *Acer monspessulanum* (Montpelier maple) and *Corylus avellana*; its associated fauna (section 8.4) suggests an Ipswichian age (Bramwell 1964; Briggs and Burek 1985).

A peat raft discovered within Devensian till at Hutton Henry near Hartlepool, County Durham, has been examined by Beaumont *et al.* (1969). A number of pollen assemblages have been identified, with high *Carpinus* frequencies in one suggesting that they all could belong to the Ipswichian. At Scandal Beck near Ravenstonedale, Cumbria, analysis of an organic deposit found occupying a hollow in Carboniferous Limestone has produced a palynological record divisible into three zones and considered characteristic of an interglacial. The organic deposit is overlain by sand and gravel and till of Devensian age, and is probably Ipswichian (Carter *et al.* 1978). Letzer (1981) has identified a till, probably a Wolstonian deposit, below this organic material (Chapter 7).

In Scotland, the upper of two palaeosols at Kirkhill, Buchan, with a pollen record characterized by substantial *Alnus* frequencies together with *Betula* and coryloid (*Corylus* or *Myrica*) grains, has been equated with the Ipswichian (Connell *et al.* 1982; Connell and Romans 1984). A palynological spectrum in which Gramineae, *Alnus* and *Plantago* (plantain) predominate, and obtained from sands associated with palaeosol features lying below a till at Teindland, Morayshire, has also been ascribed an Ipswichian age (FitzPatrick 1965; Edwards *et al.* 1976).

From islands offshore of Scotland, Sutherland and Walker (1984) have reported peat (in excess of 47 000 BP on radiocarbon-dating evidence) located beneath head and raised-beach material at Toa Galson on Lewis. Its pollen record implies that the vegetation at this time changed from open grassland to ericaceous heath. Low tree-pollen frequencies (including *Betula*, *Pinus* and *Picea*) are probably the result of long-distance transport of these grains from wooded localities without the island. Peat found below till at Sel Ayre on Mainland Shetland (Mykura 1976) has been investigated by Birks and Peglar (1979). Palaeobotany has revealed a vegetation succession from open grassland with abundant ferns to dwarf-shrub (ericaceous) heath, then open grassland again. Arboreal pollen frequencies (5% of total pollen and spores) include *Picea*, *Abies* and *Carpinus* and are most likely due to long-distance transport. Comparisons with the palaeobotanical record from Fugla Ness on the same island

(Chapter 6), where *Carpinus* is not encountered, have lent support to an Ipswichian date for the Sel Ayre organic material.

The controversy regarding the age of the Gortian deposits in Ireland is discussed in Chapter 6. If these deposits are equated with those of Hoxnian age in other parts of the British Isles (Mitchell *et al.* 1973a), sediments yielding botanical data referable to the Ipswichian have been very rarely encountered and are of equivocal status (McCabe 1987). Indeed, this interglacial in Ireland has been described as the 'Hidden Interregnum' by Mitchell (1976). However, he has suggested that a freshwater mud located between two tills (below the lower of which a Gortian sequence has been recorded) at Baggotstown, County Limerick, and which contains a pollen spectrum reflecting temperate woodland that included alder, pine, oak, hazel, holly and yew, could belong to the Ipswichian. At Shortalstown, County Wexford, pollen of Ipswichian affinity has been obtained from estuarine sediments lying between two tills (Colhoun and Mitchell 1971; Mitchell 1976).

The piecemeal information above serves to emphasize both the temporal and spatial discontinuities which exist in the botanical record of Ipswichian environments. However, West (1980b) has considered that the plant biostratigraphic information available from sites in East Anglia and from Wing enables a plausible composite representation of the Ipswichian interglacial cycle to be established (Figure 8.3). Moreover, it has been stated that lithostratigraphic and biostratigraphic evidence from these localities is such that their allocation to more than one temperate stage can not be supported. West has summarized the vegetation succession as follows. Substage Ipswichian I opened (Ip Ia) with a decline in the herbaceous vegetation characteristic of late Wolstonian (l Wo) time and an increase in *Betula* woodland. During Ip Ib, *Pinus* occurrence increased and *Ulmus*, *Quercus* and *Acer* were present. Thus, during Ip I, birch woodland was replaced by birch–pine forest, with indications that mixed-oak forest trees had immigrated and begun to spread by its close. Substage Ip II witnessed the emergence of temperate forest. During Ip IIa, a peak of *Pinus* frequency occurred. This taxon thereafter declined as *Quercus* came to dominate and *Acer*, *Alnus*, *Corylus* and *Fraxinus* (ash) were present. Ip IIb was a *Quercus–Pinus* dominated part of the substage. *Corylus* increased in importance, and in different parts of the area, *Acer*, *Taxus* and *Tilia* became better represented. *Alnus* also exhibited variability in its pollen representation over that part of the British Isles considered. Substantial alder frequencies are a likely consequence of its presence in local carr and it does not seem to have been a significant component of regional forest. Increased frequencies of non-arboreal pollen in Ip II could reflect herbaceous vegetation that had been encouraged by the grazing and trampling activities of large mammals. An increase in *Carpinus* values is characteristic of Ip III, and it has been thought possible that the tree was of most significance in the forests at the time, with other warmth-demanding arboreal taxa of considerably reduced importance. *Picea* was now likely to have been a forest component. Although its pollen frequencies are low it produces relatively few grains, and Phillips (1974) has argued for its presence in the British Isles during the Ipswichian. Substage Ip IV commenced with *Pinus* succeeding *Carpinus* as the major aboreal component. An expansion of open-habitat plant communities that had begun during Ip III continued, with an increase in non-aboreal pollen frequencies at the conclusion of Ip IV signifying the opening of the Devensian Cold Stage.

On the basis of palaeobotanical data from Sel Ayre, Fugla Ness and Murraster (Johansen 1975) (Chapter 10), Birks and Peglar (1979) have suggested an interglacial vegetational cycle for northern oceanic localities, such as the Shetland Isles. Climatic characteristics, notably low summer temperatures and exposure to wind, accompanied by the effects of salt spray, have led to minimal tree growth. In the early part of the cycle, open arctic–alpine grassland seems to have been replaced by basic sward with tall herbs and ferns. Such vegetation persisted into the middle of the cycle and until the oceanic climate caused

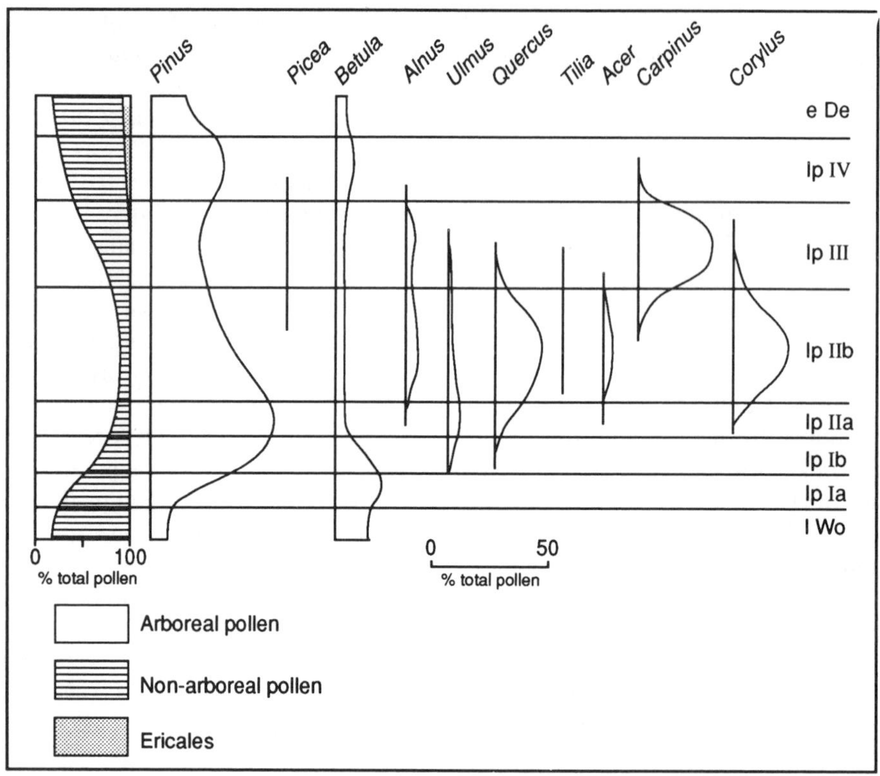

Figure 8.3 Pollen diagram of the Ipswichian. Palynological data for the late Wolstonian and Early Devensian substages are also shown. From West (1980b).

soil deterioration (mainly through podsolization) and the accumulation of mor humus and peat. This led to the onset of a retrogressive vegetational succession, with acid heath and mire communities, likely to have included *Bruckenthalia* and *Empetrum* during the Ipswichian, emerging later on. A reversion to open grassland with arctic–alpine plants completed the cycle.

8.2 Non-marine Mollusca

Freshwater and land snails have been recovered from numerous sites of Ipswichian age. At the Bobbitshole type locality (Sparks 1957, 1964), a mainly aquatic fauna and some terrestrial taxa are indicative of gently moving water flanked by marsh. An increase in the number of stenotopic species up-profile suggests that the climate was ameliorating. However, of especial interest is a re-expansion of tolerant taxa near to the top of the sequence which hints at a more harsh environment; all these trends being evident prior to the climatic optimum of the interglacial. Notable records are of *Belgrandia marginata* and *Vallonia pulchella* var. *enniensis* (Vallonia enniensis) which are extinct in the British Isles today and possess more southerly and continental distributions. The Histon Road sediments contain *Corbicula fluminalis* (Hollingworth *et al.* 1949). The reinvestigation of these deposits by Sparks and West (1959) has enabled a complex sequence of local environmental to be worked out from the molluscan succession. This complexity probably resulted from the migration of a meandering stream across its floodplain. Wretton (Sparks and West 1970) has yielded 85 species of non-marine Mollusca. The largely freshwater fauna suggests the existence of a slow-moving stream, with the terrestrial taxa dominated by those from adjacent marsh. Significant records here are of *Belgrandia marginata*, *Corbicula fluminalis* and *Clausilia pumila*, the latter also characteristic of present more continental climates than that of Norfolk. The Stutton brickearth (Sparks and West 1963) has also yielded over 80 species, seven of which are currently extinct in the British Isles. In addition to the taxa noted for Wretton, *Potomida littoralis*, *Bithynia troscheli* and *Azeca menkeana* (*Azeca goodalli*) have been recorded. This fauna possesses a southerly distributional aspect today and the behaviour of certain of its components (*Corbicula*, for example) suggests that a climatic deterioration was in progress through the sequence. The latter, probably belonging to the second part of the Ipswichian, was incorporated in aggradational sediments of the River Stour. At Wortwell (Sparks and West 1968) *Planorbis laevis* and *Planorbis crista* are dominant in deposits of Ip Ia age, and the molluscan fauna indicates the existence of open conditions in the vicinity at this time. Later Ipswichian deposits here exhibit *Bithynia tentaculata* and *Valvata piscinalis*, consonant with an overall picture of a riverside-broad environment. The organic deposits at Swanton Morley also contain species characteristic of Ipswichian molluscan assemblages (Sparks 1976). *Belgrandia* and *Corbicula* had been recorded from the upper part of the organic sediments and lower part of the brickearth at Ilford prior to the work of West *et al.* (1964). In their investigation, the fauna, dominated by *Bithynia* and *Valvata cristata*, indicated the existence of either a small meandering stream or a fair-sized pond. A list of molluscs from Aveley (Cooper 1972) has been supplemented by Holyoak (1983a). The assemblage, typical of those of Ipswichian type, is particularly characterized by *Belgrandia*, *Anisus vorticulus* and *Gyraulus laevis*. The silt at Little Thurrock, Grays (West 1969b) has produced *Corbicula*, which also occurs as a component of temperate molluscan faunas in brickearths at Erith and Crayford in the Lower Thames Valley (Kennard 1944; West *et al.* 1964). At Northfleet, freshwater interglacial sediments observed intercalated with periglacial deposits (Zeuner 1959) have produced terrestrial molluscs indicative of open habitats, probably referable to Ipswichian Substage IV (Kerney and Sieveking 1977). Mollusca from the Ipswichian IIb deposit at Thatcham (Holyoak 1983a) include *Belgrandia*, *Gyraulus laevis*, *Vallonia enniensis* and *Discus ruderatus*; the two last-named currently having central and southern, and boreal and alpine ranges in Europe respectively (Kerney and Cameron 1979). In the Upper Thames Valley, the Eynsham Gravel in Oxford numbers *Corbicula fluminalis* among its fauna (Thew 1985).

Ipswichian freshwater deposits at Selsey have produced *Bithynia*, *Corbicula* and *Valvata piscinalis* (West and Sparks 1960), while cave sediments of this age at Bacon Hole, Gower, contained a non-marine molluscan assemblage dominated by *Clausilia pumila* and including *Discus rotundatus* (Stringer *et al.* 1986). The Ipswichian Zone II organic channel-fill at Galley Hill (Preece and Ventris 1983) contains a terrestrial fauna mostly associated with calcareous grassland. Of especial interest is the occurrence of the extinct *Candidula crayfordensis*, also recorded from Ipswichian contexts at Ilford, Erith and Crayford. The land snails from Maxey (French 1982) reflect the existence of dry, open grassland in this area of Cambridgeshire during the first part of Ipswichian time. *Corbicula* is also a component of faunas from the March Gravels of Cambridgeshire (Baden-Powell 1934; West 1987; Keen *et al.* 1990). The Ipswichian localities at Tattershall have produced characteristic molluscan faunas (Holyoak and Preece 1985). They include *Bradybaena fruticum*, a land snail currently extinct in the British Isles but widely distributed in central and eastern Europe, and recorded solely from Ipswichian contexts in the British Isles. The terrestrial fauna of Ip IIb deposits at Tattershall Castle includes the closed-forest inhabitant *Spermodea lamellata*. Silt below these deposits contains an appreciable number of taxa (including *Pupilla muscorum* and *Vallonia excentrica*) diagnostic of dry, open terrain, which probably existed in earlier Ipswichian time.

In the western Midlands of England, sediments examined below Terrace 3 of the River Avon at

Eckington near Evesham, Worcestershire (Keen and Bridgland 1986) and at nearby Cropthorne (Bridgland *et al.* 1989; Maddy *et al.* 1991) have produced broadly similar faunas including *Belgrandia marginata*, *Bithynia tentaculata* and *Valvata piscinalis*, thus supporting the assignation of an Ipswichian age to this terrace (Tomlinson 1925; Shotton 1985) (Figure 7.6). In the Kenn area near Clevedon, Somerset, freshwater sediments found associated with deposits indicative of a marine transgression (section 8.9) have *Belgrandia*, *Corbicula* and *Vallonia enniensis* among their fauna (Gilbertson and Hawkins 1978).

8.3 Marine Mollusca

At Selsey and Stone (West and Sparks 1960; Brown *et al.* 1975), the incidence of *Hydrobia* spp., *Scrobicularia plana* and *Phytia myosotis* points to the existence of brackish inter-tidal conditions in an estuarine environment. The Ipswichian IIb deposits at Wretton contain *Hydrobia* (Sparks and West 1970), as do those belonging to this part of the same substage at Tattershall Castle, where *Pseudamnicola confusa*, also tolerant of brackish water, has been recorded (Holyoak and Preece 1985). In the latter locality, deposition of sediment just above the main brackish influence of spring tides has been postulated. *Pseudamnicola confusa* has also been identified in Ip III deposits at Stutton (Sparks and West 1963) and from Aveley (Holyoak 1983a).

In Somerset, the Burtle Beds (Chapter 7) include *Macoma balthica*, *Cardium* (*Cerastoderma*) *edule*, *Littorina* and *Trochus* (Bulleid and Jackson 1937, 1941; Gilbertson and Beck 1974). Deposits with an interglacial inter-tidal fauna have been located to their east, at Kenn in the Clevedon area (Gilbertson and Hawkins 1978; Andrews *et al.* 1984). The overall composition of these faunas is indicative of estuarine conditions. The March Gravels in the Eye, Wimblington and Manea areas of Cambridgeshire have yielded *Cerastoderma* and other molluscan indicators of a marine environment (Baden-Powell 1934; West 1987; Keen *et al.* 1990).

Raised beaches of Ipswichian age (section 8.9) have provided molluscan evidence of marine conditions. While there are inter-site variations in these faunas consequent upon local coastal environments, they usually possess a number of taxa in common. *Littorina* spp. are frequent, while *Macoma balthica*, *Patella vulgata*, *Mytilus edulis*, *Ostrea edulis* and *Cerastoderma* are often found. The Portland East Raised Beach has yielded 62 taxa, overall indicative of a rocky nearshore environment close to low-water mark, in a sea whose surface temperatures were at least as warm as those of today (Headon Davies and Keen 1985). The beaches at Thatcher Rock and Shoalstone in Torbay have produced mainly littoral molluscs which now live in the English Channel, implying a marine environment similar to that of today (Mottershead *et al.* 1987). The Belle Hougue Raised Beach in Jersey (Keen *et al.* 1981) has produced *Astralium rugosum*. The northern limit of this species is currently some 300 km south of the island, and its occurrence indicates a sea-surface temperature 3–4°C higher than prevails at present. Other temperate faunas from raised beaches of Ipswichian age have been recorded at Minchin Hole, Gower (George 1932; Sutcliffe and Currant 1984) (Figure 7.5), Swallow Cliff, Middlehope, Somerset (Gilbertson and Hawkins 1977), Sewerby, Yorkshire (Lamplugh 1891; Catt 1987a,c) and Easington, County Durham (Woolacott 1920, 1922).

8.4 Vertebrates

As noted in Chapter 7, a fundamental tenet of the argument against an Ipswichian (*sensu* Mitchell *et al.* 1973a) age for all deposits assigned to this stage has been variation in vertebrate (especially mammalian) faunal assemblages (Sutcliffe 1975, 1976). However, Stuart (1976), while confirming and accepting certain of the differences pointed out by Sutcliffe, has refuted the notion of separate stages for some of the assemblages. Indeed, Stuart has postulated that sufficient fossils have been recovered from sites at which parts of the Ipswichian interglacial cycle have been identified using palaeobotanical criteria, to allow a detailed reconstruction of mammalian faunal history during the stage. There has been a dearth of information concerning vertebrates during Substage Ip I, while Ip III and Ip IV sediments (frequently brickearths) have often yielded sparse pollen, which has militated against close correlation of the two lines of evidence. Information has been better on large mammals than small ones, with *Hippopotamus amphibius* regarded as the animal diagnostic of an Ipswichian age. The records which follow are derived from Stuart (1976, 1982), unless otherwise stated.

From the l Wo–Ip Ia deposits at Selsey (West and Sparks 1960), remains of *Equus ferus* have been recorded, while in those of Ip Ib–early IIb age, *Palaeoloxodon antiquus*, *Dicerorhinus hemitoechus* (narrow-nosed rhinoceros) and *Castor fiber* are among mammal fossils, and vestiges of the reptile *Emys orbicularis* occur. At Bobbitshole in l Wo–Ip IIb sediments (West 1957), *Castor fiber* is part of a fauna that contains numerous voles, including the extinct *Arvicola cantiana*. *Emys orbicularis* is present here too, and in Ip Ib–early IIb deposits at Mundesley (Phillips 1976), where *Palaeoloxodon* is found. At Swanton Morley, Ip IIa sediments (Coxon *et al.* 1980) have yielded shrew and vole fossils (including *Arvicola cantiana*) plus remains of *Crocuta crocuta*, *Dama dama*, *Cervus elaphus* (red deer) and *Bos primigenius*. The European pond tortoise is also represented. At Beetley (Phillips 1976) and Galley Hill (Forbes and Cambridge 1967; Preece and Ventris 1983) Ip IIa–IIb deposits have yielded *Palaeoloxodon* and *Megaceros*. Substage Ip IIb sediments at Beetley, and at Barrington (Gibbard and Stuart 1975), Trafalgar Square (Franks *et al.* 1958; Franks 1960), Stone (West and Sparks 1960; Brown *et al.* 1975), Wortwell (Sparks and West 1968; Coxon 1982), Swanton Morley (Phillips 1976; Coxon *et al.* 1980) and Aveley (West 1969b) possess elements of a middle interglacial, large-mammal fauna which includes *Hippopotamus amphibius*, *Palaeoloxodon antiquus*, *Megaceros giganteus*, *Dama dama*, *Cervus elaphus* and *Bos primigenius*. *Hippopotamus* is also associated with early Ip III deposits at Swanton Morley, while those of a smiliar age at Aveley have yielded remains of *Mammuthus primigenius* and *Equus ferus*. The Ip III and perhaps Ip IV sediments at Stutton (Sparks and West 1963), those of probable Ip III date at nearby Harkstead (Stuart 1982) and analogous ones at Ilford (West *et al.* 1964) lack hippopotamus, but are otherwise similar in large-mammal content to Aveley and Swanton Morley. The possible Ip IV faunas from Lexden near Colchester, Essex (Shotton *et al.* 1962), Stoke Tunnel (Layard 1912, 1920; Turner 1977; Wymer 1985) and Crayford (Kennard 1944) are analogous to those of Ip III, with *Lemmus lemmus* and *Dicrostonyx torquatus* (arctic lemming) components of the fauna at the latter locality.

There have been numerous sites with faunal remains of Ipswichian type but without corroborative palaeobotanical evidence. The hippopotamus, straight-tusked elephant, narrow-nosed rhinoceros, spotted hyaena, red deer, fallow deer and brown bear contributing to a rich fauna from talus deposits at Joint Mitnor Cave, Buckfastleigh in Devon (Sutcliffe 1960) were characteristic members of animal communities that existed during the warmest part of Ipswichian time (Sutcliffe 1974b). The Hyaena Stratum at Tornewton Cave nearby (Sutcliffe and Zeuner 1962; Sutcliffe 1974a) is of like age. In Minchin Hole, Gower, the mammalian fauna from deposits immediately above the *Patella* Beach lacks hippopotamus but is otherwise typically Ipswichian (Sutcliffe and Currant 1984; Sutcliffe *et al.* 1987). At Bacon Hole (Stringer *et al.* 1986) in a comparable stratigraphic context (Figure 9.6), an analogous fauna has been discovered. Of particular interest is a record of *Calonectris diomeda* (Cory's shearwater) which currently nests on the coastline of southern Europe (Harrison 1977). Kirkdale Cave north of Kirbymoorside, North Yorkshire, first investigated by the Reverend W. (Dean) Buckland (1822) then reported by Boylan (1972, 1977a), has produced a typical Ipswichian climatic-optimum fauna, as has the Lower Cave Earth from Victoria Cave, Settle (Boylan 1977b), and one associated with travertine at Elder Bush Cave, Staffordshire (Bramwell 1964; Briggs and

Burek 1985). Interglacial deposits at Sewerby have yielded a comparable faunal assemblage (Boylan 1967; Catt 1987c), while the Upper Channel at Marsworth, Buckinghamshire (Green *et al.* 1984) (Figure 7.8), whose animal remains include *Hippopotamus amphibius* and *Dicerorhinus hemitoechus*, is also of Ipswichian age. At East Mersea in north-east Essex, two localities contain organic deposits. *Hippopotamus* has been collected from both and straight-tusked elephant and narrow-nosed rhinoceros from one of them. An analogous situation to that at Trafalgar Square, further up the Thames Valley, has been suggested by Bridgland *et al.* (1988). The Upper Thames Valley has produced both *Hippopotamus* and *Palaeoloxodon* from the Eynsham Gravel around Oxford (Sandford 1924; Currant 1985b). Terrace 3 of the Warwickshire–Worcestershire Avon has yielded hippopotamus from a number of localities (Strickland 1835; Whitehead 1989). At Alvaston near Derby, a typically Ipswichian faunal assemblage has been extracted from the Allenton Terrace of the River Derwent (Jones and Stanley 1974). A similar fauna has been reported from deposits marking a former course of the River Witham at Fulbeck in Lincolnshire and ascribed a probable Ip IIb/III age (Brandon and Sumbler 1988). At Brundon in the Stour Valley, Suffolk, the faunal assemblage (Moir and Hopwood 1939; Spencer 1970) is comparable with those of the later part of the Ipswichian elsewhere. However, Wymer (1985) has equated the Brundon fauna with that of the Stoke Tunnel Bone-bed in Ipswich, and has referred it to earlier Late Pleistocene time on archaeological and geochemical dating evidence (Chapter 7).

Zeuner (1946) has reported remains of a dwarfed form of *Cervus elaphus* from the 8 m raised beach at Belle Hougue Cave in Jersey (section 8.9). Lister (1989b) has confirmed this and pointed to the presence of full-sized ancestors on the island. It has been estimated that a six-fold reduction in body weight of the red deer occurred over about 6000 years during the Ipswichian Interglacial when Jersey was isolated from mainland France. The re-emergence of a land connection with France at the end of the Ipswichian is thought to have led to the demise of the dwarf form, with predation, competition and hybridization likely causes.

Stuart (1976, 1982) has provided information on Ipswichian faunal history and ecology. In Ip Ia, the presence of *Equus ferus* is consistent with the grassland vegetation inferred from palaeobotanical data. The latter have shown how between Ip Ib and Ip IIa, boreal forest was replaced by one of mixed-deciduous type (section 8.1). *Castor fiber*, *Palaeoloxodon antiquus* and *Dicerorhinus hemitoechus* were members of the fauna at this time. The mixed-oak forest of Ip IIb possessed an assemblage of animals that inhabit this sort of environment today; *Clethrionomys glareolus* (bank vole), *Apodemus sylvaticus*, *Meles meles* (badger) and *Dama dama* providing examples. In open vegetation such as existed in river valleys, *Panther leo*, *Crocuta crocuta*, *Dicerorhinus hemitoechus* and *Hippopotamus amphibius* would have lived, the latter in proximity to water. The less-closed, *Carpinus*-dominated forest of Ip III carried some of the animals of Ip IIb which required open areas in order to move and feed. *Mammuthus primigenius* and *Equus ferus* were present, while *Palaeoloxodon* continued in the fauna. During Ip IV, the vegetation was largely herbaceous but some boreal forest existed. *Dicerorhinus*, *Mammuthus*, *Equus* and *Panthera* were resident. Animals today characteristic of tundra and steppe communities (*Dicrostonyx torquatus*, *Lemmus lemmus*, *Spermophilus* (ground squirrel) and *Ovibos moschatus* (musk ox), for example) now appeared.

While general climatic inferences have been made from the overall faunal compositions, some members are thought to indicate particular parameters in this regard. For example, *Hippopotamus amphibius* is today restricted to Africa. During the Pleistocene it occurred in western Europe, supporting the notion gained from other fossil evidence of mild winters and warm summers in this region. The possible climatic significance of *Emys orbicularis* is noted in Chapter 1. This reptile, resident in the British Isles by early in the Ipswichian, is currently unable to survive beyond the 18°C July isotherm, needs higher temperatures in order to breed, and thus is distributed in central, southern and eastern Europe (Stuart 1976, 1982; Arnold and Burton 1978).

The proposals of Currant (1989) regarding the grouping of Pleistocene mammalian faunas of the British Isles are noted in earlier chapters. The Group 1 Assemblage of this scheme (Table 4.1) comprises the typical *Hippopotamus* fauna of Ipswichian age. The presence of *Castor fiber* at Bobbitshole has led to the suggestion that this fauna may belong to Group 2.

8.5 Insects

Ipswichian insect faunas have been investigated at a limited number of sites. However, Coope (1974a) has reported on Coleoptera from organic silt (equivalent to Substage Ip II) at Bobbitshole. Twenty-one species and a number of genera suggest the existence of a local environment consisting of a shallow, nutrient-rich lake fringed by marsh. Of particular interest are records of dung beetles from drier, more distant habitats where large mammals must have been present. Six species found, *Oodes gracilis*, *Cybister laterali-marginalis*, *Airaphilus elongatus*, *Caccobius schreberi*, *Onthophagus opacicollis* and *Valgus hemipterus*, are now extinct in the British Isles but are widely distributed in central and southern Europe, with the northern limit of many in southern Scandinavia. *Onthophagus opacicollis*, a dung beetle, now has a Mediterranean range. The overall implication of the fauna is that July temperatures averaged *circa* 20°C (about 3°C higher than today in this area) (Figure 9.3), with winter values probably little different from those of today. This fauna corresponds closely with a substantial one from Trafalgar Square (Franks *et al.* 1958) which also includes numerous species currently resident in central and southern Europe and diagnostic of a warmer climate (Coope 1977a). At Austerfield, organic silt of probable Ip III age has produced an insect fauna that includes *Bembidion elongatum*, *Bothrideres contractus*, *Brachytemnus submuricatus* and *Scolytus carpini*. All are now extinct in the British Isles and possess ranges in southern and central Europe. *Scolytus carpini* exists largely on *Carpinus*, the presence of which is indicated in the palaeobotanical record. The fauna is different from those at Bobbitshole and Trafalgar Square, but this has been considered a function of its later occurrence during the interglacial (Gaunt *et al.* 1972). Insect fossils have been obtained from sediment of Ip IIb age at Tattershall (Girling 1974, 1977). A large beetle fauna is indicative of well-developed, primarily deciduous forest and of marsh and stream habitats and is also suggestive of higher summer temperatures than those at present characteristic of Lincolnshire. Tree-dependent species (for example, *Rhynchaenus quercus* and *Dryophthorus corticalis*) were abundant. *Isorhipis melasoides* and *Pycnomerus terebrans*, currently extinct in the British Isles, inhabit mature deciduous forests within Europe (Coope 1979). It seems relevant to restate here that Shotton *et al.* (1983) have pointed to a dissimilarity between the Ip II insect assemblages from Bobbitshole and Trafalgar Square and those of assumed similar age at Aveley (Chapter 7).

A sparse insect assemblage occurs in the Ip IIb deposit at Four Ashes, Staffordshire, but is not diagnostic of the climatic environment (A. Morgan 1973).

8.6 Other fossils

At Tattershall Castle (Holyoak and Preece 1985), an ostracod fauna provides evidence of a slow-flowing stream with a muddy substrate. Of particular interest in the fauna are *Cyprideis torosa*, *Heterocypris salina* and *Candona angulata* which denote a brackish-water influence. This phenomenon is also evidenced by certain Mollusca (section 8.3), and by the foraminifera *Haynesinia germanica* and *Elphidium articulatum*. Ostracods and foraminifera indicative of brackish conditions occur in the March Gravels (Keen *et al.* 1990). Foraminifera identified from the Portland East Raised Beach (Haynes cited in Headon Davies and Keen 1985) suggest sea temperatures similar to the present English Channel.

At Langham and Armthorpe (Gaunt *et al.* 1974), deposits, of probable Ipswichian IIb and

III/IV ages, respectively, have yielded dinoflagellate cysts dominated by *Spiniferites* spp. At Langham, *Lingulodinium machaerophorum*, tolerant of weakly saline conditions, has been recorded. The overall inference from the assemblages is of an estuarine environment in a cool-temperate climate, the latter not dissimilar to that of the seas around the British Isles today.

8.7 The presence of man

Evidence of human occupation in the British Isles during the Ipswichian is sparse. Wymer (1985, 1988) has reserved an Ipswichian context for sites with faunas including hippopotamus, straight-tusked elephant, red deer and fallow deer, but excluding horse and mammoth (Chapter 7). Levallois flakes and Mousterian-type hand axes have been recorded from the Selsey Raised Beach (Wymer 1988). Industries known as Mousterian of Acheulian Tradition (Figure 6.4) occur in marine and estuarine deposits of assumed similar ages to Selsey, at Christchurch near Bournemouth (Calkin and Green 1949) and at Great Pan Farm in the Isle of Wight (Shackley 1973) (section 9.1.6). The exact provenance of a hand axe from Victoria Cave (Roe 1968) in relation to dated Ipswichian deposits there (section 8.10) is unknown. A Mousterian of Acheulian Tradition Industry from the Loamy Cave Earth at Kent's Cavern has an associated pollen spectrum depicting mainly open vegetation, thought by Campbell and Sampson (1971) and Campbell (1977) to be of Devensian age. However, Wymer (1981), while noting an insecure relationship between the palynological and artifactual evidence, has postulated a possible late Ipswichian context. Crayford has Levalloisian flakes and hand axes (Chandler 1914) which have been ascribed to either early (Wymer 1977) or late (Stuart 1982) Ipswichian time. The East Mersea locality has produced one flake, as has Barrington; that from the latter locality in a rolled condition and possibly extraneous (Wymer 1988). The age of artifacts of Levallois type found at Brundon, Maidenhall, Stoke Tunnel and Stutton, Suffolk, is controversial. Wymer (1985, 1988) has preferred a post-Hoxnian/pre-Ipswichian date, while Stuart (1982) has allocated them to the Ipswichian.

Wymer (1988) has emphasized the fact that palaeoliths from terrace deposits could have been derived. The Eynsham Gravel of the Summertown–Radley Terrace of the River Thames has yielded both *in situ* and derived hand axes of Middle/Late Acheulian type (Roe 1976; MacRae 1985). Terrace 3 of the Warwickshire–Worcestershire Avon has also produced palaeoliths whose provenance has been questioned by Wymer (1988). Overall, however, the quantity of such material is small. As Wymer has observed, unless a substantial number of the Palaeolithic artifacts contained in terrace sediments of Devensian age were originally deposited on Ipswichian land surfaces, then reworked, a convincing demonstration of the occupation of the British Isles during the latter stage is not possible.

There has been more evidence of human occupation during the early and late parts of the Ipswichian than in its middle sector. During these times, Palaeolithic man appears to have favoured the lower portions of substantial river valleys in lowland England. Here habitation took place in an environment with slow-moving water, often adjoining which was marsh, with drier, herb-rich floodplain grassland further away. Such a mosaic of habitats would have been conducive to a hunter–gatherer culture. Considerable quantities of mammalian bones have been found in such areas but very few in archaeological contexts (Wymer 1981).

8.8 Pedogenesis

Red mottling observed in soils on Harwood Dale Moor near Scarborough, in an area unaffected by Devensian glaciation, has been thought to be at least as old as the last interglacial by Bullock *et al.* (1973). A reddened, weathered soil horizon located beneath solifluction deposits formed during the last cold stage, at Sherburn-in-Elmet near Leeds (Smith 1977), has been cautiously referred to the Ipswichian by Catt (1977a). In South Wales, terra rossa and terra fusca palaeosol features may be of Ipswichian age (Ball 1960; Bowen 1970). Deeply weathered soils containing red-mottled palaeoargillic B horizons have been reported from the Needwood Forest area of Staffordshire (Jones 1983) and the Coventry and Kenilworth districts of the West Midlands and Warwickshire (Beard 1984). These have been suggested by Avery (1985) to have originated during the Ipswichian, as probably did the reddening in deeply weathered soils over Anglian till in Essex, identified by Sturdy *et al.* (1979). In the latter case a Hoxnian age is possible for the rubification, but Avery has drawn attention to indications of warmer conditions during the Ipswichian which would have favoured such a process. Micromorphological investigations of red-mottled palaeoargillic horizons in the Northern Drift of the Cotswolds (Bullock and Murphy 1979) and the Chiltern Drift (Avery *et al.* 1982) have revealed at least two phases of reddening, the last probably during the Ipswichian. The Troutbeck Palaeosol in Cumbria possesses features denoting gleying and clay translocation (Boardman *et al.* 1981; Boardman 1985). This deeply weathered soil developed in pre-Devensian glacigenic sediments, and formed over long periods with humid temperate conditions, one of which was almost certainly the Ipswichian Interglacial. At Northfleet in Kent, a reddened soil recognized in Wolstonian loess is likely to be of Ipswichian age (Catt 1986). Pedogenetic features consistent with a temperate environment have been identified in sediments associated with a raised beach of Ipswichian date in La Cotte de St. Brelade, Jersey (van Vliet-Lanoë 1986). The Teindland (FitzPatrick 1965; Edwards *et al.* 1976) and Kirkhill (Upper) (Connell *et al.* 1982) palaeosols in Scotland are also thought to be Ipswichian.

Brown *et al.* (1975) and Green and Keen (1987) have suggested a Devensian age for gravel and brickearth (loess) above the Ipswichian deposits at Stone. However, Reynolds (1987) has reported that a Lower Brickearth above the gravel at Lepe Cliff in this area exhibits palaeoargillic features consistent with a pre-Devensian stage, probably the Ipswichian Interglacial. If this interpretation is correct, the interglacial deposits earlier in the sequence belong to a pre-Ipswichian temperate episode.

8.9 Rivers and sea-levels

As noted above, the majority of sites with lithostratigraphic and biostratigraphic evidence of Ipswichian environments are in river valleys. In the middle and lower sectors of the Thames Valley, a thin post-Taplow gravel was laid down at the conclusion of the Wolstonian. During the Ipswichian, both in the Thames and its tributary valleys in this area, the deposition of sands, silts, clays and organic material took place (Gibbard 1985, 1988b; Bridgland 1988). Data from Ilford (West *et al.* 1964) and Aveley (West 1969b) suggest that the second part of the Ipswichian was characterized by significant aggradation, with alluvium spread up to 15 m OD in the Lower Thames Valley. The upper part of the Summertown – Radley Terrace of the Upper Thames Valley (Figure 7.7) accumulated during the Ipswichian (Briggs *et al.* 1985).

Histon Road (Sparks and West 1959), Stutton (Sparks and West 1963), Wretton (Sparks and West 1970) and Swanton Morley (Phillips 1976) have yielded evidence of Ipswichian valley alluviation in East Anglia. In the English Midlands, Terrace 3 of the Warwickshire – Worcestershire Avon (Tomlinson 1925; Shotton 1986; Maddy *et al.* 1991) (Figure 7.6), the Kidderminster Terrace of the River Severn (Wills 1938; Stephens 1970a), the Beeston Terrace of the River Trent (Clayton 1953, 1957b, 1977, 1979), the Allenton Terrace of the River Derwent (Jones and Stanley 1974) and the Fulbeck Terrace of the River Witham (Brandon and Sumbler 1988) have been ascribed to the Ipswichian.

Some of the interglacial sediments at Bobbitshole are below present sea-level but lack clear lithostratigraphic and biostratigraphic indications of marine influence (West 1957; Sparks 1957). It appears that in this locality, the backing-up of fresh water and flooding by it occurred as sea-level rose during the first part of the Ipswichian. As a result, wetland developed to *circa* 1 m OD.

Such an elevation is problematical, however, as sea-level during the Ipswichian is thought to have risen higher than this. West (1957) has invoked subsequent subsidence of the area to explain this apparent anomaly. Downwarping has also been advocated by Sparks and West (1963) to account for the height of Ipswichian deposits at nearby Stutton, which also lack clear signs of marine influence. Subsidence of the southern North Sea Basin is noted in Chapter 1. In Germany and the Netherlands, the upper limit of marine deposits belonging to the same interglacial as the mainly freshwater ones at Bobbitshole and Stutton varies from about −4 to −10 m as a result of downwarping (Woldstedt 1952; Zagwijn 1961; Oele and Schüttenhelm 1979). Attention has also been drawn by Sparks and West (1963) to the different ages of the Ipswichian deposits at Bobbitshole and Stutton in broadly similar height contexts. Rejecting differential warping over such a short distance to account for this phenomenon, they have suggested that base-level rise and aggradation were interrupted between substages Ip IIb and Ip III. A temporary fall in base level took place and erosion was heightened, before base level rose once more and riverine alluviation recommenced during Ip III. Possible causes of such changes could involve a sequence of Ipswichian environments analogous to those proposed for Denmark and north-west Germany during the Eemian by Jessen and Milthers (1928). They have gathered palaeobotanical data which they have interpreted as depicting lower and upper temperate phases during this interglacial, between which was an episode with sub-arctic conditions. It should be noted, however, that Andersen (1957) has refuted such an interpretation, arguing that the upper temperate phase was not genuine because the deposits containing the floristic information have been reworked from those of similar type lower in the sequence.

It has also been postulated that there may have been two episodes of high sea-level during the Ipswichian, the Main Monastirian and Late Monastirian stands, to *circa* 18 m and 8 m above present datum, respectively (Woldstedt 1952; Zeuner 1959). An examination of Ipswichian sea-levels, mainly on the basis of localities within the Lower Thames Valley, by Hollin (1977), has provided a further insight into such stands of the sea. It has been concluded that sites at Crayford, Little Thurrock and Purfleet indicate aggradation of tidal sand, silt and clay to in excess of 11 m OD early in the interglacial. This could be related to a sea-level of about 7 m OD, as seen in a number of raised beaches. In the middle of the Ipswichian, sea-level was reduced to below present datum (0 m), with freshwater fossils in a West Thurrock sequence supporting this change. Sites at Ilford, Aveley, Little Thurrock, West Thurrock and Crayford have demonstrated significant aggradation of brickearth to around 14 m OD during the late Ipswichian. This aggradation may have been related to a second and greater rise in sea-level to *circa* 16 m OD. The second event could have been the result of an ice surge in the Antarctic. Wilson (1964) has suggested that such surges were responsible for the onset of Pleistocene ice ages. Zeuner (1959) has proposed that the Main Monastirian sea-level (+18 m) was the earlier one. However, Hollin (1977) has contended that the first rise, up to 7.5 m OD, took place in a temperate environment, with the second, higher one occurring in cool conditions that were generated by the surge.

At Stutton, there is an indication of brackish conditions at about 1 m OD from a component of an Ip III molluscan fauna (section 8.3). These conditions have been thought by Sparks and West (1963) to delimit the highest sea-level of the interglacial. Plant and molluscan evidence has identified a brackish horizon from −1.95 to −1.20 m OD at Wretton. This is of Substage Ip IIb age but signs of subsequent marine conditions are lacking (Sparks and West 1970). At Block Fen, Chatteris, Cambridgeshire, brackish or marine horizons occur in Ip III (−6.1 to −5.5 m OD) and Ip II (−3.7 m OD) contexts, and at Somersham nearby, are referable to Ip II (−2.7 to −0.4 m OD) (West 1987). These localities are close to the area covered by the March Gravels, marine deposits of possible Ipswichian age, that have been located up to *circa* 3 m OD (Baden-Powell 1934; West 1987; Keen *et al.* 1990). At Tattershall Castle, the Ip IIb deposits from −1.8 to −0.2 m OD accumulated under brackish conditions (Holyoak and Preece 1985). Sites in the environs of Goole and Doncaster (Gaunt *et al.* 1974) reflect a substantial rise in sea-level during the early and middle parts of the Ipswichian. Estuarine deposits of this age occur from −12 to −6 m OD at Langham. The depth of these deposits also points to considerable incision early in the interglacial, when sea-level, and hence base level, were low.

At Selsey, brackish influence in a sequence of sediments associated with the Ipswichian marine transgression and regression is first detected at −1.76 m OD, early in Ip IIb (West and Sparks 1960). Here the brackish conditions were followed by marine ones and a raised beach was deposited. At Stone, estuarine conditions prior to the Ip IIb sea-level maximum have been recorded by West and Sparks (1960). These authors have suggested that a gravel located higher in the Stone sequence is of raised-beach type, but this has been refuted by Brown *et al.* (1975). The raised beach at Selsey has been correlated on altimetric grounds with one at Black Rock, Brighton (*circa* 7.5 m OD) and at other localities along the Sussex coast (West and Sparks 1960; Shephard-Thorn and Wymer 1977). A raised beach at Bembridge Foreland on the Isle of Wight has an Ipswichian context (Preece and Scourse 1987; Preece *et al.* 1990) (section 8.11). Mottershead (1977) has pointed out that raised-beach deposits along the south coast of England which occur at similar elevations have traditionally been referred to the same interglacial. This has been the case at Portland (Baden-Powell 1930; Arkell 1947) but Headon Davies and Keen (1985) have concluded that only the one at Portland East (its base from 6.95 to 10.75 m OD) is of Ipswichian age. The shore platforms and raised beaches of Devon and Cornwall have all been ascribed to the Ipswichian by Zeuner (1959) and Kidson (1971, 1977). However, Mitchell (1960), Stephens (1966), and Mitchell and Orme (1967) have suggested that beaches of both Hoxnian and Ipswichian age are present, with those representing the latter stage the least well developed. Mottershead *et al.* (1987) have demonstrated that of the marine features at 8–12 m OD in Tor Bay, the raised beaches at Thatcher Rock and Shoalstone belong to the Ipswichian. Uranium-series ages and amino-acid data from Belle Hougue Cave, Jersey (Keen *et al.* 1981) have enabled the 8 m raised beach of the Channel Islands (Keen 1978) to be referred to the Ipswichian (section 8.11). Andrews *et al.* (1984) have discussed Ipswichian marine deposits around the Severn Estuary, concluding that they occur from *circa* −4 to 12 m OD. A raised beach at Swallow Cliff near Weston-super-Mare and a buried one at Llanwern close to Newport have been included in this category on the basis of amino-acid racemization evidence. The Outer (*Patella*) Beach and associated beds at Minchin Hole (Figure 7.5) and other deposits with similar amino-acid ratios (Headon Davies 1983) around the Gower Peninsula are of Ipswichian age according to geochronometric dates (section 8.11).

The marine beach at *circa* 2 m OD at Sewerby in eastern Yorkshire is of Ipswichian age on faunal evidence (Catt and Penny 1966; Catt 1987a,c). An associated buried cliff has been traced along the western side of Holderness and the Lincolnshire Marsh (Catt 1977a, 1987a) (Figure 9.17). At Easington in County Durham, a raised beach at *circa* 18 m OD (Woolacott 1920, 1922) possesses a temperate molluscan fauna of Ipswichian affinity. Its height is difficult to reconcile with that of the Sewerby beach. However, its current elevation may be a response to post-Ipswichian tectonic movement (Shotton 1981). Alternatively, the beach may represent one of two high sea-levels during the Ipswichian (see above). Around Morecambe Bay, wave-cut notches observed in cliffs of Carboniferous Limestone, from *circa* 4.9 to 6.4 m OD, might demonstrate Ipswichian marine activity (Tooley 1977). Certain elevated rock-platforms and shorelines found in the Inner Hebrides (McCann 1968) may reflect Ipswichian sea-levels. However, these features, with heights of up to 40 m OD, have been thought likely to be composite and to have been subjected to isostatic uplift, by Jardine (1977).

In the Isle of Man a fossil cliff identified at Ballure, buried by a complex of Devensian deposits, has been regarded as Ipswichian by Thomas (1977), and Dackombe and Thomas (1985). The controversy concerning the Gortian deposits and Courtmacsherry Raised Beach of Ireland is explored in Chapter 6. Warren (1985) has considered them both of last interglacial age, while Synge (1985) has equated the beach at 4–6 m above sea-level around the coast of Ireland with the Ipswichian.

8.10 The offshore sequence

Garrard and Dobson (1974) and Garrard (1977) have reported the existence of marine sands and gravels containing layers of clay and silt, beneath Devensian and above Wolstonian till of Irish Sea type in St. George's Channel of the southern Irish Sea. These sediments have yielded a rich fauna. The ostracods and foraminifera are of boreal affinity and are indicative of slightly cooler conditions than now exist in the area.

According to Cameron *et al.* (1987), marine sediments of Ipswichian age have been widely encountered in the British sector of the southern North Sea, east of 2°E. These sediments are up to 20 m in thickness and consist of beach, intertidal and shallow-marine sands and clays in the west, with more fully marine facies developed in the east (Long *et al.* 1988) (Figure 1.1). Biostratigraphic evidence (pollen, molluscs and foraminifera) from Germany (Lafrenz 1963; Hinsh 1985; Knudsen 1985; Menke 1985) suggests that during the Eemian marine transgression, the climate ameliorated from sub-arctic or high boreal, to one with temperatures at, or above, their current values (Long *et al.* 1988) (Figure 1.1).

In the central North Sea, valleys eroded in late Wolstonian time have been found to contain Ipswichian interglacial sediments (Jansen and Hensey 1981; Stoker *et al.* 1985a,b; Cameron *et al.* 1987). Ipswichian interglacial beds have also been located in the northern North Sea. Some of these deposits have yielded temperate foraminifera and dinoflagellates, and have exhibited magnetic polarity reversal characteristics consistent with the Blake Event, dated to between 75 000 and 125 000 BP (Skinner and Gregory 1983) (Figure 1.1).

8.11 Correlation, age and duration

Within the British Isles, Bowen *et al.* (1985) have employed D/L ratios (Chapter 1) in marine Mollusca from raised beaches in south-west England and South Wales to establish a Pennard (D/L) Stage. The stratotype of this stage is the Outer Beach (Sutcliffe and Bowen 1973), *Neritoides* Beach (George 1932) and part of the Earthy Breccia Series (Sutcliffe 1981), designated beds 3, 4a, 4b and 4c by Henry (1984), at Minchin Hole. At nearby Bacon Hole, the Sandy Breccio-Conglomerate and Shelly Sand have been shown to have belonged to the same stage, along with a number of other beach deposits around the Gower Peninsula. The Portland East and Hope's Nose beaches have also been included in the Pennard Stage, as has that overlying a raised platform at Trebetherick Point in north Cornwall (Arkell 1943). Raised beaches at Thatcher Rock and Shoalstone, and at Baggy Point and Saunton in north Devon have been referred to an Unnamed (D/L) Stage, located between the Pennard (D/L) Stage and the Minchin Hole (D/L) Stage, the stratotype of the latter being the Inner Beach at Minchin Hole (Chapter 7). The Minchin Hole Stage has been correlated with Oxygen Isotope Stage 7. With reference to the oceanic record of climatic fluctuations and to geochronometric dates, Bowen *et al.* have postulated that both the Pennard and Unnamed stages could be equated with Oxygen Isotope Substage 5e (the latter quite firmly correlated with the Ipswichian), and that the Minchin Hole Stage could be equated with Oxygen Isotope Stage 7. However, they have also hypothesized that the Pennard Stage could equate with Oxygen Isotope Substage 5c or Substage 5a, the Unnamed Stage with Substage 5e and the Minchin Hole Stage with Stage 7. A final permutation involves the Pennard Stage correlating with Substage 5e, the Unnamed Stage

Table 8.1 Age estimates for the (D/L) stages. Model 1a does not conform to the 'Barbados model' of sea-level change (for example, Cronin 1983). Ages are from SPECMAP (Imbrie *et al.* 1984), with some approximations (5a and 5e), and gross estimates for substages of Stage 7. From Bowen *et al.* (1985). Reprinted with permission from *Quaternary Science Reviews* **4**, D. Q. Bowen *et al.*, Amino acid geochronology of raised beaches in south west Britain; © 1985 Pergamon Press Plc.

(D/L) Stage	*D/L Ratio*	*Oxygen Isotope Stage and Age (ka BP)*		
		Model 1a	*Model 1b*	*Model 2*
Pennard	0.105 ± 0.016		5a(*ca.* 80) 5c(*ca.* 100)	5e(122)
		5e(122)		
Unnamed	0.135 ± 0.014		5e(122)	7(194?)
Minchin Hole	0.175 ± 0.014	7(186–245)	7(186–245)	7(216?)

and Minchin Hole Stage being representative of Oxygen Isotope Stage 7 (Table 8.1).

Bowen *et al.* (1989), using isoleucine epimerization in non-marine Mollusca accompanied by geochronometric dating, have identified a distinctive Ipswichian Stage in the British Isles (Figure 1.15). D/L ratios from the type site at Bobbitshole are similar to those from Bacon Hole. The latter locality has produced a U-series date of *circa* 122 000 BP on a relevant stratum which has enabled its correlation with Oxygen Isotope Substage 5e. The Trafalgar Square, Tattershall Castle, Tattershall Thorpe and Block Fen sites could also be referred to the Ipswichian on the basis of D/L ratios. Preece *et al.* (1990) have reported thermoluminescence dates of around 115 000 BP from sand lenses located within the Bembridge Raised Beach, thereby reinforcing its correlation with the Ipswichian.

The most significant early attempts at correlation of the Ipswichian deposits in the British Isles with others in mainland Europe were on botanical grounds. Hollingworth *et al.* (1949) have related the flora from Histon Road with zones f and g of the Eemian Interglacial as described from Denmark and north-west Germany by Jessen and Milthers (1928) (Table 1.1). West (1957) has drawn parallels between the pollen stratigraphy at Bobbitshole and the same sequence, correlating the Suffolk record with its zones b–f. The Selsey palaeobotanical evidence has also been considered comparable with that delimited by these Eemian zones (West and Sparks 1960). The vegetational history of the Ipswichian/Eemian Interglacial in the British Isles and continental Europe has been discussed by Phillips (1974). The Ipswichian has been correlated with the Eemian of the Netherlands (Zagwijn 1975, 1989; de Jong 1988) (Figure 9.9).

Kellaway *et al.* (1975) have equated the raised beach at Brighton with the Normannien II Beach in France, an exposure of which has been described at Sangatte close to Calais. They have postulated that tectonics have played little part in the disposition of this shoreline at *circa* 8 m around the English Channel. Balescu and Haesaerts (1984) have suggested that the Sangatte Raised Beach is referable to a Middle Pleistocene high sea-level, which Haesaerts and Dupuis (1986) have postulated may have been during Oxygen Isotope Stage 7. However, thermoluminescence dating of fossil beach material from Sangatte has yielded an age of 130 000 ± 17 000 BP (Balescu *et al.* 1991). Destombes *et al.* (1975) have proposed the existence of a constricted Strait of Dover during the last interglacial, probably in Substage Ip IIb and Substage Ip III.

Flowstone from within the Lower Cave Earth at Victoria Cave has produced U-series dates ranging from 135 000 ± 8000 to 114 000 ± 5000 BP. From the same locality, speleothems covering a characteristic Ipswichian mammal fauna (section 8.4) have been aged to 120 000 ± 6000 BP, and correlation of the animal remains with Oxygen Isotope Substage 5e has been suggested (Gascoyne *et al.* 1981). Speleothem dates from the Craven District of north-west England have been presented by Gascoyne *et al.* (1983). A period of abundant speleothem growth has been distinguished between 90 000 and 135 000 BP. Within this timespan, the Ipswichian Stage has been defined (on evidence from Victoria Cave) as occurring between 135 000 and 115 000 BP. A speleothem sample from Uamh an Claonaite near Inchnadamph, Sutherland, dated by $^{230}Th/^{234}U$ to 122 000 ± 12 000 BP, has provided an age for the Ipswichian Interglacial in Scotland (Lawson 1981). Uranium-series determinations on travertine obtained from the raised beach at Belle Hougue Cave have given an age of $121\,000^{+14\,000}_{-12\,000}$ BP. This has allowed the deposit to be referred to Oxygen Isotope Stage 5 (probably Substage 5e) of the oceanic record (Keen *et al.* 1981). Flowstone found resting upon the *Patella* Beach at Minchin Hole has given U-series ages of 127 000–107 000 BP (Schwarcz 1984a). At Bacon Hole, the Shelly Sand has produced stalagmite fragments that provided U-series ages ranging from 129 000 to 116 000 BP (Schwarcz 1984b; Stringer *et al.* 1986; Sutcliffe *et al.* 1987). On the basis of peaks in speleothem growth and U-series ages, Gordon *et al.* (1989) have distinguished three temperate episodes: Ip3, 124 000 BP; Ip2, 105 000 BP and Ip1, 90 500 BP. The likelihood of these episodes each representing separate interglacials has been postulated (Chapter 9).

Shells of *Cepaea nemoralis* which occur within calcareous silt of probable early Ipswichian age at Tattershall Castle have given U-series dates between 115 000 and 75 000 BP. A thermoluminescence age of 114 000 ± 16 000 BP has been obtained from the calcareous silt. Amino-acid racemization data from Mollusca indicate that the Tattershall Castle and nearby Tattershall Thorpe interglacial deposits are of similar age (Holyoak and Preece 1985) (see above).

8.12 Conclusion

Clear recognition of an Ipswichian interglacial cycle has been hampered by both the limited number, and fragmented nature, of similar lithostratigraphic and biostratigraphic sequences. Additionally, the issue of whether certain deposits should be assigned to either this, or to an earlier post-Hoxnian temperate stage, has been controversial. However, the existence of an interglacial climate has been established between *circa* 135 000 and 115 000 BP. About the latter time this climate began to deteriorate, an event which marked the onset of the next, and in the orthodox succession the last, cold stage.

9

The Devensian Cold Stage

The type locality for this stage, which consisted of stadial and interstadial environments, has been designated as Four Ashes in southern Staffordshire (Mitchell *et al.* 1973a) (Figures 9.1 and 9.11). The stage name was derived from that of the Devenses, a tribe of ancient Britons which inhabited the region around Four Ashes (Shotton 1977b). The orthodox view is that the Devensian followed the Ipswichian Interglacial and began around 115 000 BP (Table 1.1). It has been formally divided into Early (pre-50 000 BP), Middle (50 000–26 000 BP) and Late (26 000–10 000 BP) periods (Mitchell *et al.* 1973a). However, as Shotton (1977b) has pointed out, the start of the Devensian has not been easy to delimit. This age is beyond the range of radiocarbon assay, although some U-series dates and amino-acid ratios have provided pointers to the commencement of Devensian time. Moreover, biostratigraphic evidence of the transition from the Ipswichian to the Devensian is scanty. Indeed, as Lowe and Walker (1984) have stated, if evidence in localities outside the British Isles is considered (notably from deep-sea and ice cores, loess stratigraphy, and pollen diagrams) (Chapter 1), a more complex sequence of environmental change is suggested, with the timing of the onset of the Devensian equivocal. It appears that a deterioration in climate occurred about 115 000 BP, probably equivalent to Oxygen Isotope Substage 5d. This was followed by a climatic amelioration of either interstadial or interglacial rank *circa* 100 000 BP (Substage 5c), then a cooling episode around 90 000 BP (Substage 5b). An interstadial or interglacial environment ensued once again about 80 000 BP (Substage 5a). A significant cooling of climate is recorded for 80 000–65 000 BP, which may equate with Stage 4 (Figure 1.2(a)). Thus, the Devensian could have begun about 115 000 BP, with the temperate episodes noted above being interstadials. Alternatively, these climatic ameliorations could have been post-Ipswichian and pre-Flandrian interglacials, and the last cold stage could have commenced with the marked cooling *circa* 80 000 BP (section 9.1.11).

During the Early and Middle Devensian, the Polar Front (oceanic and probably atmospheric) in the North Atlantic Ocean lay mainly north of the British Isles. However, a move to the south of them around 75 000 BP has been postulated (McIntyre *et al.* 1972). In the north-east Atlantic Ocean, annual surface-water temperatures were about 5°C lower than today (McIntyre and Ruddiman 1972). This cold water and variable continental weather patterns are likely to have exerted a controlling influence on the climate of the British Isles (West 1977b). During the Late Devensian, the Polar Front was considerably to the south of the British Isles *circa* 20 000 BP. It subsequently began to move northwards, and by about 13 000 BP lay just north-west of Iceland. Around 11 000 BP it migrated south of the British Isles once again (Ruddiman and McIntyre 1973,

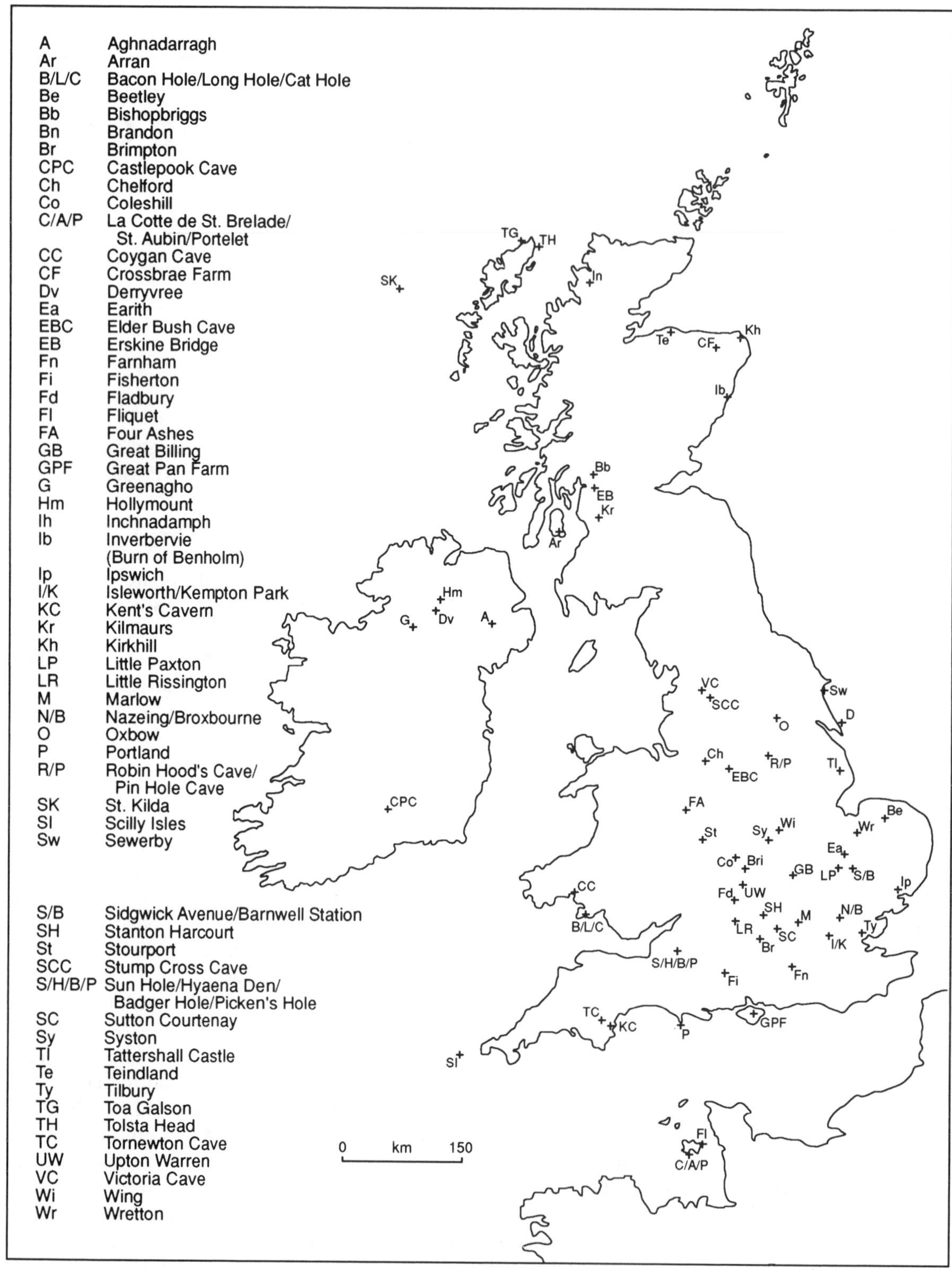
A Aghnadarragh
Ar Arran
B/L/C Bacon Hole/Long Hole/Cat Hole
Be Beetley
Bb Bishopbriggs
Bn Brandon
Br Brimpton
CPC Castlepook Cave
Ch Chelford
Co Coleshill
C/A/P La Cotte de St. Brelade/ St. Aubin/Portelet
CC Coygan Cave
CF Crossbrae Farm
Dv Derryvree
Ea Earith
EBC Elder Bush Cave
EB Erskine Bridge
Fn Farnham
Fi Fisherton
Fd Fladbury
Fl Fliquet
FA Four Ashes
GB Great Billing
GPF Great Pan Farm
G Greenagho
Hm Hollymount
Ih Inchnadamph
Ib Inverbervie (Burn of Benholm)
Ip Ipswich
I/K Isleworth/Kempton Park
KC Kent's Cavern
Kr Kilmaurs
Kh Kirkhill
LP Little Paxton
LR Little Rissington
M Marlow
N/B Nazeing/Broxbourne
O Oxbow
P Portland
R/P Robin Hood's Cave/ Pin Hole Cave
SK St. Kilda
SI Scilly Isles
Sw Sewerby
S/B Sidgwick Avenue/Barnwell Station
SH Stanton Harcourt
St Stourport
SCC Stump Cross Cave
S/H/B/P Sun Hole/Hyaena Den/ Badger Hole/Picken's Hole
SC Sutton Courtenay
Sy Syston
Tl Tattershall Castle
Te Teindland
Ty Tilbury
TG Toa Galson
TH Tolsta Head
TC Tornewton Cave
UW Upton Warren
VC Victoria Cave
Wi Wing
Wr Wretton
0 km 150

Figure 9.1 Localities with evidence of Early and Middle Devensian/Midlandian environments.

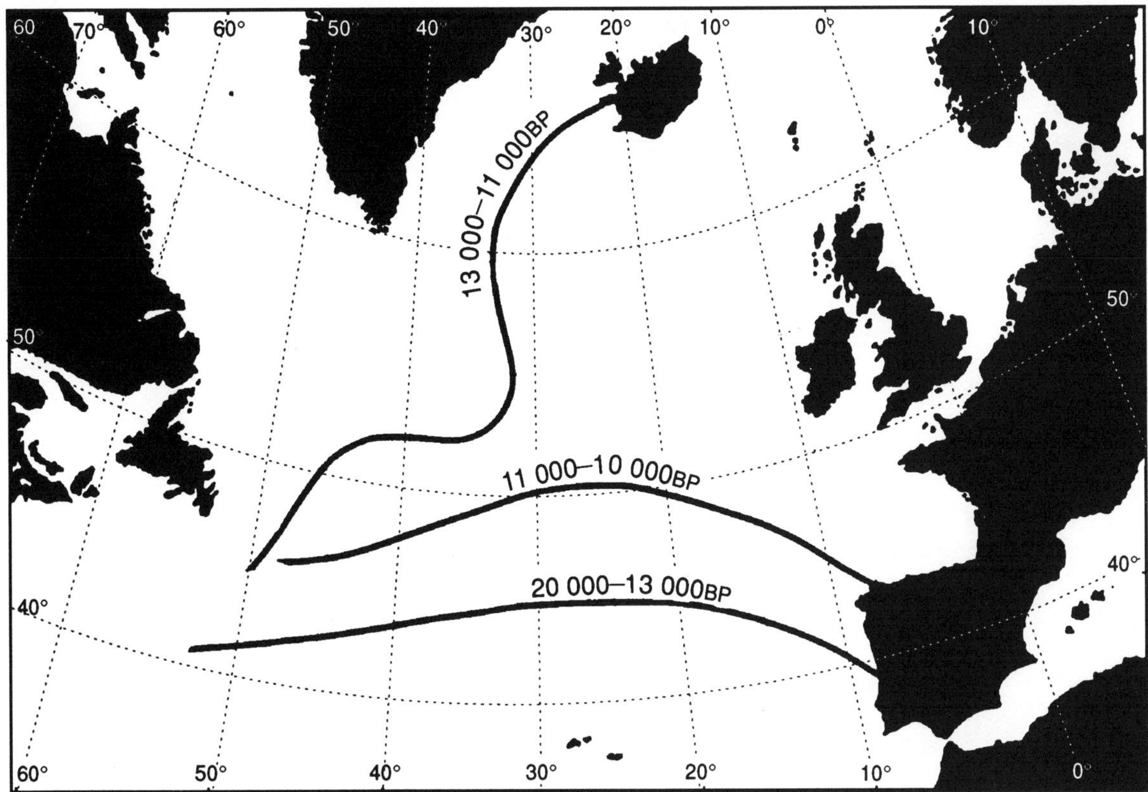

Figure 9.2 Deglacial retreat and readvance of the North Atlantic Polar Front. Positions shown were occupied during the major parts of the intervals indicated, but transitions between positions were not necessarily instantaneous. From Ruddiman and McIntyre (1981).

1981; Ruddiman *et al.* 1977) (Figure 9.2). Therefore, the North Atlantic Ocean contained a substantial component of polar water during the Late Devensian.

9.1 The Early Devensian

As noted above, the Early Devensian has traditionally been considered to be a period before 50 000 BP, which may have extended back to 115 000 BP. Evidence of glacial, periglacial and interstadial environments for this timespan has been forthcoming from various parts of the British Isles.

9.1.1 The Cheshire–Lancashire lowlands

Two tills separated by sands have been recognized across much of this area (Hull 1864; Hull and Green 1866; Pocock 1906). The palaeobotany of bands of organic mud within the sands at Chelford, south of Alderley Edge, Cheshire, has been described by Simpson and West (1958). The pollen record, in which arboreal taxa are dominant, indicates the presence of forest with *Betula*, *Pinus* and *Picea*. Macrofossils from these trees have been found. The forest also contained dwarf shrubs, and is closely comparable with that currently classed as continental conifer–

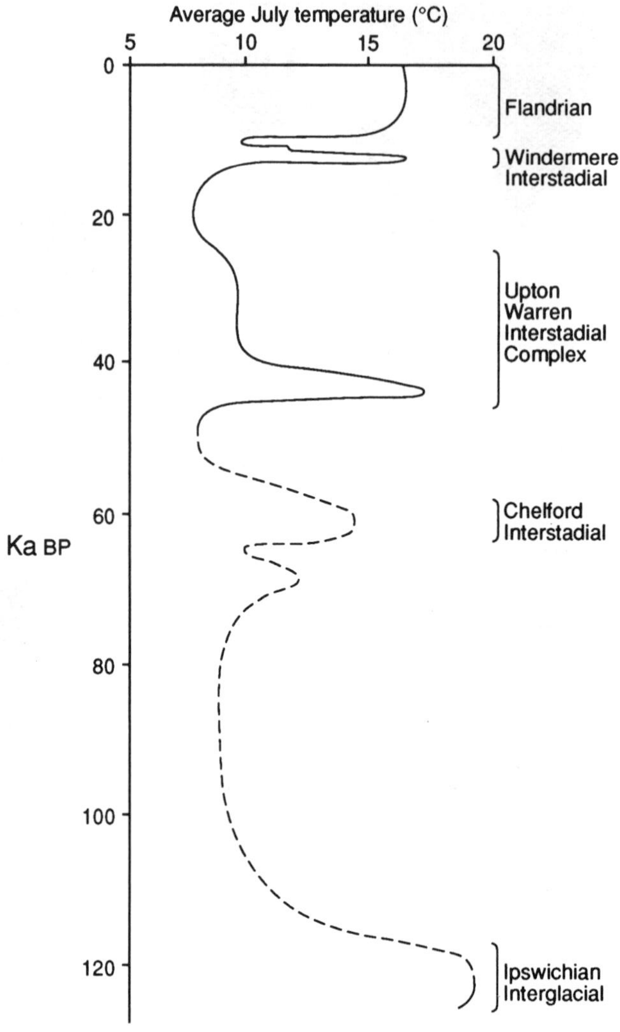

Figure 9.3 Variations in the average July temperatures in lowland areas of the southern and central British Isles since the Last (Ipswichian) Interglacial, indicated by fossil Coleoptera. From Coope (1977a).

birch type in Fennoscandia. The palaeobotanical evidence is consistent with a late Ipswichian age for the organic sediments. However, a radiocarbon date of *circa* 57 000 BP on *Picea* wood has led to the suggestion of their deposition during an Early Devensian interstadial (Figure 9.9). The insect fauna of the organic muds has been reported by Coope (1959, 1977a). It includes *Boreaphilus henningianus* and *Simplocaria metallica*, now widely distributed in boreal Europe but absent from the British Isles. The presence of several relatively thermophilous taxa hints at a substantial improvement in climatic conditions during this episode. A current distributional overlap of characteristic northern and southern taxa found in the insect assemblage occurs in southern Finland and western Russia. On this basis, it has been estimated that mean temperatures during February were around −11°C, and those of July *circa* 15°C (Figure 9.3). The climate was continental, with long, dry, cold winters.

Boulton and Worsley (1965) have questioned the lithostratigraphy at Chelford. They have dismissed the existence of a lower till, indicating that the Middle (Chelford) Sands Formation rests directly on Keuper Marl, and was succeeded by glacigenic sediments making up the Stockport Formation of Devensian age (Worsley 1966, 1967). In this succession, the interstadial occurred within a periglacial rather than glacial environment. However, Pleistocene sediments beneath the Chelford Sands have been described by Worsley *et al.* (1983). These include the Oakwood Till (Worsley 1978) (Chapter 7) below which are gravels and silts. Plant, molluscan, insect and ostracod assemblages from the latter suggest a stadial environment. The pollen record is dominated by grasses and sedges, and Mollusca present include *Vertigo genesii*, currently an arctic–alpine inhabitant. The insect fauna, of boreal/boreo-montane character, contains *Diacheila arctica*, *Pelophila borealis* and *Helophorus jacutus* among its components. The inference from the insects is of July temperatures not above 10°C and a continental climatic regime. The gravels and silts may be of Early Devensian age, as would be the overlying Oakwood Till; the latter being stratigraphically inferior to the Chelford interstadial sediment. A number of conflicting radiocarbon assays have been obtained from interstadial material at Chelford, and its age is now considered to be in excess of 60 000 BP (Vogel and Zagwijn 1967; Worsley 1980).

9.1.2 The Midlands of England

At the type locality, Four Ashes, most of the organic material was deposited *in situ* in a braided-stream environment (A. V. Morgan 1973) (Figure 9.4). Organic sediments in contact with bedrock have yielded a pollen assemblage including *Pinus*, *Picea* and *Betula*, comparable with that from Chelford (A. Morgan 1973; Andrew and West 1977b). The insect fauna from these sediments is also analogous to that from Chelford (A. Morgan 1973). It includes *Blastophagus piniperde*, a conifer-dweller and *Scolytus ratzburgi*, an obligate of *Betula*. The fauna is overall of continental character, and one found today in central Fennoscandia. The oldest stadial deposits (gravels with finer-grained inclusions) at Four Ashes could either pre-date or post-date the Chelford episode. One locality with these sediments has produced a radiocarbon age in excess of 43 500 BP. Their beetle fauna includes numerous inhabitants of present-day tundra, for example, *Bembidion hasti*, *Amara alpina* and *Diacheila polita*. The insect evidence implies a sparsely vegetated landscape containing numerous small water-bodies. Trees were lacking, with grasses and mosses dominating the flora. The climate was continental; winters were very cold and July temperatures around 10°C. Corroboration of a cold environment is provided by frequent records of *Lepidurus arcticus*, a crustacean now confined to the arctic. Palynological data from these sediments have revealed high non-arboreal frequencies. Grasses and sedges dominate the spectrum and the suggestion is of herbaceous vegetation with perhaps some localized birch and willow scrub (Andrew and West 1977b). The combined evidence has led Morgan and Morgan to prefer a post-Chelford age for these deposits.

Underlying the Main Terrace of the River Severn at Stourport, Worcestershire, Dawson (1988) has identified glacigenic diamict deposits older than Late Devensian. If these are not pre-Devensian (Chapter 7), they may have been emplaced early in the last cold stage. In the Avon Valley, gravels below the surface of Terrace 4 have been assigned to the late Ipswichian or Early Devensian (Tomlinson 1925; Shotton 1953, 1968, 1973). However, Bridgland *et al.* (1989) and Maddy *et al.* (1991) have ascribed some of these sediments to a pre-Devensian cold stage, and have noted the presence of temperate deposits within the gravel body (Figure 7.6).

At Wing, Leicestershire, a late Ipswichian pollen assemblage (Chapter 8) passes upwards into a Gramineae–Ericales–*Pinus* biozone, identified within several metres of silty, stony

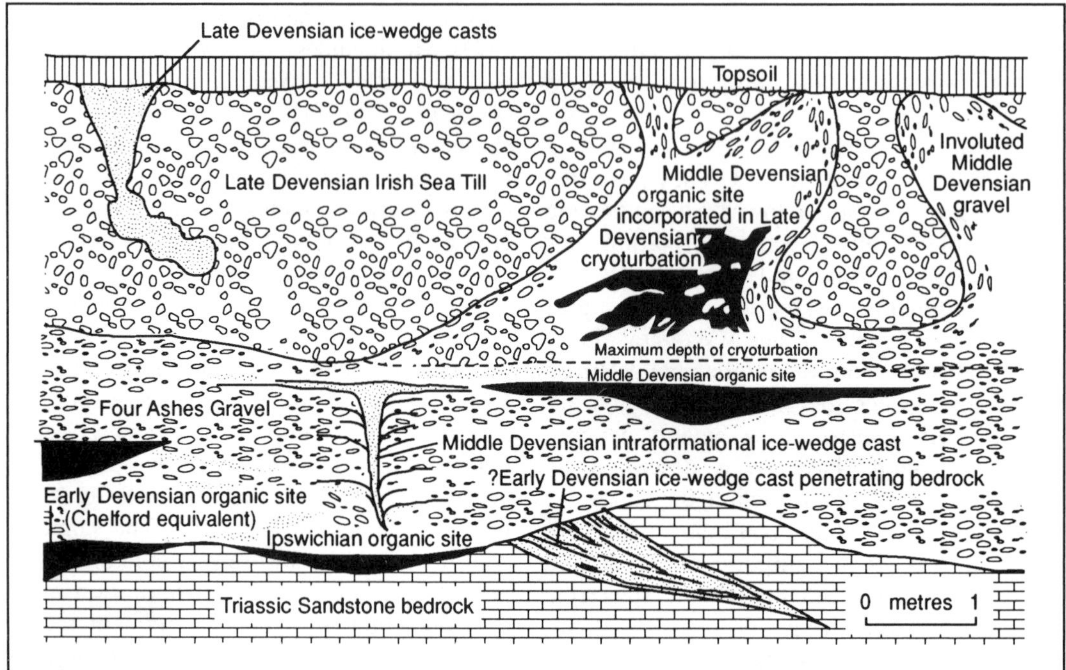

Figure 9.4 General stratigraphy of the Four Ashes pit. From A. V. Morgan (1973).

clays. Early Devensian tundra-like vegetation has been inferred, with peatland as well as herb-rich grassland around a lake basin (Hall 1980).

9.1.3 Yorkshire, Lincolnshire, the Peak District, northern Dukeries and East Anglia

Straw (1979a,c) has advanced the hypothesis that two discrete episodes of glaciation occurred in eastern England during the Devensian. He has also suggested that the first of these glaciations was the more extensive and took place in the Early Devensian (Figure 9.5). Morphological and lithostratigraphic evidence from eastern Lincolnshire has enabled Straw (1957, 1958, 1961) to recognize these episodes. A Lower Marsh Till has been referred to an Early Devensian glaciation by Straw (1969). However, mineralogical studies of tills in Lincolnshire, Holderness and north Norfolk have led Madgett and Catt (1978) to postulate their deposition during a single Late Devensian glaciation (section 9.3.2). Straw (1979a,c) has contended that in eastern Yorkshire, topographic differences observable west and east of the River Hull could have resulted from two glaciations, and that such duality is also identifiable in the morainic landforms about the eastern end of the Vale of Pickering. On the west side of the Vale of York, the Linton–Sutton Gravels, located in the vicinity of Leeds (Edwards *et al.* 1950), have been equated by Straw with an Early Devensian glaciation which was more widespread than that delimited by the York–Escrick moraines (section 9.3.2). The environs of Selby have been postulated as the southern limit of the Early Devensian ice.

To the west of, and at a higher elevation than, gravels in the lower Bain Valley in Lincolnshire (section 9.2.2), older gravels have been recognized. According to Straw (1979a,c) the latter were laid down during the Early Devensian on the fringes of a proglacial lake, termed Lake Fenland. Straw has also reassigned strandlines from 12 to 24 m OD around a former Lake

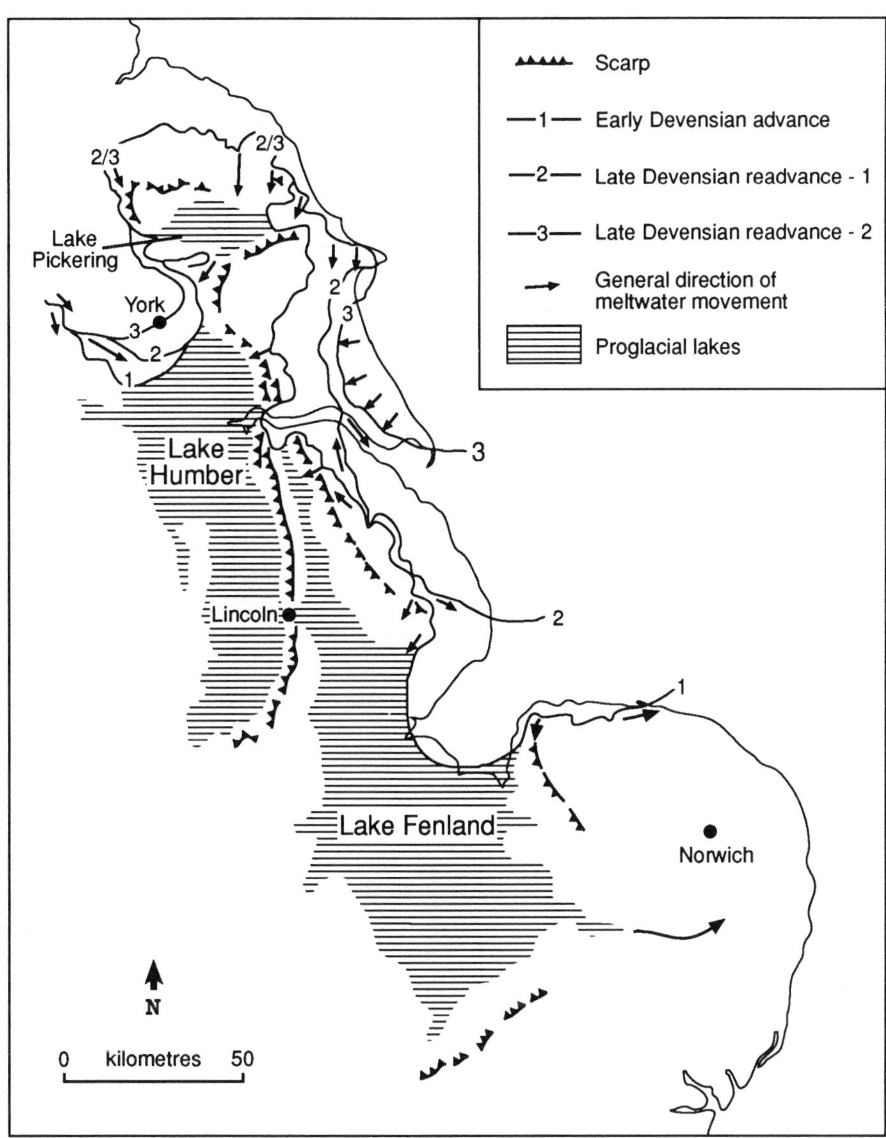

Figure 9.5 Postulated limits of Devensian ice advances into eastern England, and maximum extent of proglacial lakes. From Straw (1979a).

Humber in the southern part of the Vale of York to the Early Devensian. Gaunt *et al.* (1972) have indicated that these strandlines are younger than lacustrine deposits at *circa* 8 m OD related to Lake Humber, but had formed during the Late Devensian (section 9.3.2). Straw has intimated that Lake Fenland and Lake Humber existed in the Early Devensian (Figure 9.5), with the latter then impounded to its highest level.

At Sewerby in eastern Yorkshire, colluvium and blown sand identified above an Ipswichian raised beach (Chapter 8) could be of Oxygen Isotope Substage 5d to Substage 5a age (Catt 1987a). In the southern part of the Vale of

York a ground surface including ice-wedge casts, ventifacts and probable thermokarst features, developed on Ipswichian river gravel and beneath Late Devensian glacigenic deposits, is likely to have been formed in the Early Devensian (Bisat 1946; Gaunt 1970, 1976, 1981; Gaunt *et al.* 1972, 1974; Catt 1977a).

In the Peak District, Elder Bush Cave in the Manifold Valley south of Buxton has produced a vertebrate fauna transitional between those of the Ipswichian and Devensian. Woolly rhinoceros, cave bear, bison, cave lion and hyaena are among its components (Bramwell 1964; Bramwell and Shotton 1982). Extensive cave sedimentation occurred in the Peak District during the Early Devensian (Briggs and Burek 1985). Speleothem growth ceased in these caves between about 90 000 and 75 000 BP (Ford *et al.* 1983). The Hope Terrace of the Derwent Valley, an Ipswichian feature, received a cover of soliflucted material in the Early Devensian (Waters and Johnson 1958; Briggs and Burek 1985).

In the northern Dukeries, Robin Hood's Cave has yielded Mousterian artifacts and Early Devensian pollen spectra from its basal stratigraphy (Campbell 1977; Roe 1981). At Pin Hole Cave, in the same Creswell Crags complex north of Bolsover, two separate Middle Palaeolithic industries and many vertebrate remains, including mammals (notably *Crocuta crocuta*), amphibians, birds and fish, have been discovered in a likely Early Devensian context (Campbell 1977; Jenkinson 1984; Jenkinson *et al.* 1985).

In the Craven District around the Yorkshire–Lancashire border, speleothem growth was low from 90 000 to 45 000 BP (Gascoyne *et al.* 1983). Atkinson *et al.* (1978) have reported speleothem growth at Gavel Pot in north-west Yorkshire about 60 000 BP, a time which may relate to the Chelford Interstadial (see above). However, Gascoyne *et al.* have not been able to confirm this activity. Their results have suggested a tundra-like climate during the Early Devensian in this part of north-west England, where permafrost was deep but not continuous. Watson (1977) has suggested that permafrost formed in the British Isles during the latter part of the Early Devensian and became continuous after the Chelford Interstadial, when a polar-desert type environment existed.

Stump Cross Cave in northern Yorkshire, about 20 km from Victoria Cave, Settle, has provided an insight into the sequence of environmental changes in the timespan normally associated with the Early Devensian (Sutcliffe *et al.* 1984, 1985). Flowstone, dated by U-series to 83 000 ± 6000 BP (within Oxygen Isotope Substage 5b), encases faunal remains that include *Rangifer tarandus* and *Gulo gulo* (wolverine). Thus, a cold environment existed at this time. Wolverine is especially diagnostic of such conditions, possessing a current circumboreal distribution which reaches the tundra.

In Norfolk, the mostly fluviatile sands and gravels of the Low Terrace of the River Wissey have fossiliferous organic deposits below, and within, them at Wretton (West *et al.* 1974). A complex sequence of sediments above Ipswichian deposits (Chapter 8) has been assigned an Early Devensian age. The terrace gravels resulted from a braided-river environment in a periglacial climate. Structures characteristic of the latter are present in the gravels, and it is likely that certain of the organic deposits accumulated in slight hollows which originated by the melting of ground-ice mounds. The plant remains suggest an initial cold episode when the vegetation was mainly herbaceous. Grasses, sedges and a range of heliophytes including *Armeria* (thrift), *Artemisia* (mugwort/wormwood) and *Helianthemum* (rockrose) are especially prominent. This is followed by a biozone indicative of birch–pine woodland and heath. Grassland is indicated by the subsequent biozone, with the next a further woodland one in which pine, birch and spruce were the most important taxa, and heath plants were also present. The second woodland episode has been correlated with that of the interstadial at Chelford, the first assigned to a new Wretton Interstadial (Figure 9.9). The pollen spectra in the sediments following those of the Chelford Interstadial are again diagnostic of grassland. All of the herb biozones at Wretton contain taxa analogous to those comprising the vegetation of present-day tundra. West *et al.* have suggested that the lack of trees may have been caused by July temperatures of less than 10°C, which is the threshold for their growth. Alternatively, aridity may have prevented tree growth, so that summer temperatures could have been 15°C or higher. The periglacial phenomena indicate that winter temperatures were of the order of −10°C.

Mollusca have been obtained from the sediments corresponding to the first and last herb biozones. The presence of *Belgrandia marginata* in the former probably reflects its survival from the Ipswichian Interglacial (the corresponding plant record also contains likely thermophilous vestiges from this stage), but *Columella columella*, *Pupilla muscorum* and *Vertigo parcedentata* indicate cold conditions. In the second assemblage, enhanced frequencies of *Pupilla muscorum* denote the preponderance of cold, dry and open habitats.

Coleoptera have also been recorded from Wretton (Coope 1974b). Those sediments indicative of grassland from their plant fossils have produced no insect obligates of woodland. *Otiorrhynchus arcticus*, *Otiorrhynchus nodosus* and *Notaris aethiops* occur, implying a cool but not tundra climate. More equivocal, however, is the insect evidence from the sediments yielding pollen indicative of woodland. Here only beetles of open habitats are represented. They include the tundra-dwellers *Amara alpina* and *Helophorus jacutus*; the overall assemblage suggesting mainly barren, sandy terrain, with some reed swamp.

Early Devensian interstadial deposits (sandy muds with a wood peat near to their base) have been reported by Phillips (1976) from Roosting Hill, Beetley, Norfolk. The sediments occupy a depression in the Hungry Hill Gravels (Chapter 7), probably caused by the collapse of a ground-ice mound. Also at this locality, Ipswichian interglacial deposits (Chapter 8) are succeeded by a water-lain clay containing involutions, also probably of Early Devensian age. Plant remains from the organic deposits suggest an initial grass-dominated vegetation. Of particular interest is the incidence of charcoal in the stratigraphy, which has led to the suggestion that burning may have been responsible for woodland removal and grassland development. This episode was followed by one when coniferous woodland (*Pinus–Betula–Picea*) developed, with subsidiary heath (including *Calluna* (heather), *Empetrum* and *Bruckenthalia*) and wetland, the latter containing *Rubus chamaemorus* (cloudberry) and *Sphagnum* (bog moss). The most common plant macro-remains are from *Picea*, and although its pollen frequencies do not exceed 4% of total land taxa, it is considered an important woodland constituent. A subsequent *Pinus–Betula–Picea–*Ericales–*Sphagnum* biozone indicates that the forest opened up, with heath and herbaceous vegetation expanding in its place. The final biozone is a herb one, depicting vegetation in which grasses and sedges were of considerable importance. Phillips has suggested three possible interpretations for this sequence. Firstly, it could demonstrate the demise of Ipswichian interglacial woodland and the spread of herbaceous plant communities in the Early Devensian. Secondly, it may reflect the termination of wooded interstadial conditions and their replacement by open vegetation. Thirdly, and perhaps the more likely, a complete interstadial may be represented. There is a close similarity between the woodland biozone and that at Chelford, and its equivalent at Wretton. At Beetley the full duration of this interstadial has been recorded: an opening herb phase, then a woodland one, with a further herb phase at its close.

At Sidgwick Avenue, Cambridge, organic deposits underlying the Intermediate Terrace of the River Cam have produced plant macrofossil and molluscan assemblages comparable to those at Wretton, which have been designated as early glacial by Lambert *et al.* (1963). At Earith north-west of Cambridge, organic material is present in the lower part of the terrace gravels of the River Ouse, next youngest to those of Flandrian age (Bell 1970). Ice-wedge casts and involutions occur in the gravels, the former having affected the earliest organic material. This was deposited in shallow pools on the floodplain and has given a radiocarbon age in excess of 45 000 BP. The vegetational history has been derived mainly from macrofossils which include *Salix herbacea* (dwarf willow) and *Betula nana* (dwarf birch). In addition to scrub, there is a rich terrestrial herb flora. Of especial interest within this is a marked halophyte component including *Glaux maritima* (sea milkwort), *Suaeda maritima* (herbaceous seablite) and *Triglochin maritima* (sea arrow-grass). The climatic indications from the flora are of July temperatures *circa* 16°C and of cold winters, but a snow cover appears to have been either absent or infrequent. The closest modern analogue for this vegetation occurs on floodplains in Siberia where plants characteristic of steppe, tundra and saline environments are located. The climate is highly continental. Winter–summer temperature variations of up to 40°C have been recorded and permafrost is present.

Wymer (1985) has ascribed Early Devensian contexts to Mousterian of Acheulian Tradition (Figure 6.4) and Levalloisian industries recorded at Little Paxton, Huntingdonshire (Paterson and Tebbutt 1947) and in terrace gravels of the River Gipping at Bramford Road, Ipswich.

9.1.4 The Thames Valley

In the Upper Thames area, some of the gravels composing the Summertown–Radley Terrace (Figure 7.7) contain periglacial features and have been identified as Early Devensian by Goudie and Hart (1975). Silts in these terrace gravels at Stanton Harcourt have given thermoluminescence dates ranging from 93 000 to 91 000 BP which supports this contention (Seddon and Holyoak 1985). Gravels of this affinity have been identified in the Middle (Gibbard 1985) and Lower (Bridgland 1988) Thames areas, and represent cold-climate accumulations of Early Devensian time.

9.1.5 Wales

Bowen *et al.* (1986) have stated that no evidence exists of an Early Devensian glaciation in Wales. However, at Broughton Bay on the Gower Peninsula a raised beach, ascribed to Oxygen Isotope Substage 5e by the amino-acid ratios of shells contained within it (Chapter 8), can be observed overlain by a till. Shells from the till have given similar amino-acid ratios to those of the raised beach and an Early Devensian age for the glacigenic material has been hypothesized (Campbell *et al.* 1982). Wood from the till has been radiocarbon dated to *circa* 68 000 BP (Campbell and Shakesby 1985). This implies a Chelford Interstadial age for the tree growth and could mean later (Late Devensian) incorporation of the wood into the till. None the less, in view of the uncertainties surrounding radiocarbon dates of such antiquity (Worsley 1980) (section 9.1.1), the wood could be of greater age, possibly Ipswichian, and the till, Early Devensian.

At Long Hole, Gower, a palynological sequence covers the Ipswichian–Devensian transition, and includes a *Picea–Pinus–Betula* assemblage, the latter probably equating with the Chelford Interstadial. Mousterian artifacts have also been recorded from this part of the cave stratigraphy (Campbell 1977). Archaeological material of Mousterian–Levalloisian affinity has also been recovered from Coygan Cave, Carmarthenshire (Wymer 1977).

A complex transition from Ipswichian to Devensian time has been proposed, mainly on faunal evidence, by Stringer *et al.* (1986) and Sutcliffe *et al.* (1987) after investigations at Bacon Hole, Gower. They have regarded the accepted model relating to faunal changes at the conclusion of a temperate stage and the commencement of a cold one as elementary. In this, members of the former fauna were supplemented by those of the latter. Faunal succession has been considered to be more complex, and the lack of late interglacial faunas superimposed on earlier ones containing characteristic taxa has been noted. As observed at the beginning of this chapter, the oxygen isotope record points to two cold and two temperate episodes between Substage 5e and Stage 4. At Bacon Hole and elsewhere in the British Isles, Substage 5e has been established (Chapter 8). However, substages 5d to 5a have not been pin-pointed here. The Bacon Hole stratigraphy (Figure 9.6) contains stalagmite overtopping the Upper Cave Earth. Aged to *circa* 81 000 BP by U-

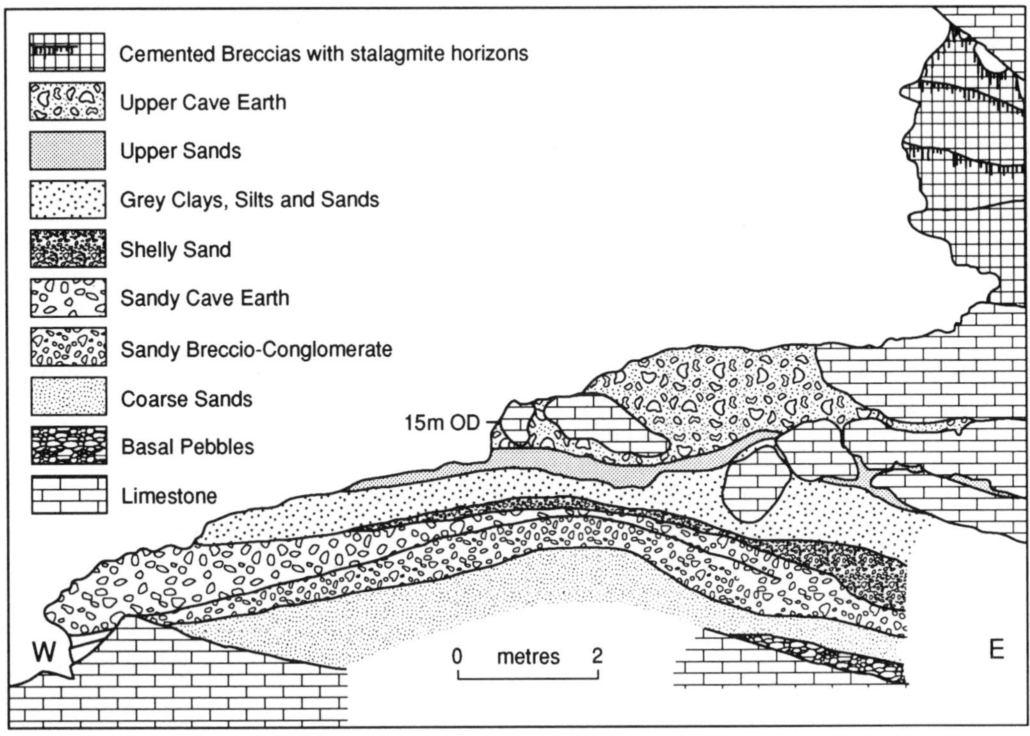

Figure 9.6 Bacon Hole. Representative transverse (E–W) section of platform deposits. From Sutcliffe *et al.* (1987).

series analysis, this stalagmite may equate with Oxygen Isotope Substage 5a. Above the Shelly Sand of Substage 5e age, the lithostratigraphy consists of clays, silts and sands whose mineralogical and chemical properties imply humid conditions accompanying falling temperatures and a lowering sea-level. Biostratigraphic data from these sediments indicate that *Mammuthus* was present in South Wales, as were two birds, *Anser fabalis* (bean goose) and *Clangula hyemalis* (long-tailed duck) currently winter visitors to the British Isles. Above these sediments is aeolian sand indicative of a drier environment. The Upper Cave Earth was deposited in temperate, fairly dry conditions, when sea-level was lower than at the time of accumulation of an earlier Sandy Cave Earth. It contains a limited mammalian fauna of interglacial type. In context, this fauna is most likely to be that of an Early Devensian interstadial during which *Palaeoloxodon antiquus* and *Dicerorhinus hemitoechus* were present and *Equus ferus* was absent.

9.1.6 England south of the Lower Severn and Thames valleys

Synge (1977) has postulated a Devensian ice shelf extending southward in the Irish Sea (section 9.3.5). Till on the coast of north Devon and in the Scilly Isles (Chapters 5 and 7) could have been emplaced from such a feature, perhaps prior to the Late Devensian (Synge 1985).

Periglacial deposits (especially head and Coombe Rock), some of which could be Early Devensian in age, have been widely recognized over southern England (Stephens 1974a; Shephard-Thorn and Wymer 1977). Much of the evidence concerning Early Devensian environments in this area has come from river-terrace deposits. From Brimpton at the confluence of the

Enborne and Kennet valleys in Berkshire, fine-grained fossiliferous sediments interbedded with terrace gravels have been reported by Bryant *et al.* (1983a). The earliest pollen assemblage is indicative of regional birch–pine woodland, and a flora lacking taxa currently confined to high latitudes and altitudes is present. This episode was probably an interstadial, which may have been equivalent to the Wretton Interstadial. If it was not an interstadial, floristics and vegetational history at the close of the Ipswichian Interglacial may be portrayed. This episode was followed by a stadial, when open vegetation existed, in which sedges, grasses and other herbs were dominant, low shrubs were present and trees were absent. A woodland biozone has been recognized from the succeeding sediment, with birch, pine and spruce of most importance. Macroremains of *Betula pubescens* (birch), *Pinus sylvestris* (Scots pine) and *Picea abies* (Norway spruce) occur. Heath plants were also present in the vegetation, which overall bears close resemblance to that of the Chelford Interstadial, with which it has been correlated. A further herb biozone (notably with sedges and grasses) reflects an open, stadial environment. This is replaced by a boreal forest pollen assemblage with *Betula* and *Pinus* regional dominants. The name Brimpton Interstadial has been given to this episode (Figure 9.9), for which an Early Devensian context has been suggested as dated Middle Devensian stadial deposits overlie its sediment.

Molluscan and insect faunas from these Early Devensian sediments both support the sequence of events deduced from the botanical data. The first (interstadial) deposit contains *Vallonia pulchella*, *Trichia hispidia* and *Pisidium supinum*, molluscs of thermophilous Devensian contexts (Holyoak 1982). The second (stadial) deposit possesses only three land (*Pupilla muscorum*, *Catinella arenaria* and *Oxyloma pfeifferi*) and three freshwater molluscs. The third (interstadial) deposit has six land snails (including *Discus ruderatus*, *Nesovitrea hammonis* and *Vertigo substriata*) characteristic of current boreal forest and scrub habitats. Insect remains in this sediment, where *Gymnusa variegata*, *Trechus rivularis* and *Notaris aethiops* occur, are of taxa now mainly boreo-montane in distribution; their ranges overlapping in central and southern Fennoscandia which experiences a January mean temperature of −7.5°C and a July one of 15°C. The fourth (stadial) deposit contains Mollusca which today inhabit dry, open localities (*Pupilla muscorum*, for example), and live under herbs or dwarf shrubs (*Columella columella*, for example). Insect evidence is scarce but supportive of the habitat inferences made from the molluscs. The final (interstadial) deposit contains a fairly diverse and reasonably temperate suite of Mollusca, including *Myxas glutinosa*, *Cochlicopa lubrica*, *Vallonia pulchella* and *Trichia hispida*.

At Fisherton near Salisbury in Wiltshire, colluvial terrace deposits of the River Nadder contain a molluscan fauna including *Pupilla muscorum* and an ostracod assemblage with *Ilyocypris gibba*. These animals have been interpreted as cool-temperate and near-interstadial in character, and the deposit of Early Devensian age (Green *et al.* 1983).

Silts filling a channel cut in Lower Cretaceous Folkestone Beds at Farnham, Surrey, are associated with the Wrecclesham Gravel. The latter, a braided-river deposit, has the Railway Terrace of the River Wey developed on it (Bryant *et al.* 1983b). The silts have yielded a fairly consistent pollen flora in which Cyperaceae and Gramineae are dominant, and numerous other herbaceous taxa (notably Compositae, Chenopodiaceae, *Plantago* and *Rumex* (sorrel/dock) species) occur. The vegetation contained few trees, but *Juniperus* and *Salix* pollen and macrofossil *Betula nana* are indicative of a better developed shrub component. The macrofossil plant record also includes the arctic–alpine mosses *Encalypta ciliata*, *Scopidium turgescens* and *Timmia norvegica*. A molluscan fauna containing *Pupilla muscorum* and *Columella columella*, together with an insect assemblage of cold affinity, have also been obtained from the silts. Bones of *Mammuthus primigenius* have been recovered from the channel, and it has been suggested that some of the open vegetation in the area may have been perpetuated by the grazing and trampling activities of mammoth and other large herbivores. The silts have been dated by thermoluminescence to 107 000 ± 15 000 and 106 000 ± 11 000 BP, the time of Oxygen Isotope Substage 5c. However, plant macrofossil material from the silts has given a radiocarbon age of about 36 000 BP. A subsequent radiocarbon assay (*circa* 37 000 BP) on a *Mammuthus* tusk fragment from the gravel has led Gibbard *et al.* (1986a) to suggest a probable Middle rather than Early Devensian context for these deposits (section 9.2.4).

Head observed overlying the Portland West Raised Beach (Chapters 7 and 8) is likely to be of Early Devensian age. Its land-snail fauna includes *Pupilla muscorum* and *Helicella itala*, indicative of dry grassland and a cool climate (Keen 1985).

Hand axes, including Mousterian types, and Levallois flakes (Figure 6.4), obtained from the Christchurch Gravels of Hampshire (Calkin and Green 1949) and Great Pan Farm in the Isle of Wight (Shackley 1973) are probably in Early Devensian rather than Ipswichian contexts (Chapter 8) according to Wymer (1988).

From Sun Hole in the Mendip Hills of Somerset, Campbell (1977) has presented palynological data that have been interpreted as marking the vegetation changes at the end of the Ipswichian Interglacial and the start of the Devensian Cold Stage, with a Chelford-type interstadial event recorded by the flora. A pollen sequence from Hyaena Den, Wookey, in the same area (Tratman *et al.* 1971) commences with heathland taxa, substantial *Betula* frequencies and the presence of some thermophilous trees. It has been inferred that this could also represent late Ipswichian or Early Devensian vegetation. A subsequent palynological assemblage possesses higher herbaceous values indicative of a colder episode but evidence of the Chelford event is lacking.

9.1.7 Jersey

A radiocarbon date of 47 000 ± 1500 BP, possibly contaminated with recent humic material (Vogel and Waterbolk 1963) has been obtained from black organic material, perhaps associated with a Mousterian Industry, at La Cotte de St. Brelade (McBurney and Callow 1971; Callow 1986b,c). Teeth (and possibly bone) of *Homo sapiens neanderthalensis* of Late Pleistocene affinity (Zeuner 1940; McBurney and Callow 1971; Stringer and Currant 1986) have been found in an Early Devensian context (Callow 1986d) at this rock shelter overlooking St. Brelade's Bay (Callow and Cornford 1986), to which Palaeolithic hunters could have migrated from France across tundra-like terrain (see below). Pedogenetic (van Vliet-Lanoë 1986) and palynological data (Campbell and Pohl 1971; Jones 1986) indicate a complex post-Ipswichian sequence of environments consisting of interstadials or interglacials separated by cold episodes with gelifluction.

Peat found resting on the shore platform at Fliquet in the north-east of the island, has produced evidence of the late Ipswichian/Early Devensian environment (Coope *et al.* 1980). The earlier of two palynological assemblages accompanied by plant macrofossil remains indicates local wetland (including salt-marsh) beyond which were pine–birch woods and shrub heath, each with a substantial herbaceous component. A later pollen and spore assemblage reflects reduced woodland cover. Increased herbaceous pollen frequencies, including Saxifragaceae (saxifrage) species, and lesser quantities of salt-marsh taxa, together with the incidence of *Lycopodium* (*Huperzia*) *selago* (fir clubmoss) spores, is indicative of a colder climate and lower sea-level. An insect fauna from this peat is also divisible into two biozones. The older supports the idea of local wetland and is of cool-temperate aspect; the younger contains a sub-arctic assemblage including *Diacheila arctica*, *Helophorus sibiricus*, *Pycnoglypta lurida* and *Boreaphilus nordenskioeldi*. The present climatic tolerances of these species indicate that July temperatures in Jersey were 11–12°C and the climate continental. The peat, aged by radiocarbon assay to older than 25 000 BP, is unlikely to pre-date the Ipswichian, as the high sea-level of that stage would have almost certainly effected its removal from the foreshore. While the timing of peat deposition is uncertain, the most feasible options seem to be either at the conclusion of the Ipswichian Interglacial/start of the Devensian Cold Stage, or during the middle and late parts of an interstadial during the Early Devensian.

Further insight into Early Devensian environments has come from another foreshore peat revealed in St. Aubin's Bay on the south coast (Coope *et al.* 1985). Only single pollen and insect assemblage biozones have been delimited in this sediment. The pollen record is diagnostic of dominantly herb-rich vegetation, with occasional birch–pine woodland and dwarf-shrub heath. Many of the herbaceous plants are consistent with those comprising present-day arctic tundra and steppe vegetation. The coleopteran fauna includes *Agonum exaratum*, *Pynoglypta lurida*, *Boreaphilus henningianus* and *Simplocaria metallica*. All of these insects are today absent from the British Isles and have boreal or boreo-montane ranges. *Agonum exaratum* does not live below the tree line, and the inference from the fauna is of arctic conditions, with July mean temperatures of 10°C or less during an Early Devensian stadial.

9.1.8 Scotland

Close to the current coastline, in southern Arran and at Inverbervie in Kincardineshire, for example, elevated beds of marine shells have been located. The shells are mainly of arctic type but include some temperate taxa. Sutherland (1981a) has considered these shell beds to be *in situ* rather than ice transported, having formed as a result of a marine transgression. Crustal loading and hence downwarping, due to the build up of the last Scottish ice sheet, made this transgression possible. Infinite radiocarbon dates from the shells have suggested a pre-Late Devensian age. The build up of ice has been thought to have occurred during the Early Devensian, probably about 75 000 BP, as indicated by oceanic data (Shackleton and Opdyke 1973; Ruddiman *et al.* 1980). At the Burn of Benholm (Inverbervie), peat lenses found in till above the shelly material have yielded a dominantly herbaceous pollen flora, whose composition (including the currently arctic/sub-arctic *Koenigia islandica*) has been interpreted as being from tundra grassland with some low shrubs. A radiocarbon date of over 42 000 BP on the peat means that it is most likely to be of Early Devensian age (Donner 1979). The high-level rock platform in the Hebrides (McCann 1968; Sissons 1976; Jardine 1977) could have been eroded by an Early Devensian marine transgression (Sissons 1981a, 1982).

The upper tills and sand and gravel, and an underlying head, part of the complex sequence at Kirkhill in Buchan (Chapters 5 and 7), may be Early Devensian deposits (Connell 1984; Hall 1984a). At Crossbrae Farm near Turiff, Buchan, the lower of two tills separated by an interstadial deposit (section 9.2.6) is probably of Early Devensian age (Hall 1984a).

An organic deposit at Toa Galson on Lewis (Sutherland and Walker 1984) (Chapter 8) has produced a pollen flora and a radiocarbon age (>47 000 BP) consistent with the Early Devensian (Lowe 1984). In south-west Scotland and the Midland Valley, faunal remains located either beneath or within sediments of the last cold stage have possible Early Devensian interstadial contexts (Sutherland 1984). For example, an antler of *Rangifer tarandus*, radiocarbon dated to in excess of 40 000 BP, has been obtained from above a marine layer and below a till at Kilmaurs, Ayrshire (Shotton *et al.* 1970). A borehole at Erskine Bridge on the River Clyde west of Glasgow has encountered a pre-Late Devensian till (Browne 1980), probably of Early Devensian age (Sutherland 1984).

9.1.9 Ireland and the Isle of Man

The last cold stage in Ireland has been termed the Midlandian (Mitchell *et al.* 1973a), while Warren (1985) has proposed the use of Fenitian (Table 1.1). At Aghnadarragh on the eastern side of Lough Neagh in County Antrim, McCabe *et al.* (1986a, 1987) have identified a sequence beginning with a Lower Till, above which are periglacial deposits that contain remains of *Mammuthus primigenius*, the latter radiocarbon dated to over 44 330 BP. Next youngest is a woody detritus peat, whose botanical remains denote the existence of interstadial woodland with birch, pine and spruce. Insects from this peat include *Rhyncolus strangulatus* and *Rhyncolus elongatus*, both now entirely dependent on *Picea* or *Abies* and extinct in the British Isles. The beetle evidence is indicative of July mean temperatures between 15 and 18°C, with those in January from −11 to 4°C. Wood from the peat has been radiocarbon dated to over 48 180 BP. The biostratigraphic evidence is similar to that from Chelford, and the age of the deposit means that the Lower Till and periglacial sediments can be assigned to an Early Midlandian stadial. Above the interstadial peat are sands, gravels and diamictons, together with organic mud (radiocarbon dated to in excess of 46 850 BP) and sand of a cold episode, then an Upper Till. Plant remains in the organic mud have been interpreted as coming from treeless vegetation in which grasses and sedges were predominant and heath plants well represented. Macroremains of *Salix herbacea* and *Selaginella selaginoides* (lesser clubmoss) have been recorded, implying that a cold climate was in existence. Insect remains support vegetational evidence of the latter. The assemblage, including *Diacheila arctica*, denotes a climatic regime at the end of the Early Midlandian with July temperatures 11–13°C and those of January −18 to −7°C.

At Hollymount, County Fermanagh (McCabe *et al.* 1978), freshwater organic silts have been discovered between two tills. Plant macrofossils from the silts include taxa (*Saxifraga oppositifolia* (purple saxifrage), for example) from open vegetation in high latitudes today. Their insect fauna incorporates *Arpedium brachypterum*, currently a northern and montane species. Organic detritus from the silts has given a radiocarbon age in excess of 41 500 BP, implying an Early Midlandian age for the older till. A radiocarbon date in excess of 34 000 BP from organic mud found between two tills at Greenagho, County Fermanagh, has suggested a similar relationship (Dardis *et al.* 1985).

At Derryvree in the Clogher Valley, and close to Hollymount, a Middle Midlandian interstadial deposit (Colhoun *et al.* 1972) (section 9.2.7) is underlain by a till similar in character to that beneath the silts at the latter locality (Bowen *et al.* 1986).

The sequence of Early Midlandian events has been summarized by McCabe (1987) who also has set out the terminology for them (Figure 9.7(a)). While the establishment of a clear geochronometric framework for these events has been hampered by the imprecision of the radiocarbon dates, with the likelihood that most are infinite (McCabe *et al.* 1987), they have permitted a gross timescale to be formulated. The initial episode was the Fermanagh Stadial. This occurred prior to 44 330 BP and included a lowland glaciation, which covered most of Ulster and emplaced lodgement tills. The erratic content of these tills has indicated their deposition by coeval ice moving from central Ulster and from Scotland (Figure 9.7(b)). It is probable that the lower members of multiple till sequences in other parts of Ireland are of Early Midlandian age. Deglaciation was followed by periglaciation of the deglacial sediments. The second event was the temperate interval, named the Aghnadarragh Interstadial, which took place before 48 180 BP (providing a further marker for the age of the Fermanagh Stadial) and included the development of boreal forest. Thirdly came the Hollymount Cold Phase, which was in progress earlier than 41 500 BP at Hollymount and 46 850 BP at Aghnadarragh, when open environments were present.

On the Isle of Man, local drift identified in the uplands has been thought, by Mitchell (1972a) and Thomas (1976, 1977), to be a product of severe periglaciation that took place on a nunatak throughout the Devensian. Head deposits of almost identical character to the insular drift have been seen to underlie Late Devensian foreign glacigenic sediments at the margin of the island (section 9.3.7), and probably reflect earlier periglaciation there, probably within the last cold stage (Thomas 1977; Thomas and Hardy 1985).

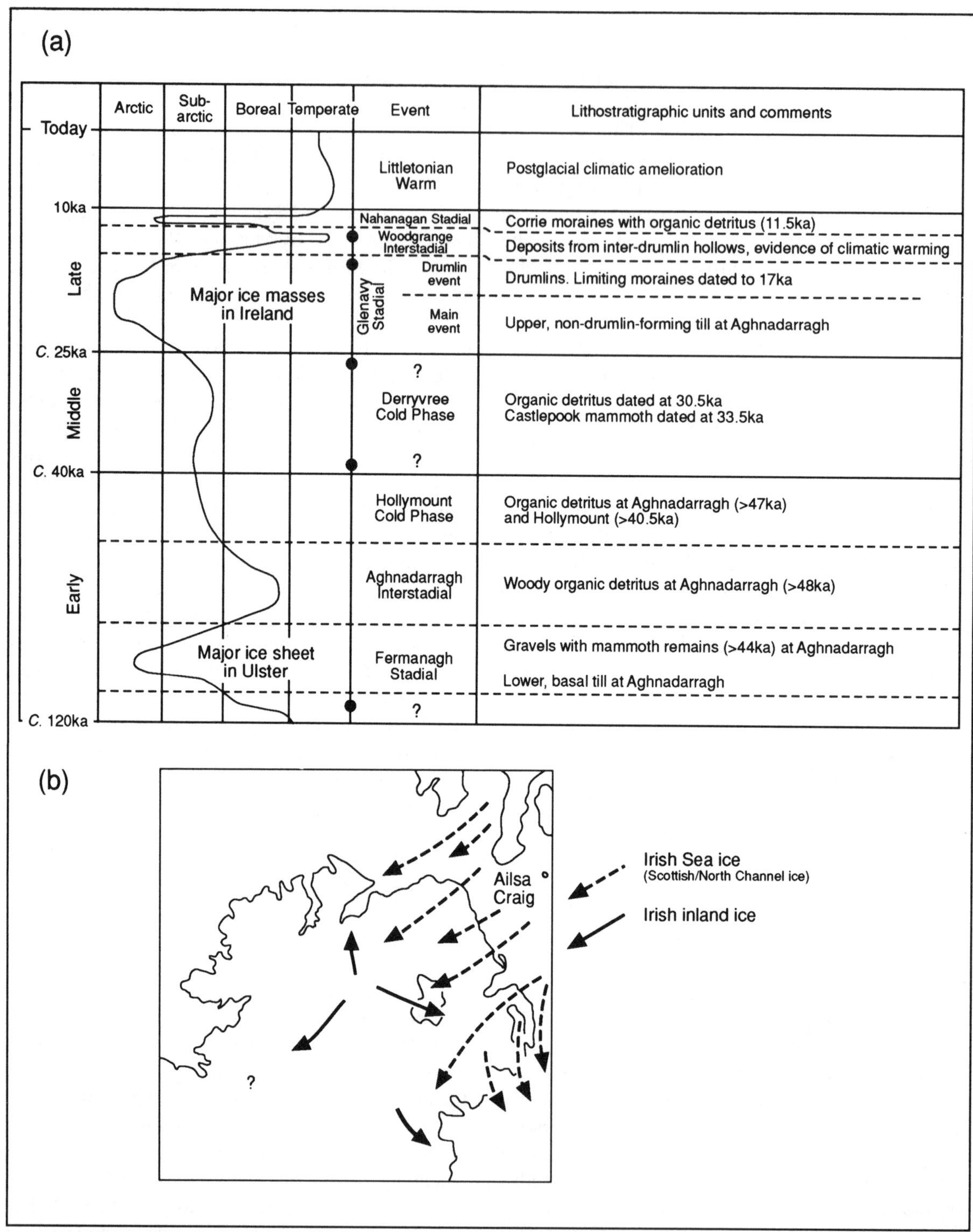

Figure 9.7 The Midlandian Cold Stage. (a) Curve to indicate general temperature oscillations and major events. Dots indicate the position of known unconformities. From McCabe (1987); modified from Mitchell (1976). (b) General ice movements in Ulster during the Early Midlandian. From McCabe (1987). Reprinted with permission from *Quaternary Science Reviews* **6**, A. M. McCabe, Quaternary deposits and glacial stratigraphy in Ireland; © 1987 Pergamon Press Plc.

9.1.10 The offshore sequence and sea-levels

Organic material from a core through the infill of the Fosse Dangeard in the English Channel has been analysed for pollen by Morzadec-Kerfourn (1975). A probable Early Devensian interstadial has been identified on this basis and tentatively correlated with that of Brørup (section 9.1.11) (Figure 9.9). Of especial interest was evidence of saline wetland vegetation at *circa* −90 m NGF which has afforded information concerning Early Devensian sea-level in this area.

According to Cameron *et al.* (1987) the Middle and Late Pleistocene sediments of the North Sea have provided evidence of only three major episodes of glaciation, ascribable to the Anglian, Wolstonian and Devensian cold stages (Figure 1.1). Jansen *et al.* (1979) have suggested that the level of the North Sea was reduced to at least 110 m below that of today by the start of the Late Devensian (section 9.3.8). Ipswichian deposits found in the southern North Sea are succeeded in numerous localities by brackish-marine lagoonal clays belonging to the Brown Bank Formation. These sediments were deposited during a marine regression, when sea-level fell to about 40 m below that of today. They have been extensively bioturbated and locally cryoturbated, and have yielded pollen assemblages characteristic of the late Eemian and Early Weichselian in the Netherlands (Cameron *et al.* 1989b).

In the central and northern North Sea, glaciomarine sedimentation led to the infill of pre-existing valleys during the Early Devensian (Cameron *et al.* 1987). On the inner continental shelf to the west of Scotland, the glaciomarine Stanton Formation is possibly of Early and/or Middle Devensian age (Figure 9.8). It is most prominent in the Sea of the Hebrides and the Malin Sea, and has been located directly below an erosion surface thought to equate with the Main Late Devensian Glaciation (section 9.3.8). The deposit may have had its provenance in calving ice generated from grounded glaciers on the Scottish mainland and the Inner Hebrides (Davies *et al.* 1984).

9.1.11 Age and correlation

According to Mitchell *et al.* (1973a), the Devensian could be correlated with the Weichselian Cold Stage of the Netherlands and the western part of north-west Europe. West (1977b) has equated the Early Devensian with the Early Weichselian plus the first part of the Middle Weichselian (the Lower Pleniglacial) of the Netherlands (Zagwijn 1961, 1974b, 1989; de Jong 1988) (Figure 9.9). As stated earlier, time prior to 50 000 BP has been assigned to the Early Devensian by Mitchell *et al.* (1973a), who have placed the Chelford Interstadial at around 60 000 BP. This interstadial has been correlated with the Brørup event in Jutland (Andersen 1957, 1961) by Simpson and West (1958). The Wretton Interstadial (West *et al.* 1974) has been tentatively equated with the Amersfoort Interstadial of the Netherlands (Zagwijn 1961).

Bryant *et al.* (1983a) have discussed Early Devensian interstadials in the context of their Brimpton evidence (section 9.1.6). They have pointed out that the correlation of Early Devensian interstadials in the British Isles with Early Weichselian/Wurmian ones on mainland Europe has been based on two major suppositions. The first, that the Ipswichian was the last interglacial; the second, that radiocarbon dates pertaining to the interstadials are sound. Both of these suppositions they have suggested to be insecure. Palynological information from Grand Pile in the southern Vosges Mountains of north-eastern France (Woillard 1978) (Figure 1.2(d)) has indicated that two further temperate episodes of interglacial rank (St. Germain I and II) followed the Eemian. The Brørup and Odderade (Averdieck 1967) interstadials, but not the Amersfoort Interstadial, have also been recognized at Grand Pile. Bryant *et al.* have hypothesized that the Wretton and Amersfoort events could relate, in part, to one of the two additional interglacials. A key member of the biota at this time may have been *Picea*, whose pollen was plentiful in the late Eemian, in the two extra interglacials and in the Brørup equivalent at

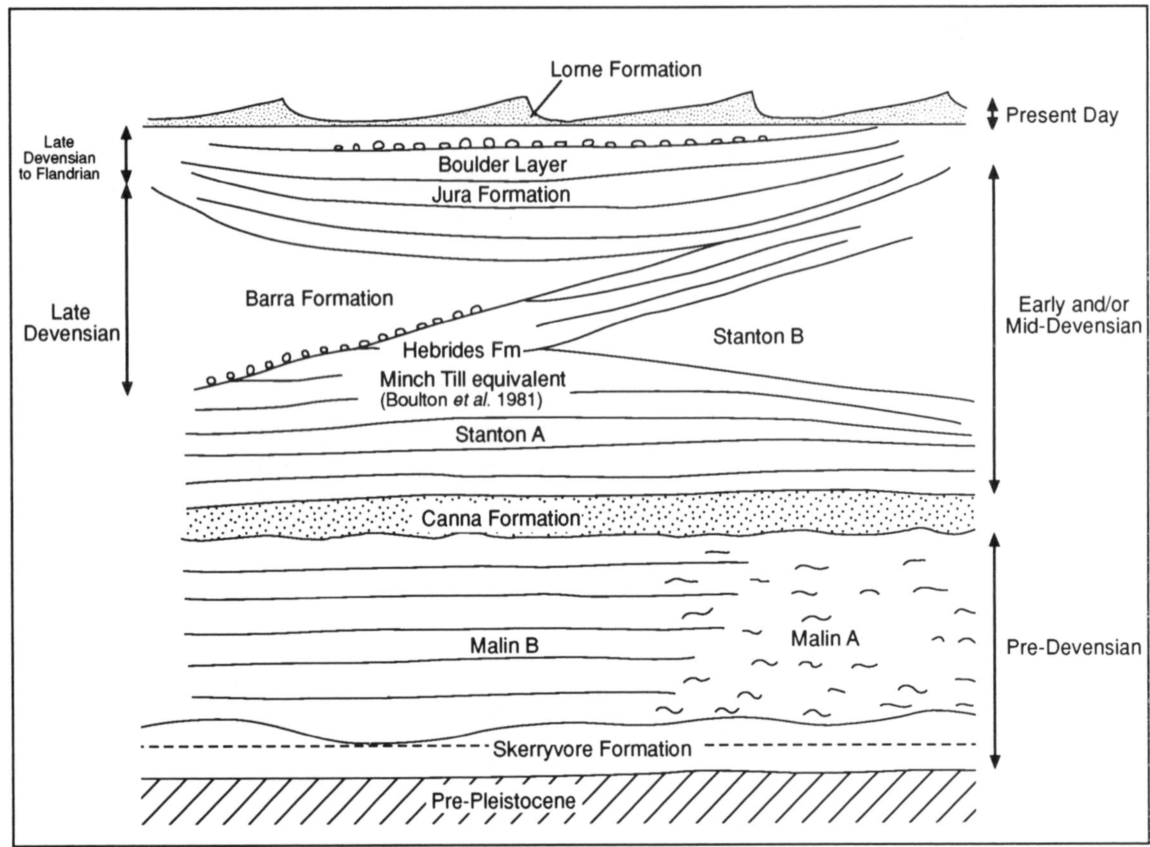

Figure 9.8 A schematic illustration of the composite Pleistocene seismic stratigraphy in the Malin–Hebrides sea areas with the suggested age for each formation. From Davies *et al.* (1984). Reproduced with permission of *Boreas* **13**, 1984; A revised seismic stratigraphy for Quaternary deposits; H. C. Davies *et al.*; Fig. 3, p. 54.

Grand Pile. In addition to *Picea abies* pollen, the Amersfoort and Brørup localities have yielded smaller grains referred to *Picea omorikoides* type (Zagwijn 1961). According to Holyoak (1983b) the closest analogue for this taxon is *Picea omorika* (Serbian spruce), currently native only in central Yugoslavia (Jalas and Suominen 1973). Only *Picea abies* occurs at Chelford (Holyoak 1983b) and thus correlation between this interstadial and that at Brørup could be questioned, according to Bryant *et al.* (1983a). However, they have stated that if *Picea omorikoides* was unable to immigrate into the forest vegetation of the British Isles, and the Chelford and Brørup events were equivalent, the Brimpton Interstadial could be correlated with that of Odderade. The ambivalence of the Chelford radiocarbon dates has been noted (section 9.1.1). Against this backdrop, it is possible that the boreal environment depicted at Chelford relates to a post-Ipswichian interglacial rather than to an Early Devensian interstadial. Finally, in the context of these problems of age and correlation, it should be noted that at Les Echets, near Lyon, France, pollen data for temperate episodes equated with those of St. Germain I and II at Grand Pile probably reflect interstadial conditions. Correlation of two episodes within St. Germain I with the Amersfoort and Brørup events, and of one during St. Germain II with the Odderade, has been suggested (de Beaulieu and Reille 1984). St. Germain I and II have been equated with Oxygen Isotope Substage 5c and Oxygen Isotope Substage 5a, respectively, by Guiot *et al.* (1989).

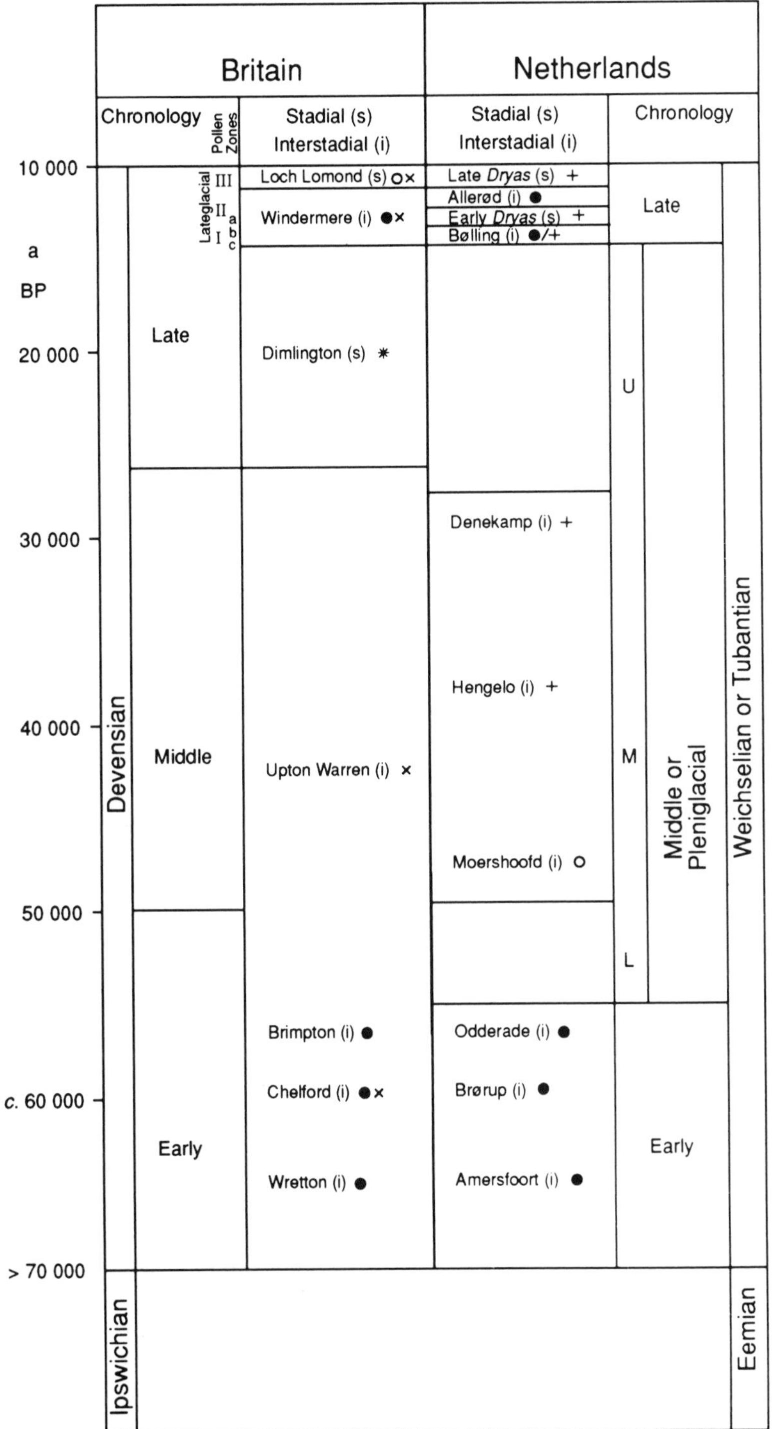

Figure 9.9 The Devensian sequence in Britain and its correlation with that of the Weichselian in the Netherlands. *, insect and plant macrofossil evidence of stadial conditions; ○, pollen-based stadial or interstadial: tundra, +, pollen-based stadial or interstadial: shrub tundra; ●, pollen-based interstadial: woodland, x, insect-based stadial or interstadial. From West (1977b); additional sources, Bryant *et al.* (1983a), Coope and Pennington (1977), Gray and Lowe (1977), Rose (1985, 1989a), van Staalduinen *et al.* (1979), van der Hammen (1951), van der Hammen *et al.* (1967), Zagwijn (1961, 1974c).

The U-series ages of speleothems (section 9.1.3) have assisted in the clarification of the pattern of environmental change over the timespan usually assigned to the Early Devensian, especially around 80 000 BP. Gascoyne *et al.* (1983) have reported low growth of speleothems in north-west England 90 000–45 000 BP, inferring the existence of a non-glacial tundra-like climate. Gordon *et al.* (1989) have pointed to a clear fall in speleothem growth frequency in the British Isles after a peak at 90 500 BP (Chapter 8), with a minimum at 80 000 BP considered to mark the boundary between the last interglacial and last cold stage. Peaks of speleothem growth at 76 000, 57 000 and 50 000 BP have been proposed as indicators of Devensian interstadials.

Bowen *et al.* (1989) have concluded that amino-acid ratios in shells from Upton Warren interstadial deposits, traditionally regarded as of Middle Devensian age (section 9.2.1), rank nearer in time to those of Ipswichian date than do the radiocarbon assays from deposits of the interstadial. Oxygen Isotope Stage 5 (probably Substage 5a, *circa* 80 000 BP) is implied for the Upton Warren event on this basis (Figure 1.15). As such, its correlation with the Odderade and St. Germain II episodes (Woillard 1978; Woillard and Mook 1982; Guiot *et al.* 1989) seems feasible. The Chelford event was earlier than that of Upton Warren and probably occurred during 5c, when the Brørup (Behre 1989) and St. Germain I (Guiot *et al.* 1989) episodes are also thought to have taken place (Bowen *et al.* 1989).

It thus will be evident that the number, order and timing of events over this timespan is both imperfectly known and controversial.

9.2 The Middle Devensian

The time involved in this part of the Devensian, *circa* 50 000–26 000 BP (Mitchell *et al.* 1973a), is within the upper limit of radiocarbon dating. However, as is observed below, many of these dates have been infinite, with some ages based on sample size, and their standard errors have often been large. It thus appears that like those for the Early Devensian, many Middle Devensian radiocarbon age determinations have been at best, imprecise and at worst, wrong.

The overall impression from lithostratigraphic and biostratigraphic records is of a substantial period with cold conditions interrupted by a climatic oscillation which gave rise to an interstadial complex.

9.2.1 The Midlands of England

At Four Ashes, sandy detritus peat and organic clay found within sands and gravels (Figure 9.4) have yielded radiocarbon ages of *circa* 42 000–30 000 BP (A. V. Morgan 1973). These sands and gravels, which have also been identified south of the Late Devensian ice limit in this area (section 9.3.1), were deposited by a braided river and were penetrated by ice-wedge casts during the Middle Devensian. The first cold episode at Four Ashes took place prior to 43 500 BP and almost certainly after the Chelford Interstadial (section 9.1.2). Biological (mainly insect) evidence has identified this cold episode (A. Morgan 1973), which may have caused the disappearance of the boreal forest of Chelford time. The insect fauna contains many cold stenotherms, a few eurytherms and lacks warm stenotherms. Included among current northerly distributed species are *Helophorus sibiricus*, *Boreaphilus nordenskioeldi* and *Pterostichus blandulus*. The overall impression from the fauna is of a continental climate with cool summers. An indication of shifts in the cover and diversity of the herbaceous vegetation is provided by the relative abundances of phytophagous insects. A sparse plant cover of low diversity (grasses, sedges and dwarf willows) in cooler periods is denoted by a limited number of species, with warmer intervals having a greater variety of plants making a more complete ground cover indicated by an increase in plant-feeding insects.

Between about 42 500 and 38 500 BP the insect fauna at Four Ashes underwent significant changes. Warm stenotherms (including *Georyssus crenulatus*, *Crypticus quisquilicus* and *Otiorrhynchus ligneus*) came to dominance, while continental and northern species were lacking.

The inference is of a climate of reduced continentality, with warmer summers than previously. It was an interstadial-type climate, and one in which tree growth would have been possible. However, palynological investigations of the sediments have revealed an absence of tree pollen and no wood-eating insects have been discovered. Of climatic significance was the occurrence of *Calanthus melanocephalus* with a red protonum. Today, insects with the latter (as opposed to a black one) are largely distributed lower than 260 m OD in Scotland (Greenslade 1968). After about 38 000 BP, the numbers of northern and eastern stenothermic insects increased at Four Ashes, a situation which prevailed until *circa* 30 000 BP. The landscape was tundra, whose climatic characteristics included July mean temperatures of around 10°C towards the conclusion of this timespan (Figure 9.3). Colder polar-desert conditions ensued, and the Late Devensian ice advance took place later than 30 000 BP in southern Staffordshire.

The interstadial recognized at the Devensian type site was first identified by Coope *et al.* (1961) within terrace deposits of the River Salwarpe (a tributary of the River Severn) at Upton Warren, between Bromsgrove and Droitwich in Worcestershire. The episode subsequently became known as the Upton Warren Interstadial. In the lower part of the terrace gravels at Upton Warren are cryoturbated lenses of organic sandy silt that accumulated in shallow ponds. Radiocarbon assays on plant-rich organic material have given ages of 41 900 ± 800 and 41 500 ± 1200 BP. The terrace passes into the Main Terrace of the River Severn (Wills 1938) and has also been equated with Terrace 2 of the Warwickshire – Worcestershire Avon sequence (Tomlinson 1925; Maddy *et al.* 1991) (Figure 7.6). The molluscan fauna of the organic beds includes the temperate forms *Lymnaea peregra* and *Planorbis laevis*, but also *Pisidium obtusale lapponicum*, an arctic subspecies. In an ostracod assemblage diagnostic of shallow water, the incidence of *Cyprinotus salinus* is of especial interest. This oligohaline or slightly brackish-water species is unlikely to have been living in tidal water, as estimates from terrace heights imply a sea-level analogous to that of today. A more feasible explanation for its presence is the existence of brine springs in the Permo-Triassic Keuper Marl (Mercian Mudstone) that occurs beneath the gravels. The insect fauna from Upton Warren contains eight species presently found only in the higher latitudes of Scandinavia and 13 currently resident in northern parts of the British Isles. However, there is also a substantial group of Coleoptera with ranges now almost wholly south of 60°N, and which inhabit the British Isles. The ranges of the northern and southern groups are mutually exclusive in many cases. The fauna also possesses a continental element, including *Carabus hortensis* and *Philonthus linki*. Reworking of sediment, leading to the incorporation of both northern and southern forms, is unlikely as the fossils are in good condition. A more plausible explanation is that a resident cold fauna became mixed with a warmer one which colonized in response to a rapid climatic amelioration. Plant remains from the site consist of macrofossils and pollen. There are substantial frequencies of Cyperaceae, Gramineae, Compositae and Caryophyllaceae (pink family) species and little evidence of trees. Of particular botanical interest is the existence of a halophyte element in the macrofossils which includes *Glaux maritima*, *Triglochin maritima* and *Plantago maritima* (sea plantain). The flora is mainly temperate in character and of open habitats but with a notable boreal and arctic–alpine element. *Thalictrum alpinum* (alpine meadow rue), *Draba incana* (hoary whitlow grass), *Salix herbacea* and *Viscaria alpina* (red alpine catchfly) are among those plants present that now inhabit high latitudes or altitudes. In a climate probably akin to that of southern Sweden today, a largely thermophilous assemblage of herbs seems to have immigrated into this part of the Severn Basin. However, these plants were not followed by shrubs and trees, as would be expected during a temperate episode. Remains of *Mammuthus primigenius*, *Coelodonta antiquitatis*, *Bison priscus* (extinct bison) and *Rangifer tarandus* (animals also typically found fossilized in the Severn Main Terrace and Avon

Terrace 2) have been recorded at Upton Warren. It has been suggested that grazing by these large mammals could have inhibited the development of a tree and shrub cover.

The base of Avon Terrace 2 at Fladbury, Worcestershire, contains peat-like material which has been radiocarbon dated to 38 000 ± 700 BP (Coope 1962). Insect evidence from the organic deposit shows that it accumulated in a floodplain pool around which sedges and willows were growing. Beyond the wetland, the vegetation was sparse and the landscape sandy. A similar large-mammal fauna to that of Upton Warren has been obtained from this site. The dung beetle *Aphodius* is present in profusion, suggesting use of the pool by these animals for drinking. The insect assemblages from Fladbury and Upton Warren are similar. However, at Fladbury, *Diacheila arctica*, *Boreaphilus henningianus* and *Symplocaria metallica* occur, demonstrating the existence of harsher climatic conditions.

The Syston Terrace of the Soar Valley, Leicestershire (Rice 1968), has been regarded as equivalent to Avon Terrace 2, the Low Terrace of the River Tame (see below) and the Floodplain Terrace of the River Trent (Clayton 1953). Cryoturbated sand and gravel located within the Syston Terrace contain organic layers which have been described by Bell *et al.* (1972). Plant remains from one of the layers have been radiocarbon dated to 37 420 $^{+1670}_{-1390}$ BP. No pollen has been recorded and plant macrofossils are mainly of herbs now lacking southerly distributions in Europe. Tree remains have not been encountered, although *Betula nana* and *Salix herbacea* existed at this time. The insect assemblage, including arctic – alpine taxa, resembles that from Fladbury.

Coope and Sands (1966) have described involuted, peaty sand lenses from the base of gravels of the Low Terrace, at three localities around Minworth and Coleshill, in the Tame Valley, Warwickshire. One of these organic deposits has been radiocarbon dated to 32 160 $^{+1780}_{-1450}$ BP. The insect fauna, including *Boreaphilus nordenskioedi*, indicates a cold, continental climate and a vegetation cover lacking trees. July mean temperatures have been estimated at *circa* 13°C, those of January may have been as low as −17°C. The deposits in the Tame Valley, plus those at Fladbury and Upton Warren have been linked in what is termed the Upton Warren Interstadial Complex.

At Brandon near Coventry, cryoturbated deposits of Avon Terrace 2 have produced organic material aged by radiocarbon assay to between 32 000 and 29 000 BP (Shotton 1968). Mollusca present in this sediment include the cold-climate forms *Columella columella* and *Pupilla muscorum*. Remains of *Dicrostonyx torquatus* are also present. The macroflora (no pollen has been obtained) has been described by Kelly (1968). *Betula nana*, *Salix herbacea*, *Polygonum viviparum* (alpine bistort) and *Thalictrum alpinum* were present in a sub-arctic or low-arctic environment with bare ground, grassland, heath and shrubland. The insects from Brandon have been reported by Coope (1968a). No tree-dependent taxa have been recorded but *Rynchaenus foliorum*, an obligate of *Salix*, is abundant. *Carabus arvensis* and *Amara alpina* denote barren heath in drier areas, while *Bembidion aeneum* and *Ochthebius viridus* hint at saline conditions in the locality. A tundra-like climate, harsher than that under which the Tame Valley organic deposits were laid down, has been postulated. July temperature values of *circa* 10°C (Figure 9.3), and January ones of about −20°C, with a mean annual value of at most −5°C, have been indicated for the Coventry area at this time.

At Great Billing, in the Nene Valley near Northampton (Morgan 1969), periglacial terrace gravels contain organic clays radiocarbon dated to *circa* 28 000 BP. Insects from these clays include *Pelophila borealis*, *Diacheila arctica* and *Boreaphilus nordenskioeldi*; molluscs include *Pupilla muscorum*, *Pisidium casertanum* and *Oxyloma pfeifferi*; and plants include *Carex bigelowi* (stiff sedge), *Thalictrum alpinum* and *Betula nana*. Remains of the large mammals *Mammuthus primigenius* and *Coelodonta antiquitatis* have also been recorded from this locality, where fossil evidence suggests an arctic tundra environment.

9.2.2 Yorkshire, Lincolnshire, the northern Dukeries and East Anglia

In the valley of the River Aire, at Oxbow between Leeds and Castleford, Gaunt *et al.* (1970) have observed sand and gravel succeeded by silt. *Mammuthus primigenius* remains from the silt have been dated to 38 600 $^{+1720}_{-1420}$ BP by radiocarbon assay. Botanical data indicate the presence of tundra, whose components included *Betula nana*, *Thalictrum alpinum*, *Arenaria ciliata* (fringed sandwort) and *Potentilla crantzii* (alpine cinquefoil), in addition to species of Gramineae and Cyperaceae. The insect fauna contains the arctic inhabitants *Diacheila polita*, *Amara alpina* and *Feronia blandulus*, supporting the concept of an open, treeless landscape. July temperature values of around 10°C and a continental climatic regime has been suggested.

By means of U-series age determinations, Gascoyne *et al.* (1983) have deduced overall low speleothem growth in the Craven District of north-west England from 85 000 to 35 000 BP. For the period 44 000–34 000 BP, a minor increase in the number of dated speleothems occurs, which may signify episodes of milder climate. No ages have been obtained for 34 000–13 000 BP. This has been interpreted as demonstrating the existence of continuous permafrost then ice in the area. However, Atkinson *et al.* (1986) have produced age data from speleothems in the same locality which suggest their intermittent growth from 40 000 to 26 000 BP. Such a situation would have meant that discontinuous rather than continuous permafrost was present over this timespan (Chapter 1).

From Tattershall, in the lower Bain Valley, Lincolnshire, two organic horizons which occur in sands and gravel above Ipswichian deposits (Chapter 8) at the Castle locality have provided insect evidence regarding the age and characteristics of the environments just prior to, and during, the Upton Warren temperate episode *sensu stricto* (Girling 1974, 1977). The lower layer has given radiocarbon ages of 44 300 $^{+1600}_{-1300}$ and 42 100 $^{+1400}_{-1100}$ BP, and contains a meagre insect assemblage of tundra affinity that includes *Diacheila arctica* and *Olophorum boreale*. The radiocarbon ages of the demonstrably younger, upper organic deposit, overlap those of the older one. However, the upper layer possesses a diverse and thermophilous insect fauna, with a component (represented by *Hydrochus flavipennis* and *Aphodius bonvouloiri*) that now inhabits southern Europe. These insects were able to survive in Lincolnshire during what appears to have been a short thermal maximum within the interstadial complex. Vertebrate remains from the Middle Devensian deposits at Tattershall Castle and Tattershall Thorpe have been discussed by Rackham (1978). At the former, an upper silt contains bison and reindeer. At the latter, there is only one silt, radiocarbon dated to 34 800 BP, whose mammal fauna, dominated by woolly rhinoceros, mammoth and horse, indicates a subsequent colder and more continental environment. The insect fauna from this silt is of arctic affinity but different in composition to that recorded by Girling from the older (lower) silt at the Castle locality. The Middle Devensian fluviatile gravels of the lower Bain Valley have revealed abundant periglacial structures. The latter have also been identified in gravels of comparable type and age in other parts of the Fen Basin (Straw 1979a) (Figure 1.8).

At Pin Hole in the Creswell Crags Gorge, Earlier Upper Palaeolithic artifacts have been recorded from the Upper Cave Earth (Campbell 1977; Jenkinson 1984; Jenkinson *et al.* 1985). At Robin Hood's Cave in the same area, Earlier Upper Palaeolithic artifacts have also been found (Mellars 1974) (Figure 9.10), and a radiocarbon date of 28 500 $^{+1600}_{-1300}$ BP has been obtained from bone of *Ursus arctos* (Campbell 1977). Hunters and their prey thus seem to have been resident in northern Derbyshire and Nottinghamshire during the Middle Devensian.

In the Cambridge region, one of the plant beds at Earith (Bell 1970) (section 9.1.3) has been radiocarbon dated to 42 140 $^{+1890}_{-1530}$ BP. The fossil flora indicates treeless steppe vegetation that included willow scrub and many herbs. Some of the latter (*Arenaria ciliata*, *Saxifraga oppositifolia* and *Draba incana*, for example) have northern

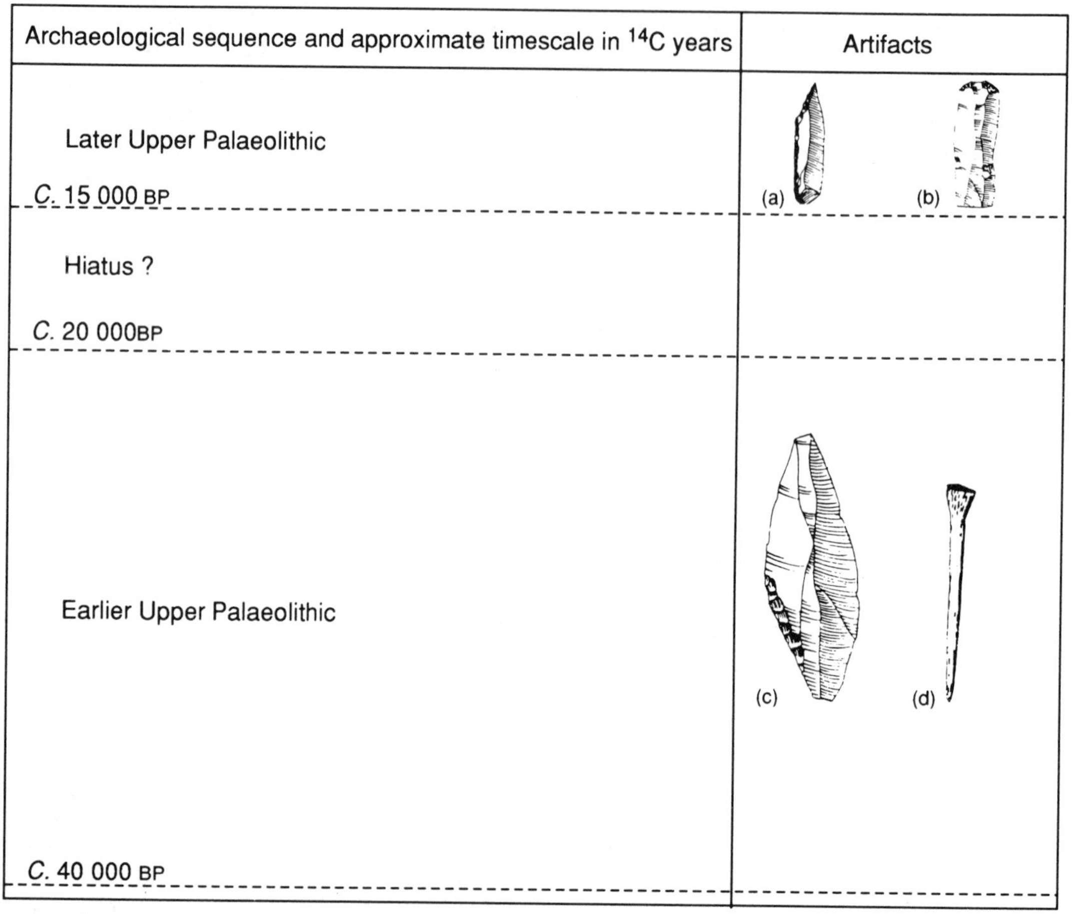

Archaeological sequence and approximate timescale in ^{14}C years	Artifacts
Later Upper Palaeolithic *C.* 15 000 BP	(a) (b)
Hiatus ? *C.* 20 000BP	
Earlier Upper Palaeolithic *C.* 40 000 BP	(c) (d)

Figure 9.10 The Upper Palaeolithic archaeological sequence and approximate timescale. From Campbell (1977). Reproduced by permission of Oxford University Press. Artifacts. (a) Creswell point, Dead Man's Cave, North Anston. From Mellars (1974). (b) End scraper, Mother Grundy's Parlour, Creswell. From Mellars (1974); source, Armstrong (1925). (c) Bifacially-worked 'leaf point', Robin Hood's Cave, Creswell. From Mellars (1974); source, Garrod (1926). (d) Bone pin, Kent's Cavern. From Mellars (1974); source, Garrod (1926).

ranges today. There were also halophytes including *Suaeda maritima*, *Glaux maritima* and *Juncus gerardii* (mud rush), which probably grew in depressions where salt accumulated as a result of high evaporation rates during summer months. *Najas flexilis*, *Lycopus europaeus* (gipsy-wort) and *Groenlandia densa* have been recorded, indicating July mean temperatures of around 16°C. Winters were cold and permafrost was likely to have been present.

At Barnwell Station near Cambridge, an insect fauna with an arctic–alpine element, obtained from peats within sands of the Second Terrace of the River Cam, has been referred to the closing phase of the Upton Warren Interstadial Complex by Coope (1968b).

Warren (1912), Reid (1949) and Godwin (1956) have described an Arctic Plant Bed in the Broxbourne–Nazeing areas of the Lea Valley, about the border of Hertfordshire and Essex. *Juniperus* and *Betula nana* remains have been found, but no remains of trees. There are abundant herbs, and a sedge-grass tundra at the time of deposition is indicated. Mollusca from these organic deposits include *Columella columella* and *Pupilla muscorum*, and a mammalian fauna associated with them numbers mammoth, woolly rhinoceros, reindeer and lemming among its components. A radiocarbon date of 28 000 ± 1500 BP has been obtained from the organic material (Godwin and Willis 1960).

9.2.3 The Thames Valley

An organic silt close to the base of the stratigraphy in the Upper Floodplain Terrace of the River Thames (Figure 3.6) at Isleworth, Middlesex, has yielded a substantial insect fauna (Coope and Angus 1975). Remains of mammoth, woolly rhinoceros, reindeer and bison have also been recorded in this stratigraphic context. The gravels above the silt contain ice-wedge casts, while plant remains from the organic deposit have given a radiocarbon age of 43 140 $^{+1520}_{-1280}$ BP. A reasonable sized pond fringed by sedge-dominated wetland was present locally. Moist grassland and drier, less-vegetated areas existed further away. No beetles dependent on trees have been discovered and dung-feeding and carrion-feeding taxa occur. The general characteristics of this fauna are analogous to those of the insect assemblage in southern England today. However, a mix of northern and more temperate taxa existed, and 22 species now extinct in the British Isles but occurring on mainland Europe were present. Such an assemblage of insects can currently be found in the lowlands of northern Germany. Here January mean temperatures are about 0°C, with those in July close to 18°C. These climatic characteristics are representative of the thermal maximum of the Upton Warren Interstadial Complex (Figure 9.3). As at Upton Warren, trees could have grown in the Thames Valley but did not. Coope and Angus have pointed out that tundra was in existence for about 15 000 years prior to the onset of interstadial conditions. Forest had then retreated from the British Isles, probably into southern Europe. The beginning of the Isleworth–Upton Warren temperate episode was seemingly rapid. Such a situation would have enabled thermophilous insects to immigrate quickly from southerly refugia. Rates of plant migration would have been slower, with herbs the first to arrive in the British Isles. The less efficient seed dispersal, longer generation time and more precise edaphic requirements of trees would have militated against their colonization over a time-span of perhaps *circa* 1000 years (Coope 1979), before tundra developed once again. It is also possible, as in the Severn Basin, that herbivore grazing and trampling could have inhibited the spread of woodland to the Thames Valley. Reindeer and bison are migratory and could have ranged widely to the south via land links at this time (section 9.2.8), thereby impeding the northward advance of trees.

The deposits at Isleworth have also been studied by Kerney *et al.* (1982). The gravels have been established as equivalent to those at Kempton Park (see below). Botanical evidence indicates regional temperate grassland and little or no tree growth. No distinctively arctic–alpine plants or molluscs have been recorded. Among the latter, *Pupilla muscorum* is dominant in the terrestrial assemblage, implying a dry, calcareous, treeless landscape. *Anisus vortex* and *Pisidium moitessierianum*, also present, have normally been associated with interglacial environments in the British Isles.

At Kempton Park, Sunbury, Surrey, gravel (the Kempton Park Gravel) beneath the Upper Floodplain Terrace of the River Thames contains clayey silt with organic remains (Gibbard *et al.* 1982). Plant macrofossils, notably *Betula nana* and *Salix herbacea*, indicate a cold-stage deposit. The only terrestrial mollusc is *Pupilla muscorum*. The incidence of *Candona* cf *lozeki* within the ostracod sequence suggests a worsening climate during silt deposition. This climatic deterioration has also been detected by means of the insect fauna. *Diacheila polita* and *Colymbetes dolabratus*, recorded near to the top of the silt, are currently circumpolar dwellers at high latitude and altitude, respectively. A shift to colder and continental conditions is indicated. Organic material from the silts, dated by radiocarbon, shows that this episode concluded at 35 230 ± 185 BP. Thermoluminescence dating of the silts has given a mean age of 36 600 ± 6400 BP (Southgate 1984). These silts were thus deposited during the cooling phase of the Upton Warren Interstadial Complex that followed the Isleworth–Upton Warren thermal maximum episode.

According to Bridgland (1988) the East Tilbury Marshes Gravel of the Lower Thames Valley can be equated with the Kempton Park Gravel. Wymer (1988) has reported Mousterian-type axes from both these gravels. Bell (1969) has described organic silt in Thames floodplain gravel at Marlow, Buckinghamshire. This silt has produced a herb-dominated treeless plant assemblage including southern, steppe and halophytic taxa. A radiocarbon age in excess of 31 000 BP (Shotton and Williams 1971) has been obtained from the deposit, which may belong to one of the cooler parts of the Upton Warren Interstadial Complex.

The earlier deposits of the Floodplain Terrace in the Upper Thames Valley (Figure 7.7) have been ascribed a Middle Devensian age by Briggs and Gilbertson (1980). They are typically braided-river sediments containing periglacial features together with organic horizons. A mammalian fauna including reindeer, woolly rhinoceros and mammoth has been associated with the deposits, on the bedrock surface below which Levalloisian artifacts have been found *in situ*. Occasional finds of Mousterian implements have been made within the gravel. Thus, a limited presence of man at this time seems likely (Briggs *et al.* 1985). Radiocarbon ages of 39 300 ± 1300 BP from Queensford, 34 500 ± 800 BP from Little Rissington, 33 190 ± 3450 BP from Sutton Courtenay, 29 500 ± 300 BP from Standlake Common and 28 400 ± 100 BP from North Lechlade have confirmed the Middle Devensian context of the organic deposits. Plant evidence from Sutton Courtenay and North Lechlade indicates open herb-dominated vegetation and a cold climate. Mollusca from Sutton Courtenay, North Lechlade, Standlake, Bourton-on-the-Water and Abingdon are typical of many Middle Devensian localities. The insect faunas are characteristic of those associated with cool episodes during the latter part of the Upton Warren Interstadial Complex. A barren, exposed tundra-like landscape has been inferred from coleopteran evidence. Beetles present include *Tachinus arcticus*, now of Siberian distribution, and abundant *Aphodius* species, which feed on mammal dung. A continental climate, in which July mean temperatures have been estimated at around 10°C and those of January at least −10°C and perhaps down to −25°C, has been inferred.

Possible Early Devensian deposits at Stanton Harcourt are discussed above (section 9.1.4). However, Seddon and Holyoak (1985) have reported a radiocarbon age of 34 730 $^{+440}_{-420}$ BP on plant remains from a channel-fill at this site, which is at variance with thermoluminescence dates from silt in a superior stratigraphic position. The pollen assemblage is mainly herbaceous, and macrofossil plant fossils include the northern forms *Arenaria ciliata* and *Salix herbacea* and the southerly distributed *Groenlandia densa* and *Linum perenne* (perennial flax). The terrestrial molluscan fauna is dominated by *Pupilla muscorum* and the freshwater molluscan fauna is dominated by *Oxyloma pfeifferi*. Mammoth, horse and bison remains have also been recorded in this stratigraphic context. The upper sediments at Stanton Harcourt contain evidence of deposition when continuous permafrost existed. The lower sediments (containing the organic material) were assigned, mainly on biological evidence, to the later part of the Upton Warren Interstadial Complex.

9.2.4 Wales, and England south of the Lower Severn and Thames valleys

Little evidence of Middle Devensian environments has been found in Wales. Long Hole and Cat Hole caves on the Gower Peninsula have produced palynological sequences indicative of cold episodes either side of an interstadial of Upton Warren type. The associated sediments at Long Hole have yielded Earlier Upper Palaeolithic artifacts, together with remains of reindeer, horse and fox (Campbell 1977). A radiocarbon date of 38 684 $^{+2713}_{-2024}$ BP has been obtained from reindeer remains in Coygan Cave, Laugharne, on the coast of Carmarthenshire (Stuart 1982).

The Brimpton locality in Berkshire (Bryant

et al. 1983a) (section 9.1.6) also provides evidence of Middle Devensian environmental conditions. Deposition of the Upper Gravel commenced at this time. Organic material within it has produced radiocarbon ages of 29 500 ± 460 and 26 340 ± 1210 BP. Plant fossils from the organic material indicate treelessness but *Salix*, *Juniperus* and *Empetrum* suggest a shrub component in the vegetation. Herbaceous taxa are well represented and include *Arenaria ciliata*, *Cerastium alpinum* (alpine mouse-ear chickweed) and *Thalictrum alpinum*. All members of an abundant molluscan fauna, which include *Columella columella*, have current ranges extending into the arctic. Insects, among which *Tachinus jacutus* (now an inhabitant of eastern Siberia and North America) are notable, include phytophages and dung feeders and reinforce the picture of a cold climate.

The equivocal nature of the geochronometric dating evidence from Farnham (Bryant *et al.* 1983b; Gibbard *et al.* 1986a) (section 9.1.6) means that the organic deposits at this locality could reflect an environment that followed the thermal optimum of the Upton Warren Interstadial Complex.

Campbell (1977) has presented pollen spectra of Middle Devensian type from Badger Hole, Wookey in the Mendip Hills. The associated stratigraphy contains Earlier Upper Palaeolithic artifacts and bones of *Ursus*, *Coelodonta*, *Equus* and *Dicrostonyx*. At nearby Hyaena Den, where Earlier Upper Palaeolithic artifacts have also been found, a palynological record shows an episode of shrub (especially juniper and willow) resurgence in an otherwise herb-dominated flora. The shrub phase has been interpreted as part of the Upton Warren event. The latter has also been observed in a pollen sequence from Sun Hole, Cheddar (Campbell 1977). At Picken's Hole, Compton Bishop, Somerset, a cave earth has produced a bone assemblage including arctic fox (*Alopex lagopus*), hyaena, mammoth, woolly rhinoceros and horse, from which a radiocarbon age of 34 265 $^{+2600}_{-1950}$ BP has been obtained (Tratman 1964; Stuart 1977, 1982). The data of Atkinson *et al.* (1986) relating to speleothem growth (section 9.2.2) are also applicable to the Mendip region (Figure 1.10). During the Middle Devensian, sporadic growth took place until *circa* 30 000 BP, probably in conditions with intermittent permafrost. Later than 30 000 BP, no speleothem formation occurred, so that continually frozen ground must have existed.

Part of the Reindeer Stratum in Tornewton Cave, Devon (Sutcliffe and Zeuner 1962; Sutcliffe 1974a) and associated Earlier Upper Palaeolithic flint artifacts are likely to represent Middle Devensian activity (Campbell 1977). The Loamy Cave Earth at Kent's Cavern, Torquay, contains Earlier Upper Palaeolithic artifacts (Figure 9.10), and bones radiocarbon dated to 38 270 $^{+1470}_{-1240}$ BP (*Equus*), 28 720 ± 450 BP (*Ursus arctos*), 28 160 ± 435 BP (*Coelodonta antiquitatis*) and 27 730 ± 350 BP (*Bison*). The pollen record from this sediment has juniper, willow, grasses, sedges and other herbs among its components (Campbell and Sampson 1971; Campbell 1977).

In the Scilly Isles, organic deposits located below glacigenic and periglacial sediments have produced pollen assemblages that are indicative of tundra grassland. One of these deposits has been radiocarbon dated to 34 500 $^{+880}_{-885}$ BP (Scourse 1986).

9.2.5 Jersey

Rousseau and Keen (1989) have described a cold-climate molluscan fauna (including *Columella columella* and *Pupilla muscorum*) from loess of Middle Devensian age in Portelet Bay on the south coast of the island. Changes in the assemblage of Mollusca have been interpreted as a response to progressively wetter and colder conditions.

9.2.6 Scotland

A soliflucted organic palaeosol horizon found beneath glacial sediments at Teindland, Morayshire, has been radiocarbon dated to 28 140 $^{+480}_{-456}$ BP (FitzPatrick 1965). The Teindland sequence has been considered in more detail by Edwards *et al.* (1976) who have produced a pollen record dominated by Gramineae, Rubiaceae (bedstraw family) and *Calluna* species from the organic horizon. Caseldine and Edwards (1982) have reported further radiocarbon assays from this material of 40 710 ± 2000 and 38 400 ± 1000 BP, which suggest an age within the Upton Warren Interstadial Complex.

At Tolsta Head, Island of Lewis, an organic lake deposit discovered below till has been radiocarbon dated to 27 333 ± 240 BP. The pollen assemblage from this deposit contains substantial frequencies of Gramineae, *Calluna* and Compositae. *Juniperus* and *Salix* have also been recorded, as has *Koenigia islandica* (von Weymarn and Edwards 1973). The implication of the evidence is of an ice-free, possible interstadial environment in this part of Scotland *circa* 27 000 BP. The organic sediments at this site have also been examined by Birnie (1983). Two pollen assemblage zones have been delimited, the first dominated by Gramineae, the second by Cyperaceae. *Salix herbacea* macrofossils have been found. This, together with diatom records, indicate an episode during which conditions were cold enough for solifluction and the climate deteriorated.

There are a number of other sites in Scotland with organic deposits that have yielded radiocarbon ages from the conclusion of Middle Devensian time to the commencement of Late Devensian time, and which appear to reflect interstadial conditions. Remains of woolly rhinoceros in sands overlain by till at Bishopbriggs in the Kelvin Valley, Lanarkshire, have been dated to 27 550 $^{+1370}_{-1680}$ BP (Rolfe 1966; Sissons 1967b). Reindeer antler from Reindeer Cave, Assynt, has produced ages of 25 360 $^{+810}_{-740}$ BP and 24 590 $^{+790}_{-720}$ BP (Lawson 1984). At Crossbrae Farm, Buchan, peat underlain by probable Early Devensian till (section 9.1.8) and with soliflucted Late Devensian till above, has been aged to 26 400 ± 170 and 22 380 ± 250 BP. Pollen and plant macroremains of interstadial type have been obtained from the peat (Hall 1984a). An organic sand that has been found below head on Hirta in the St. Kilda archipelago has produced a date of 24 710 $^{+1470}_{-1240}$ BP. A pollen flora dominated by grasses, sedges and composites, but with a limited representation of shrub and heath plants, has been obtained from the sand (Sutherland *et al.* 1984).

Uranium-series ages (38 000 ± 6000; 30 000 ± 4000; 26 000 ± 3000 and 26 000 ± 2000 BP) obtained by Atkinson *et al.* (1986) from speleothems in Assynt indicate that if, as Sutherland (1981a) has proposed (section 9.1.3), an ice sheet had been building up in Scotland since *circa* 75 000 BP, it was to the south of this area.

9.2.7 Ireland

At Derryvree, County Fermanagh (Colhoun *et al.* 1972), freshwater sediments located between two tills (the lower deposited during the Fermanagh Stadial) (section 9.1.9) have been radiocarbon dated to 30 500 $^{+1170}_{-1030}$ BP. Plant remains (mainly grasses, sedges and mosses) from these organic deposits suggest a treeless landscape, although macrofossils of *Dryas octopetala* (mountain avens) and *Salix herbacea* denote a dwarf-shrub component in the tundra-like vegetation. The insect fauna is analogous to that at Brandon (section 9.2.1). The environment represented has been referred to the Derryvree Cold Phase and the Derryvree Interstadial Complex (Mitchell 1976; McCabe 1987) (Figure 9.7(a)).

A mammoth tooth from Castlepook Cave, County Cork, has been radiocarbon dated to 35 000 ± 1200 BP and a hyena bone to 34 300 ± 1800 BP. The mammal fauna also includes lemming, arctic fox and reindeer. However, in common with other Middle Midlandian localities in Ireland, woolly rhinoceros and bison are lacking (Scharff *et al.* 1918; Mitchell 1981a; Stuart 1982, 1985).

9.2.8 The offshore sequence and sea-levels

Kellaway *et al.* (1975) have suggested that a major part of the floor of the English Channel was laid bare during the Middle Devensian. Braided rivers crossed this area as sea-level continued to fall to its Late Devensian minimum. This fall in sea-level also caused land to be exposed in the North Sea (Long *et al.* 1988). In the northern North Sea, Middle Devensian sedimentation was mainly glaciomarine (Stoker *et al.* 1985c). Radiocarbon ages of 31 150 ± 1200 BP (Milling 1975) and 30 190 ± 360 BP (Rise and Rokoengen 1984; Rokoengen and Rise 1984) have been obtained from organic material found within possible littoral or fluvioglacial sediments in the south of this area (Cameron *et al.* 1987). Part of the glaciomarine Stanton Formation identified on the continental shelf west of Scotland was probably deposited in the Middle Devensian (Davies *et al.* 1984) (Figure 9.8).

9.2.9 Age and correlation

The suggestion of Bowen *et al.* (1989) regarding the age of the Upton Warren episode, whose thermal maximum has been placed at 43 000–40 000 BP on radiocarbon evidence, and which normally has been assigned to Oxygen Isotope Stage 3 (Figure 1.2(a)), is noted above (section 9.1.11). This alternative chronology serves to heighten the observation made at the outset of this section that numerous radiocarbon determinations of this magnitude seem to have been either imprecise or wrong, with many affording minimal ages (Bowen *et al.* 1989).

West (1977b) has considered that the Middle Devensian approximated to the Middle Pleniglacial of the Netherlands (Zagwijn 1974c) (Figure 9.9). It has been concluded by A. V. Morgan (1973) that correlation of the Middle Devensian climatic fluctuations deduced at Four Ashes with localities outside the British Isles is difficult. The fullest and most securely dated vegetational histories over this timespan are from the Netherlands (Zagwijn 1961, 1974c; van der Hammen *et al.* 1971; de Jong 1988). Here there were three interstadials between *circa* 50 000 and 26 000 BP: the Moershoofd (50 000–43 000 BP), Hengelo (39 000–37 000 BP) and Denekamp (32 000–29 000 BP). As Morgan has pointed out, the Upton Warren Interstadial has been based on insect evidence. Given the quicker response of insects than plants to environmental modifications, the thermal maximum of the Upton Warren event (*circa* 41 000 BP on radiocarbon evidence) could be equivalent to the pollen-based Hengelo Interstadial, dated to around 39 000 BP.

The problem of environmental interpretation based on different lines of biological evidence is further discussed at the conclusion of this chapter.

9.3 The Late Devensian

Most authorities have concluded that the maximum expansion of Devensian ice took place within the last part of the cold stage (Figure 9.11). According to Rose (1985), one cycle of climatic decline and improvement is likely, although fluctuations seem to have taken place during the recovery phase. The cycle occurred *circa* 26 000–13 000 BP, constituting a chronozone and a stadial, for which the term Dimlington Stadial has been proposed (Figure 9.9). As already observed, Four Ashes has been designated the type locality for the Devensian. Rose, however, has pointed out that at Four Ashes the lithostratigraphy contains scant evidence of the glaciation. It has also been noted that the biostratigraphic information and the radiocarbon dates which have given the glaciation a minimum age at the locality are not closely connected with the episode. On the other hand, the Dimlington locality in eastern Yorkshire (section 9.3.2) possesses a distinctive suite of glacigenic sediments which are underlain by radiocarbon-dated fossiliferous organic material. Moreover, close to Dimlington, this glacigenic material is overlain by radiocarbon-dated fossiliferous interstadial deposits (Figure 9.12). The latter began accumulating about 13 000 BP and continued until *circa* 11 000 BP during the Windermere Interstadial (Coope and Pennington 1977) (section 9.3.3). Above the interstadial deposits in this area are those of a further stadial. This took place between *circa* 11 000 and 10 000 BP and has been termed the Loch Lomond Stadial (Gray and Lowe 1977) (section 9.3.6). The conclusion of the Loch Lomond Stadial marked the termination of the Devensian Late-

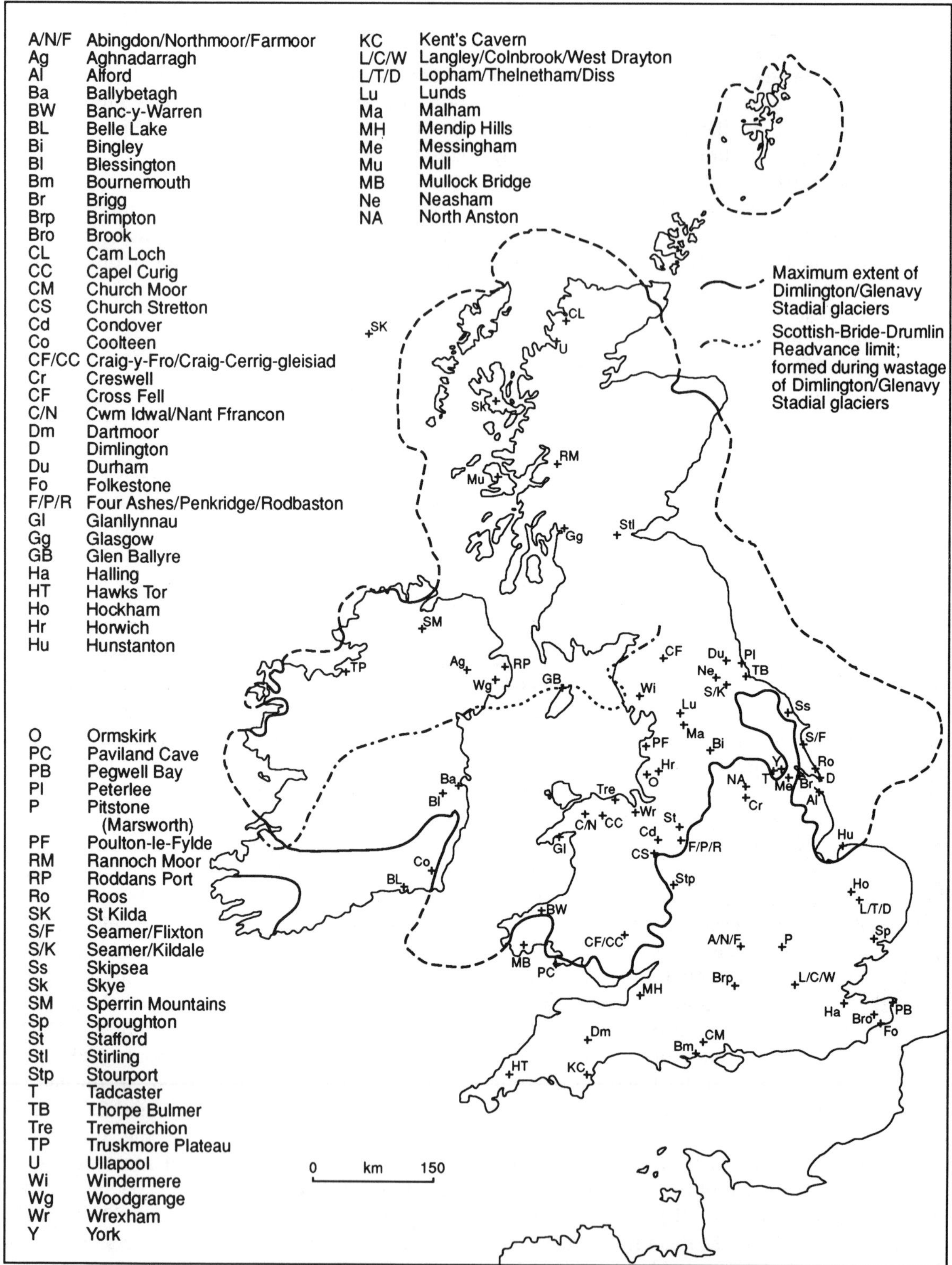

Figure 9.11 Localities with evidence of Late Devensian/Midlandian environments. The maximum extent of the Dimlington/Glenavy Stadial glaciers, and the limit of a possible readvance during their wastage (sources, Mykura and Phemister 1976, Bowen *et al.* 1986, Cameron *et al.* 1987) are also shown.

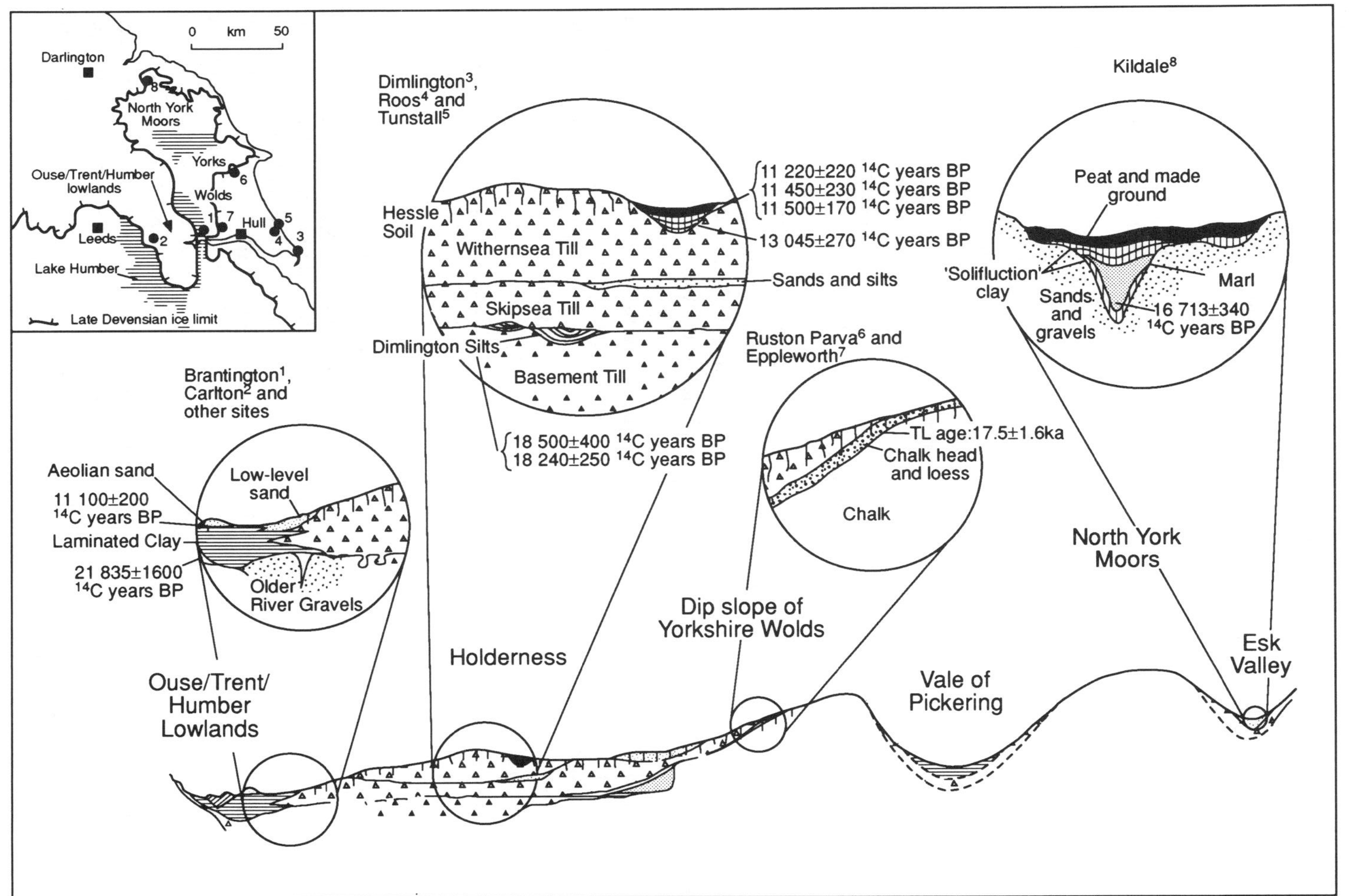

Figure 9.12 Diagrammatic representation of stratigraphic evidence for the Dimlington Stadial in eastern England. An inset map shows the site locations in the region from the Ouse/Humber/Trent lowlands in the south, to the North York Moors in the north, and the relation of the sites to the Late Devensian ice limit (Madgett and Catt 1978). Formational names and lithological descriptions are given for the Dimlington, Roos and Tunstall evidence, and are represented by equivalent symbols at the other sites, except for the Devensian Till north of the Yorkshire Wolds, which is not differentiated into Skipsea and Withernsea types. From Rose (1985). Reproduced with permission of *Boreas* **14**, 1985, The Dimlington Stadial/Dimlington Chronozone, J. Rose; Fig. 2, p. 228.

glacial period and Cold Stage, and the start of the Flandrian Interglacial (Table 1.1).

9.3.1 The Midlands of England, Cheshire–Lancashire lowlands and northern Borders of Wales

At Four Ashes the youngest radiocarbon date obtained is 30 500 ± 400 BP from organic material within the sands and gravels (Figure 9.4) (A. V. Morgan 1973). The absence of younger organic material prior to the ice advance has been attributed to the onset of polar-desert conditions by A. Morgan (1973). Above the sands and gravels, till was emplaced. This is a typical Irish Sea type deposit, containing flint from Chalk outcrops on its bed, and marine shells. Erratics from the Lake District and Scotland are also found within the till, and the overall picture is of ice movement into southern Staffordshire from north-west to south-east. Fluvioglacial deposits and meltwater channels have been identified in association with the till. The ice advance ceased in the vicinity of Wolverhampton (Figure 9.11) where thicker till has been observed but no clear morainic landforms have been seen. Ice-wedge casts and involutions found in the glacigenic sediments, together with patterned ground and solifluction deposits, are indicative of a subsequent periglacial environment. Ventifacts have also been recorded, implying the existence of a barren, sandy landscape and a climate in which powerful winds were operative.

At Stafford, basal organic deposits occupying a depression in fluvioglacial gravels have been radiocarbon dated to 13 490 ± 375 BP, indicating that this area was then free of ice (A. V. Morgan 1973). Faunal evidence demonstrates that at this time the climate was arctic (Morgan 1970). The Devensian Lateglacial palaeobotany and further radiocarbon assays of the Stafford sediments have been presented by Bartley and Morgan (1990). Dates of 13 250 ± 300 and 12 070 ± 220 BP relate to the Windermere Interstadial, towards the conclusion of which, tundra with scattered birch trees existed. A second pollen assemblage biozone and plant macrofossils (notably *Betula nana*) suggest more open tundra, with fewer or no tree birches, which developed under colder conditions during the Loch Lomond Stadial. At Pillaton Hall south of Stafford, peat above sand has been radiocarbon dated to 11 660 ± 250 BP; at nearby Penkridge an age of 11 580 ± 140 BP has been obtained from organic material in gravels; peat at Rodbaston, about 2 km north of Four Ashes, has produced dates of 10 670 ± 130 and 10 300 ± 170 BP (A. V. Morgan 1973). Ashworth (1969) has described coleopteran assemblages from these three localities. At Pillaton Hall and Penkridge, Windermere Interstadial deposits (*circa* 11 500 BP) contain insects whose summer temperature requirements are now a few degrees below those of Staffordshire. A fairly open landscape was present, although patches of willow and birch occurred. The beetle faunas between 10 700 and 10 300 BP at Rodbaston contained arctic stenotherms consonant with a cold environment during the Loch Lomond Stadial (A. V. Morgan 1973).

Worsley *et al.* (1983) and Worsley (1985) have presented detail on the succession of glacigenic sediments in the Cheshire Plain. The Stockport Formation, including till, outwash and lacustrine deposits, was the product of Late Devensian Irish Sea ice. These deposits are equivalent to the majority of the former Lower Boulder Clay–Middle Sands–Upper Boulder Clay sequence. In the Horwich area close to the eastern margin of the Lancashire lowlands, outwash sediments have been identified (Hibbert *et al.* 1971). At nearby Red Moss, Devensian Lateglacial lake deposits have produced assemblages of Coleoptera which demonstrate the replacement of thermophilous taxa with more cold-tolerant ones. Radiocarbon dating has established that this had taken place by 12 160 ± 140 BP (Ashworth 1972; Coope 1977a).

The pathways of Late Devensian ice into and across Cheshire, Shropshire and Staffordshire have been summarized by Shotton (1977c). Ice originated in western Scotland and the Lake District and then entered the Irish Sea. At this time, ice masses were also emanating from Ireland and North Wales, so that the Scottish–Lake District ice deviated down the Cheshire

Plain. The ice limit ran from the environs of Church Stretton to those of Wolverhampton, then northwards to the edge of the Southern Pennines (Figure 9.11). Ice retreat across Cheshire and northern Shropshire appears to have been a complex process, involving halts and perhaps minor readvances. Hummocky moraine has been identified extending in two segments: a western one from the neighbourhood of Wrexham via Ellesmere to Whitchurch (Figure 1.3), and an eastern one from the latter locality to Woore and Bar Hill. This feature has been interpreted mainly as the product of an earlier glaciation by Poole and Whiteman (1961), and alternatively as an end moraine of the Late Devensian Glaciation by Boulton and Worsley (1965). A third explanation has envisaged it as being associated with either a readvance or an equilibrium phase of the Late Devensian ice (Peake 1961, 1981; Worsley 1970) (Figure 9.13). Thomas (1985c, 1989) has examined the glacial succession in the western part of the Cheshire–Shropshire lowlands, where Irish Sea and Welsh ice progressed, joined and separated. A basal till was related to ice advance. The Wrexham–Ellesmere–Whitchurch moraine was probably the product of retreating Irish Sea ice. While agreeing with Peake (1961) on the distribution of Welsh till, Thomas (1989) has indicated that this was not emplaced by a readvance of Welsh ice subsequent to the disappearance of Irish Sea ice. It has been suggested that Welsh ice may have been the first to reach this area. Although Welsh till has been demonstrated overlying Irish Sea till, the converse is also true, with penecontemporaneity the most likely explanation. Sedimentation of the Wrexham Delta Terrace (Figure 1.5) has been considered by Thomas (1989) to have occurred at the western margin of a glacial lake, and to have been started by meltwater draining the Alyn river system during deglaciation.

Beales (1980) has reported a vegetational history from Crose Mere within the Ellesmere–Bar Hill Moraine that covered zones I, II and III of Lateglacial time (Figure 9.9). A clay, detritus mud and silt/sand lithostratigraphic sequence has produced a palynological record at first dominated by grasses and sedges, and treeless. During Zone II, tree birches expanded in the area but substantial frequencies of herbaceous taxa suggest the absence of closed woodland. The tree-birch pollen curve exhibits a temporary fall in Zone II and a deterioration in climate has been suggested (section 9.3.3). In Zone III, frequencies of herbaceous pollen increase, notably *Artemisia* and *Rumex*, and grassland vegetation came to dominance.

At Condover near Shrewsbury, clay in a kettle hole within the limit of Late Devensian ice has yielded remains of *Mammuthus primigenius*, radiocarbon dated to 12 920 ± 390 and 12 700 ± 160 BP. The clay contains an interstadial pollen flora and insect fauna (Coope and Lister 1987). The mammoth remains are of especial interest because the species had previously been thought to have become extinct in the British Isles earlier in the Late Devensian, as a result of the intense cold associated with the glacial episode.

Among the river valleys of the Midlands, deposits below the surface of Avon Number 1 Terrace have produced a cold fauna and are of Late Devensian age (Shotton 1977c). The lower (Main, Worcester and Power House) terrace deposits of the River Severn date from the Late Devensian glacial maximum (Dawson 1985, 1989). The magnitude of these terraces indicates that they marked a main drainage route from the ice sheet. However, Dawson and Gardiner (1987) have calculated that the mean annual flood discharges of the Severn at this time were analogous to those of today. The Main Terrace of the River Severn has been considered by Wills (1938) and Shotton (1977b,c) to relate to the Irish Sea Glaciation on erratic and faunal contents. Dawson (1989) has identified glacigenic diamicts inter-bedded with Main Terrace sediments near Bridgnorth and has suggested accumulation of the terrace *circa* 25 000–18 000 BP. The Worcester Terrace occurs at a lower altitude and is distinguished in the area between Bewdley and Tewkesbury. At Bridgnorth, Irish Sea erratics have been found in this terrace, an outwash feature of Late Devensian age (Shotton 1977c; Shotton and Coope 1983). It may have been

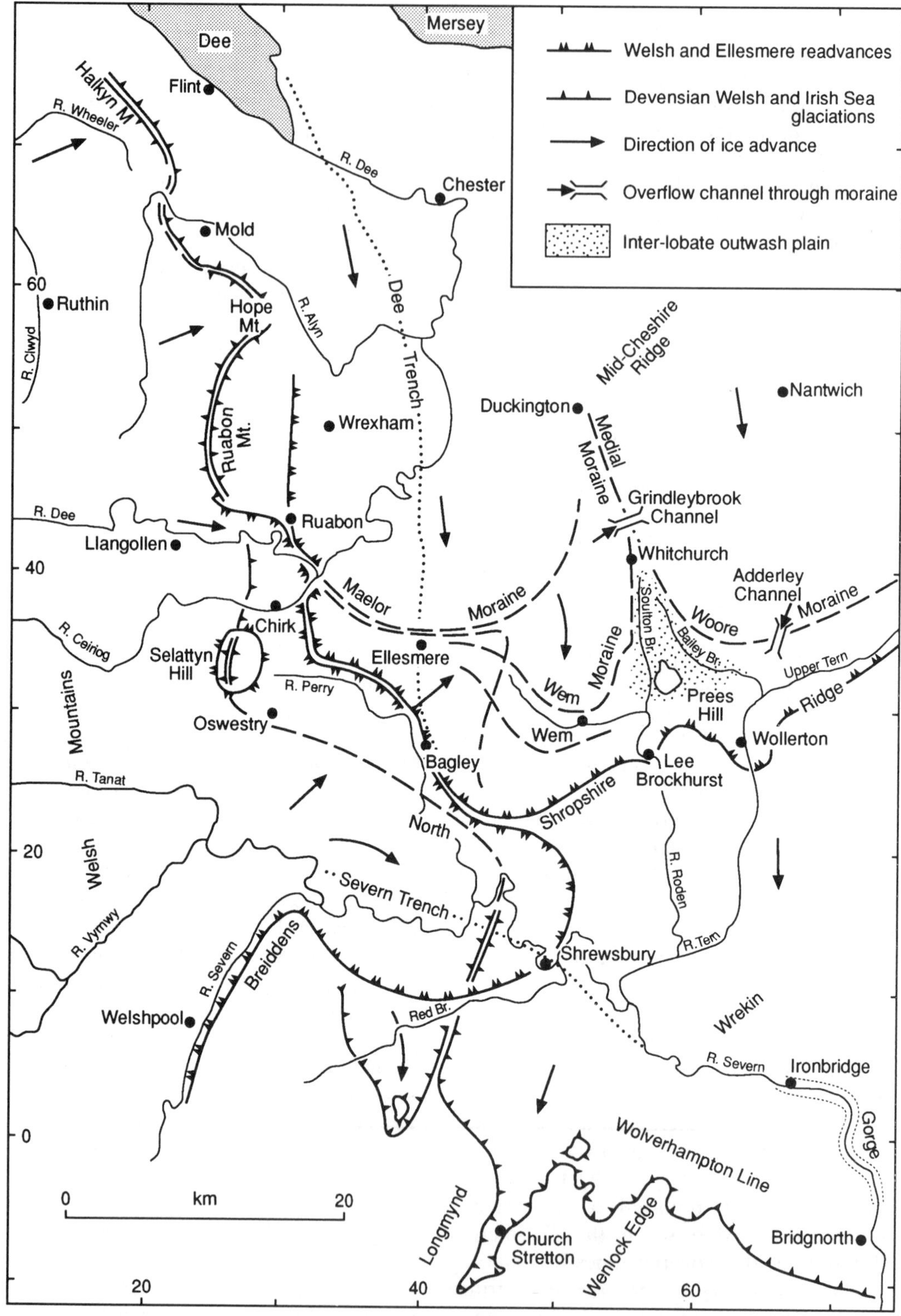

Figure 9.13 The Devensian Glaciation on the north Welsh border. From Peake (1981).

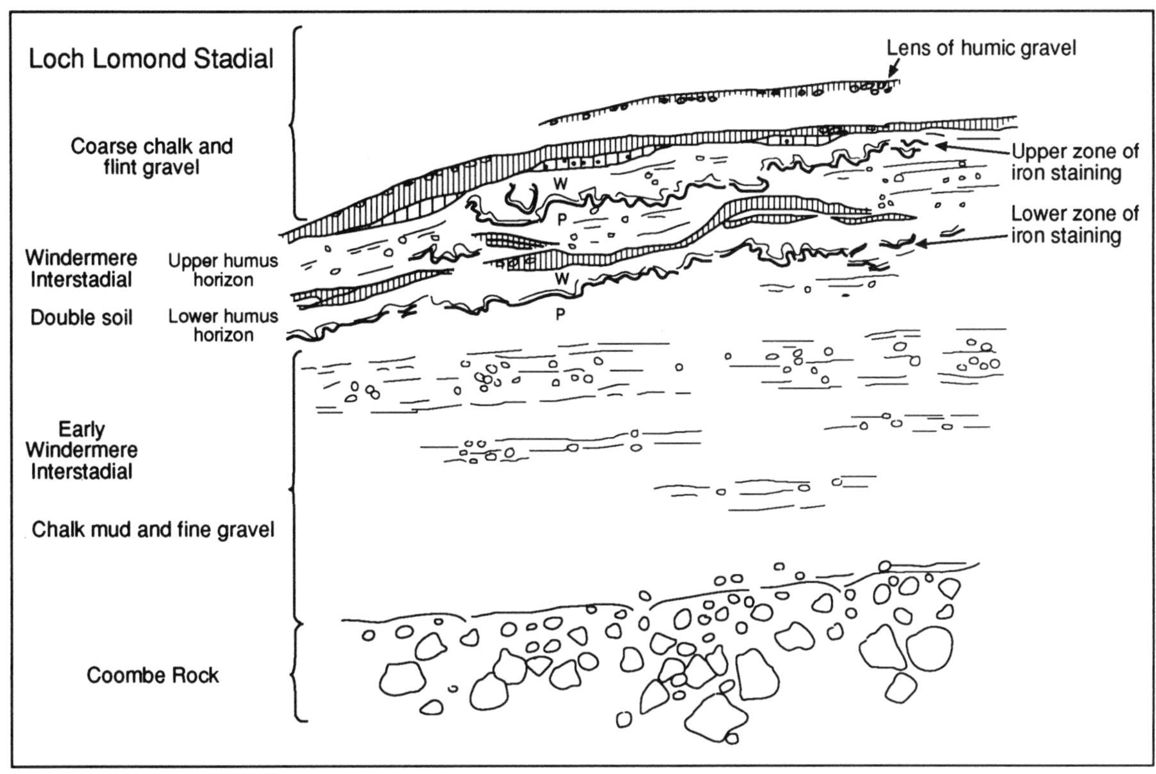

Figure 9.14 The Pitstone Soil at Pitstone, Buckinghamshire. From Rose *et al.* (1985b); source, Evans (1966).

associated with the Ellesmere Readvance (Beckinsale and Richardson 1964) (see above). The Power House Terrace of the Lower Severn Valley has been recognized as several intermittent surfaces beneath the level of the Worcester Terrace. A radiocarbon date of 12 570 ± 220 BP from an organic deposit above gravel in an equivalent terrace of the Stour Valley near Stourport has established a Lateglacial age for this material (Shotton and Coope 1983). Plant and insect remains from the organic deposit indicate a temperate episode early in the Windermere Interstadial, together with the cold conditions of the Loch Lomond Stadial. In the Trent Valley, the Floodplain Terrace deposits (Clayton 1953; Posnansky 1960) have been considered as Late Devensian (Shotton 1977c).

Well to the south of the Late Devensian ice limit, at Pitstone (Marsworth, Chapter 7) on the Buckinghamshire–Hertfordshire border, Coombe Rock has been observed overlain by chalk rubble and mud. This rubble and mud formed the parent material for soil development during the Lateglacial Interstadial. Two profiles have been recognized, the lower the less well developed, but both with an assemblage of terrestrial Mollusca characteristic of open ground. Between the two profiles, fine rubble and chalk mud are present. Above them, a coarse gravel consisting of flint and chalk occurs, having been laid down during the last, short cold phase of the Devensian (Evans 1966). According to Rose *et al.* (1985b) this is the most completely documented soil from the Windermere Interstadial and they have recommended that it should be taken as a characteristic profile (Figure 9.14). It has been suggested that a harsher environment may have prevailed during the first phase of soil formation at Pitstone than during the second phase, and that inorganic deposition upon the lower soil was probably due to heightened erosion of the surrounding terrain. The double soil could reflect the Bølling and Allerød climatic oscillations which are thought to have occurred during the Lateglacial Interstadial (Figure 9.9).

9.3.2 North-eastern England and East Anglia

Late Devensian ice originated in southern Scotland and northern England. This ice advanced down the Vale of York, abutting, but not crossing, the higher parts of the North Yorkshire Moors and encroached upon near-coastal sections of Yorkshire, Lincolnshire and north-west Norfolk (Figure 9.11).

The type locality proposed for the Dimlington Stadial (Rose 1985, 1989a) is in Holderness, eastern Yorkshire (Figure 9.12). Bisat (1939, 1940) has recognized Drab and Purple clays above a Basement Clay in Holderness. Catt and Penny (1966) have proposed a sequence involving Devensian Drab, Purple and Hessle tills above an earlier Basement Till (Chapter 7) in the same area. Hollows in the Basement Till at Dimlington contain silts, above which is the Drab Till. Moss remains in the silts have been radiocarbon dated to 18 500 ± 400 and 18 250 ± 250 BP (Penny *et al.* 1969; Catt 1987b). Plant and animal fossils from the silts indicate their deposition in a shallow lake bearing little vegetation, around which was open ground including moss patches. The moss dated is *Pohlia wahlenbergii* var. *glacialis* which currently inhabits glacial meltwater. In places the silts are succeeded by sands of aeolian type, indicative of dry, cold conditions prior to glaciation. Before ice advanced into eastern Yorkshire and Lincolnshire, periglaciation was intense. A thin layer of loess was laid down in many places (Catt 1987a) (Figure 9.15). This loess possesses a similar mineralogy to the Late Devensian glacial deposits of the area and was probably derived from proglacial outwash in the North Sea Basin (Catt *et al.* 1974). A solifluction deposit below Drab (Skipsea) Till (see below) on the Yorkshire Wolds at Eppleworth, west of Kingston-upon-Hull, incorporates loess which has been dated by thermoluminescence to 17 500 ± 1600 BP (Wintle and Catt 1985; Catt 1987a) (Figure 9.12).

Particle-size analyses and petrological investigations in eastern Yorkshire, Lincolnshire and north Norfolk have led Madgett and Catt (1978) to conclude that only two Devensian tills exist. These they have termed the Skipsea (Drab) and Withernsea (Purple) tills. The Hessle Till has been dismissed as a postglacial weathering profile of whichever of the other two tills occurs at the surface. The most widely distributed till is the Skipsea, occurring throughout most of eastern Yorkshire and Lincolnshire. It is pertinent at this point to recall the views of Straw (1969, 1979a,c) (Figure 9.5) that there are Lower Marsh drifts of Early Devensian age in Lincolnshire and a direct correlative of them in north Norfolk, and a Late Devensian Upper Marsh Till in the former area. The Hunstanton Till of Norfolk is an equivalent of the Skipsea Till according to Catt (1980).

In Holderness the Withernsea Till has been shown to be restricted to an area about the south-east coast. The number of ice advances that produced the till sequence is uncertain. However, a lack of weathering between the two tills has suggested a single episode to Catt (1987a). Further north, in Filey Bay, tills resembling both the Skipsea and Withernsea have been identified (Edwards 1981). In County Durham and Northumberland, most of the Pennines and the valleys traversing them (Tynedale, Weardale and Teesdale, for example) were affected by the passage of Late Devensian ice (Smith and Francis 1967). The origin of till in Teesdale can be traced to Pennine ice which crossed Stainmore and separated into two parts near Darlington. One part continued down the Tees Valley and around the North Yorkshire Moors to the coast (Francis 1970). Here it coalesced with coastal ice moving southwards from Northumberland and Durham. The Devensian ice-sheet model of Boulton *et al.* (1977) (Figure 9.16) was accompanied by a suggestion that a surge down the east coast of England resulted from an unstable ice front in the Tees lowlands. According to Catt (1987a), such a

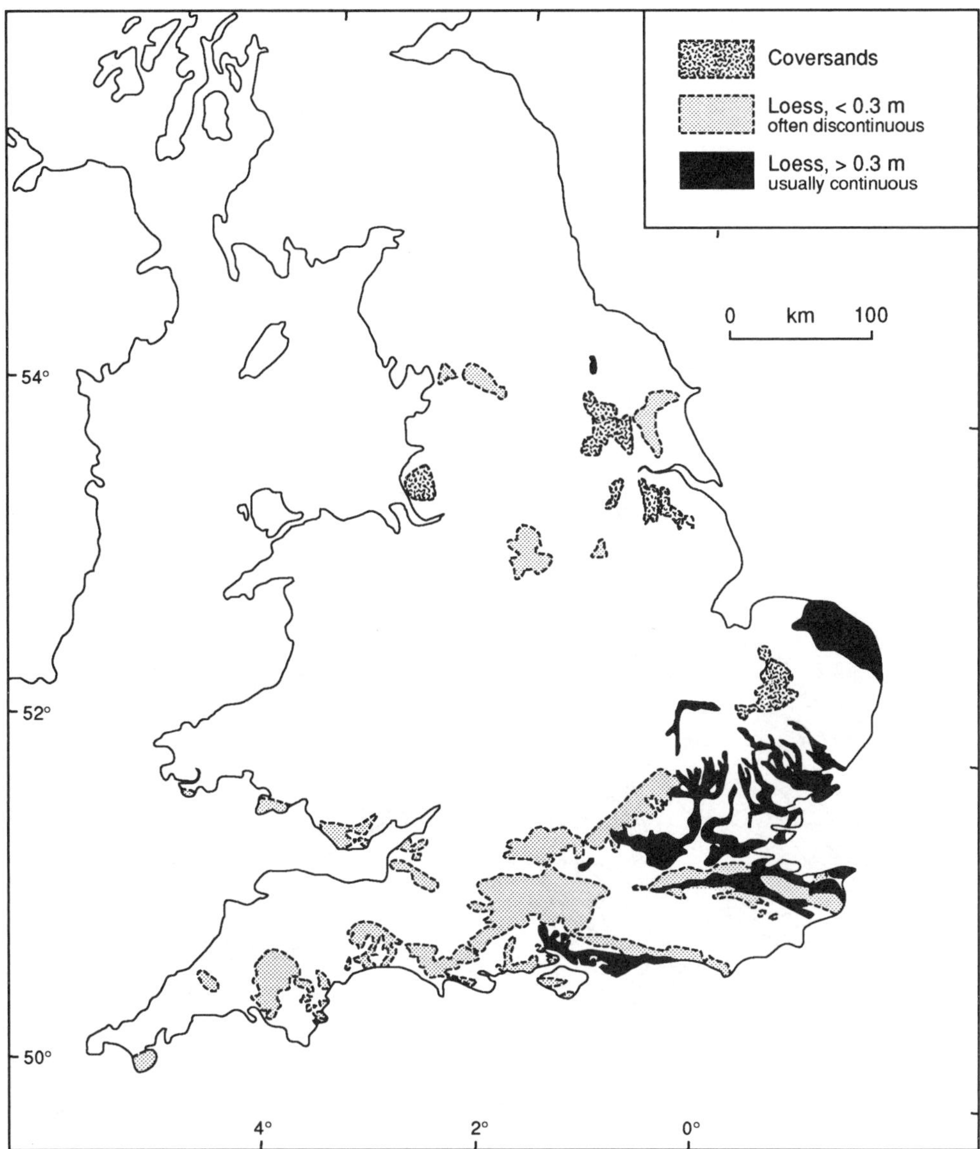

Figure 9.15 Distribution of loess and coversands in England and Wales. From Catt (1977c). Reproduced by permission of Oxford University Press.

surge could have started if Stainmore ice overrode coastal ice. This phenomenon would also have provided a mechanism for deposition of the Withernsea Till.

A considerable thickness of Late Devensian glacigenic sediments has been identified in the valleys and coastal area of lowland Durham (Francis 1970; Beaumont 1971; D. B. Smith 1981). Outwash was deposited above the Wear and Blackhall tills, products of the coastal ice stream that deposited the Skipsea Till. A substantial amount of outwash has been recognized in the environs of the City of Durham and the town of Peterlee. Loess above Wolstonian till

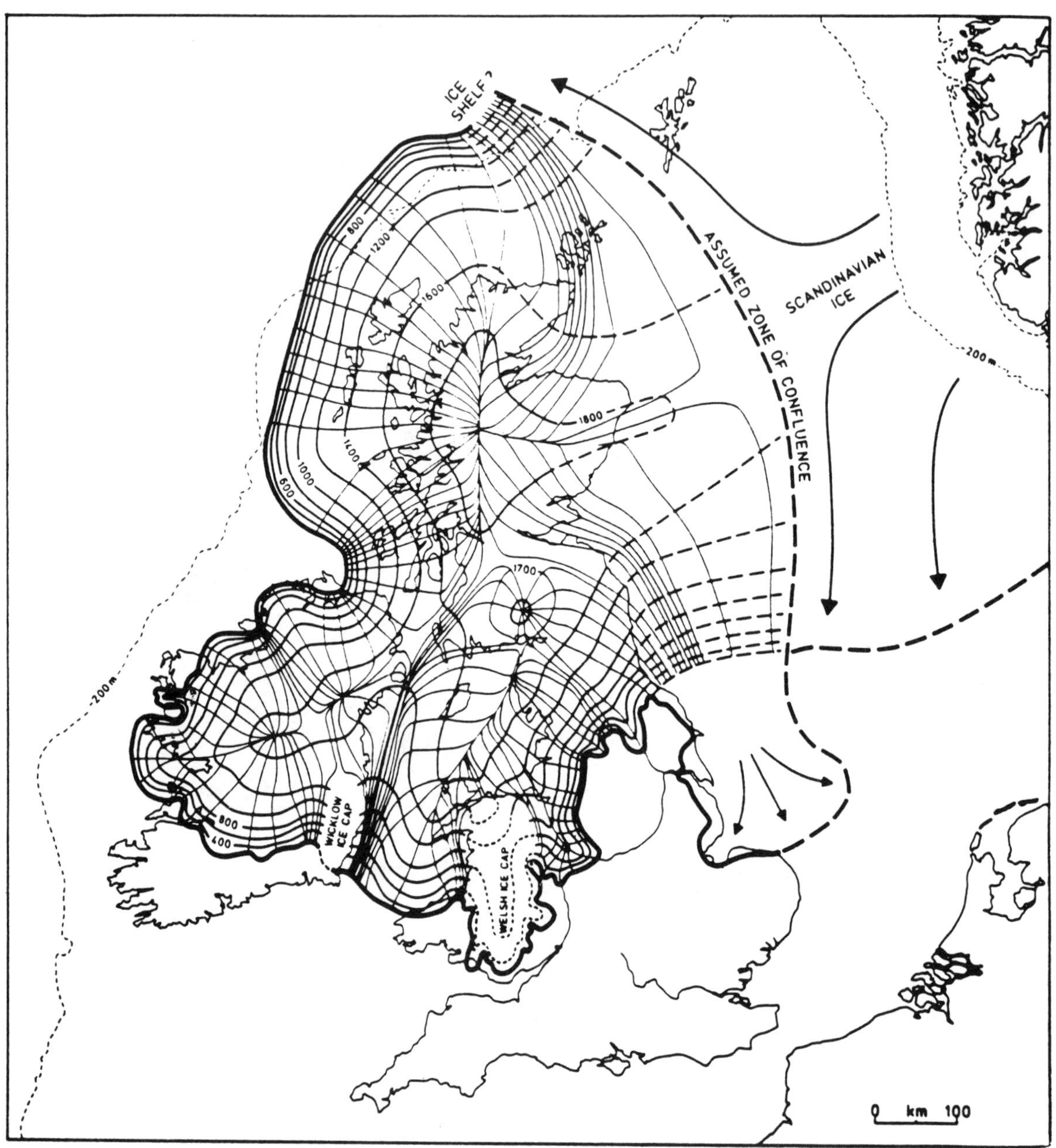

Figure 9.16 The modelled surface topography and flow lines of the Late Devensian ice-sheet. From Boulton *et al.* (1977). Reproduced by permission of Oxford University Press.

on the Durham coast (Chapter 7), and in soils developed over Magnesian Limestone outcrops in the county, could have been emplaced in an extraglacial, early Late Devensian environment (Catt *et al.* 1974).

Till and supervening outwash in the Northumberland coastal district were deposited by ice which probably came from the Cheviot Hills (Clapperton 1970, 1971) and the Tweed Valley. A later, North Sea ice advance was able to penetrate the coastal lowlands of Northumberland when the influence of western ice declined (D. B. Smith 1981).

The second part of the Stainmore ice (see above) proceeded down the Vale of York. Ice engulfed the northern part of the North Yorkshire Moors. Much of the meltwater from ice in this area drained sub-glacially eastwards (Gregory 1962, 1965). Both ends of the Vale of Pickering were sealed by ice. Meltwater from ice marginal to the North Yorkshire Moors, much of which probably flowed through Newtondale, helped sustain Lake Pickering (Kendall 1902). The latter overflowed into the Vale of York via the Kirkham Abbey Gorge in the Howardian Hills (Catt 1977a) (Figure 9.17).

The Late Devensian sequence in the Vale of York–Humber region has been detailed by Gaunt (1976, 1981). Ice moved in from the north and west. The York and Escrick moraines (Figure 9.17) were deposited more or less parallel to the ice front on its retreat. To the south of the ice margin, Lake Humber (section 9.1.3) developed up to 33 m OD, and sand and gravel mark its former margins. Bone from either below or within these high-level lake deposits has a radiocarbon date of *circa* 27 000 BP (Gaunt 1974). During deglaciation a low-level lacustrine phase ensued. Lake Humber now extended northwards, with its surface 10–4 m OD in the environs of Doncaster and York. Lacustrine clay and marginal sand (the 25′ Drift) were deposited during this episode. Silting up was the most likely cause of the lake vanishing (Gaunt 1981). A palaeosol on laminated clay at West Moor near Doncaster has been radiocarbon dated to 11 100 ± 200 BP (Gaunt *et al.* 1971) (Figure 9.12), thereby providing a minimum age for disappearance of the lake (Gaunt 1981). The subsequent development of river courses towards the Humber led to the formation of sand levées (Gaunt 1981). Near Selby, levée deposition ceased by 10 469 ± 60 BP, the radiocarbon age of peat above levée sand (Jones and Gaunt 1976). An aeolian phase, probably initiated during the Loch Lomond Stadial, led to coversand deposition over a wide area (Gaunt *et al.* 1971) (Figure 9.15). The West Moor palaeosol was overlain by blown sand. Peat obtained from the lower part of blown sand near York has given a radiocarbon age of 10 700 ± 190 BP. Organic material above this blown sand has been dated to 9950 ± 180 BP. More sand was emplaced over the organic material (Matthews 1970). Sand mobilization in the Loch Lomond Stadial may have been assisted by a reduction in the groundwater table (Gaunt 1981). It is possible that a brief and intense episode of fluvial incision, caused by a quick and substantial fall in base level in the Humber area, began during this period (Gaunt *et al.* 1971; Gaunt 1981). An Upper Periglacial Land Surface has been detected in the Vale of York. It developed in the top of Late Devensian glacigenic sediments, probably during the Loch Lomond Stadial, and contains ice-wedge casts, ventifacts and desert pavement (Gaunt 1981).

Beyond the limit of Devensian ice in East Anglia, periglaciation was intense (Straw 1960). Nivation, cryoturbation and solifluction took place, and patterned ground was well developed (Williams 1964, 1968, 1975) (Figure 9.18). In north-east Norfolk, widespread and relatively thick loess of Late Devensian age has been identified (Catt *et al.* 1971). Loess of this affinity has also been recognized in south-east Essex (Catt 1977c) (Figure 9.15). In the Nar Valley, the Pentney Sand and Gravel possesses ice-wedge casts and was deposited by a braided Late Devensian river. Organic material within these deposits has a radiocarbon age of *circa* 20 000 BP (Ventris 1986). At Walton Common in this area, pingoes collapsed during the Windermere Interstadial but reformed in the Loch Lomond Stadial (Sparks *et al.* 1972).

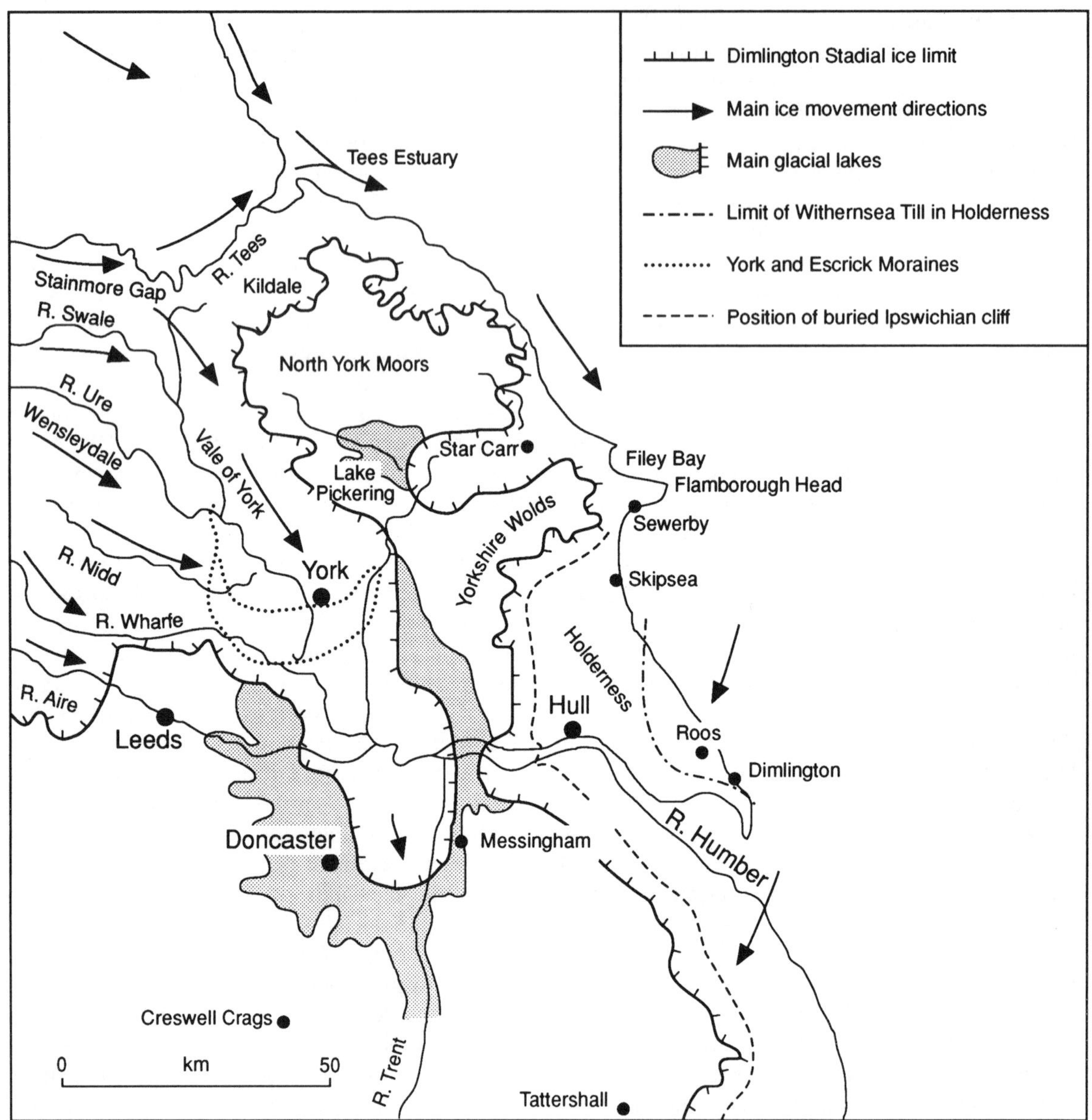

Figure 9.17 Devensian glacial features of east Yorkshire and nearby areas, and localities mentioned in the text. From Catt (1987a).

As noted (section 9.3), radiocarbon-dated Lateglacial deposits have established the time before which Late Devensian ice disappeared from Holderness. Such deposits have been found over quite a wide area of north-eastern England and East Anglia, and have also yielded a considerable amount of biostratigraphic and archaeological data.

A kettle hole in Withernsea Till at Roos, Holderness, possesses an infill of lake deposits

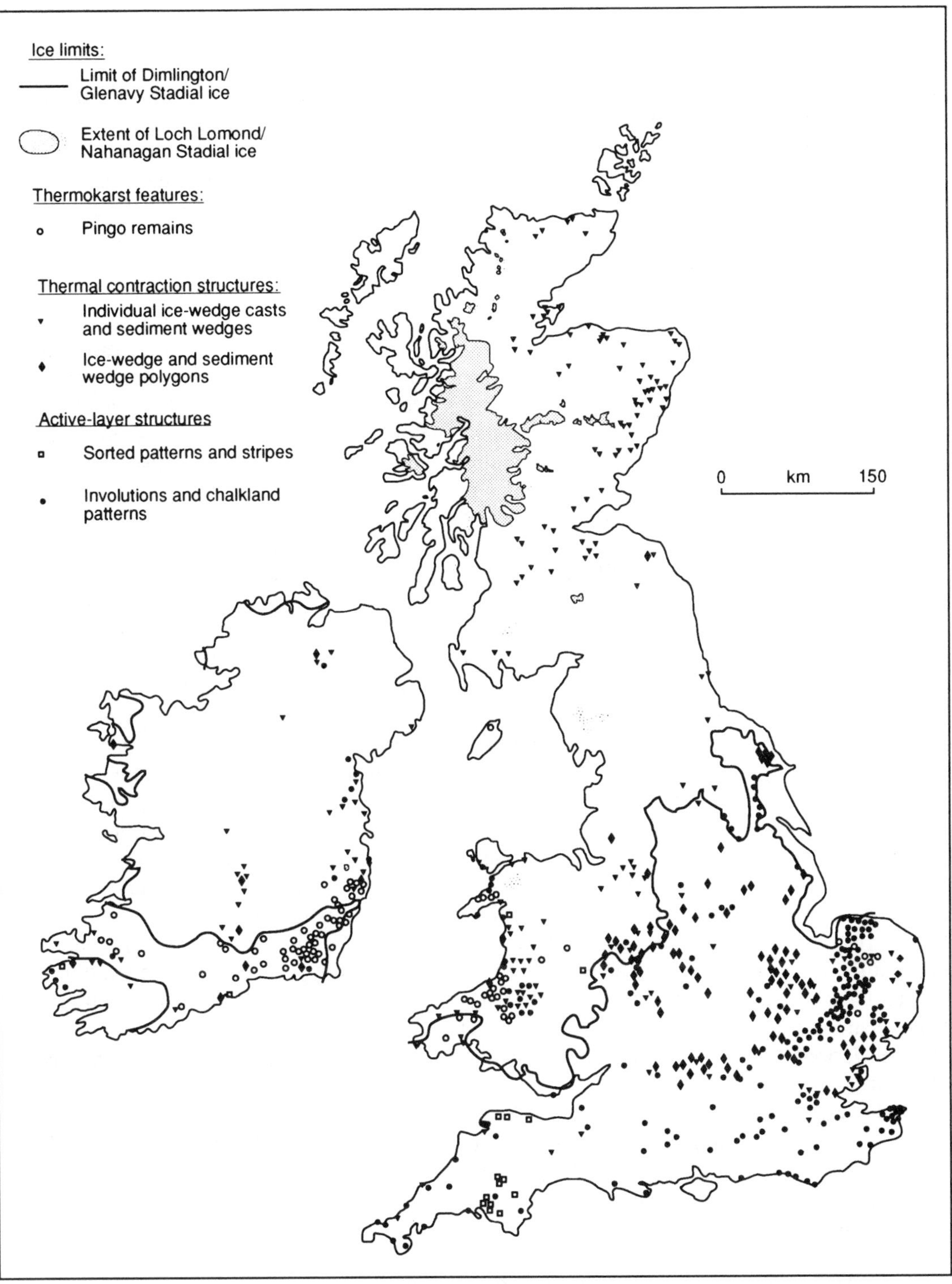

Figure 9.18 Periglacial features. Late Devensian soil structures in Britain. From Rose *et al.* (1985b); sources, West (1977a), Watson (1977), with additional information from Gaunt (1976), Girling (1977), Greig (1981), Douglas (1982), D. G. Sutherland (personal communication). Late Devensian thermokarst features in Britain. From Watson (1977). Periglacial features in Ireland. From Lewis (1985); sources, Lewis (1978), P. O'Callaghan (unpublished) – pingos, Hoare and McCabe (1981) – features in east-central Ireland. © *Irish Journal of Earth Sciences*. Ice limits; source, Bowen *et al.* (1986).

(Beckett 1981) (Figure 9.12). The oldest radiocarbon date, 13 045 ± 270 BP, is associated with the earlier and smaller of two peaks in tree-birch pollen. The second peak, *circa* 11 500–11 200 BP, is separated from the first by a phase in which herbaceous pollen frequencies increase. The end of the Loch Lomond Stadial has been dated to 10 120 ± 180 BP. The stadial was characterized by open vegetation including *Betula nana*, *Juniperus* and numerous herbs among which *Artemisia* was prominent. Skipsea Withow Mere in the same area has produced plant and molluscan evidence of Lateglacial environmental changes (Hunt *et al.* 1984), together with an Upper Palaeolithic flint blade of pre-Windermere Interstadial age (Mellars 1984). At Willow Garth in the Great Wold Valley of the Yorkshire Wolds, just within the Devensian ice limit, Bush and Flenley (1987) and Bush and Ellis (1987) have described deposits referable to the later part of the Windermere Interstadial and to the Loch Lomond Stadial on plant fossil evidence. Cold-climate grassland was present, in which *Gentiana verna* (spring gentian), *Diphasium* (*Lycopodium*) *alpinum* (alpine clubmoss) and *Saxifraga oppositifolia* grew. Seamer Carr and Flixton in the Vale of Pickering have Lateglacial vegetational histories (Walker and Godwin 1954), the latter site also yielding Later Upper Palaeolithic flint artifacts and horse bones (Moore 1954; Campbell 1977). Star Carr in the same area has produced an open occupation site including a stone hearth and Later Upper Palaeolithic flint artifacts. The latter are from peat, radiocarbon dated to 11 300–10 200 BP, which occurs beneath aeolian sand (Schadla-Hall 1987).

In the northern part of the Vale of York, Seamer Carrs, occupying an inter-drumlin depression near Stokesley, shows a Windermere Interstadial vegetation sequence with two birch pollen peaks, the earlier the larger, and the later radiocarbon dated to 13 042 ± 140 BP (Jones 1976a). At Kildale Hall in the nearby Leven Valley, a kettle hole in outwash has produced a radiocarbon age of 16 713 ± 340 BP on basal organic material (Figure 9.12), which implies early deglaciation of this locality (Jones 1977a). Ostracod evidence indicates that the first of two mild episodes during the Windermere Interstadial (also identified on palynological and molluscan criteria) was the more genial. The Loch Lomond Stadial had concluded by 10 350 ± 200 BP (Keen *et al.* 1984). At Neasham near Darlington, an intra-till hollow contains lake sediments which have recorded Lateglacial vegetational history. A skeleton of *Alces alces* (elk) has been recovered from Zone II muds, the latter radiocarbon dated to 10 851 ± 630 BP (Blackburn 1952). In south-east Durham, a kettle hole in till at Thorpe Bulmer near Hartlepool has produced a pollen stratigraphy indicative of two *Betula* peaks in a Lateglacial Interstadial context. The earlier of these peaks is the biggest, and the sequence has been correlated with that at Low Wray Bay in the Lake District (section 9.3.3) (Bartley *et al.* 1976). Lake deposits in outwash at Bamburgh in Northumberland have revealed that birch woodland was absent from this area during the Windermere Interstadial (Bartley 1966).

Lake clays and muds occupying a hollow in the Escrick Moraine near Tadcaster accumulated throughout the Lateglacial, and show a double oscillation in the *Betula* pollen curve which Bartley (1962) has suggested may be a reflection of the Bølling and Allerød interstadials. At Cawood near Selby, an organic deposit on levée sand and beneath blown sand, radiocarbon dated to *circa* 10 500 BP (see above), contains a pollen spectrum typical of open vegetation growing in cold conditions (Jones and Gaunt 1976; Gaunt 1981).

From Dead Man's Cave, North Anston, in the northern Dukeries region about the borders of Derbyshire, Nottinghamshire and Yorkshire, Campbell (1977) and Jenkinson (1984) have reported deposits referable on their pollen content to Zone II and Zone III of the Lateglacial and depicting interstadial birch woodland. Creswellian artifacts (Figure 9.10), and vertebrate remains including a *Rangifer tarandus* antler radiocarbon dated to *circa* 10 000 BP, have also been recorded (Mellars 1969, 1974). Similar archaeological and biostratigraphic evidence has been forthcoming from Mother Grundy's Parlour

(Figure 9.10) and Robin Hood's Cave, Creswell Crags (Jenkinson 1984). At Robin Hood's Cave, Later Upper Palaeolithic implements, together with human and animal skeletal remains, have been recorded, some of the latter radiocarbon dated to around 10500 BP (Campbell 1977). At Pin Hole Cave in the same area, Later Upper Palaeolithic artifacts and human remains have been discovered (Jenkinson *et al.* 1985). Uranium-series dating at Robin Hood's Cave has confirmed that speleothem growth took place during the Lateglacial Interstadial (Rowe and Atkinson 1985).

At Messingham near Scunthorpe, peat below coversands (Figure 1.7) contains a Later Upper Palaeolithic flint artifact and an insect fauna of arctic aspect. The peat has been radiocarbon dated to 10280 ± 120 BP (Buckland 1976). An age of 10550 ± 250 BP has been obtained from organic material 2 m higher in this sequence. The overlap between the standard deviations of the dates suggests that the coversands may have been deposited fairly quickly (Buckland 1982). However, hard-water error (Shotton 1972) has also been taken into account as the possible cause of the older date from the later stratigraphic layer. Also, insect evidence has led Coope (cited in Buckland 1984) to infer that the radiocarbon age of the lower peat could be about 500 years too young, and that the Messingham Sands above it were deposited during the coldest episode of the Lateglacial. At Aby Grange near Alford, Lincolnshire, a kettle hole in Devensian moraine has sediments from which Lateglacial (zones I–III) pollen spectra have been obtained (Suggate and West 1959). At Sturton near Brigg in the Ancholme Valley, Lincolnshire, a calcareous sand, whose molluscan fauna includes *Columella columella*, *Catinella arenaria* and *Vertigo genesii*, was probably deposited during the Lateglacial (Preece and Robinson 1984).

In East Anglia, Wymer *et al.* (1975) have described the association of Upper Palaeolithic barbed points of antler and bone and mammalian remains (*Rangifer tarandus* and *Equus ferus* – the latter dated by radiocarbon to 11300–9880 BP) in sands and gravels of the River Gipping at Sproughton near Ipswich. Rose *et al.* (1980) have presented a detailed analysis of the stratigraphy at this locality. Prior to 12200 BP, head was laid down, after which a palaeosol developed in chalk mud. The mud contains fossils of shrubs and grasses indicative of a cool-temperate climate. A lacustrine silt developed in a backswamp includes evidence of tree-birch growth about 11800 BP. However, the emergence of a northern insect fauna denotes a climatic deterioration at this time. River incision under arctic conditions took place *circa* 11300–11000 BP. After this time, and before the onset of the Flandrian, braided-river sedimentation and gelifluction took place, and permafrost was probably in existence. A beetle fauna of arctic–alpine type occurs in the corresponding deposits.

Sands and gravels of the River Waveney floodplain at Lopham Little Fen, Norfolk, contain organic remains in hollows, whose age has been established as Lateglacial on botanical evidence by Tallentire (1953). At Thelnetham in the Little Ouse Valley, Suffolk, a short distance away, pollen of *Betula*, *Pinus*, *Salix*, *Juniperus*, *Hippophaë*, *Helianthemum* and *Thalictrum* in organic remains has been assigned a Devensian Lateglacial age (Coxon 1978). Nearby Diss Mere has been investigated by Peglar *et al.* (1989). Here a small depression in chalky till may be either a thermokarst feature of Devensian age, or a doline. A succession of clays, silts and sands containing organic detritus has been recorded. The pollen flora, incorporating substantial frequencies of *Betula*, together with high values for grasses, sedges and numerous other herbs (including *Artemisia* and *Rumex acetosa*) is indicative of an age between *circa* 12000 and 10000 BP. An incomplete cover of birch woodland, as envisaged at Lopham, was present at this time. Data on Lateglacial vegetational history have also come from Hockham Mere near Thetford (Godwin and Tallantire 1951; Bennett 1983). In this area of the Breckland, Devensian coversand was emplaced above a till from an earlier cold stage (Perrin *et al.* 1974; Catt 1977c) (Figure 9.15). Bennett has demonstrated that at the start of the Windermere Interstadial, sedimentation was mainly organic (radiocarbon dated to 12 620 ± 85 BP), and scanty, open birch woods were in existence. A shift to largely inorganic (aeolian sand) deposition ensued but little change in vegetation composition *circa* 11 000 BP is indicated. However, the woodland cover was likely to have been reduced early in the Loch Lomond Stadial. At Sea Mere, close to Hockham, the pollen record from organic-poor rubble, sand, silt and clay indicates initial vegetation of open character which included *Dryas octopetala*, *Helianthemum* and *Artemisia* (Hunt and Birks 1982). A subsequent chalky silt has revealed the first of a three-phase grassland biozone with a pollen flora of open, species-rich type. Detritus mud above the silt denotes a more closed and extensive plant cover of the same kind, and a decreasing organic content has been associated with more open grassland vegetation. If these changes were not local, their correspondence with the Bølling and Older *Dryas* episodes of the Windermere Interstadial (Figure 9.9) has been postulated. A further detritus mud includes high counts of *Betula*, denoting woodland development after grassland and shrub successional stages. Overlying silts, clays and mud are well endowed with Cyperaceae and *Artemisia* pollen, while frequencies of *Betula* and *Filipendula* (dropwort/meadow-sweet) decline. This biozone (sub-divided into three) is probably the representative of the Loch Lomond Stadial.

9.3.3 North-western England

Valley glaciers moved down Swaledale, Wensleydale, Nidderdale and Wharfedale in the Late Devensian (Raistrick 1926, 1931), thereby augmenting the Vale of York ice on its western side. In Airedale the glacier halted north-west of Leeds (Catt 1977a, 1987a) (Figure 9.17). During the glaciation a number of the Pennine summits may have been nunataks where intense physical weathering took place. Many periglacial deposits and landforms, which may date from either the time of the Main Devensian Glaciation, or the time of the Loch Lomond Stadial, have been identified in the Knock Fell, Cross Fell and Dun Fell areas (Tufnell 1985). Tors such as the Brimham Rocks, Nidderdale (Linton 1964), could also have formed in one or both of these episodes. After their maximum extent, the Pennine glaciers retreated up-valley, leaving moraines and drumlins (Raistrick 1927, 1932; Rose 1980a). In numerous places, Lateglacial sediments have been recognized occupying depressions in these deposits. Keen *et al.* (1988) have described a Lateglacial stratigraphy from a kettle hole in outwash at Bingley Bog, within the limits of the Aire Valley glacier. Mollusca and Ostracoda define a rapid climatic amelioration at the start of the Windermere Interstadial. This was followed by a short, less-temperate episode registered by both animal and plant fossils. While a partial climatic recovery was made after this event, the fauna signal an overall decline in environmental conditions towards those of the Loch Lomond Stadial. During the latter, the fauna was extirpated and the flora changed from that of birch woodland to tundra. Lateglacial deposits have also been identified at Lunds in the upper part of the Ure Valley (Walker 1955) and at Malham (Pigott and Pigott 1963). Victoria Cave, Settle, was the site of Later Upper Palaeolithic occupation *circa* 12 000–11 000 BP (Campbell 1977; Wymer 1981).

If the estimate of Boulton *et al.* (1977) of an ice thickness of around 1600 m in the region was actual (Figure 9.16), the whole of the Lake District would have been glaciated in the Late Devensian. In Cumbria there appear to have been three major episodes during the Late Devensian glaciation (Huddart 1972, 1977; Evans and Arthurton 1973; Huddart *et al.* 1977; Tooley 1977). During the first of these, proglacial deposits and till of local origin accumulated in Lake District valleys. Local ice was then over-run by ice from Scotland. Drumlins in the Solway lowlands, Eden Valley and Furness district were a product of this ice, which wasted *in situ*. In the second episode, Irish Sea ice readvanced into the Low Furness area, then decayed *in situ*. A third episode may have involved a further readvance of Irish Sea ice, the Scottish Readvance (Trotter and Hollingworth 1932) (Figure 9.11), into the Cumberland lowlands, to produce the youngest glacial deposits. This event has been supported on palynological grounds by Walker (1966), but Pennington (1970) has disputed this as evidence of two stadials and one interstadial during Zone I.

In the north-eastern Lake District, the Late Devensian Threlkeld Formation contains till, sands, gravels and laminated sediments (Boardman *et al.* 1981). Many drumlins were developed in material of this formation (Evans and Arthurton 1973). The Wolf Crags Formation (Boardman *et al.* 1981) resulted from a corrie glacier at this locality during the Loch Lomond Stadial (Pennington 1978; Sissons 1980a). This glacier led to the deposition of moraine and outwash gravel in the upper Mosedale Valley, west of Penrith. Ice occupied numerous other Lake District cirques during the Loch Lomond Stadial, and in some localities advanced to the heads of valleys, depositing hummocky moraine at its limits (Sissons 1980a). The Skiddaw Formation (scree) in the Keswick area has been observed overlying a Late Devensian till and is of Loch Lomond Stadial age (Boardman 1978). In the north-western Lake District (in Keskadale and Coledale, for example), protalus ramparts and rock glaciers were formed beyond the limits of Loch Lomond Stadial ice (Sissons 1980a).

Devensian Lateglacial environments have been well illustrated by biostratigraphic (especially palynological) studies in the Lake District, notably the Windermere area. The lake has provided the type locality for the Lateglacial Interstadial (Coope and Pennington 1977) (section 9.3). At Blelham Bog about 1 km west of Lake Windermere (Pennington 1970, 1975a; Pennington and Bonny 1970), *Betula* pollen frequencies have been shown to have attained their Lateglacial maximum *circa* 13 000–12 000 BP during a climatic amelioration correlated with the Bølling Interstadial, identified at Bøllingsø in Denmark (Iversen 1954) (section 9.3.9) At Blelham, the *Betula* pollen curve shows that during a subsequent interstadial episode equated with the Allerød of Denmark (Iversen 1954), rather than ameliorating, the climate of the Lake District was deteriorating as the Loch Lomond Stadial approached.

At Low Wray Bay, Windermere, the Lateglacial vegetational history, which confirms that at Blelham, has been reported in detail by Pennington (1977a) (Figure 9.19). Prior to 14 500 BP, seasonal snow-melt from declining Late Devensian valley glaciers led to the deposition of a coarsely laminated clay, from which no floral evidence has been forthcoming. From *circa* 14 000 to 13 600 BP, chionophilous (snowbed) and/or fell-field plant communities existed in a periglacial environment. Cyperaceae, *Salix herbacea* and *Lycopodium selago* were prominent in the vegetation. Between about 13 600 and 13 000 BP, the plant record (now including *Rumex*, Gramineae, *Betula pubescens*, *Filipendula* and *Typha latifolia* (great reedmace)) signifies a climatic amelioration, but one unaccompanied by a significant expansion of woody plants. From *circa* 13 000 to 11 000 BP, the Windermere Interstadial occurred. An organomineral mud possesses a higher concentration of pollen than the underlying sediments. The palynological record shows that the interstadial complex began with vegetation in which *Juniperus* thickets were important. Lateglacial pollen sedimentation reached its maximum during this episode, which took place 13 000–12 500 BP. From the latter time until 12 000 BP, a *Betula*-dominated assemblage indicates a tree-birch cover of around 50%, which probably succeeded that of *Juniperus*. For the period from *circa* 12 000 to 11 800 BP, *Betula* and *Rumex* characterize the pollen spectrum. Frequencies of grasses, sedges and other herbs were now higher than previously in the interstadial; a situation that was subsequently sustained. During a *Betula–Juniperus* zone from about 11 800 BP to just before 11 000 BP, birch woodland with ferns was prominent in the landscape, where grasses also assumed importance. The inference is of more open vegetation during this woodland episode than the earlier one. Disintegration of the plant cover, the likely cause of which has been thought to be increased snowfall, is marked by a Cyperaceae–*Selaginella* biozone, in which *Betula* and *Juni-*

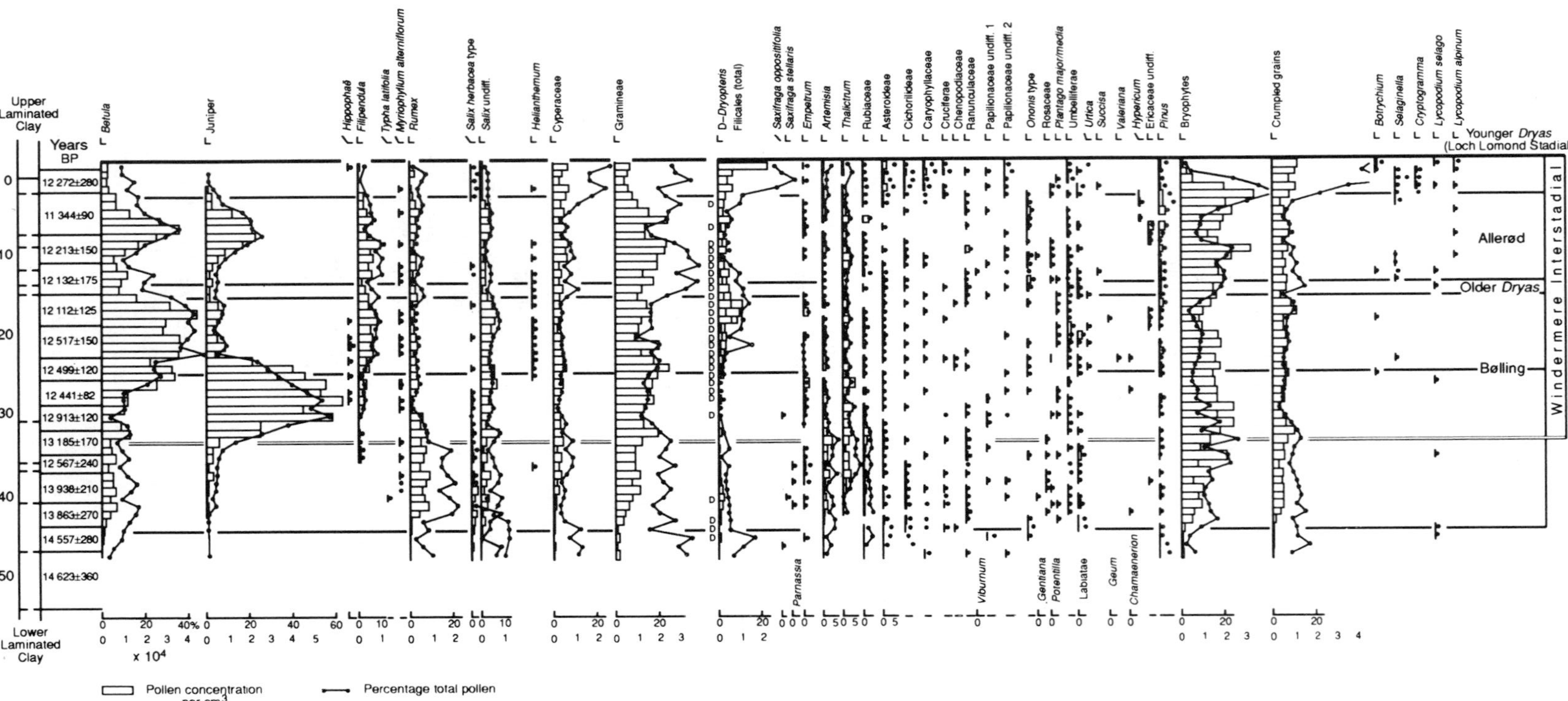

Figure 9.19 A percentage and absolute pollen diagram from Low Wray Bay, Windermere. From Pennington (1977a).

perus frequencies are reduced, reflecting the vegetation around 11 000 BP. Thus, the interstadial consisted of two episodes which could be chronostratigraphically correlated with those of Bølling and Allerød, between which a worsening of conditions took place (Mangerud *et al.* 1974) (section 9.3.9). The period from about 11 000 to 10 500 BP at Windermere witnessed the deposition of paired annual varves, consequent upon glacial activity during the Loch Lomond Stadial. Glaciation, and periglaciation beyond the ice limits, resulted in a sparse vegetation cover and hence sparse pollen output. However, from *circa* 10 500 to 10 000 BP, pollen influx to the lake increased, with pioneer vegetation dominated by Gramineae, Cyperaceae, *Artemisia* and *Rumex* developing as the climate began to ameliorate towards the conclusion of the stadial.

The insect fauna of the sediments at Low Wray Bay has been described by Coope (1977a). No insects have been recovered from the basal clay. However, in the lowest interstadial deposit, obligate arctic–alpine taxa are lacking. Species characteristic of the Lake District today are in evidence, together with a group of thermophiles whose current ranges are mainly southern European. Both the rise to, and fall from, a thermal maximum early in the interstadial appear to have been rapid. The decline probably involved a reduction of *circa* 3°C in mean temperatures of the warmest summer month (Atkinson *et al.* 1987) (Figures 9.3 and 10.16(a)). About 12 000 BP, perhaps coincident with the Older *Dryas* cooling event (Figure 9.19), the thermophiles disappeared and were replaced by species that now possess more northerly distributions. The insect evidence does not signal a climatic improvement during the Allerød episode. For the period from *circa* 12 200 to 11 000 BP, the Coleoptera indicate a cool-temperate environment. Insects have not been recorded from the Loch Lomond Stadial deposits in Windermere.

In south-west Lancashire, coversands (the Shirdley Hill Sand Formation) (Figure 9.15) were deposited during the Loch Lomond Stadial. Peat below these sands at Clieves Hills, north of Ormskirk, contains a pollen flora dominated by grasses, sedges and dwarf shrubs and has given a radiocarbon age of 10 455 ± 100 BP (Tooley and Kear 1977). At Poulton-le-Fylde, an elk skeleton bearing weapon marks, and Proto-Maglemosian barbed points, have been discovered in lake mud radiocarbon dated to 12 200 ± 160 and 11 665 ± 140 BP. The pollen assemblage from the mud indicates that birch woodland was present in the area (Hallam *et al.* 1973). Elk bone from this locality has been radiocarbon dated to 12 400 ± 300 BP (Jacobi *et al.* 1986).

Late Devensian glacigenic deposits have not been recorded in the Peak District (Briggs and Burek 1985). Periglaciation was intense during this timespan, with landsliding and scree formation taking place in valleys such as Edale (Redda 1985).

9.3.4 The Thames Valley

The Floodplain Terrace in the Upper Thames Valley (Figure 7.7) was partly formed during the Late Devensian (Briggs and Gilbertson 1980). While it is uncertain whether deposition or channel incision occurred during the glacial episode, silty and peaty layers within the terrace have been established as cold-stage accumulations. The inference from molluscan assemblages (including *Columella columella* and *Pupilla muscorum* and lacking *Ancylus fluviatilis*) is of cold conditions, and a shallow, mainly slow-flowing river. A cryoturbated surface observed within the terrace could be related to the glacial episode. Above this surface, Lateglacial deposits have been identified and radiocarbon dated to *circa* 13 500 BP at Abingdon, 11 000 BP at Northmoor and 10 500 BP at Farmoor (Briggs *et al.* 1985). The Abingdon sequence has been detailed by Aalto *et al.* (1984). The flora includes *Ranunculus platanifolius* (large white buttercup), now an inhabitant of alpine meadows. The insect fauna possesses *Pycnoglypta lurida*, today extinct in the British Isles but common here during the Lateglacial. The radiocarbon age of this deposit indicates a pre-Lateglacial Interstadial context. However, the insect evidence is more compatible with a later phase when the climate was deteriorating from the thermal maximum of the interstadial. The Farmoor deposit, of Loch Lomond Stadial age, contains *Helophorus glacialis*, an arctic and high-montane insect, whose main habitat in present-day Europe is drainage water from melting snow patches (Briggs *et al.* 1985).

In the Middle Thames Valley, gravel (the Shepperton Gravel, for example) and silt (the Langley Silt Complex) accumulated during the Late Devensian (Gibbard 1985, 1988b). The Langley Silt Complex has given a thermoluminescence age of *circa* 17000 BP (Gibbard *et al.* 1987). Braided-river sediments of Late Devensian age have been described from West Drayton and Colnbrook in west London (Gibbard and Hall 1982). In the Coln Valley, organic deposits within channels have been radiocarbon dated to 13405 ± 170 and 11230 ± 120 BP. Plant macrofossils of cold-climate affinity have been obtained from the organic material.

9.3.5 Wales, and England south of the Lower Severn and Thames valleys

Numerous views have been expressed concerning the extent of the Main Devensian Glaciation within Wales (Bowen 1969b; Lewis 1970a) (Figure 9.20). Charlesworth (1929) has postulated that most of the country was covered by ice, the limit of which was an end-moraine in South Wales. Wirtz (1953) has suggested that the lowland behind Cardigan Bay was bypassed by the ice, while Mitchell (1960) has postulated that Irish Sea ice had no influence south of the Lleyn Peninsula, with a large part of central Wales remaining unglaciated. Synge (1963) has considered that only a small sector of the North Wales coast was affected by ice at this time. Bowen (1969a,b, 1970, 1973a,b, 1974, 1981) has outlined the limits of Late Devensian glaciation in Wales, where both local and Irish Sea ice were juxtaposed (Bowen *et al.* 1986). The key stratigraphic markers have been raised beaches of Ipswichian (Oxygen Isotope Substage 5e) age (Chapter 8), above which glacigenic and periglacial deposits were emplaced. A radiocarbon date of 18000 ± 1200 BP on woolly mammoth bone, part of a cold-climate faunal assemblage discovered beneath Irish Sea till at Tremeirchion Caves in the Vale of Clwyd (Rowlands 1971), has provided a maximum age for the glaciation in this part of Wales (Shotton 1977a). Shells within glacial deposits in south-west Dyfed (see below) and northern Gower (Campbell *et al.* 1982) have amino-acid ratios which support the concept of marine transgressions, probably correlating with Oxygen Isotope Substage 5e and Stage 3 (Bowen 1984). The youngest shells have provided a maximum age for the glaciation, considered to have taken place during Oxygen Isotope Stage 2 (Bowen 1984; Bowen and Sykes 1988).

In north-west Wales, glacial deposits on the Lleyn Peninsula have been interpreted as indicative of two episodes of ice activity (Whittow and Ball 1970), the latter one perhaps a readvance (Saunders 1968; Bowen *et al.* 1986). The Loch Lomond Stadial witnessed the re-establishment of glaciers in the cwms of Snowdonia (Seddon 1957; Gray 1982, 1990; Gray and Lowe 1982). Within the innermost moraines associated with these cwms, for example, Cwm Idwal (Godwin 1955; Tipping 1990), Cwm Cywion and Cwm Llydaw (Ince 1983), only sediments containing Flandrian (Postglacial) pollen assemblages have been discovered, thus indicating that the glacigenic material was produced in the Loch Lomond Stadial. Away from the cwms, periglaciation was intense across the uplands of north-west Wales at this time. Outside the innermost moraines associated with the cwms, in Nant Ffrancon (Seddon 1962), for example, complete Lateglacial vegetational histories have been obtained. Elsewhere in Snowdonia, pollen and diatom records from lake deposits near Capel Curig exhibit biozones I, II and III of the Lateglacial (Crabtree 1972). At Glanllynnau, near Criccieth on the Lleyn Peninsula, a kettle hole in outwash contains an infilling of sediments covering the Lateglacial period (Simpkins 1974). Zone I at first has sparse pollen. Its quantity then builds up, and *Juniperus* and *Rumex* frequencies increase, before the *Betula* curve rises during Zone II. In

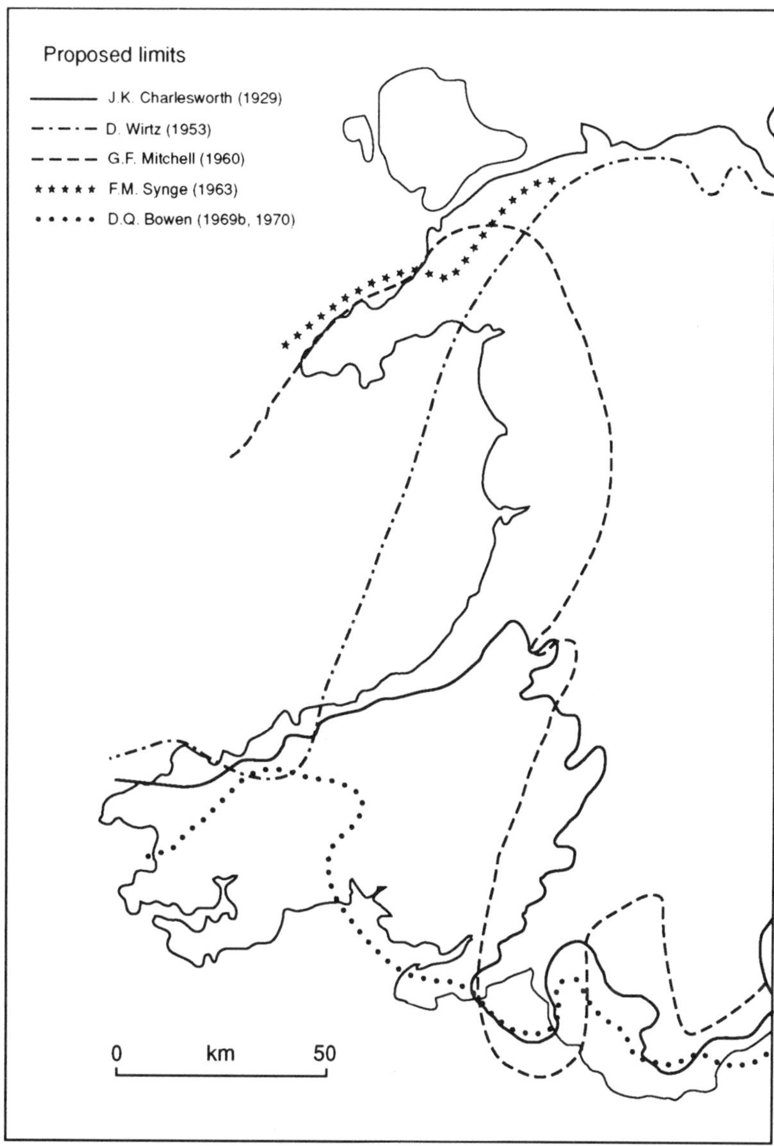

Figure 9.20 Views on the extent of Devensian glaciation in Wales. From Bowen (1969b); additional sources, Bowen (1970), Lewis (1970a).

Zone III the vegetation became open once again. One radiocarbon date of 14 468 ± 300 BP has been obtained from the base of the Lateglacial sequence, and another of 12 556 ± 230 BP from organic material just prior to that whose pollen assemblage indicates the commencement of the birch woodland phase (Shotton and Williams 1971). The Lateglacial insect fauna from Glanllynnau has been described by Coope and Brophy (1972), who have also compared the palynological and coleopteran sequences over this timespan. The insects indicate a different climatic history to that suggested by the plant evidence. About 13 000 BP an intensely cold, continental environment, with July temperatures of 10°C or less, and an insect fauna including *Amara alpina* and *Helophorus sibiricus*, was suddenly replaced by one whose thermal characteristics were equivalent to or warmer than those of North Wales today. Among the insects present were *Feronia macra*, now an inhabitant of central Europe, and *Bembidion grisvardi*, currently resident in south-west Europe. July mean temperatures of at least 17°C are indicated. This insect assemblage corresponds to the palynological one dominated by *Juniperus* and *Rumex* that indicates a scarcity or the absence of trees. For the period from about 12 000 to 11 000 BP, the pollen data depict birch woodland growth in the area. The corresponding insect fauna, including the boreo-montane *Pycnoglypta lurida* and *Arpedium brachypterum*, signals a deteriorating climate. This deterioration continued and worsened, with the reappearance of such taxa as *Helophorus sibiricus* marking the onset of the Loch Lomond Stadial.

As the insects from Glanllynnau denote quite a warm climate *circa* 13 000 BP, it has been suggested by Coope and Brophy that sparse, pioneer vegetation at the time may have reflected a lack of competition from other plants and unsuitable soils for the development of closed communities, rather than the incidence of an arctic climate. They have also hypothesized that warmth-demanding plants were slower to immigrate than thermophilous insects following the climatic amelioration, and that the former had no difficulty establishing in the prevailing environment on their later arrival.

Welsh ice from Snowdonia, the Arenigs and the Berwyns, and Irish Sea ice, were present in north-east Wales during the Late Devensian. However, large sectors of upland Denbighshire were probably devoid of ice, with a local ice cap developed over only the highest localities in this area (Embleton 1970). In mid-Wales, Watson (1970) has concluded that the Cardigan Bay lowlands experienced periglacial rather than glacial conditions in the Late Devensian. Moore (1970) has described the Lateglacial vegetational history of the Elan Valley in the mid-Wales uplands. During Zone I, tall herbs and dwarf-shrub heath were dominant; in Zone II, birch scrub was dominant; in Zone III, tundra was dominant.

According to John (1970), Irish Sea ice was present in both north and south Pembrokeshire during the Late Devensian glaciation. At Mullock Bridge near Milford Haven, marine mollusc fragments from outwash of this glaciation have given a radiocarbon date of *circa* 37 000 BP (John 1965). In the Banc-y-Warren area of south Cardiganshire, sands and gravels contain wood radiocarbon dated to 33 750 $^{+2500}_{-1900}$ BP (John 1967) and organic mud radiocarbon dated to 31 800 $^{+1400}_{-1200}$ BP (Brown *et al.* 1967). Organic material from this locality has produced pollen, mainly *Pinus*, and may have formed during an earlier interstadial (John 1970).

In some parts of central and south-east South Wales, the periglaciation that had characterized much of earlier Devensian time continued during the Late Devensian, with head a major product. Ice moved from further north in Wales to other parts of these areas. Its limit extended from eastern Gower across Swansea Bay and the Vale of Glamorgan, with the glaciers just getting offshore between Cardiff and Newport (Bowen 1969b, 1970) (Figure 9.20). At Paviland Cave on the Gower Peninsula, *Homo sapiens sapiens* remains, of a young male, radiocarbon dated to 18 460 ± 340 BP, have been found (Oakley 1968). Earlier Upper Palaeolithic artifacts have been found associated with these human remains (the so-called Red Lady because of red ochre adhering to the bones of what their excavator Dean Buckland (1823) thought a female), which according to Campbell (1977) could mean that the cave was occupied during the last glacial maximum. The habitation might have been earlier, however. In this context, a radiocarbon date of 25 650 ± 1300 BP from a *Bos* bone at the locality may be relevant (Molleson and Burleigh 1978).

The limit of the main Late Devensian ice has been traced in the middle Usk and upper Wye valleys (Lewis 1970b) and across the Herefordshire lowlands (Luckman 1970) (Figure 9.11). Glaciers appear to have developed on the Brecon Beacons and Fforest Fawr but not the Black

Mountains, which were also not surmounted by ice. In the Brecon Beacons–Fforest Fawr region, arcuate moraines and protalus ramparts have been identified in cwms and beneath steep escarpments (Lewis 1970b). Palynological data from Craig-y-Fro and Craig-Cerrig-gleisiad in the Brecon Beacons (Walker 1980, 1982) have established that many of these features were produced during the Loch Lomond Stadial, although some (the outer moraines in the latter locality, for example) may have been of earlier origin. Both of these basins contained ice during the Loch Lomond Stadial. The initial pollen assemblage of each includes substantial frequencies of *Juniperus* and Gramineae, before *Betula* values rise. At Craig-y-Fro the basal organic material has a radiocarbon age of 10 030 ± 100 BP. At Traeth Mawr, on Mynydd Illtydd 3 km to the north, a dead-ice hollow at *circa* 330 m OD contains sediments with a complete Lateglacial vegetational sequence (Walker 1982). The first pollen spectra, of early interstadial age, depicts vegetation consisting of grasses, sedges and other pioneer herbs such as *Rumex* and *Artemisia*, together with some dwarf shrubs. This was succeeded later in the interstadial; first by juniper – willow shrubland, then *Betula* woodland interspersed with open grassland and tall-herb communities, the latter including *Thalictrum* and *Filipendula*. Of especial interest are two phases within the interstadial when woodland cover appears to have been reduced. One of these has been considered likely to represent the Older *Dryas* event. The start of organic matter accumulation (later than the onset of the interstadial) has been radiocarbon dated to 11 600 ± 140 BP and the commencement of the Loch Lomond Stadial to 10 620 ± 100 BP. During the latter episode, montane grassland and alpine vegetation were widespread on Mynydd Illtydd. Solifluction took place and high frequencies of *Artemisia* pollen suggest the incidence of disturbed soil. The stadial concluded at 9970 ± 115 BP.

Most authorities have concurred that Devensian ice did not impinge on the South-West Peninsula of England or its offshore islands (Figure 9.11). However, Synge (1977, 1979, 1981, 1985) has suggested that floating shelf-ice filled most of the Celtic Sea between Ireland, Wales and Cornwall during the Devensian (section 9.1.6). This ice may thus have emplaced the glacigenic deposits recognized at Fremington in Devon, Trebetherick in Cornwall and in the Scilly Isles (Chapters 5 and 7). Support for Devensian glaciation in the Scilly Isles has been provided by Scourse (1986). Above a raised beach in the islands, solifluction breccia (Scourse 1987), with organic layers below and within it, has been encountered. Palynological investigation of these organic deposits, and radiocarbon dating, indicate the existence of tundra grassland *circa* 34 000–21 000 BP (section 9.2.4). In the northern Scillies, till has been observed above the breccia, and in the southern isles, loess. The latter has been dated by thermoluminescence to around 18 600 BP (Wintle 1981). Scourse (1986) has suggested that ice advanced as far as the northern isles of Scilly during the Dimlington Stadial.

Outside localities of possible ice-contact, periglaciation was widespread in south-western England during the Late Devensian. At Kenn in the Bristol area, coversands, ice-wedge casts and involutions of this age have been identified (Gilbertson and Hawkins 1978). Head formed on the Mendip Hills (Findlay 1965) and Brean Down (ApSimon *et al.* 1961; ApSimon 1977) in Somerset. Waters (1964, 1974) has postulated that on Dartmoor, an Upper Head, the majority of which consists of blockfields (clitters), was produced during the Devensian. The partial destruction of tors (Chapter 7) by periglacial processes during this period has been thought to have provided material for the head. In north Devon, Stephens (1970b) has recognized an Upper Head of Devensian age (Figure 1.4), in addition to a Lower Head (Chapter 7).

Sedimentation in a Lateglacial lake at Hawks Tor on Bodmin Moor, Cornwall (Conolly *et al.* 1950), began just before 13 000 BP (Brown 1977). Palaeobotanical studies have revealed initial snow beds and flushes, with open grass heath present. This was succeeded by juniper scrub, after which tree-birches colonized sheltered localities and

Empetrum heath occurred on hillsides. About 11 000 BP, wetland dominated by grasses and sedges emerged in the locality. Solifluction took place in a cold but oceanic climate during the Loch Lomond Stadial which terminated prior to 9700 BP (Brown 1977).

Later Upper Palaeolithic artifacts have been recorded from Kent's Cavern, Devon, and Gough's Cave and Sun Hole, Somerset. Cave earths at these sites have yielded arctic/boreal pollen floras consistent with those of Lateglacial type. At Kent's Cavern and Sun Hole, large-mammal remains of this age include *Crocuta crocuta*, *Canis lupus*, *Vulpes vulpes* (red fox), *Alopex lagopus*, *Ursus arctos*, *Coelodonta antiquitatis*, *Cervus elaphus*, *Megaloceros giganteus*, *Rangifer tarandus*, *Bison priscus* and *Equus przewalskii* (*ferus*) (horse) (Campbell 1977). *Ursus arctos* bone from the Black Band at Kent's Cavern has produced a radiocarbon age of 14 275 ± 120 BP (Campbell and Sampson 1971), while bovid vertebrae from the same layer (which also contains implements) have been dated to 11 570 ± 410 BP (Jacobi *et al.* 1986). Bones of *Equus ferus*, some with human cut-marks, from Gough's Cave, have been radiocarbon dated to *circa* 12 000 BP (Burleigh *et al.* 1985).

South-eastern England was also unaffected by Devensian glaciation. As in south-western England, periglacial features, some of Late Devensian age, have been widely recognized (Catt 1977c; Williams 1968, 1975, 1987; Worsley 1977, 1987) (Figures 9.15 and 9.18). At this time, solifluction, cryoturbation and river incision took place in the Weald, while slope and coombe deposits formed on the Downs (Shephard-Thorn and Wymer 1977). Sarsens located within dry valleys in the chalklands are likely, at least in part, to be products of Late Devensian solifluction (Small *et al.* 1970). Loess was deposited widely on a variety of substrates across southern England as far west as Devon in the Late Devensian (Catt 1977c). Up to 4 m of loess has been observed at Pegwell Bay in north Kent (Pitcher *et al.* 1954; Shephard-Thorn 1977) (Figure 1.6) where it has been dated by thermoluminescence to about 14 800 BP. Thermoluminescence dates from other loesses in southern England have ranged from *circa* 18 800 to 14 500 BP (Wintle 1981). At Broadstairs, two fossil soils, probably of Bølling and Allerød age, have been found above loess (Kerney 1965). A rendzina soil between two chalk silts has been recognized in numerous localities over the chalklands of southern England. Near Halling, Kent, this soil contains birch charcoal. Organic material within a solifluction deposit below the base of the soil has been radiocarbon dated to 13 190 ± 230 BP (Kerney 1963). These lithostratigraphic units possess a characteristic molluscan biostratigraphy. The lower silt, of Zone I age, includes *Pupilla muscorum* and *Vallonia costata*. The interstadial soil possesses a relatively thermophilous fauna in which *Abida secale* is most abundant. In the upper silt, deposited during Zone III, *Trichia hispida* is especially well represented (Kerney 1977c) (Table 10.3). At the Devil's Kneadingtrough near Brook, Kent, the lower part of the fill of a coombe in the Chalk escarpment is rubble and silt of periglacial origin, which also occur on the plain beyond. Beneath these deposits is a palaeosol and an organic mud, the latter radiocarbon dated to 11 900 ± 160 BP (Kerney *et al.* 1964). The Lateglacial molluscan and plant stratigraphy of calcareous deposits located at Folkestone and near Maidstone, Kent, has been described by Kerney *et al.* (1980). In the former locality, basal gravels and silts of Zone III age contain a mainly herbaceous flora but also *Betula nana*, *Juniperus*, *Empetrum* and *Arctostaphylos* (bearberry) which were components of tundra vegetation. The corresponding molluscan assemblage includes *Pupilla muscorum*, *Columella columella*, *Catinella arenaria*, *Vertigo genesii*, *Trichia hispida* and *Abida secale*.

At Brimpton, Berkshire, a terrestrial molluscan fauna of Windermere Interstadial age that has been obtained from silty material lying above the terrace gravels of the rivers Kennet and Enborne includes *Trichia hispida* and *Vallonia pulchella* (Bryant *et al.* 1983a). During the Loch Lomond Stadial, when the River Kennet was braided and possessed a mainly gravel bed, the flora in its environs was dominated by Cyperaceae,

Gramineae and other herbs. The equivalent molluscan fauna included *Pupilla muscorum*, *Columella columella*, *Helicella itala* and *Vallonia excentrica* (Holyoak 1983c).

Some of the more recent terrace deposits of the Hampshire Avon (terraces 1–4), which have been seen in the Bournemouth–Fordingbridge area, are of Late Devensian age (Clarke and Green 1987). From Church Moor in the New Forest, Hampshire, Clark and Barber (1987) have described Lateglacial plant macrofossil and pollen assemblages. An interstadial peat, radiocarbon dated to 12 440 ± 600 BP near to its top, contains high birch-pollen frequencies, and both tree and dwarf *Betula* species have been recorded as macrofossils. A fairly well-developed Windermere Interstadial birch forest was replaced by herbaceous communities during the Loch Lomond Stadial. In the Isle of Wight, sandy deposits have produced palynological data which indicate that dwarf-shrub and short-turf vegetation was important during the Loch Lomond Stadial (Scaife 1982, 1987).

9.3.6 Scotland

Till has been demonstrated overlying Middle Devensian deposits at only a small number of localities. Thus, the presence of a Late Devensian ice sheet is unproven in many parts of Scotland. Moreover, the chronology of ice-sheet expansion has not been established (Bowen *et al.* 1986). Radiocarbon dates for Lateglacial deposits in Scotland have led Sissons (1981b) to the conclusion that deglaciation was widespread by *circa* 13 000 BP. Boulton *et al.* (1977) (Figure 9.16) and Price (1983) have postulated that the whole of Scotland was glaciated. However, Sutherland (1984) has proposed a restricted distribution of Late Devensian ice, with areas of the mainland and some offshore islands being unglaciated (Figure 9.11). Charlesworth (1956) and Synge (1956) have concluded that much of the Buchan plateau in north-east Scotland escaped Devensian glaciation. Proglacial lakes and meltwater channels identified in this area have been considered the product of coastal ice by Synge. In contrast, Clapperton and Sugden (1975, 1977) have asserted that all of north-east Scotland was subjected to Devensian glaciation, when ice crossed the area from the Moray Firth, eastern Grampians and Strathmore. In this context, it is worth recalling the recognition of an interstadial deposit dated 26 000–22 000 BP at Crossbrae Farm in central Buchan (section 9.2.6). Above the interstadial deposit is a soliflucted till, which represents possible evidence of ice-free conditions during the Late Devensian (Hall 1984a). Such conditions may have been more widespread, occurring for example, in other parts of Buchan and over much of Caithness and Orkney (Sutherland 1984). There was an unglaciated area in Lewis (Sutherland and Walker 1984), while St. Kilda possessed a small valley glacier (Sutherland *et al.* 1984). The inference from evidence in these north-western islands is of mainland ice halting east of the Outer Hebrides, with land exposed over parts of the continental shelf (section 9.3.8). It should be noted that Flinn (1978, 1980) and von Weymarn (1979) have concluded that a separate ice cap existed in the Outer Hebrides. Mull, about 40 km west of Rannoch Moor, possessed an ice cap on its southern mountains from which glaciers diverged. The northern and eastern parts of Mull may also have been reached by mainland ice (Walker *et al.* 1985). In the northern isles, Shetland developed its own ice cap, according to Mykura and Phemister (1976).

The Western Highlands and Southern Uplands were sources of ice during the Late Devensian glacial maximum. Here large amounts of precipitation assisted the formation of corrie glaciers. Of particular importance were the mountains west of Rannoch Moor, where an ice cap developed from which valley glaciers radiated. Ice-moulded bed forms, tills, outwash, drumlins and meltwater channels recorded in the Glasgow region resulted from the Main Late Devensian Glaciation (Price *et al.* 1980).

The general pattern of deglaciation in Scotland seems to have involved retreat until the ice was mainly land based in the eastern coastal area, while on the west coast an ice dome existed about the Firth of Clyde (Bowen *et al.* 1986). The ice

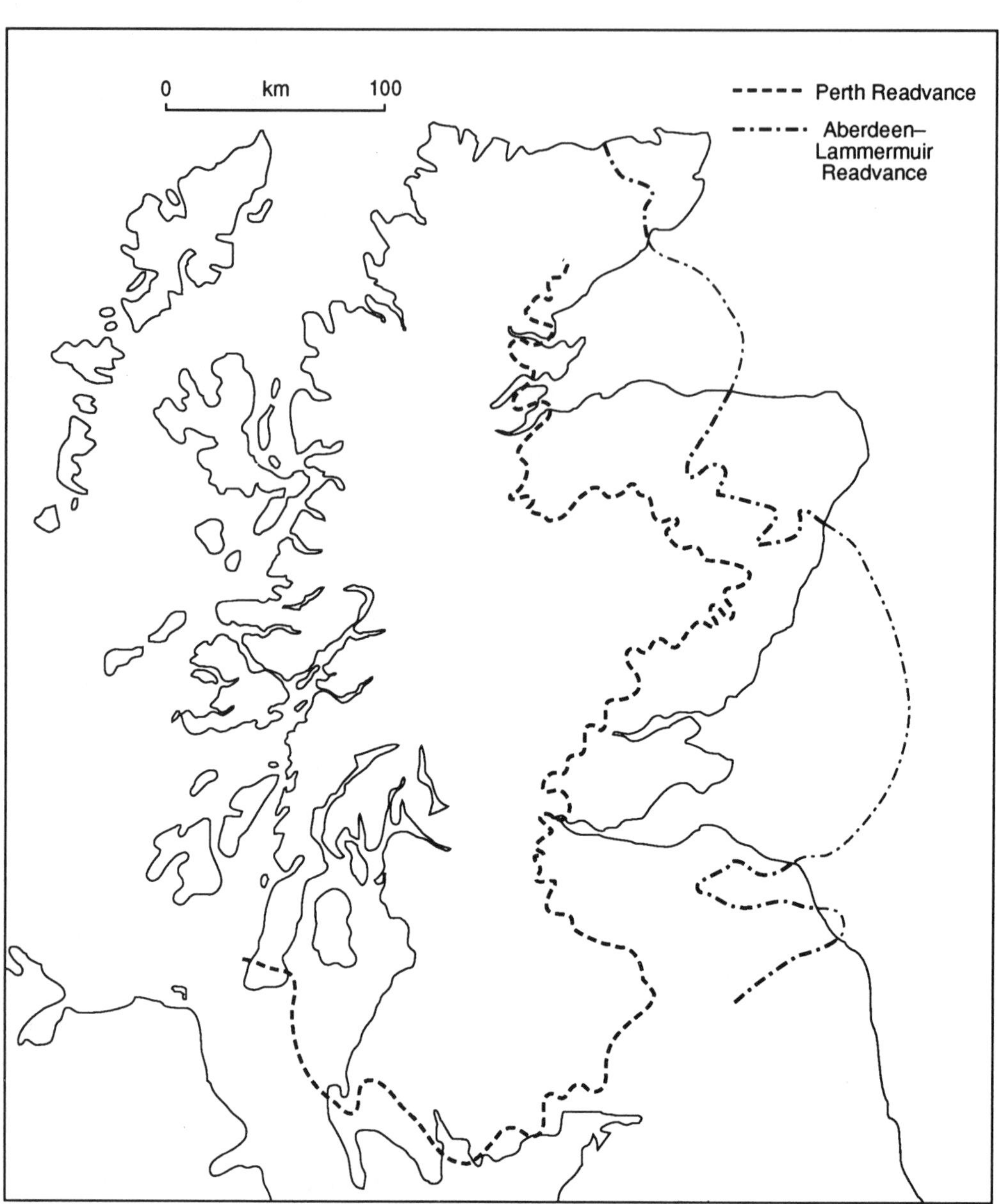

Figure 9.21 The Aberdeen–Lammermuir Readvance and Perth Readvance limits in Scotland. From Sissons (1967b).

eventually retreated to within the limits of the Loch Lomond Stadial glaciers, but whether it wasted entirely during the Windermere Interstadial has not been resolved (Sissons 1976; Sutherland 1984).

The Aberdeen–Lammermuir Readvance (Synge 1956; Sissons 1965) and the Perth Readvance (Sissons 1963, 1964, 1967b) (Figure 9.21) have been refuted and doubted, respectively, by Sissons (1974a). The Main Perth Shoreline in the estuaries of the Tay and Forth (section 9.3.8) has been purported to mark a high sea-level associated with the Perth Readvance. In the Stirling area, a rapid fall in sea-level has been

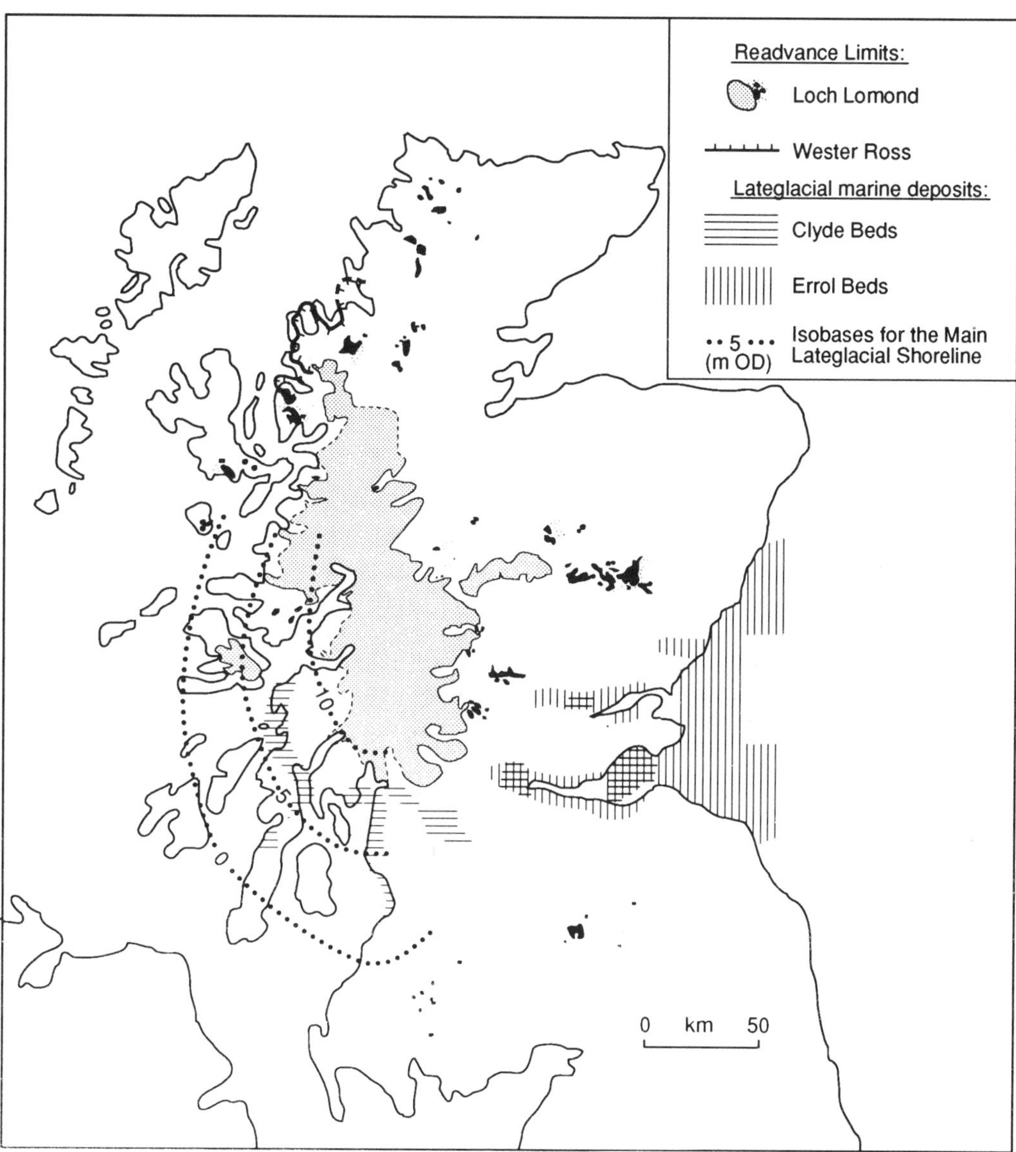

Figure 9.22 Limits of the Wester Ross Readvance (source, Robinson and Ballantyne 1979) and the Loch Lomond Readvance (source, Sissons 1980b) in Scotland. The distribution of Lateglacial marine deposits in Scotland and offshore (source, Peacock 1975, 1981). Isobases for the Main Lateglacial Shoreline in western Scotland (source, Dawson 1984). Reprinted with permission from *Quaternary Science Reviews* **3**, A. G. Dawson, Quaternary sea-level changes in western Scotland; © 1984 Pergamon Press Plc.

considered the consequence of retreat of this readvanced ice (Sissons and Smith 1965).

Moraines that were related to a readvance of ice after the Main Late Devensian Glaciation and before the Loch Lomond Stadial Glaciation have been identified in the environs of Loch Carron and Ullapool, in the North-Western Highlands of Wester Ross (Robinson and Ballantyne 1979) (Figure 9.22). The timing of this oscillation in the ice margin during a phase of overall climatic amelioration is uncertain, but the most likely period was 13 500–13 000 BP, when the Oceanic

Polar Front was moving north of the coast of western Scotland (Ruddiman and McIntyre 1973, 1981), (Figure 9.2). Milder conditions resulting from this movement could have generated more precipitation which led to an enhanced build up of ice (Bowen *et al.* 1986). At Cam Loch in Sutherland, north of Wester Ross, a basal Lateglacial radiocarbon date of 12 960 ± 240 BP (Pennington 1975a) has demonstrated that deglaciation had been accomplished in this area. The moraines of the proposed Wester Ross Readvance have been linked to a shoreline (the Main Wester Ross Shoreline) by Sissons and Dawson (1981).

The country around Glasgow had been vacated by ice of the Main Late Devensian Glaciation by about 12 500 BP (Peacock 1971b; Browne *et al.* 1977). An extremely cold environment in this area just after 12 000 BP has been identified by Dickson *et al.* (1976). At this time, valley glaciers were expanding in the mountains north of the Midland Valley. In these mountains, corries and steep-sided troughs were enlarged and moraine was deposited. These glaciers progressed to beyond the southern shore of Loch Lomond during an ice readvance (Figure 9.22) in a stadial which has been named after the locality (Gray and Lowe 1977) (section 9.3). The Loch Lomond Readvance glaciers attained their maximum extent in the area between the heads of the Firth of Clyde and the Forth Valley *circa* 10 900–10 700 BP (Sutherland 1984). Radiocarbon dating of the deglacial phase of the stadial has proved difficult, with diverse results (Lowe and Walker 1980, 1984; Sutherland 1980, 1984).

In Mull the Loch Lomond Stadial has been investigated by Gray and Brooks (1972). An ice cap was present in the south-east of the island, with isolated glaciers in a number of other locations. Sissons (1976, 1979a, 1980b), using data on firn lines and ice limits, has suggested that inter-area climatic variations during the Loch Lomond Stadial gave rise to glaciers of differing dimensions. In Mull during the Loch Lomond Stadial, mean temperatures of under −7°C in January and around 5°C in July have been postulated by Walker *et al.* (1985). Within the limits of Loch Lomond Stadial ice on Mull, only Flandrian pollen stratigraphies occur; beyond them, Lateglacial ones have been found (Walker and Lowe 1982).

Sugden (1970) has maintained that ice may not have readvanced in the Cairngorms during the Loch Lomond Stadial and that the area could have possessed an ice cover throughout the Lateglacial. However, Sugden and Clapperton (1975) have concluded that some corrie glaciers were likely to have existed in these mountains at this time. Sissons (1972) has proposed a plateau ice cap in the south-east Grampians, from which valley glaciers emanated during the Loch Lomond Stadial.

Lateglacial environments in Scotland have been considered by Gray and Lowe (1977), and Lowe and Walker (1984), while Walker (1984a) has examined evidence for vegetational history over this timespan. The earliest sediments are dominated by pollen and spores of open-habitat plants, notably Gramineae, Cyperaceae, Caryophyllaceae, Chenopodiaceae, Compositae, *Rumex* and *Lycopodium* species. A paucity of shrub and arboreal pollen indicates that woodland was either absent or minimal during this pioneer stage. Radiocarbon ages of *circa* 13 000 BP (early Windermere Interstadial time) have been obtained from sediments with these palynological charcteristics in the central Grampians (Sissons and Walker 1974) and Mull (Walker and Lowe 1982). Differing environmental factors meant that after the pioneer stage the interstadial vegetation was varied in its composition across Scotland. For example, in the south-west (Moar 1969a) and Argyllshire (Rymer 1977), stands of *Salix*, *Juniperus*, *Empetrum* and probably tree-birch, accompanied grassland. Heath, grassland and *Betula* woods have been thought to have been present during the Lateglacial Interstadial on Skye by Birks (1973). However, Walker *et al.* (1988) have demonstrated that the majority of palaeobotanical evidence from Skye assumed to be Lateglacial by Birks relates to the early Flandrian. The Lateglacial vegetation of Skye has been shown to be analogous to that of nearby Hebridean islands, especially Mull (Lowe and Walker 1986), and adjacent mainland areas.

In these localities the pioneer vegetation was followed by an extension of heathland with *Juniperus* and *Empetrum* as the Windermere Interstadial proceeded. The onset of climatic worsening, which reached its acme during the Loch Lomond Stadial, caused the heathland vegetation to decline. It was replaced by tundra in which grasses, sedges, *Artemisia*, *Rumex* and *Lycopodium* were prominent. In Sutherland (Pennington *et al.* 1972), Caithness (Peglar 1979), Orkney (Moar 1969b) and Shetland (Birnie 1981), *Empetrum*-dominated dwarf-shrub heath was juxtaposed with open grassland, and *Juniperus* fluctuated in its contribution to the plant cover. Tree-birches may have been present in Caithness but are likely to have been absent from Sutherland (Pennington 1975a). The interstadial vegetation of the central Grampians was shrub tundra in which *Empetrum*, *Salix* and *Betula* were present (Walker 1975a; Birks and Mathewes 1978). Sheltered parts of valley bottoms and lower slopes in the southern and eastern Grampians carried patches of birch woodland. In more exposed areas, closed grassland with juniper, dwarf birch and willow was present, while at greater elevations, impoverished grassland and moss heath dominated (Walker 1975b; Lowe and Walker 1977; Lowe 1978). The Midland Valley developed grassland and birch copses during the Windermere Interstadial (Donner 1957; Newey 1970). Similar vegetation then existed in Fife (Donald 1981) and the south-eastern Borders (Webb and Moore 1982).

Coleopteran assemblages from south-western Scotland point to an interstadial climatic amelioration prior to 13 000 BP. The highest interstadial temperatures were attained around 13 000 BP, following which they underwent a progressive decline (Bishop and Coope 1977). Pollen diagrams from north-west Scotland (Pennington 1977b) exhibit a short episode, radiocarbon dated to 12 000–11 800 BP, when frequencies of woody plant taxa (*Empetrum*, *Juniperus* and *Betula*, for example) decreased and those of open ground (*Rumex* and *Artemisia*, for instance) increased. The corresponding sediments contain an enhanced quantity of minerogenic material. Such vegetational trends (section 9.3.3) have also been recognized in the Dee Valley, Aberdeenshire (Vasari and Vasari 1968; Vasari 1977), the eastern Cairngorms (Clapperton *et al.* 1975) and south-eastern Grampians (Walker 1977; Caseldine 1980).

Birks (1973) has suggested that the plant communities which existed during the Loch Lomond Stadial on Skye were analogous to those of present-day tundra, and of low-alpine shrubland in Scandinavia. The incidence of such taxa as *Artemisia* and *Rumex* in pollen records of this age has been considered indicative of widespread bare ground and disturbed soil. *Artemisia* will not now thrive where there is a snow cover (Pennington 1980). Low frequencies of *Artemisia* pollen in the eastern Grampians during Zone III have been correlated with a widespread snow blanket by Walker (1975b). High values for this pollen taxon in the Spey Valley have led Birks and Mathewes (1978) to postulate a drier local environment here during the Loch Lomond Stadial.

Insect evidence from south-west Scotland indicates near polar-desert conditions during the Loch Lomond Stadial (Bishop and Coope 1977). Radiocarbon dates for the onset of the stadial range from over 11 000 BP (Vasari 1977) to 10 700 BP (Walker and Lowe 1982). Sites within the limits of Loch Lomond Stadial ice have produced radiocarbon ages from basal organic sediments which indicate a climatic amelioration *circa* 10 500 BP (Lowe and Walker 1976, 1977, 1980, 1984; Walker and Lowe 1979, 1982).

9.3.7 Ireland and the Isle of Man

Little organic material has been found below glacigenic deposits of Late Midlandian age in Ireland. Such age determinations as exist have indicated an ice sheet developing later than 30 000 BP, with its maximum extent *circa* 24 000–20 000 BP (McCabe 1987). Radiocarbon dates of 17 300 and 16 900 BP from *Macoma calcarea*, which occurs in raised glaciomarine deposits in County Mayo, have demonstrated that about one third of the ice sheet had wasted by that time (McCabe *et al.* 1986b).

Aghnadarragh has been recommended by McCabe (1987) as the type locality for the Glenavy Stadial (Figure 9.7(a)), with the Upper Till distributed around Lough Neagh, the lithostratigraphic representative of Late Midlandian glaciation in this area. It has been stated by McCabe that two important ice domes were present, in north-central and west-central Ireland. Thus, in contrast to the majority of the British Isles, the ice centres were at low elevations. From here ice dispersal took place across the majority of the lowlands, with some mountains, such as the Sperrins (Colhoun 1971a) also becoming engulfed. Certain mountainous areas, the Wicklows for example, also generated ice. The maximum extent of Late Midlandian ice was the South Irish End Moraine, which has been identified crossing the island in the vicinities of Limerick, Tipperary and Wicklow (Figure 9.23). In some districts, the Wicklow Mountains (Synge 1973) for instance, the ice limits were complex due to the influence of both upland and lowland ice. There were ice-dammed lakes (at Blessington, for example) in the Wicklow Mountains during the Late Midlandian (McCabe and Hoare 1978).

It is noted above (section 9.3.5) that Synge (1977, 1979, 1981, 1985) has proposed that Midlandian ice occupying the Celtic Sea was a floating shelf that came into contact with the northern coast of Devon and Cornwall and the northern Scilly Isles. The Ballycroneen Till, which has been recognized along the southern coast of Ireland, has been designated as marine and a product of this shelf ice by Devoy (1983). This interpretation has been disputed by McCabe (1987), who has categorized the glacigenic deposit as a glaciomarine diamictic mud which was generated by debris flows from tidewater glaciers (McCabe 1985). The diamicton has been identified overlying the Courtmacsherry Raised Beach, and McCabe (1987) has concluded that the latter was not of interglacial origin (Chapter 8) but of glacial origin, having been produced by a sea whose level was about that of the present day. In support of this hypothesis, McCabe has pointed to the existence of angular and erratic clasts within the beach, which he has suggested became submerged and covered by glaciomarine sediments during the Late Midlandian. The area was isostatically depressed by ice, which when it retreated enabled the sea to transgress (see below).

Mitchell *et al.* (1973a), Herries Davies and Stephens (1978) and Warren (1985) have regarded shelly (Irish Sea) tills located in south Donegal, north Mayo and along the eastern coast as having resulted from ice located on land, and their stratigraphic complexity to be connected with advance and retreat episodes. Colhoun (1973) and Warren (1985) have associated such tills with the presence of Scottish ice. However, Eyles and McCabe (1991) have stated that they were glaciomarine sediments. They have postulated that the mechanism outlined above – isostatic depression because of ice loading, which permitted subsequent marine transgression – was in operation.

Synge (1969) has postulated that a readvance of ice occurred during the Late Midlandian (Figure 9.11). In northern and western Ireland, drumlins, formed about 17 000 BP, have been considered by Dardis (1985) to be the result of floating and surging ice on a deformable substrate. The drumlin swarms, fronted by moraines, have been assigned the same age by McCabe (1985). The issue of the rates of deglaciation in the Late Midlandian has been discussed by McCabe (1987). In this context, the fact that organic sediment had been lacking has been noted, and it has been suggested that this could have been a response to rapid ice wastage, whereby biogenic material was prevented from accumulating.

Certain parts of Ireland appear to have escaped glaciation in the Late Midlandian (Figure 9.23). Considerable evidence of periglaciation has been found in Ireland (Lewis 1985) (Figure 9.18). For example, relict periglacial features referable to this period have been identified on the Truskmore Plateau, near the border of Sligo and Leitrim (Coxon 1988). In some areas, the summit plateau of the Sperrin Mountains (Colhoun 1971b), for

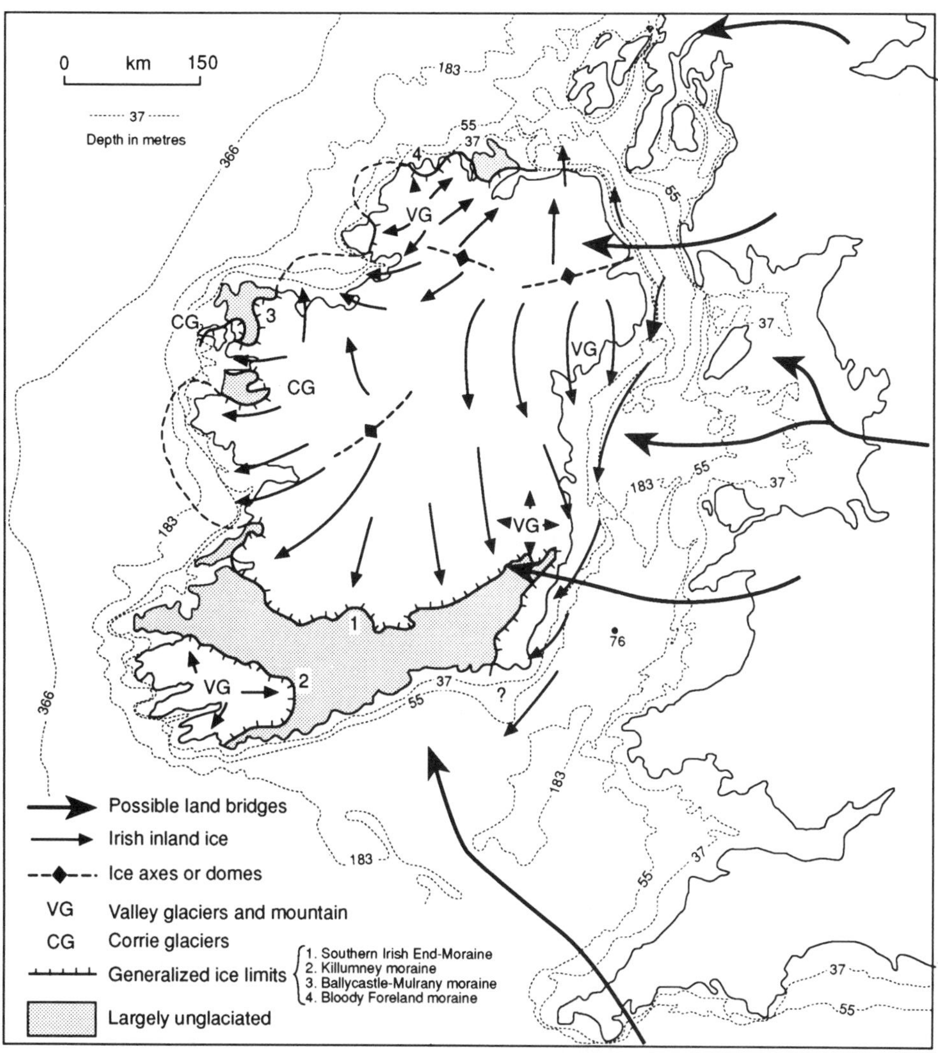

Figure 9.23 General direction of ice-sheet movement and generalized ice limits in Ireland during the Late Midlandian. From McCabe (1985); sources, the published works of Charlesworth, Creighton, Colhoun, Farrington, McCabe, Mitchell, Stephens and Synge. Bathymetry and location of postulated Late Pleistocene landbridge routeways between Britain and Ireland. From Devoy (1985a). Bathymetry; sources, Royal Irish Academy (1979) and Bickmore and Shaw (1963). Reprinted with permission from *Quaternary Science Reviews* **4**, R. J. N. Devoy, The problems of a late Quaternary landbridge between Britain and Ireland; © 1985 Pergamon Press Plc.

example, periglacial features formed subsequent to Late Midlandian deglaciation have been recognized.

Mitchell (1976) has stated that the Midlandian Lateglacial period in Ireland contained the Woodgrange Interstadial and the Nahanagan Stadial (Figure 9.7(a)), which could be equated with the Windermere and Loch Lomond events in the remainder of the British Isles. Sites with important Lateglacial sequences include Roddans Port (Morrison and Stephens 1965), Woodgrange (Singh 1970), Coolteen and Belle Lake (Craig 1978), in the eastern part of the island. Watts (1977, 1980, 1985) has detailed the vegetational history of Ireland during the Midlandian Lateglacial. Prior to *circa* 13 000 BP, sparse herbaceous plant communities were present. From 13 000 to 12 400 BP the floral assemblages, dominated by *Rumex* and containing *Salix herbacea*, were representative of those comprising current snow-patch vegetation. A first *Juniperus* phase occurred from 12 400 to 12 000 BP, when this shrub was extensive in scrub, especially in eastern Ireland. *Empetrum* was now an important component of the flora in western Ireland. A short erosion phase followed (12 000–11 800 BP). Inorganic sedimentation increased and juniper pollen frequencies fell during a climatic deterioration that led to soil instability. From 11 800 to 10 900 BP, a Gramineae-dominated episode was in progress, during which birch copses were also present.

This timespan included the *Betula* (Zone II) period identified by Jessen (1949) in a scheme of Lateglacial vegetation changes for Ireland. Subsequent investigations have established that birch was of only limited, local abundance during the Lateglacial Interstadial (Watts 1985). During the grassland phase of this interstadial, *Megaloceros giganteus* and *Rangifer tarandus* were abundant. Both animals subsequently became extinct, the latter in the Nahanagan Stadial (Mitchell 1976; Stuart 1982, 1985). The first botanical analysis of Lateglacial deposits in the British Isles, at Ballybetagh, County Dublin, was undertaken by Jessen and Farrington (1938). The scheme of Lateglacial vegetational history subsequently provided for Ireland by Jessen, also included Older (Zone I) *Salix herbacea* and Younger (Zone III) *Salix herbacea* periods. The scheme presented by Watts (1985) contained an *Artemisia* phase coincident with the Nahanagan Stadial, 10 900–10 000 BP. *Salix herbacea* was present at this time, together with a diverse herb flora, in tundra-like vegetation. Cirque moraines developed in the Wicklow Mountains during the Nahanagan Stadial. Organic material in ice-pushed lacustrine clay beneath moraine at Lough Nahanagan contains a pollen flora dominated by *Juniperus* and *Empetrum*, and has given radiocarbon ages of 11 600 ± 260 and 11 500 ± 550 BP, thereby placing the stadial after this time. The snow line in the area was around 470 m above sea-level during the stadial, and on this basis a fall in mean temperature of about 7°C has been calculated (Colhoun and Synge 1980).

According to Thomas (1976), foreign glacigenic material, confined to lower altitudes in the Isle of Man, was deposited during the Devensian (Figure 9.11). Pleistocene sediments at higher altitudes have been thought to be heads developed on a nunatak during the Devensian. Eyles and Eyles (1984) have suggested that the foreign suite of glacigenic material had a glaciomarine origin, an interpretation questioned by Thomas and Dackombe (1985), who have favoured a terrestrial provenance. Evans and Arthurton (1973) and Bowen (1973b) have suggested that the high-level deposits were till, laid down by Late Devensian ice whose thickness reached at least 1.6 km (Boulton *et al.* 1977) (Figure 9.16).

At Glen Ballyre on the north-west coast of the Isle of Man, sediment filling a kettle hole in outwash has revealed a Lateglacial vegetational sequence. *Dryas*, *Salix* and *Rumex* were prominent plants during Zone I; Gramineae and *Betula* were prominent in Zone II. A radiocarbon date of 12 150 BP has been obtained from interstadial mud (Mitchell 1965a; Dickson *et al.* 1970). At the same locality a basal moss layer above the outwash has been radiocarbon dated to between 18 900 and 18 400 BP (Shotton and Williams 1971, 1973; Shotton 1977a), thus indicating the time by which ice of the Main Late Devensian Glaciation had disappeared (Dackombe and Thomas 1985; Rose 1985). Insect assemblages from the Glen Ballyre deposits show a lower one of northern aspect and pre-Windermere Interstadial age in which the Eurasian arctic species *Bembidion lapponicum* occurs. This is followed by an early interstadial one including *Asaphidion cyanicome*, now an inhabitant of southern and central Europe where July temperatures are at least 17°C. The insect assemblage indicates the existence of grassland vegetation and sporadic tree growth. The sediments of the later part of the interstadial at first contain an insect fauna including *Amara torrida* and later *Diacheila arctica*. The indications from the insects are of cool conditions and abundant vegetation at the outset of this phase. Subsequently it became very cold and acid bog developed in the vicinity of the site (Joachim *et al*. 1985).

9.3.8 The offshore sequence and sea-levels

By the Late Devensian, glacial or periglacial environments existed over the whole of the southern North Sea. Off eastern Yorkshire a lobe of Late Devensian till, mainly of Withernsea type (Catt 1987a), has been identified on the floor of the North Sea to about 3°E, whence it passes into fluvioglacial sands and gravels. This lobe extends as far south as 53°N and has valleys eroded into it, many of which contain fluvioglacial and glaciolacustrine sediments (Cameron *et al*. 1987).

The Devensian glacigenic deposits in the North Sea (Figure 1.1) indicate that ice from the British Isles was not confluent with Scandinavian ice (Cameron *et al*. 1987). Till has not been recorded beyond 100 km offshore of eastern Scotland. The Wee Bankie Formation, including moraine-like ridges, occurs nearest the coast. Beyond this the shallow-water glaciomarine Marr Bank Formation has been found (Thomson and Eden 1977). Mostly outside the ice limit and mainly trending north–south are valleys cut during the Devensian and filled with Late Devensian glaciomarine sediments. Valleys also occur within the ice limit. These formed after till deposition and below an extensive cover of sea ice that may have extended as far south as 53°N. A climatic amelioration and consequent rise in sea-level (but not to beyond −60 m OD) may have assisted this valley incision (Cameron *et al*. 1987). Coeval Scandinavian till has not been found west of the Norwegian Trench. That part of the north-central North Sea between the Scottish and Scandinavian ice sheets was a shallow marine embayment with arctic conditions and probably a thick cover of sea ice (Stoker and Long 1984; Cameron *et al*. 1987). In the northern North Sea (above 60°N), till-like Late Devensian sediments have been identified (Stoker *et al*. 1985c). The Shetland ice cap (section 9.3.6) extended some 60 km east of the archipelago but did not come into contact with Scandinavian ice. Here also, till passing eastward into fluvioglacial deposits has been identified. Valleys are located in this area, mainly trending north-west to south-east. Some of these valleys are outside the ice limit; others were cut and filled after till was laid down during the Late Devensian. Some of the area between the two ice masses seems to have been land, while a marine embayment also existed (Long and Skinner 1985; Cameron *et al*. 1987).

Shards of volcanic glass correlated with those of the Vedde Ash Bed of western Norway, dated to 10 600 ± 600 BP (Mangerud *et al*. 1984), have been recognized in sediments from the central North Sea (Long and Morton 1987). Biostratigraphic and lithostratigraphic information from these sediments has indicated a Devensian Lateglacial age (Long *et al*. 1986). The shards occur in the deposit of a short cold period in the stratigraphic sequence.

West of the Scottish mainland, Davies *et al*. (1984) have established the existence of glaciomarine sediments of the Stanton Formation (Figure 9.8). These were associated with pre-Late Devensian ice whose locus was the Inner Hebrides (section 9.1.10). During the Main Late Devensian Glaciation, this ice advanced over the continental shelf west of Scotland, thereby forming a widespread erosion surface. Indications of ice-flow directions imply the existence of an independent ice cap in the Outer Hebrides, as postulated by Flinn (1978, 1980) and von Weymarn (1979) (section 9.3.6). Jardine (1977) has suggested that around 18 000 BP, the marine shoreline was located west of the Outer Hebrides. South of the Outer Hebrides platform, glacial erosion continued to the margin of the continental shelf, with the ice then calving into the distal, deeper water. The ice is likely to have been thin, as such a body erodes actively and deposits little till. A thin boulder layer or diamicton, plus an erosion surface constitute the Hebrides Formation. Above the latter is a complex of glaciomarine deposits making up the Barra Formation. These accumulated rapidly under stadial conditions, probably subsequent to ice retreat from the shelf margin and to a rise in sea-level. The Jura Formation is the next youngest, also complex and up to 300 m thick (Figure 9.8). It consists of glaciomarine and marine sediments, and contains faunal evidence of both cold and temperate episodes (Binns *et al.* 1974). Accumulation of this formation probably began *circa* 13 500 BP. Thus, included layers of cobbles and boulders could have been emplaced during the Loch Lomond Stadial (Davies *et al.* 1984).

The Jura Formation may, therefore, correlate in part with the Clyde Beds (Peacock 1975, 1981) (Figure 9.22). These marine and estuarine silts and clays have produced foraminifera, Ostracoda and Mollusca (notably *Arctica islandica*). The fauna is of boreal affinity and analogous to that around the Faroe Islands today. The Clyde Beds were laid down during a marine incursion that began *circa* 13 000 BP, during the Main Late Devensian deglaciation phase, when the ice was retreating westward along the Clyde Estuary. Sissons (1974a) has postulated a marine limit of about 25 m OD in the Glasgow area at this time. Radiocarbon dates from shells in the Clyde Beds signal their deposition in a marine environment between 13 150 and 11 800 BP, over an area similar to that of the current Glasgow–Paisley district (Peacock 1971b; Rose 1975).

The evidence for Late Devensian sea-level movements in western Scotland has been assembled by Dawson (1984). High-level (18–51 m OD) rock platforms, frequently 500 m in width and backed by a cliff not far short of 100 m high, have been recognized in this area. Sissons (1981a) has hypothesized that the eastern extremity of these platforms marked the western limit of the last Scottish ice sheet. Therefore, they have been interpreted as the result of periglacial coastal erosion during the last (and previous) cold stages; variations in their height being a response to glacio-isostatic inclination. During deglaciation, Lateglacial shorelines developed. These have risen isostatically, and have been demonstrated to increase in height as the centre of uplift in the Western Highlands is approached (Figure 9.22). The well-marked raised shoreline features of the Inner Hebrides (up to *circa* 36 m OD in western Mull (Gray 1974), for example) and the weakly developed ones in the Outer Hebrides, are testimony to the distance of the latter area from the centre of uplift. In the south-west Highlands, where substantial reductions in the marine limit have been correlated with either cessations in ice-sheet decay or small readvances, a minimum of seven Lateglacial shorelines has been identified (Sutherland 1981b). The most prominent of these, formed around 13 000 BP, is a likely correlative of the Main Perth Shoreline in eastern Scotland (Sissons *et al.* 1966). In Mull the disappearance of the ice of the Main Devensian Glaciation led to rapid isostatic uplift, so that relative sea-level was reduced below 12 m OD (Walker *et al.* 1985).

The overall pattern of relative sea-level movement in western Scotland during the Lateglacial Interstadial involved a quick decline from 13 000 to 12 000 BP, then a slower one from 12 000 to 11 000 BP (Dawson 1984). Marine erosion was heightened during the Loch Lomond Stadial. Along the western seaboard of Scotland, the Main Lateglacial Shoreline was cut by a combination of wave action and frost shattering (Sissons 1974b; Gray 1978; Dawson 1980). This shoreline has been shown to reach its maximum height (10–11 m OD) in the neighbourhood of Oban, descending below OD in Islay, Colonsay, Mull, Kintyre and Arran as the centre of uplift becomes more distant (Dawson 1984) (Figure 9.22). Eustatic sea-level during the Loch Lomond Stadial is likely to have been between 35 and 50 m below that of today (Sissons 1979b). The reduction in relative sea-level at this time has not been clearly established. Data presented by Rose (1980b) have implied a relative sea-level of *circa* 2 m OD, or less, in the Clyde Estuary, while Browne (1980) has postulated a fall to about −13 m OD in this area.

Lateglacial raised shorelines have also been recognized in east-central Scotland. The existence of the Main Perth Shoreline in the Tay and Forth estuaries has been noted (section 9.3.6) and the Main Lateglacial Shoreline has been identified in the Forth Valley (Sissons 1969, 1974b). In the Earn–Tay lowlands, Cullingford (1977) has discerned five Lateglacial shorelines. The highest and best developed of these is the Main Perth Shoreline, which did not form during a readvance, but when the decaying ice margin was situated close to Perth. The present elevation of this feature has been considered to be a consequence of the interaction between glacio-isostasy (including rapid uplift during deglaciation) and eustatic movement of sea-level. If as projected, the Main Perth Shoreline formed *circa* 13 500 BP, the associated sea-level was rising but still some 60 m lower than that of today (Mörner 1969). As the shoreline is currently found up to 30 m OD, uplift of about 90 m must have taken place (Cullingford 1977).

Marine clays, silts and sands containing ice-rafted stones and possessing a cold molluscan fauna of low diversity (including *Portlandia arctica*), have been identified above the Clyde Beds (which occur below OD) in the Forth and Tay estuaries (Peacock 1975). These sediments, the Errol Beds (Figure 9.22), were deposited during ice retreat, and have produced a more impoverished fauna than that of the Clyde Beds. The latter, as noted above, began to accumulate *circa* 13 500 BP at the beginning of the Windermere Interstadial and continued until about 11 800 BP. The Errol Beds thus seem likely to have recorded the climatic deterioration evident throughout the later part of the interstadial, and the cold climate of the Loch Lomond Stadial.

Raised Lateglacial shorelines have been found in northern and eastern, but not western, Ireland (Stephens and McCabe 1977; McCabe 1987). The absence of shorelines in the west could be linked to isostatic activity. A forebulge mechanism (Walcott 1970, 1973; Sutherland 1981a) may have operated, so that sufficient uplift took place in the west to put this area above marine influence (McCabe 1987).

The possibility of a Late Midlandian land bridge between Ireland and the remainder of the British Isles has been discussed (Figure 9.23). Synge (1985) has considered that eustatic sea-level was lowest *circa* 15 000 BP. Then the floor of the southern Celtic Sea would have been exposed, with Ireland connected to Cornwall and Brittany. Subsequent eustatic rise would have led to the narrowing of this land bridge until it was restricted to a mainly morainic ridge that has been traced between the coast of Wicklow and the Lleyn Peninsula in North Wales. The final breach of this connection probably occurred during the Woodgrange Interstadial, about 12 000 BP. Devoy (1985a) has dismissed the notion of such a land bridge, suggesting that if a link had existed, one via peninsulas and islands across the Malin Sea from Islay to Donegal is most feasible. Isostatic uplift could have played a rôle in the creation of this type of routeway, for which a fall in relative sea-level to at least 40 m below OD would have been necessary. Radiocarbon-dated indicators of sea-levels (Carter 1982) suggest that a link of this sort could not have formed prior to 12 000 BP and that a timing of between 11 400 and 10 200 BP is most likely. Rapid isostatic uplift ceased about 10 000 BP in northern Ireland, at which time sea-level was *circa* −13 m OD in this area (Synge 1981, 1985). Such a tenuous and transient feature would not have afforded an easy migratory route for plants and animals. For instance, although present adjacent to the land bridge on the mainland of the British Isles, woolly rhinoceros, elk, vole and man were absent from Ireland during the Late Midlandian. Against the backdrop of a difficult connection across the Irish Sea, these absences may mean that only the best-adapted animals were able to make the crossing (Stuart 1985).

The Melville Formation south of the Scilly Isles has been considered by Pantin and Evans (1984) to be partly of Late Devensian age. Mainly sandy tidal deposits formed ridges after the shoreline moved over the middle shelf in the marine transgression soon after the glacial maximum. These features have associated ice-rafted material that points to their partial or entire growth when icebergs existed. The shallowest ice-rafted material encountered is at *circa* −113 m OD, which provides a minimum height for sea-level at that time. No evidence of actual glaciation during the Devensian has been forthcoming from this part of the Celtic Sea.

9.3.9 Age and correlation

Geochronometric (especially radiocarbon) dating has provided numerous ages for Late Devensian events. However, the date of commencement of this period, together with the timings of the onset and termination of its main glaciation, have not been well established in many parts of the British Isles. Organic deposits, mostly associated with tills and kettle holes, have indicated that glaciation began around 28 000 BP, with the ice reaching its maximum extent *circa* 18 000 BP (Shotton 1977b). The absence of speleothem growth in north-west Yorkshire and north-west Scotland from 26 000 to 15 000 BP has been ascribed to glaciation by Atkinson *et al.* (1986). According to Climap Project Members (1976) the greatest volume of global ice was present about 18 000 BP. Subsequent deglaciation appears to have been rapid and was widely accomplished by approximately 14 000 BP, when the Lateglacial period began.

Watts (1980) has observed that the Lateglacial was a complex climatic event, unparalleled prior to earlier interglacial cycles and has postulated that Lateglacial vegetation possessed interglacial characteristics, with a false start being made in its development. The latter had to begin again in the Flandrian, following reversion during the Loch Lomond Stadial.

Insect evidence has identified an abrupt and substantial climatic amelioration at 13 500–13 000 BP which marked the opening of the Windermere Interstadial. Until about 12 200 BP the Lateglacial climate was analogous to that of the present-day British Isles. Around 12 200 BP a sudden decline from the thermal maximum of the interstadial occurred, and *circa* 12 000–11 000 BP a cool-temperate climate existed. From about 11 000 to 10 000 BP a tundra or polar-desert climate was present over most of the British Isles during the Loch Lomond Stadial (Coope 1977a).

Uranium-series ages of speleothems have denoted a revival in their growth in caves in north-west England during the Windermere Interstadial. This growth continued during the Loch Lomond Stadial, when temperatures in this area could not have been reduced sufficiently to inhibit groundwater seepage (Gascoyne *et al.* 1983). Data from Yorkshire, Somerset and Assynt have led Atkinson *et al.* (1986) to conclude that speleothem growth commenced in the Late Devensian *circa* 15 000 BP.

The D-alloisoleucine/L-isoleucine ratios in non-marine molluscs from Halling, where a Lateglacial deposit has a radiocarbon age of *circa* 13 000 BP (section 9.3.5), are consistent within the aminostratigraphic framework of Devensian events presented by Bowen *et al.* (1989) (Figure 1.15).

West (1977b) has correlated Late Devensian time prior to about 13 000 BP with the Upper Pleniglacial of the Netherlands, and the Lateglacial (*circa* 13 000–10 000 BP) with the Late Weichselian of that region (van der Hammen *et al.* 1967; van Staalduinen *et al.* 1979) (Figure 9.9). Lithostratigraphic markers such as corrie sediments of Loch Lomond Stadial age have been found in the British Isles and coversands of Late (Younger) *Dryas* age recognized in the Netherlands and Belgium (Maarleveld 1960; Vandenberghe 1985). However, the majority of correlations between Lateglacial episodes in the British Isles and elsewhere in north-west Europe have been made using biological data. Watts (1980) has reviewed regional variations in vegetation and climate during the Lateglacial in this area. The cold climate of *circa* 11 000–10 000 BP has been well illustrated by botanical evidence from the Netherlands and northern Germany. Here during the Younger *Dryas* Stadial, birch and pine forest became more open and *Empetrum* heath developed. In Denmark and Sweden the pollen record indicates a considerable reduction in tree growth at this time. Either the Bølling event or the Allerød event or both of them have been detected in Lateglacial vegetational sequences from Scandinavia (Mangerud *et al.* 1974; Berglund 1979), the Netherlands (van der Hammen 1951) and Germany (Usinger 1975). The climatic deterioration 12 000–11 800 BP established in the British Isles has found little expression in botanical records from Denmark and Germany.

Radiocarbon ages of *circa* 13 500 BP, equating with the time of deglacial warming of the North Atlantic Ocean envisaged by Ruddiman *et al.* (1977), have been obtained from marine organisms in southern Sweden (Berglund 1976). In western Norway this warming had occurred by 13 000 BP and a Bølling oscillation (marked by an expansion of *Betula* pollen) took place about 12 650 BP (Mangerud 1970).

Late Devensian vertebrate faunas in mainland Europe and the British Isles were broadly similar. However, in northern Germany and in Denmark, *Bos primigenius*, *Sus scrofa* and *Alces alces* were present during the Younger *Dryas*; a reflection of the continentality of these areas according to Stuart (1982). The evidence from Shropshire (section 9.3.1) has supported that indicating the survival of *Mammuthus primigenius* elsewhere in Europe until near to the end of the Devensian (Berglund *et al.* 1976).

9.4 Conclusion

This chapter covers a variety of evidence accumulated for the environments of at least 80 000 years. The task has been hindered for several reasons which have been noted by West *et al.* (1974). The first of these is the existence of localized lithostratigraphic detail which has often not been easy to correlate on a regional basis. Second is the presence of biostratigraphies based on different organisms (notably plants and insects) which has sometimes led to conflicting climatic interpretations. Pollen-based and insect-based interstadials (Figure 9.9), which lack synchroneity, have been designated (Coope *et al.* 1971). These contradictions have not been easy to resolve but seem to reflect differing rates of response by insects and plants to environmental change. West *et al.* (1974) have suggested that in such circumstances, it may be prudent to designate interim interstadials based on a particular type of evidence. The formal definition of the event should await the emergence of sufficient corroborative data from as many sources as possible. A third difficulty is geochronological. The reliability of radiocarbon ages in Early and Middle Devensian contexts has been questioned, and the technique reaches its limit late in the former period. Alternative methods of absolute age determination for these timespans, such as U-series and thermoluminescence, have not been without problems in their application.

In spite of these hazards, environmental changes during the Devensian have been reconstructed more clearly than those of any previous cold stage.

10

The Flandrian Temperate Stage

Appropriate nomenclature for the most recent Pleistocene stage has generated more discussion than that pertaining to the naming of any other division of this rank (Mangerud *et al.* 1974; Paepe *et al.* 1976; Hyvarinen 1978; Mangerud and Berglund 1978; West 1979). The present interglacial has been termed the Flandrian after a marine transgression within its timespan which has been identified on the Flemish coastal plain. With a lower limit of 10 000 radiocarbon years BP, the Flandrian usually has been considered equivalent to the formally designated Holocene Series and the informal Postglacial episode. The use of Flandrian has been recommended in the British Isles, although for Ireland the term Littletonian has been presented as synonymous (Mitchell *et al.* 1973a) (Table 10.1).

It is standard geological practice to select a type locality for the deposits of a stage. However, West (1979) has suggested that this is unnecessary in the case of the Flandrian, whose deposits, unlike those of earlier Pleistocene stages, are widely distributed and amenable to correlation by radiocarbon dating. In previous interglacials, biozonation on the basis of pollen assemblages has led to the recognition of Pre-temperate, Early-temperate, Late-temperate and Post-temperate zones (Turner and West 1968) (Chapter 1), and to the designation of type sites within the British Isles. During the ten millennia that have so far elapsed in the Flandrian, the vegetational characteristics of the first three of these zones appear to have been encountered (Table 10.1). Although Hibbert *et al.* (1971) have suggested that the Flandrian deposits of Red Moss, Lancashire (from which a series of pollen assemblage biozones and chronozones have been obtained) (Table 10.2) may be suitable as the type locality, one has not been proposed for England, Wales and Scotland by Mitchell *et al.* (1973a). However, Littleton Bog, County Tipperary (Mitchell 1965b) and Roddans Port, County Down (Morrison and Stephens 1965) have been cited for Ireland in this context by these authors. Hibbert and Switsur (1976) have referred to Red Moss as the Flandrian type site, possessing the standard stratigraphic sequence of chronozones.

As Turner and West (1968) have pointed out, the large number of studies accomplished over a wide area of north-western Europe and the complexity introduced by anthropic influence on vegetation have led to intricacy in the biozonation of Flandrian deposits. Contrasting zonation schemes of both local and regional significance have emerged. Apart from the system of Turner and West (1968), one of the most widely adopted subdivisions of Flandrian time has been the Blytt – Sernander scheme of climatic periods (Blytt 1876; Sernander 1908). This has been derived from plant macrofossil sequences in Scandinavian wetlands, and has been related to a series of pollen zones by von Post (1916, 1918) (Davis and Faegri 1967). The pollen zones established for England and Wales by Godwin (1940a) have also

Table 10.1 Sub-divisions, terminology and chronology of the Flandrian/Littletonian Stage. Additional sources: Britain, Godwin (1975); Ireland, Edwards and Warren (1985)

^{14}C years BP	England and Wales: STAGE	England and Wales: Blytt – Sernander periods	England and Wales: Godwin (1940a) Pollen zones	England and Wales: West (1980b) Chronozones	Scotland: West (1977a) Pollen zones	Ireland: STAGE	Ireland: Blytt – Sernander/ Jessen (1949) periods	Ireland: Jessen (1949) Pollen zones	Ireland: Mitchell (1956) Pollen zones	^{14}C years BC/AD
0										2000
									X	
1000		Sub-Atlantic	VIII						IX	AD 1000
2000							Sub-Atlantic	VIII		0
									VIIIb	
3000		Sub-Boreal	VIIb	Fl III	VIII–VIIb					1000 BC
4000	FLANDRIAN					LITTLETONIAN				2000
							Sub-Boreal			
5000								VIIb	VIIIa	3000
		Atlantic	VIIa							
6000					VIIa		Atlantic	VIIa	VII	4000
				Fl II						
7000										5000
8000			VI c b a		V–VI			VI	VI	6000
		Boreal					Boreal			
9000			V					V	V	7000
				Fl I						
		Pre-Boreal	IV		IV		Pre-Boreal	IV	IV	
10 000										8 000

been extensively used in segmenting the period since the conclusion of the last cold stage in the British Isles (Table 10.1).

Terminological controversy and biostratigraphical difficulties notwithstanding, a greater amount of information has been forthcoming concerning Flandrian environments than about those of any previous stage of the Pleistocene. Thus, the account below is able to provide only the barest outline of these environments, based on the main lines of evidence. Many of the latter have been synthesized and these accounts furnish summary detail for the ensuing text.

Table 10.2 Flandrian pollen assemblage zones, radiocarbon chronology and chronozones from Red Moss. From Hibbert *et al.* (1971)

Pollen assemblage zone	*Date* BP *of commencement*	*Chronozone*
Quercus–Alnus Zone	5010 ± 80	FIII
Quercus–Ulmus–Alnus Zone	7107 ± 120	FII
Pinus–Corylus–Ulmus Zone	8196 ± 150	FId
Corylus–Pinus Zone	8880 ± 170	FIc
Betula–Pinus–Corylus Zone	9798 ± 200	FIb
Betula–Pinus–Juniperus Zone	*circa* 10 250	FIa

10.1 The presence of man

As noted above, man has had a considerable impact on Flandrian vegetation. Hence, there has been human impact on other aspects of ecosystems and this has been especially evident during the last six millennia. The course of prehistory in the British Isles and characteristics of its cultures have been examined by the contributors to Renfrew (1974) and Megaw and Simpson (1979) (Figure 10.1). Archaeological sites mentioned in the text are shown in Figure 10.2.

10.1.1 Later Upper Palaeolithic and Mesolithic

The Later Upper Palaeolithic culture continued into the earliest part of Flandrian time. Some of the cave sites of this culture in the Mendips and northern Dukeries, as well as Kent's Cavern in Devon, were probably occupied into the initial stage of the Mesolithic (Campbell 1977; Mountain 1979). In addition to caves, there appears to have been an extension of open-air sites during the last part of Later Upper Palaeolithic time. These, like the cave localities, and including Hengistbury Head, Dorset (Mace 1959), Sproughton, Suffolk (Wymer and Rose 1976) and Portland, Dorset (Palmer 1977), were concentrated in the southern and eastern sections of the British Isles (Figure 10.2).

Jacobi (1973, 1976) has pointed to marked variations in typology between Later Upper Palaeolithic and Mesolithic stone implements (Figures 9.10 and 10.1). Essentially, blades and scrapers decreased in size, with the former eventually becoming a microlith. The Mesolithic also saw the emergence of the flaked tranchet axe. Bone and antler tools were now more diverse, and wood and shell were used in artifacts.

Mountain (1979) has noted that Mesolithic implements were present *circa* 10 000 BP, and their makers must have co-existed with Later Upper Palaeolithic inhabitants, both availing themselves of the same sites in certain places. According to Jacobi (1973), all Mesolithic industries before *circa* 8500 BP possessed a limited number of microlithic shapes, but subsequently, narrow microliths were characteristic. These differences could be referred to an Earlier Mesolithic, including Proto-Maglemosian or Maglemosian, and a Later Mesolithic comprising the Sauveterrian (Figure 10.1). Mellars (1974) has stated that the persistence of larger, non-geometric microliths indicates continuity between the early and late stages of the Mesolithic in the southern part of the British Isles. Thatcham in the Kennet Valley, Berkshire, is an Earlier Mesolithic site radiocarbon dated to between *circa* 10 365 and 9480 BP (8415 and 7530 BC) (Wymer 1962; Churchill 1962). Star Carr in the Vale of Pickering, eastern Yorkshire has produced radiocarbon ages from 9800 ± 80 BP (7850

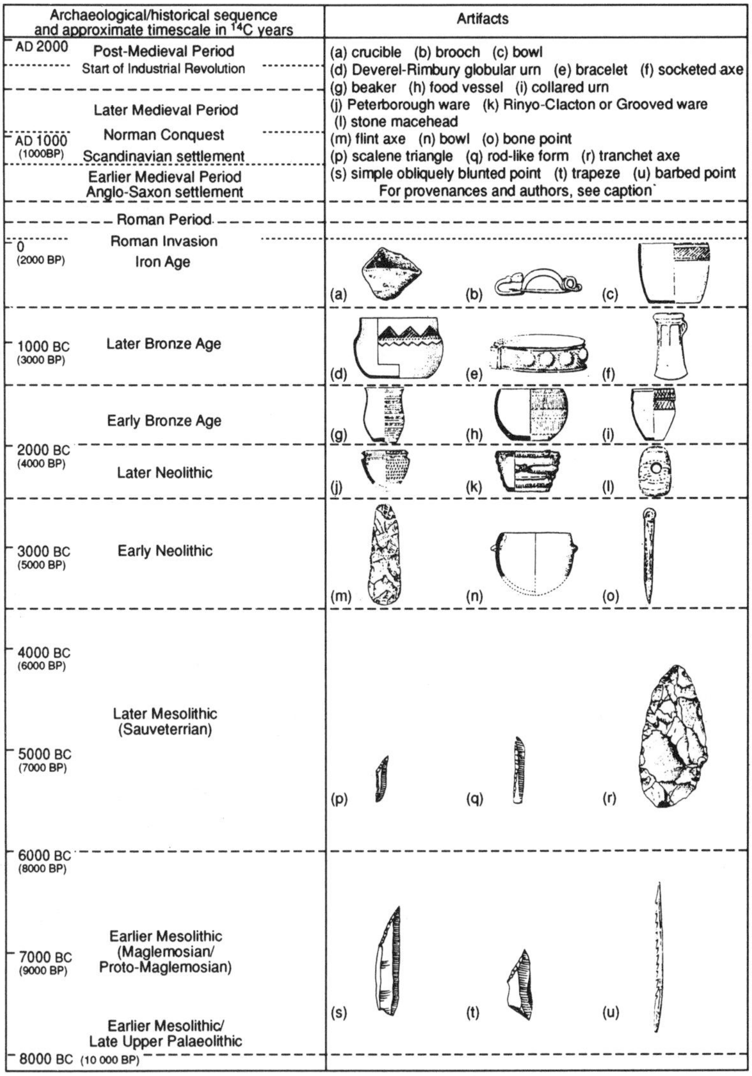

Archaeological/historical sequence and approximate timescale in ^{14}C years		Artifacts
AD 2000	Post-Medieval Period	(a) crucible (b) brooch (c) bowl
	Start of Industrial Revolution	(d) Deverel-Rimbury globular urn (e) bracelet (f) socketed axe
		(g) beaker (h) food vessel (i) collared urn
	Later Medieval Period	(j) Peterborough ware (k) Rinyo-Clacton or Grooved ware
		(l) stone macehead
AD 1000	Norman Conquest	(m) flint axe (n) bowl (o) bone point
(1000BP)	Scandinavian settlement	(p) scalene triangle (q) rod-like form (r) tranchet axe
	Earlier Medieval Period Anglo-Saxon settlement	(s) simple obliquely blunted point (t) trapeze (u) barbed point For provenances and authors, see caption
	Roman Period	
0	Roman Invasion	
(2000 BP)	Iron Age	(a) (b) (c)
1000 BC (3000 BP)	Later Bronze Age	(d) (e) (f)
	Early Bronze Age	(g) (h) (i)
2000 BC (4000 BP)	Later Neolithic	(j) (k) (l)
3000 BC (5000 BP)	Early Neolithic	(m) (n) (o)
4000 BC (6000 BP)		
	Later Mesolithic (Sauveterrian)	
5000 BC (7000 BP)		(p) (q) (r)
6000 BC (8000 BP)		
7000 BC (9000 BP)	Earlier Mesolithic (Maglemosian/ Proto-Maglemosian)	(s) (t) (u)
	Earlier Mesolithic/ Late Upper Palaeolithic	
8000 BC (10 000 BP)		

Figure 10.1 The Postglacial archaeological and historical sequence and its approximate timescale. Terminology and chronology of prehistory: sources, Jacobi (1973), Megaw and Simpson (1979). Artifact provenances (where known), and authors. (a) Meare, Somerset. From Champion (1979); source, Bulleid and Gray (1953). (b) Blandford, Dorset. From Champion (1979); source, Smith (1925). (c) Maiden Castle, Dorset. From Champion (1979); source, Wheeler (1943). (d) Thorny Down, Wiltshire. From Megaw (1979); source, Piggott (1973). (e) Norton Fitzwarren, Somerset. From Megaw (1979); source, Langmaid (1971). (f) Everthorpe, Yorkshire. From Megaw (1979); source, Langmaid (1976). (g) Roundway, Wiltshire. From Simpson (1979); source, Annable and Simpson (1964). (h) Omagh, County Tyrone. From Simpson (1979); source, Simpson (1968). (i) Largs, Ayrshire. From Simpson (1979); source, Morrison (1968). (j) West Kennet, Wiltshire. From Simpson (1979); source, Piggott (1962). (k) Woodlands, Wiltshire. From Simpson (1979); source, Stone (1949). (l) From Simpson (1979); source, Piggott (1954). (m) Windmill Hill Culture. From Simpson (1979); source, Piggott (1954). (n) Norton Bavant, Wiltshire. From Simpson (1979); source, Piggott (1954). (o) Windmill Hill Culture. From Simpson (1979); source, Piggott (1954). (p), (q) Farnham, Surrey. From Mellars (1974); sources, Clark and Rankine (1939), Clark (1955). (r) Oakhanger, Hampshire. From Mountain (1979); source, Rankine and Dimbleby (1960). (s), (t) Star Carr, Yorkshire. From Mellars (1974); source, Clark (1954). (u) Star Carr, Yorkshire. From Mountain (1979); source, Clark (1972). (a)–(o) and (u) reproduced from *Introduction to British prehistory: from the arrival of Homo sapiens to the Claudian invasion*, J. V. S. Megaw and D. D. A. Simpson (eds), 1979, by permission of Leicester University Press.

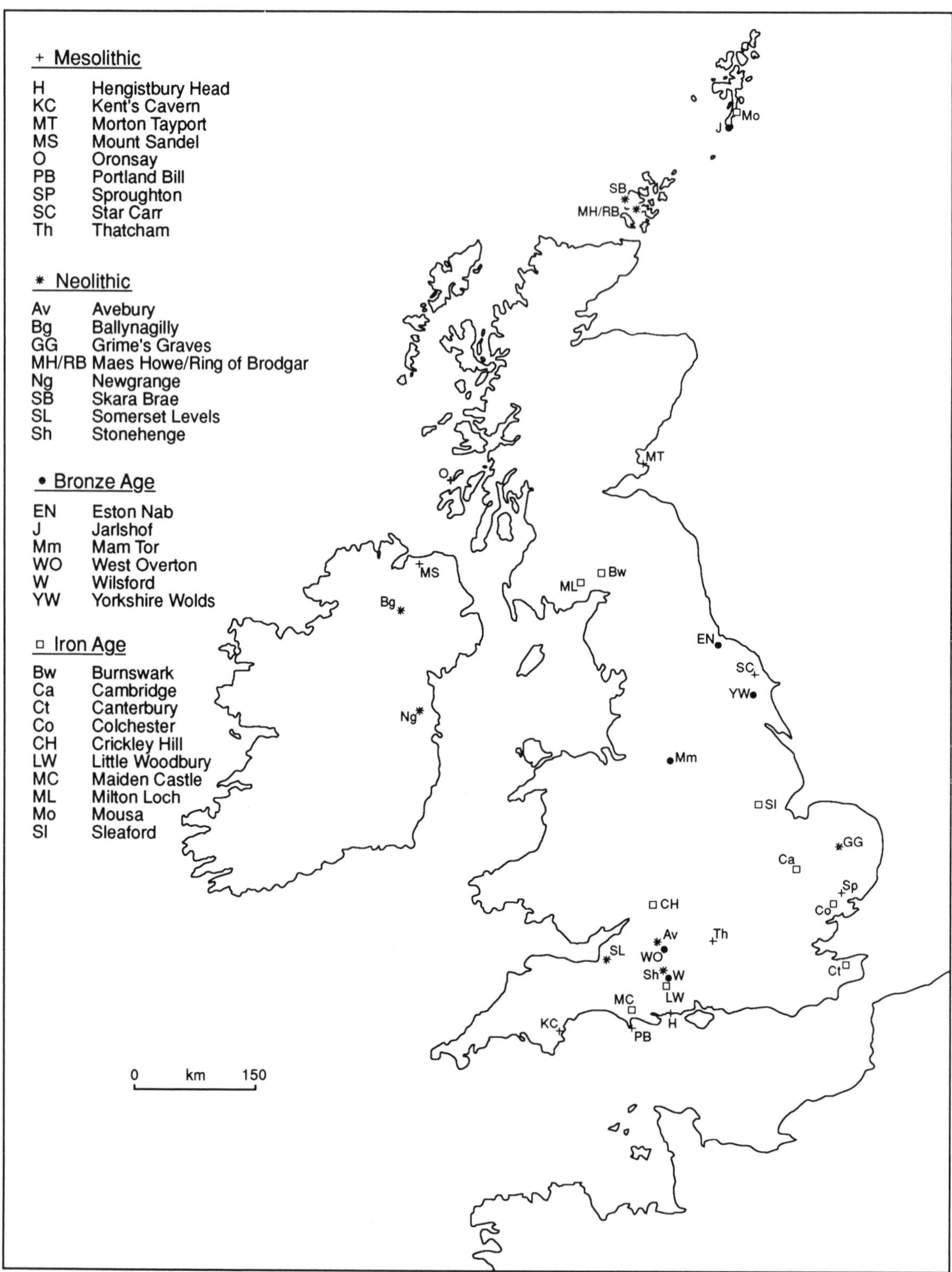
+ Mesolithic
H Hengistbury Head
KC Kent's Cavern
MT Morton Tayport
MS Mount Sandel
O Oronsay
PB Portland Bill
SP Sproughton
SC Star Carr
Th Thatcham
* Neolithic
Av Avebury
Bg Ballynagilly
GG Grime's Graves
MH/RB Maes Howe/Ring of Brodgar
Ng Newgrange
SB Skara Brae
SL Somerset Levels
Sh Stonehenge
• Bronze Age
EN Eston Nab
J Jarlshof
Mm Mam Tor
WO West Overton
W Wilsford
YW Yorkshire Wolds
□ Iron Age
Bw Burnswark
Ca Cambridge
Ct Canterbury
Co Colchester
CH Crickley Hill
LW Little Woodbury
MC Maiden Castle
ML Milton Loch
Mo Mousa
SI Sleaford
0 km 150
Mo
J
SB
MH/RB
MT
O
MS
Bg
ML
Bw
EN
SC
YW
Ng
Mm
SI
GG
Ca
Sp
Co
CH
Av
Th
SL
WO
Sh
W
LW
Ct
MC
KC
PB
H

Figure 10.2 Archaeological sites referred to in Chapter 10.

± 80 BC) to 9030 ± 100 BP (7080 ± 100 BC), and occupation has been dated to around 9600 BP (Clarke 1954, 1972; Cloutman and Smith 1988). Jacobi (1973) has observed that Earlier Mesolithic sites are concentrated in two regions, tributaries of the River Thames, and Yorkshire and Lincolnshire, but considered that subsequent valley alluviation (section 10.10) has been responsible for burial of others in the intervening area. The traditional view, based upon a number of criteria including similar tool types, has been that lowland localities such as these were inhabited during winter, with neighbouring uplands such as the North Yorkshire Moors, visited for hunting purposes in summer (Clark 1972; Jacobi 1973). However, recent research on faunal remains from the Star Carr area has indicated year-round occupation of this lakeside site (Schadla-Hall 1987).

From *circa* 10 000 to 8000 BP the southern North Sea was a plain (section 10.11) across which the estuaries and floodplains of the rivers Thames, Rhine and Meuse were developed. Thus, a land connection to mainland Europe was available for Earlier Mesolithic hunter-gatherers and moreover, one whose food resources were diverse and plentiful.

In Ireland, the Mesolithic has provided the earliest evidence of man (Woodman 1985). There was a settlement at Mount Sandel in County Londonderry *circa* 9000–8800 BP (Woodman 1981a). Microliths similar to those found in Earlier Mesolithic contexts in northern England have been recorded, and the inhabitants are thought to have immigrated from this area across the Irish Sea, via the Isle of Man, when sea-level was low.

During the Later Mesolithic (*circa* 8000–5500 BP), sea-level rose (section 10.11), thereby reducing the land area of the British Isles and establishing a geography akin to that of the present day. A reduction in land area coupled with an expanding population led to widespread colonization of the British Isles (Jacobi 1973). Coastal sites of this period have been located in south-western England and Wales where they have escaped submergence by the sea (Palmer 1977). There were occupation sites in the Cotswolds, the Weald, the Pennines and North Yorkshire Moors. In Scotland the rising sea-level reduced the width of coastal plains, which were also uplifted by isostatic readjustment. Here, a coastal midden at Morton Tayport, Fife, has been radiocarbon dated to *circa* 8050–6050 BP (Coles 1971). On Oronsay, shell middens, found grouped at about the elevation of the Main Postglacial Shoreline (section 10.11), have given radiocarbon ages ranging from *circa* 5900 to 5000 BP (Mackie 1972; Mellars 1978). In Ireland an important change in flint-implement typology took place after 8000 BP. Large, heavy blades became characteristic and microliths were lacking, but continued occupation throughout the Mesolithic is likely (Woodman 1985). The later Irish Mesolithic was termed Larnian by Movius (1953) and Mitchell (1971b). The majority of Mesolithic sites of this time were in lowlands, with fishing in both coastal and interior water-bodies of especial significance. The Mesolithic occupation of Ireland was thus more extensive in both a spatial and temporal context than that envisaged by Mitchell (1970), who considered it confined to the north-east of the island in the later part of the period.

10.1.2 Neolithic

It appears that the first farmers arrived in the British Isles by sea from mainland Europe, settling initially towards the beginning of the sixth millennium BP. This settlement was not extensive but widely distributed, with Ireland reached early on, according to radiocarbon-dating evidence (see below). In general, a sequence of settlement and farming, followed by the construction of funerary and ceremonial monuments seems to have taken place during the Neolithic (Simpson 1979).

The activities of Early Neolithic people have been most clearly illustrated in the southern part of the British Isles, notably the chalklands of southern England. Plant and molluscan data (sections 10.2 and 10.4) have indicated that many of these downs were forested prior to the arrival of Neolithic man. Thus, the presence of a reasonable depth of soil may have been a contributory factor in the selection of them for settlement. The culture has been termed Windmill Hill, after the area near Avebury in Wiltshire, where a substantial number of its remains (enclosures and long barrows, for example) have been identified. Elsewhere in the lowland zone of the British Isles, trackways connecting settlements, in the Somerset Levels (Coles *et al.* 1973), for example, were constructed, and flint mines such as Grime's Graves near Thetford in Norfolk (Sieveking *et al.* 1973) produced material for chipped and polished axes. Pottery at this time was characterized by bowls with round bases and there was an array of bone and antler tools (Figure 10.1). While wheat and barley were grown, cattle rearing appears to have been the mainstay of the economy, with the cultivated plants and domestic animals likely to have been imported by the settlers (Simpson 1979).

From the highland zone of the British Isles, there has been less information on Early Neolithic activities. None the less, the earliest evidence of Neolithic settlement has come from Ballynagilly in County Tyrone, dated by radiocarbon to 5745 ± 90 BP (3795 BC) (ApSimon 1969, 1976). There was both arable farming and stock rearing in the Early Neolithic in Ireland, and communal burial tombs were constructed by occupants who settled in numerous parts of the island (Woodman 1981b, 1985). Megalithic monuments built in the British Isles at this time possessed traits which made them distinct from those of mainland Europe (Renfrew 1973).

The Later Neolithic (*circa* 4500–3700 BP/2500–1700 BC) witnessed the increased development of native monument and artifact styles in the British Isles. Passage graves such as Newgrange in the Boyne Valley, Ireland (O'Kelly 1973) and Maes Howe on Mainland Orkney (Renfrew 1979) were constructed. On the latter island the important settlement of Skara Brae developed, the oldest radiocarbon date for which is 4470 ± 120 BP (2520 BC) (Clark 1976; Clark and Sharples 1985). Henges and stone circles, such as Avebury (Smith 1965) and Stonehenge (Atkinson 1960) in Wessex, and the Ring of Brodgar, close to Maes Howe, on Orkney (Renfrew 1979) date from this period. Pottery was now of Peterborough and Rinyo-Clacton (Grooved ware) types (Figure 10.1), and has been considered indicative of a distinctive culture (Piggott 1962; Wainwright and Longworth 1971). The centuries around the middle of the third millennium BC saw new, distinctive pottery appear in the British Isles. Burgess (1974) has attributed its presence to continental migrants known as the Beaker people. This group also made metal objects, and its importance increased at the beginning of the second millennium BC (Bradley 1984) (see below).

By the late third millennium BC, the British Isles may have possessed a mainly pastoral subsistence economy (Smith 1974). Whittle (1978) and Bradley (1978) have suggested that arable farming was markedly reduced in the Later Neolithic. However, Edwards (1979a) has cautioned against such a view on palynological evidence (section 10.2). Bradley (1984) has noted that there is an increased body of data that implies the existence of both settlement and enclosure, rather than a major abandonment of cleared areas and woodland regeneration, during the Later Neolithic. Woodman (1985) has indicated that the Later Neolithic economy of Ireland was mixed, but with an emphasis on pastoralism.

10.1.3 The Bronze Age

The period from *circa* 2000 to 1300 BC is usually considered to have been the Early Bronze Age (Simpson 1979). In the first centuries there was immigration from continental Europe to the British Isles, and as noted above, the Beaker Culture became established. The arrival of these people led to significant social and economic adjustments. Collective burial in passage graves gave way to normally single inhumation in graves or pits beneath round barrows, such as West Overton in Wiltshire (Smith and Simpson 1966). Metal tools, weapons and ornaments have been recovered from Early Bronze Age localities. A mixed-farming economy has been indicated. Pastoralism took place and cereals were grown, but inter-site variation in crop and animal types appears to have been quite marked (Simpson 1979). Well-developed cairn-fields on the North Yorkshire Moors have been attributed to arable farmers of this time by Fleming (1971a).

By the sixteenth century BC, beakers were no longer deposited in graves (a characteristic trait of this culture) in Wessex. Here, distinctive metal types of this time have been identified, and have been attributed to a Wessex Culture by Piggott (1938). A substantial amount of the evidence has come from round barrows such as that at Wilsford, Wiltshire (Annable and Simpson 1964). The culture flourished in Wessex and was not necessarily the product of immigrants (Clark 1966) or of trade (Renfrew 1968, 1973). The artifacts and monuments of this culture in Wessex have been considered indicative of a stratified warrior society. Fleming (1971b) has drawn a parallel with present-day pastoralists, who tend to possess such societies. A third episode of construction occurred at Stonehenge in this period, when other stone circles were also erected (Burl 1976).

The Early Bronze Age culture also saw the use of food vessels and urns. A distinctive type of the former has been recovered from sites in Yorkshire, especially on the Wolds (Simpson 1968). These food vessels appear to have been more or less contemporary with the Wessex cultural traits. Simpson (1979) has stressed that because beakers, food vessels and urns (Figure 10.1) have each been found with the other, it is not possible to state that they formed a chronological sequence in this order in an Early Bronze Age context. None the less, these contrasting types of pottery may demarcate different societies.

The later part of the Bronze Age occurred *circa* 1400–500 BC. About the latter time, artifacts linked with the later episodes of the Hallstatt Iron Age Culture of continental Europe emerged in the British Isles (see below). Hawkes (1960) has suggested the recognition of a Middle Bronze Age *circa* 1400–900 BC, which falls within the earlier part of Later Bronze Age time as defined by Megaw (1979).

Pottery of the Later Bronze Age possessed fairly distinctive insular traits (Megaw 1979). Urns of the Deverel–Rimbury type were characteristic in southern England (Figure 10.1). Bradley (1984) has suggested that local differences in this pottery may reflect the existence of separate social groups. Some Later Bronze Age metal work (Figure 10.1) has been related to a mainland European influence, while different compositions of metals have been identified in its early, middle and late phases (Allen *et al.* 1970). In southern England, substantial settlements, enclosures, field systems and cemeteries were made at this time. Drewett (1978) has established that Later Bronze Age inhabitants of the Sussex Downs cultivated wheat and barley, kept cattle and sheep and hunted deer. Mixed farming took place up to *circa* 450 m on Dartmoor where reaves were constructed to enclose land (Fleming 1978). Extending from the borderlands of Wales through the Midlands to the north of England were defensive hill-top settlements such as on Mam Tor in Derbyshire and Eston Nab in Cleveland. According to Challis and Harding (1975) these may have been constructed for one or more reasons including increased territorial organization, population pressure and a worsening climate (section 10.13). In the northern isles at this time, there was an important settlement at Jarlshof on the main island of Shetland (Hamilton 1956). There was a substantial amount of settlement in Ireland, where mixed farming was carried on (Woodman 1985). The dwellings were often associated with lakes, where artificial islands or crannogs were constructed (Megaw 1979).

10.1.4 The Iron Age

Around the middle of the seventh century BC, bronze objects with Hallstatt characteristics were present in the British Isles. It has been unclear whether these artifacts, mainly swords, were brought by settlers or obtained by bartering. The Iron Age lasted from about 600 BC to AD 200. In the southern part of the British Isles, including

Ireland, Iron Age settlements were numerous and the period witnessed a modification in interment customs, with burial largely superseded by cremation (Champion 1979). The Iron Age has been separated into three parts by Hawkes (1931). Iron Age A involved invasion from mainland Europe by the Hallstatt Culture; Iron Age B, movement into the British Isles of early La Tene Culture from this area; Iron Age C, the arrival of late La Tene Culture also from the continent. Although Harding (1976) has postulated that incomers in these three episodes gave momentum to the development of the Iron Age in the British Isles, others have been less convinced of ingression. Champion (1975) has indicated that there is a lack of firm evidence for immigrant people at the start of the Iron Age, when an increased requirement for industrial metal is likely to have resulted in the utilization of local iron by metal workers who had previously produced leaded bronze. Cunliffe (1978) has established intra-British Isles divisions based upon insular Iron Age traits.

Pottery and other artifacts (Figure 10.1) have been recovered in abundance from Iron Age localities, with the former employed in a chronological context in parts of the southern sector of the British Isles. Here, enclosures such as Little Woodbury, Wiltshire (Bersu 1940) and hill forts, including Maiden Castle, Dorset (Wheeler 1943) and Crickley Hill, Gloucestershire (Dixon 1976) were part of a phase of significant increase in woodland removal (section 10.2) and settlement. Field systems existed, notably in Wessex (Bowen and Fowler 1966) and on the Yorkshire Wolds (Challis and Harding 1975). Cereals were widely grown in the lowlands, with hulled barley and spelt of especial importance. Domesticated animals included cattle, sheep, pigs, horses and dogs, but little regionalization of their dominant presence has been possible (section 10.5). Ceremonial monuments were not characteristic of the Iron Age, whose people seem to have favoured riverside and wooded localities for such purposes (Champion 1979). During the last part of Iron Age time, notable economic growth occurred, coinage was used and large settlements (Canterbury, Colchester, Cambridge and Sleaford, for instance) were built up.

Piggott (1966) has analysed the Iron Age in the northern part of the British Isles, dividing this into four provinces, the Atlantic, North-Eastern, Tyne–Forth and Solway–Clyde. The economy here was dominantly pastoral but mixed farming was carried on, with transhumance probably an important activity. In the highland zone at this time, the physical and biological environment was inimical to economic development. Increased wetness led to the spread of blanket peat and to soil podsolization (section 10.9) which reduced the area available for farming (Ralston 1979). The Atlantic Province was distant and relied upon maritime connections. There were numerous small defensive sites (MacKie 1971). Brochs, such as Mousa on Mainland Shetland (Graham 1947), were a characteristic feature, as probably were burnt mounds, already in use in the Later Bronze Age on the Orkney Islands (Hedges 1975). The North-Eastern Province contained forts such as Castle Law, Abernethy (MacKie 1969), while that of Solway–Clyde had a scattering of defended settlements including Burnswark hill-fort in Dumfriesshire (Jobey 1971). There were crannogs in the latter region, for example at Milton Loch, Kirkudbrightshire (Piggott 1953).

In the northern part of the British Isles, coins were not in use before Roman times and urban growth was weak (Ralston 1979). The Roman invasion in AD 43 did not have much direct affect on this area. The Romans advanced into northern England by about AD 70, entering Scotland during the decade AD 80–90 (Hartley 1972).

According to Cunliffe (1974), although from a technological viewpoint the Iron Age began between about 750 and 500 BC, signs of a cultural change were evident earlier during the first millennium BC. Cunliffe has identified three phases related to the Iron Age in the British Isles. The first was one of conservatism *circa* 1000–750 BC. While the economy and society did not modify a great deal, there are indications of the bolstering or construction of defensive sites. *Circa* 750–500 BC witnessed a phase of innovation, including a multiplication of hill-forts possessing novel means of defence. Hallstatt metal artifacts emerged, probably stimulated by imports as part of well-developed trade about the Atlantic coast of Europe. A phase of development occurred from around 500 BC to AD 43–84 and later. This was the time when the southern part of the British Isles developed while the northern sector underwent scant modification. The La Tene Culture made an impact in the fifth century BC, with local traditions developing between the middle of the fourth century and the end of the second century BC. During the second part of the second century BC, immigration into the British Isles occurred. Trade with Gaul took place after the middle of the first century BC. The Roman invasion was started by Claudius in AD 43, following an attempted one by Caesar *circa* 55 BC. The later date marked the end of the pre-Roman Iron Age. The native government was overthrown and replaced by Roman rule. The Roman conquest had ended by AD 84 in most areas, with only the far north-western sector of the British Isles remaining outside their control. Roman influence was mostly military, and mainly upon communications and the more substantial settlements, with the rural economy experiencing little impact (Evans 1975).

10.1.5 Historic time

The Roman Period terminated *circa* AD 400 (Figure 10.1). Prior to the conclusion of the fourth century AD, Germanic people are thought to have colonized the British Isles, a process which accelerated during the next century. Initially this colonization was confined to the lowland zone, extending to fringes of the highland zone by *circa* AD 600. Beyond, in the highlands of Wales and Scotland, Celtic states remained in existence (Roberts 1979). During the ninth and tenth centuries AD, invaders from Scandinavia entered the British Isles, settling mainly in the north and east. The Norse element of these Viking pirates favoured Wales and Scotland, while eastern England became the domain of the Danes. The Norman Conquest of the British Isles took place during the eleventh century AD. The Domesday Book of these people provided a landscape inventory for a large part of Anglo-Saxon and Scandinavian England in 1086. The essentials were manors and estates with tenant farmers. Villages existed and settlement units were termed vills. The Normans perpetuated a similar social and economic system (Roberts 1979). Cultivation was by the ridge and furrow method in open fields, beyond which was grazing land. The Domesday Book recorded a substantial amount of woodland in lowland England. Areas especially noted included the Weald, Arden, Cannock, Charnwood and Delamere forests. Such areas were a source of building material and kindling wood, and were used for pannage.

The period from the middle of the eleventh to the middle of the fourteenth century AD witnessed an increase in population and settlement, together with an expansion of arable land at the expense of forest. Taylor (1978) has shown how Northamptonshire villages changed location and morphology in Anglo-Saxon and later Medieval time, suggesting that this was linked to shifts in the pattern of communications. The desertion of villages was a frequent occurrence during the eleventh and twelfth centuries AD. These desertions have been discussed by numerous people, including Beresford and Hurst (1971). Several factors, probably acting in combination, have been proposed as likely causes of this phenomenon, including those of an economic, social and political nature, disease and climatic change (section 10.13) (Roberts 1979). Later Medieval time also saw monasteries flourish, especially those of the Cistercians. Important farming systems were associated with these institutions, of which Strata Florida in western Wales and Rievaulx in northern Yorkshire were examples.

In England, land enclosure probably began during the thirteenth century AD (Hoskins 1970). However, this process took place with greater effect from the fifteenth century, and continued during the sixteenth, in the Post Medieval Period (Figure 10.1). In Scotland and Ireland, enclosure came later than in England, commencing in the sixteenth and seventeenth centuries AD. This had been preceded by a rather different pattern of settlement and economy. In Ireland, individual farmsteads rather than nucleated settlements characterized the lowlands during the first thousand years AD. From *circa* the seventh until late in the twelfth century, forests were cleared and arable farming expanded. By the latter time, a feudal economy was established. In both Ireland and Scotland, a farming system known as run-rig operated during the Later Medieval Period. This comprised an infield area which was not enclosed and always cropped, together with predominantly fallow outfield land, extending away from one or more farmsteads. Beyond these fields lay grazing land. Enclosure witnessed the end of the run-rig system and a change from mainly mixed farming to stock rearing (Evans 1975).

Industrial activities during the first millennium AD included the production of lead ingots around Wirksworth in Derbyshire during the first and second centuries (Butterworth and Lewis 1968) and Anglo-Saxon salt extraction at Droitwich, Worcestershire (Finberg 1972). The Later Medieval Period saw iron-working in the Forest of Dean and the Weald (Mitchell 1954).

There were numerous important urban centres during medieval times. Excavations beneath the current cities of Winchester, London, York (section 10.5) and Dublin, for example, have made possible detailed reconstructions of their successive economic and social activities. However, major population growth and urbanization began *circa* AD 1750, as the Industrial Revolution was initiated. Its progress, and later events in this sphere, while well documented, have also been evident in recent sediments (sections 10.7, 10.8 and 10.9).

10.2 Flora and vegetation

More information has been gathered on the vegetational history of the Flandrian/Littletonian than on any other temperate stage in the Pleistocene of the British Isles (Godwin 1975). In consequence, it is only possible to provide a brief account of this history for various parts of the territory, and to present summary comments concerning spatial and temporal variations in the flora and plant cover. Localities discussed are shown in Figure 10.3. It should be noted that the evidence below has come from the study of pollen grains, spores and macrofossil plant parts. Diatoms, also members of the plant kingdom, have provided rather different information (Chapter 1) concerning Flandrian environments. Hence, an assessment of this information is given in section 10.7, as is that derived from tree rings.

10.2.1 East Anglia

Pollen zones established in East Anglia, and typified at Hockham Mere, north of Thetford (Godwin 1940a,b; Godwin and Tallentire 1951), have been subsequently applied to other localities in England, Wales and Scotland (Table 10.1). The palaeobotany of Hockham Mere has been re-examined by Bennett (1983). Open birch woodland with a herb-rich understorey, present during the Devensian Lateglacial (Chapter 9), persisted until 9500 BP. Subsequent forest history saw a short opening phase of *Betula* dominance that lasted until *circa* 9300 BP. After this, *Pinus*, whose expansion began about 9400 BP, and *Corylus*, which increased in occurrence around 9300 BP, came to prominence, the former subsequently giving way to *Quercus*. From *circa* 7200 to 6000 BP, *Corylus* was replaced, first by *Tilia* and thereafter by *Fraxinus*. An initial decline in elm pollen (Godwin 1975) about 6000 BP has been ascribed to Dutch elm disease, with a second fall in elm *circa* 4500 BP linked to human disturbance of the forest. However, herbaceous pollen frequencies did not increase markedly until about 2500 BP. This, together with substantial amounts of cereal pollen later than 2000 BP, indicate a heightened impact by man on the landscape, where farming was in progress. A rise in *Calluna* (heather) pollen frequencies after 2500 BP denotes the formation of heath in this part of Breckland. *Fagus* reached the area *circa* 2400 BP and *Carpinus* around 1600 BP. The vegetational history of Diss Mere, Norfolk, between about 10 000 and 5000 BP (Peglar *et al.* 1989) is comparable with that of Hockham Mere. *Ulmus*

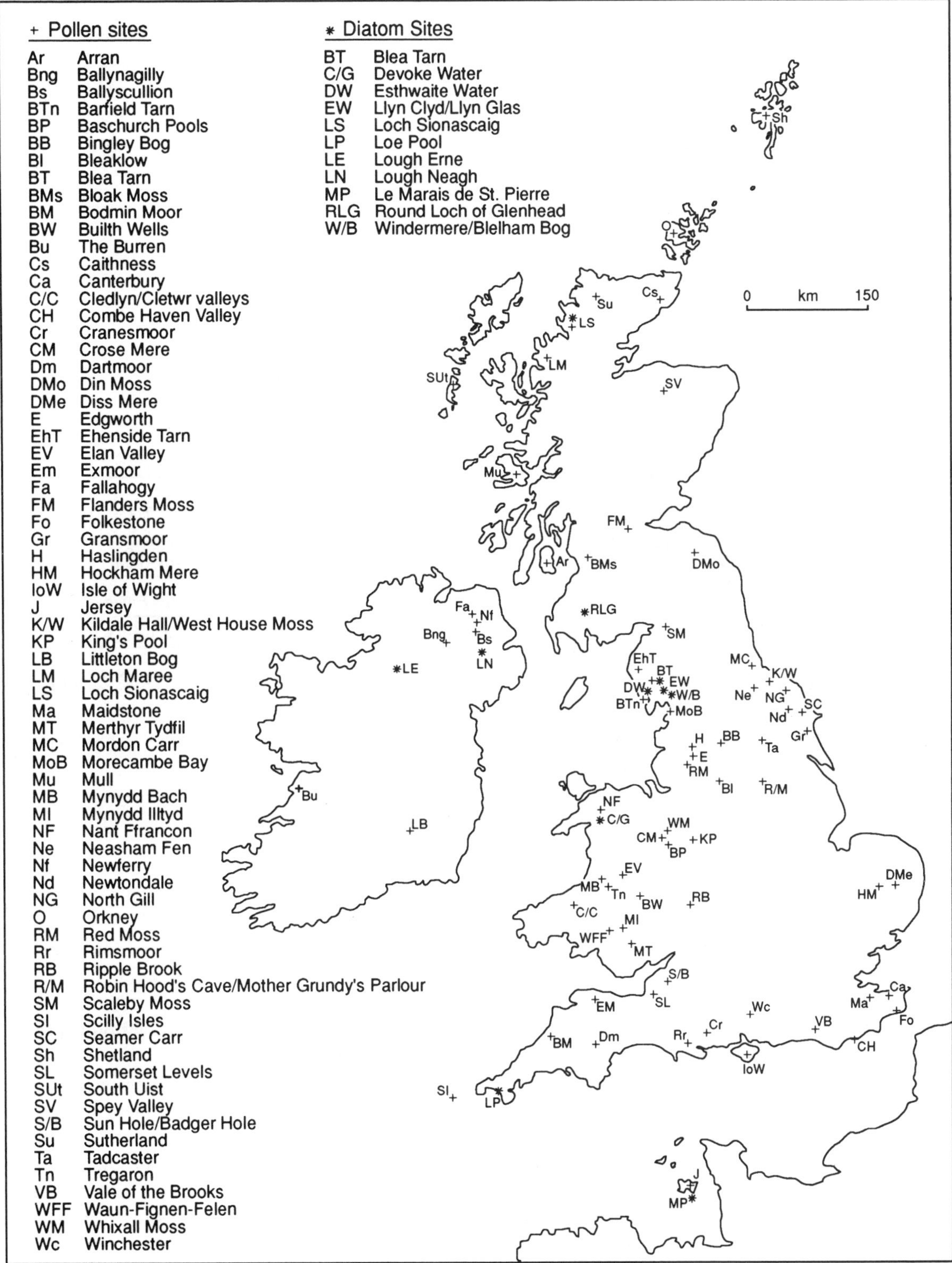

+ Pollen sites
Ar Arran
Bng Ballynagilly
Bs Ballyscullion
BTn Barfield Tarn
BP Baschurch Pools
BB Bingley Bog
Bl Bleaklow
BT Blea Tarn
BMs Bloak Moss
BM Bodmin Moor
BW Builth Wells
Bu The Burren
Cs Caithness
Ca Canterbury
C/C Cledlyn/Cletwr valleys
CH Combe Haven Valley
Cr Cranesmoor
CM Crose Mere
Dm Dartmoor
DMo Din Moss
DMe Diss Mere
E Edgworth
EhT Ehenside Tarn
EV Elan Valley
Em Exmoor
Fa Fallahogy
FM Flanders Moss
Fo Folkestone
Gr Gransmoor
H Haslingden
HM Hockham Mere
IoW Isle of Wight
J Jersey
K/W Kildale Hall/West House Moss
KP King's Pool
LB Littleton Bog
LM Loch Maree
LS Loch Sionascaig
Ma Maidstone
MT Merthyr Tydfil
MC Mordon Carr
MoB Morecambe Bay
Mu Mull
MB Mynydd Bach
MI Mynydd Illtyd
NF Nant Ffrancon
Ne Neasham Fen
Nf Newferry
Nd Newtondale
NG North Gill
O Orkney
RM Red Moss
Rr Rimsmoor
RB Ripple Brook
R/M Robin Hood's Cave/Mother Grundy's Parlour
SM Scaleby Moss
SI Scilly Isles
SC Seamer Carr
Sh Shetland
SL Somerset Levels
SUt South Uist
SV Spey Valley
S/B Sun Hole/Badger Hole
Su Sutherland
Ta Tadcaster
Tn Tregaron
VB Vale of the Brooks
WFF Waun-Fignen-Felen
WM Whixall Moss
Wc Winchester
* Diatom Sites
BT Blea Tarn
C/G Devoke Water
DW Esthwaite Water
EW Llyn Clyd/Llyn Glas
LS Loch Sionascaig
LP Loe Pool
LE Lough Erne
LN Lough Neagh
MP Le Marais de St. Pierre
RLG Round Loch of Glenhead
W/B Windermere/Blelham Bog
0 km 150

Figure 10.3 Localities from which Flandrian/Littletonian pollen and diatom data have been obtained.

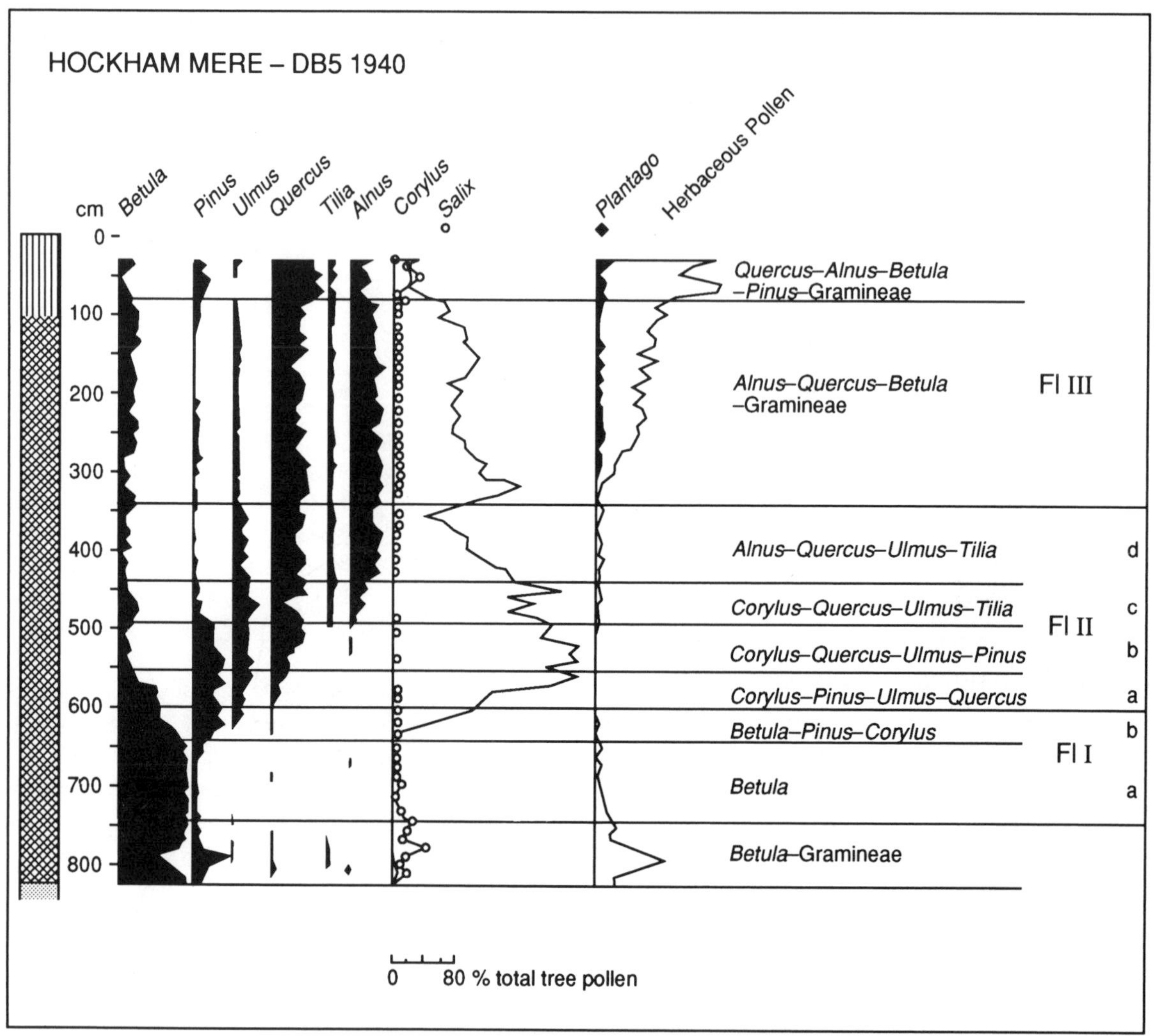

Figure 10.4 Pollen diagram of the Flandrian at Hockham. From West (1980b); source, Godwin and Tallantire (1951).

pollen declined *circa* 5000 BP and that of *Taxus* increased, perhaps reflecting the expansion of this tree in woodland clearings. A decline in *Tilia* pollen accompanied by an increase in that of herbaceous taxa and a first appearance of that of cereals around 3500 BP, is indicative of human activities. From 2500 to 1500 BP a deforestation episode has been identified, with a corresponding increase in grass pollen implying the development of pasture. The period from 1500 to 150 BP in this area is marked by more weeds and cultivars in the pollen record suggesting that cultivation was now of greater significance.

There have been many other studies of Flandrian vegetational development in East Anglia (for example, Lambert *et al.* 1960; Godwin 1968, 1978). West (1980b) has drawn together available data on its forest history within the framework of interglacial substages, referring also to the zonation scheme of Godwin (1940a) (Table 10.1; Figure 10.4). Flandrian I was initially characterized by the establishment of *Betula* forest (Fl Ia/Godwin Zone IV). However, evidence of a short phase prior to birch dominance, during which *Juniperus* was extensively distributed, is lacking. In Fl Ib (V) *Pinus* and *Corylus* assumed greater significance in the forest, with *Betula* decreasing in importance. *Betula*, *Pinus* and *Corylus* dominated this forest, in which *Ulmus* was more frequent than *Quercus*. West has pointed out that the lower boundary of substage II should be placed at the level in pollen diagrams where warmth-demanding deciduous tree genera assume importance, as had been the practise for earlier temperate stages. Thus, in East Anglia and comparable locations within the British Isles, this boundary should be at the conclusion of Fl Ib (V), and should not, as for example, had been the case at the Red Moss type site of Hibbert and Switsur (1976) (section 10.2.6) (Table 10.2), include Fl Ic and Fl Id (VI) divisions. Revision has been undertaken (Table 10.1), so that Zone VIa of Godwin has become Fl IIa. During this part of the substage *Ulmus*, *Quercus* and *Corylus* values went on rising, with oak staying inferior in representation to elm. *Pinus* frequencies remained fairly substantial, while those of *Betula* continued to decline. During Fl IIb (VIb) *Quercus* frequencies exceeded those of *Ulmus*; *Corylus* values remained high and those of pine and birch were reduced. In Fl IIc (VIc) *Tilia*, although likely to have been present since Fl Ib, became an established forest component. *Alnus* frequencies rose during Fl IIc, whose upper boundary is placed where they peak, and those of *Pinus* are low. Closed forest, of which *Quercus*, *Tilia*, *Ulmus*, *Fraxinus*, *Corylus* and *Alnus* were the most important components, was established in Fl IId (VIIa). It remained in existence for *circa* 2000 years, an initial demise being marked by a decline in elm pollen around 5000 BP, which delimits the end of Fl IId and the beginning of Fl III. During the latter, frequencies of non-arboreal pollen rose as the forest was opened and destroyed by prehistoric inhabitants. Farming took place, and heathland was created in Breckland. Although indicators of a Late-temperate substage of an interglacial (*Carpinus*, for instance) are present, they do not expand as in earlier temperate stages. This is because of removal of, and compositional adjustments within, the forest vegetation that were accomplished by man (section 10.1).

10.2.2 The Midlands of England

Data on the Postglacial vegetational history of this area have been limited and have come from its western sector. Around Crose Mere in Shropshire (Beales 1980), birch woodland developed *circa* 10 300 BP, following a phase when *Juniperus* was prominent in the vegetation. *Corylus* then became an important forest component, accompanied by *Quercus* and *Ulmus*. *Pinus* expanded about 8500 BP and was well established by *circa* 7900 BP. A phase of *Quercus–Alnus*-dominated forest ensued. A decline in elm pollen has been radiocarbon dated to 5296 ± 150 BP, but there appears to have been minimal influence on vegetational composition. The first significant episode of forest reduction took place in the Bronze Age, around 3900 BP. The pollen record indicates a mainly pastoral economy, although some cereals were being grown. Woodland is thought to have regenerated in this area during the Iron Age. Palynological data from the King's Pool, Stafford (Bartley and Morgan 1990) have revealed the earliest Flandrian vegetation to be a mosaic of open woodland, scrub and heath, of which birch, juniper, *Empetrum* and *Calluna* were members. *Betula* woodland developed *circa* 9600 BP and was succeeded by an arboreal assemblage of *Corylus*, *Quercus*, *Ulmus* and *Pinus*. During the Atlantic Period (Table 10.1), *Alnus* and *Tilia* increased their representation in the forests around Stafford, while *Pinus* declined in importance. Greig (1982) has suggested that *Tilia* may have been the most important component of mid-Flandrian deciduous forest about 10 km south of Stafford. Bartley and Morgan have considered the area around the King's Pool transitional between forest rich in *Tilia* to the south and forest poor in this tree to the north. The spread of *Alnus* and *Corylus* in the Boreal and Atlantic vegetation of this area has been thought a likely response to burning, perhaps by Mesolithic hunter-gatherers. Forest clearance, together with pastoral and arable agriculture took place from Early Neolithic to Bronze Age time, but the major human impact in this respect took place during the Iron Age, with forest clearance dated by radiocarbon to 2700–2500 BP. Close to the conclusion of the Roman Period, arable farming became increasingly important, the crops including cereals and *Cannabis* (hemp). Woodland did not regenerate subsequently around Stafford. An analogous major episode of Iron Age forest removal, after which woodland regrowth was lacking, has been reported from the Ripple Brook area, near Tewkesbury, by Brown and Barber (1985) (section 10.12). Late Flandrian pollen spectra from the Baschurch Pools near Shrewsbury have been discussed by Barber and Twigger (1987). In this area, the regional post-elm decline forest appears to have been dominated by *Quercus*. A decline in *Tilia* pollen has been identified *circa* 3200 BP at the Baschurch Pools by Barber and Twigger, who have inferred forest exploitation and small-scale clearances by man at this time in the Shrewsbury area. More extensive woodland removal followed, with regeneration of the tree cover taking place late in the Iron Age and in early Roman time. A synchronous late Flandrian decline in *Tilia* pollen and a comparable pattern of subsequent small-scale human activity has been recorded at Whixall Moss, between Whitchurch and Wem in Shropshire, by Turner (1964b).

10.2.3 Wales

A larger quantity of palaeobotanical data has been obtained for Wales than the English Midlands. Moore (1977) has summarized the Flandrian vegetational history, and this account is supplemented by more recent work below. The climatic amelioration at the opening of the Flandrian (section 10.13) was accompanied by the extension first of grassland, then juniper shrubland. A rapid expansion of *Betula* and *Corylus* followed the *Juniperus* rise. This sequence has been observed in the Elan Valley (Moore and Chater 1969; Moore 1970), on Mynydd Bach (Moore 1972), and in the Cledlyn and Cletwr valleys (Handa and Moore 1976) in west-central Wales. It has also been reported from the Wye Valley near Builth Wells (Moore 1978), and from Mynydd Illtydd, north of the Brecon Beacons, where birch–hazel woodland became established within a millennium of the conclusion of the Loch Lomond Stadial (Walker 1982) (Chapter 9). Analogous sequences have been established in the Nant Ffrancon (Hibbert and Switsur 1976; Ince 1983) and Snowdon (Ince 1983) areas. The succession to the mixed-deciduous forest of Atlantic time was similar in many parts of Wales. *Betula* declined in importance as other trees colonized in most areas. *Ulmus* enjoyed a brief period when it achieved maximum representation before being succeeded by *Quercus*. *Pinus* pollen peaked just before that of *Alnus* rose in the majority of localities. The late Atlantic forests in the uplands of southern and western Wales appear to have been dominated by *Quercus*, while the elevated sector of mid-Wales possessed much *Corylus*, perhaps forming hill-top scrub (Moore 1977). The opening of Fl III is marked by a distinct decline in elm pollen in most parts of Wales, an event radiocarbon dated to *circa* 5000 BP at Tregaron and Nant Ffrancon (Hibbert and Switsur 1976). Forest removal for agricultural purposes began in the Neolithic (Moore 1973), and by the Iron Age, tree-pollen percentages were akin to those of today over much of the country (Moore 1977).

Perturbation of the vegetation of Wales by Mesolithic man has also been identified. Chambers (1983) has inferred Mesolithic activity preceding that of Neolithic and Bronze Age peoples on moorland near Merthyr Tydfil, where peat formation (section 10.9) began *circa* 7000 BP. Smith and Cloutman (1988) have described Mesolithic artifacts and vegetation disturbance at Waun-Fignen-Felen in the Black Mountain range above the Swansea Valley, where open water is likely to have provided a focus for hunter-gatherer activity. Palaeobotanical and radiocarbon dating evidence has led to the suggestion that a small area of *Corylus*-dominated woodland was removed *circa* 8000 BP. Heath developed, and this was perpetuated by firing, remaining in some localities until *circa* 5000 BP, about which time peat began to form widely. Neolithic impact was slight, but Bronze Age inhabitants made considerable modifications to the vegetation between about 4000 and 2600 BP. Woodland regenerated early in the Iron Age, but the later part of this period witnessed its further clearance.

10.2.4 England south of the Lower Severn and Thames valleys

In south-west England, the Postglacial vegetational history of Bodmin Moor has been presented by Brown (1977). The initial hillside vegetation consisted of *Empetrum* and *Juniperus*. However, these were replaced by *Corylus* before 9000 BP. Hazel scrub was characteristic of exposed valley sides during the Boreal. At this time, oak woods were present in sheltered localities, with birch and alder dominating the arboreal associations on valley floors and around wetlands. *Pinus* was apparently absent from Bodmin Moor during the Boreal. Opening of the Boreal vegetation, and compositional changes within it, have been ascribed to burning that was either natural in a dry climate, or had been started by Mesolithic hunters. Open woodland persisted into the Atlantic Period. Slight evidence of a decline in elm pollen has been found and the amount of woodland gradually began to decline thereafter. Bodmin Moor was first settled during the Bronze Age, when pastoralism dominated. In the Iron Age further settlement took place and signs of arable farming have been detected.

On northern Dartmoor, open-habitat vegetation characteristic of the closing episode of the Lateglacial persisted into the earliest part of the Flandrian (Caseldine and Maguire 1986). Thereafter, dwarf-shrub heath with *Empetrum* and *Juniperus* developed, and was succeeded by birch woodland; a pattern also identified on southern Dartmoor (Simmons 1964; Simmons *et al.* 1983). The subsequent sequence of vegetation development across Dartmoor was the spread of *Corylus*, then *Quercus* and *Ulmus*, with an oak–hazel forest present during the Atlantic. Maguire and Caseldine (1985) have suggested that the tree line may have reached 415 m OD at this time. The Boreal and Atlantic woodlands of Dartmoor seem to have been subjected to disturbance by Mesolithic man (Simmons 1964). At Blacklane Brook in southern Dartmoor, such activity began *circa* 7760 BP (Simmons *et al.* 1983). Evidence of a falling tree-line during the Atlantic, and of blanket peat formation has been linked in part to Mesolithic man by Caseldine and Maguire (1986) (section 10.9). An elm decline has been recognized on pollen diagrams from Dartmoor, but there was little forest clearance during the Neolithic. Human impact on the vegetation became significant in the Bronze Age and intensified during the Iron Age (Simmons 1964).

On Exmoor the Atlantic forest was oak dominated. An elm decline took place *circa* 5000 BP. Neolithic people cleared areas of the forest and practised pastoral farming before allowing trees to regenerate, in a system of shifting agriculture. A similar sequence of events continued during the Bronze Age and the Iron Age. By the latter, arable agriculture had developed and the forest cover was reduced to approximately that of today (Merryfield and Moore 1974; Moore *et al.* 1984).

In the Somerset Levels, there was well-developed elm–oak–lime forest in the drier areas *circa* 5000 BP (Beckett and Hibbert 1979). The elm decline has been radiocarbon dated to about this time, and a phase of pastoral farming which lasted about 400 years, identified. This was followed by a 300-year period when regrowth of woodland took place. A further decline in elm pollen ensued over a timespan of around 400 years, during which herbaceous vegetation did not increase significantly. Woodland regeneration then occurred until *circa* 3400 BP, following which substantial Bronze Age clearance began. Elsewhere in Somerset, Badger Hole and Sun Hole caves in the Mendip Hills contain sediments which have yielded pollen spectra comparable with those of the early Flandrian elsewhere in southern England (Campbell 1977).

Around Cranesmoor, near Burley in the New

Forest, grassland and birch woodland existed at the Devensian–Flandrian transition (Clarke and Barber 1987). *Pinus* woodland emerged *circa* 9600 BP and persisted for about 1000 years, before being succeeded by *Quercus–Ulmus–Corylus* forest, which was best developed *circa* 6000 BP. A decline in oak pollen occurred at 5750 ± 60 BP and that of herbs rose. Elm pollen fell at 4550 ± 60 BP. The Atlantic forest close to Rimsmoor, near Bere Regis on the Hampshire–Dorset chalklands, was dominated by oak, elm and lime (Waton 1982; Waton and Barber 1987). Elm and other trees showed pollen declines between 5160 ± 90 and 4690 ± 70 BP. The pollen record is indicative of farming, including cereal cultivation, prior to the elm decline. However, these were only minor disturbances of the vegetation, and woodland regenerated before a clearance episode in the Early Bronze Age, *circa* 3800 BP. Regrowth of woodland again took place before major human impact in the Later Bronze Age and Iron Age. The economy was mainly pastoral at this time, but some evidence of crop growing has been discovered. However, the major cultivation phase here occurred in the late thirteenth and early fourteenth centuries AD. In the environs of Winchester, the Boreal forest included pine, hazel, oak and elm, with oak, elm, lime and ash dominating that of the Atlantic (Waton 1986). A decline in elm and other tree and tall-shrub pollen has been radiocarbon dated to 5630 ± 90 BP. Arable farming was in progress at this time, but not prior to it as around Rimsmoor.

In and around the Combe Haven Valley, between Bexhill and Hastings in eastern Sussex, oak–lime–elm closed-canopy forest existed on drier substrates *circa* 6000 BP, with alder woods in wetter areas (Smyth 1986; Jennings and Smyth 1987). Here also Neolithic people seem to have affected the vegetation little, with those of Bronze Age and Iron Age cultures having more impact on the landscape. In the Vale of Brooks and its neighbourhood in Sussex, well-developed woodland of elm and lime existed on the valley sides during the early Flandrian. The valley floor carried alder woods and the upper parts of the surrounding chalklands an oak–hazel association. Minor disturbance of woodland has been identified at *circa* 6290 BP but regeneration followed, with more extensive tree removal beginning in the Middle Bronze Age (Thorley 1971, 1981).

Pollen records from sediments at Wingham near Canterbury, and Frogholt, near Folkestone, in Kent have revealed largely deforested landscapes where farming was in progress. Deforestation had occurred by *circa* 3700 BP around Wingham, while at Frogholt, the record is of only about 500 years in the Late Bronze Age or Early Iron Age (Godwin 1962). Earlier Flandrian botanical evidence for the Maidstone and Folkestone areas has been obtained by Kerney *et al.* (1980). The pre-Atlantic woodlands of these areas contained at first birch, then hazel, which was joined by elm and oak.

10.2.5 The Scilly Isles, Channel Islands and Isle of Wight

Circa 6000 BP, oak and hazel dominated open-canopy woodland on St. Mary's in Scilly, of which elm and birch were also components. Some of this woodland was cleared, and the clearings farmed in the Early Neolithic, but the major episode of tree removal took place in the Middle and Late Bronze Age (Dimbleby *et al.* 1981; Scaife 1984, 1986b).

For the Channel Islands, only the Postglacial vegetational history of Jersey is well known (Jones *et al.* 1990). Around 9600 BP, open birch-dominated woodland with some oak and hazel existed. The pollen records show that this was succeeded by oak–hazel woodland. Removal of this woodland began *circa* 6000 BP, mainly for pastoral purposes, but some cereals were cultivated at the outset. During the Bronze Age and Iron Age, mixed farming took place on the island. From about 7000 to 4000 BP, damp substrates on the coastal plain had well-developed alder fen.

In the Isle of Wight (Scaife 1982, 1987), birch woodland was present *circa* 10 000 BP. This was followed in the early Boreal by a *Pinus*-dominated assemblage, and later in that period by oak–elm–hazel woodland. The Atlantic forest had alder, lime, oak and elm as its major components. An elm pollen decline has been dated to 4850 ± 45 BP, about when cereals were being cultivated in limited cleared areas. The Late Neolithic was a time of woodland regeneration and the Early Bronze Age one of widespread clearance and farming.

10.2.6 Northern England

A synthesis of Flandrian vegetational history in north-western England has been provided by Pennington (1970). Juniper was a significant constituent of the vegetation at the start of the Postglacial. A maximum of *Juniperus* pollen at Scaleby Moss in northern Cumberland has been radiocarbon dated to *circa* 10 3000–9800 BP (Godwin *et al.* 1957). Birch forest then developed. *Betula* was subsequently joined by *Corylus*, *Quercus* and *Ulmus*, the latter on better soils, to form closed forest that extended at least to 760 m OD (Pennington 1970). Pine was an important forest component in some areas during the late Boreal, its pollen maximum occurring later than the time of *Ulmus* and *Quercus* expansion (Oldfield 1965). *Tilia* was not present before Zone VI and then seems to have been confined to limestone terrain in the environs of Morecambe Bay. This area may have comprised part of its northerly limit in the British Isles (section 10.2.9). The Atlantic Period opened with the spread of alder woodland in depressions, valleys and hillside flushes (Pennington 1970). Pre-elm decline disturbance of the forest vegetation on the Cumberland Lowland in the later Postglacial has been reported by Walker (1966). The elm decline has been detected in southern Westmorland by Smith (1958a) and in Lowland Lonsdale by Oldfield (1960, 1963). In the latter area, a primary *Ulmus* decline, affecting only elm has been identified, together with a later episode involving more widespread tree removal and an extension of herbaceous vegetation. The expansion of grass and plantain pollen has been linked to a Landnam type of farming (Iversen 1941). Similar human activities have been recorded from northern and western Cumberland by Walker (1966) and from Barfield Tarn in south-western Cumberland (Pennington 1965, 1970). At Barfield Tarn, pollen of cereals and weeds appeared just later than 5000 BP. Both here, and at Ehenside Tarn (Walker 1966), a later clearance episode (*circa* 4200 BP) has been recognized. At Blea Tarn, Langdale, the elm decline occurred between 5300 and 5200 BP, when its foliage was probably collected for animal feed. Forest clearance ensued *circa* 5000–4000 BP. In the higher parts of the Lake District, cereal growing appears not to have commenced until about AD 200. Tree removal in preparation for this agricultural phase was responsible for the permanent deforestation of the area (Pennington 1965, 1970). Such an episode has been identified in south-western Westmorland (Smith 1959), while at the same time, lowland Cumberland witnessed heightened human impact on its vegetation (Walker 1966).

In north-eastern England the vegetational history of the North Yorkshire Moors, where in excess of 30 sites have been investigated over an area of approximately 450 km^2, has been particularly well established (Simmons *et al.* 1982). Open herb, heath and shrub vegetation, including grasses, sedges, *Empetrum* and *Salix*, was present at the opening of the Flandrian, as were scattered woods of birch and pine (Jones 1977b, 1978). The woodland gradually became more closed and during the Boreal *Pinus–Corylus*-dominated forest emerged, although at the highest elevations open plant communities appear to have been interspersed with a tree and tall-shrub cover (Simmons and Cundill 1974). At Kildale Hall in the Leven Valley, lithostratigraphic and biostratigraphic evidence of Mesolithic vegetation disturbance during the first few centuries of the Postglacial has been detected, while on moorland nearby, members of this culture are likely to have been responsible for episodes of soil erosion during the Boreal Period (Jones 1976b) (section 10.9.1). A rise in *Alnus* pollen frequencies, dated by radiocarbon to 6650 ± 290 BP at West House Moss in the Cleveland district, marked the opening of the Atlantic Period (Jones 1977b).

The Atlantic forests were mainly oak–hazel dominated, with alder important on wetter substrates. As during the Boreal, the highest parts of the moors were only lightly wooded (Simmons and Cundill 1974). During Zone VIIa, Later Mesolithic people (section 10.1) were causing perturbations in the moorland vegetation (Simmons 1969; Simmons and Innes 1981, 1988a,b; Innes and Simmons 1988). The elm decline has been radiocarbon dated to 4767 ± 60 BP at North Gill on Glaisdale Moor (Jones *et al.* 1979) and 4720 ± 90 BP at Fen Bogs in Newtondale (Atherden 1976). Neolithic forest clearance and agriculture were scarce (Spratt and Simmons 1976), but the Bronze Age witnessed substantial woodland reduction and a mixed agrarian economy. Heathland began to develop widely during the Bronze Age and brown earth soils were podsolized (Dimbleby 1962) (section 10.9.1). The Iron Age was also marked by considerable forest removal and agriculture on the moorlands of north-east Yorkshire (Spratt and Simmons 1976).

Vegetational history in the lowlands adjacent to the North Yorkshire Moors was somewhat different. In the Vale of Pickering (Walker and Godwin 1954), northern part of the Vale of York (Jones 1976a) and south and east Durham (Bartley *et al.* 1976), initial juniper shrubland was rapidly replaced by closed birch forest. *Corylus* then became important, while *Ulmus* and *Quercus* were able to become established in suitable localities during the early part of Boreal time. The late Boreal forests of these lowlands were dominated by oak, elm and birch, and those of the Atlantic by *Quercus*, *Ulmus*, *Tilia* and *Alnus*. Cloutman (1988) has suggested that in the Star Carr area of the Vale of Pickering, the expansion of alder about the transition from Boreal to Atlantic time was caused by Mesolithic man burning established vegetation. At Neasham Fen near Darlington, the elm decline has been radiocarbon dated to 5468 ± 80 BP and at Mordon Carr south of Sedgefield, to 5305 ± 55 and 5235 ± 70 BP (Bartley *et al.* 1976). However, subsequent Neolithic impact on the vegetation here, and in adjacent lowlands, appears to have been slight, with Bronze Age and Iron Age peoples responsible for much more marked effects in this context.

Bush and Flenley (1987) have demonstrated that grassland was present on the Yorkshire Wolds early in the Flandrian. An expansion of tree birches *circa* 9300 BP did not exclude grassland taxa, as neither did the subsequent immigration of other trees. Evidence has been forthcoming of Mesolithic vegetation disturbance, beginning about 8900 BP, which probably assisted the maintenance of grassland for at least a millennium. The vegetational history of Holderness has been described by Beckett (1981), Blackham and Flenley (1984) and Flenley (1984). The juniper phase found further north in north-east England has not been recognized, nor has an episode of pine dominance. *Betula* woodland developed first, then a *Corylus–Quercus–Ulmus–Alnus* forest emerged. The elm decline occurred *circa* 5100 BP at Gransmoor Quarry in northern Holderness (Beckett 1981), but as elsewhere in the region, Neolithic forest clearance was slight with Bronze Age peoples making a bigger impact.

Evidence from Tadcaster (Bartley 1962) has shown the area between Leeds and York to have first developed birch woods in the Flandrian, these later being replaced by a pine–hazel association, with mixed-oak forest emerging by the early Atlantic Period. Around Bingley in Airedale, a similar succession of vegetation took place. Birch was replaced by hazel (not accompanied by pine) as dominant in the forest during the Boreal, with mixed-deciduous trees present in Zone VIIa (Keen *et al.* 1988). In Lowland Craven, north-west Yorkshire, birch woodland was present until *circa* 9400 BP. Thereafter, *Pinus* and *Corylus* became prominent in different parts of the forest, and oak and elm colonized. Alder expanded *circa* 7500 BP, and from about 6000 to 5000 BP the forest composition was relatively stable in the upper Wharfe and Ribble valleys. Pollen and lithostratigraphic data indicate some Mesolithic forest disturbance, and just prior to 5000 BP, grazing may have been practised. The elm decline has been radiocarbon dated to 5010 ±

110 and 5080 ± 100 BP, but only limited forest clearance and farming during the Neolithic is indicated. Woodland removal increased and farming intensified on fertile soils over limestone in the Bronze Age, *circa* 3600 BP. However, it was not until about AD 600, in Anglo-Saxon times, that heavier soils developed in drift were cultivated (Bartley *et al.* 1990).

In the southern Pennines, pine woodland grew in the Bleaklow area between Manchester and Sheffield during Zone VI (Jacobi *et al.* 1976). Around 7000 BP, forest or scrub probably grew up to about 500 m OD over most of the southern Pennines; above this height, exposure may have inhibited tree growth (Tallis 1964a). Peat formation became widespread at the start of the Atlantic Period in this area (Conway 1954; Tallis and Switsur 1973), and may have been responsible for eliminating arboreal vegetation during its subsequent extension at various times (Tallis 1964b; Tallis and McGuire 1972; Bartley 1975) (section 10.9.2). However, Jacobi *et al.* (1976) have hypothesized that blanket peat formation had a minimal influence upon the amount of tree cover, which was discontinuous above *circa* 350 m OD in late Boreal/early Atlantic time. At lower elevations, a well-developed mixed-oak forest existed by *circa* 6000 BP (Hicks 1971); above this, on plateau surfaces, a mosaic of hazel scrub and herbaceous communities probably existed. Jacobi *et al.* have suggested that Mesolithic people began to utilize the southern Pennines within a few hundred years of the onset of the Postglacial, at which time a tree cover had not developed because of climatic and plant migrational characteristics (Chapter 9 and section 10.6). As afforestation took place under the influence of an ameliorating climate, the spread of trees higher than *circa* 350 m OD could have been resisted by periodic firing of the vegetation by Mesolithic man. In the Neolithic and Bronze Age, minor impermanent forest clearance occurred over the East Moor district of the north Derbyshire Pennines. Woodland removal accelerated here during the Iron Age, but pastoralism remained the dominant economic activity until Roman times. The onset of cultivation led to soil deterioration, heather moorland development and peat growth (Hicks 1971). From the Central Rossendale area of the southern Pennines, Tallis and McGuire (1972) have identified a major episode of woodland clearance in the Middle and Late Bronze Age. More trees were removed from the area between Haslingden and Edgworth during the Iron Age, but subsequently woodland regrew and was not destroyed until after *circa* AD 1200.

The proposal that Red Moss, on the western edge of the southern Pennines, be designated the type locality for the Flandrian Stage, is noted at the start of this chapter. From this site near Horwich in Lancashire, Hibbert *et al.* (1971) have defined six pollen assemblage zones, and radiocarbon dated their boundaries, together with other features within them. Three major chronozones *sensu* West (1970) have been represented (Table 10.2). It should be recalled at this point that West (1980b) has suggested revision of the location of the FlI/FlII boundary (section 10.2.1).

In the northern Dukeries area of Nottinghamshire and Derbyshire, Mother Grundy's Parlour and Robin Hood's Cave sediments have produced pollen spectra dominated by hazel, birch and pine, and of early Postglacial age (Campbell 1977). The early and middle Flandrian vegetation of the northern Pennines between the Stainmore Gap and River South Tyne has been discussed by Turner and Hodgson (1979, 1981, 1983). The Boreal forest *circa* 8800–7000 BP contained *Corylus*, *Pinus*, *Betula*, *Ulmus*, *Quercus*, *Alnus* and *Salix*, and was well developed between about 200 and 750 m OD. However, its composition varied geographically. For example, pine was important in Upper Teesdale (Turner *et al.* 1973) and in the Derwent area, but at varying elevations and on contrasting soils. Hazel and elm were more frequent at high than low altitude, with the latter not well represented in forest over Millstone Grit (Turner and Hodgson 1979). The mid-Flandrian forest exhibited more spatial than temporal variation as a result of differing climatic and edaphic conditions. The Millstone Grit country in the north-east of this region possessed forest with more birch, pine and willow, and less elm and hazel than elsewhere. *Tilia* was more abundant in the south, as was *Fraxinus*. Human disturbance of this forest was slight on the evidence of botanical data. However, a number of sites in the altitudinal range 450–600 m OD contain high enough frequencies of ash, grass and *Potentilla* (cinquefoil/tormentil) pollen for burning of the plant cover by Mesolithic people to be hinted at (Turner and Hodgson 1981, 1983).

10.2.7 Scotland

Syntheses of the vegetational history have been made by Birks (1977) and Walker (1984a), while Birks (1988) has discussed Scottish vegetation within the wider context of the ecological history of uplands in the British Isles. The earliest Flandrian pollen spectra in most parts of the country contain herbaceous taxa, notably Gramineae and *Rumex*, indicative of grassland. Following this was dwarf-shrub and scrub vegetation, varying locally and including *Empetrum*, *Juniperus* and *Salix*, then *Betula–Corylus* woodland. In areas which were glaciated during the Loch Lomond Stadial (Chapter 9), the initial herbaceous episode was lacking, with dwarf-shrub vegetation, particularly composed of *Empetrum*, the first to colonize. An early Flandrian peak in *Juniperus* pollen has also been identified in Scotland. This probably signifies increased flowering of extant juniper as a result of a climatic amelioration (section 10.13) and the spread of scrub in place of park tundra (Walker 1984a). There was a substantial amount of juniper scrub in Roxburghshire by 10 300 BP (Hibbert and Switsur 1976), and by 10 000 BP it had extended across the Grampians (Vasari 1977; Walker and Lowe 1979). In the Spey Valley (Birks and Mathewes 1978) and on Mull (Walker and Lowe 1982) this vegetation had developed by 9700–9500 BP. The initial woods were open and contained *Betula*, *Salix*, *Sorbus* (rowan) and *Populus*. Central and southern parts of Scotland had tree birches shortly after 10 000 BP (Vasari 1977; Lowe 1978), and by *circa* 9000 BP, they had reached Caithness (Peglar 1979). *Corylus* was in the Scottish Borders by around 9300 BP (Hibbert and Switsur 1976), reached the north Ayrshire coast by *circa* 9200 BP (Boyd 1988), had progressed up the western coastlands to Wester Ross by *circa* 8800 BP (Birks 1972a; Walker and Lowe 1982), and was in the central Grampians by about 8700 BP (Birks and Mathewes 1978). Southern Scotland (Newey 1968; Moar 1969a; Webb and Moore 1982), together with the south and west fringes of the Grampians (Donner 1962; Rymer 1977; Walker and Lowe 1981; Lowe 1982) saw birch–hazel woodland colonized by elm and oak, with mixed-deciduous forest emerging. In the lowlands of Aberdeenshire, pine achieved more importance than oak in the forest (Vasari and Vasari 1968), while north of the Grampians and into Ross and Cromarty, the mid-Flandrian forests were dominated by *Pinus* and *Betula* (Birks 1972a; Pennington *et al.* 1972; O'Sullivan 1974; Walker 1975c). Caithness and Sutherland appear to have had a restricted woodland cover, with birch–hazel scrub of significance in north-east Caithness (Birks 1977; Peglar 1979). The Orkney (Moar 1969b; Keatinge and Dickson 1979) and Shetland (Hawksworth 1969; Johansen 1975, 1985; Birnie 1984) islands developed areas of birch–hazel scrub in the mid-Flandrian, and it is unlikely that other trees grew there. At this time, the Outer Hebrides (Birks and Madsen 1979; Walker 1984b; Bohncke 1988), together with certain of the Inner Hebrides (Flenley and Pearson 1967), were largely devoid of arboreal vegetation. Skye possessed small areas of birch–hazel–alder woods in its northern sector, while woodland in the south of the island was better developed and included elm and oak (Birks 1973). *Pinus* was a significant component of woods in eastern Skye during the fifth millennium BP (Birks and Williams 1983). At least 50% of South Uist was covered by *Betula–Corylus* woodland, and *Quercus*, *Ulmus*, *Alnus* (all now extinct) and *Fraxinus* grew during the early Flandrian (Bennett *et al.* 1990). In Arran, birch woodland existed before 8600 BP, with hazel then joining *Betula*. Elm, oak, pine and alder thereafter achieved increased representation in forest on different parts of the island (Boyd and Dickson 1987).

It is possible that Mesolithic man caused small-scale disturbances of vegetation in Scotland (Edwards and Ralston 1984). A reduction in woodland cover took place in many areas of Scotland during the fifth millennium BP, and by *circa* 2000 BP, a substantial part of the country had been deforested (Walker 1984a). This reduction of arboreal vegetation has been attributed to both natural agencies, especially blanket-bog growth (section 10.9.2), and human activity. A decline in *Ulmus* pollen took place *circa* 5400 BP at Din Moss in Roxburghshire (Hibbert and Switsur 1976), and just prior to 5000 BP at Loch Maree in Ross and Cromarty (Birks 1972a), and at Flanders Moss west of Stirling (Smith and Pilcher 1973). Associated with the decline in elm pollen are increased frequencies of grass (including cereal), weed and heath pollen, and of *Pteridium* (bracken) spores. In Orkney, scrub clearance began *circa* 5000 BP (Keatinge and Dickson 1979), and arable farming took place during the Neolithic (Davidson *et al.* 1976). In Caithness, significant human impact on the vegetation did not take place before about 2500 BP (Peglar 1979). Around Loch Sionascaig in north-west Scotland, the initial forest destruction has been dated to *circa* 3500 BP (Pennington *et al.* 1972), about the same time that it took place in the Cairngorms, where *Calluna* heath had developed two millennia later (O'Sullivan 1974, 1976). Aberdeenshire witnessed human interference with its vegetation prior to 5000 BP (Edwards 1979b), and forest removal in Perthshire was early (Caseldine 1979). The elm decline in the Galloway Hills took place about 5000 BP, to be followed by predominantly pastoral land use (Birks 1972b). At Bloak Moss in Ayrshire, minor clearance occurred about 4000 BP, but major forest destruction was delayed until the middle of the second millennium BP (J. Turner 1970). Woodland decline on South Uist began *circa* 4300 BP as a result of grazing pressure (Bennett *et al.* 1990). On Arran, small-scale human activity is detectable at *circa* 5000 BP, with *Calluna* heath emergent 1000 years later (Boyd and Dickson 1987).

10.2.8 Ireland

Littletonian vegetational history has been discussed by Mitchell (1981b) and Watts (1985). At the start of the interglacial, there was a rapid succession of pioneer vegetation communities. The pollen record indicates that at first, grassland was important (Watts 1985). This was followed by a brief episode when *Empetrum* dominated and *Filipendula* was abundant. *Juniperus* scrub subsequently spread widely (Watts 1977; Craig 1978). The juniper phase was also short-lived and occurred prior to the development of birch–willow–aspen woodland (Watts 1985). *Corylus* subsequently invaded the birch woods to produce a more or less full cover of hazel scrub or woodland over Ireland. The peak in hazel pollen occurred *circa* 9000 BP, when *Pinus*, *Quercus* and *Ulmus* were also present. In the Burren, pine immigrated prior to hazel (Watts 1984). Not long after 9000 BP, forest dominated by oak, elm and pine had developed. This forest differed in its make-up, depending mainly upon edaphic conditions (Singh and Smith 1973, Craig 1978). *Alnus* arrived *circa* 7000 BP (Smith and Pilcher 1973) and grew profusely on damp substrates for around 2000 years.

At Newferry in County Antrim, Mesolithic activities around 7400 BP probably included forest opening and the maintenance of cleared areas by grazing (A. G. Smith 1981). Such pastoralism appears to have lasted until a decline in elm pollen, *circa* 5400 BP (Smith 1984; Edwards 1985). The incidence of cereal pollen prior to the elm decline in Ireland (and elsewhere in the British Isles) has led Edwards and Hirons (1984) to postulate cultivation at this time, probably as early as 5800 BP. Radiocarbon dates have established an average age of *circa* 5100 BP for the *Ulmus* decline in Ireland, and the phenomenon seems to have been most effective on fertile soils over limestone in the central lowlands (Watts 1985). Pollen data have indicated increased representation of grasses, cereals, plantain and hazel at the time of the elm decline. Two *Ulmus* declines have been identified in County Tyrone by Hirons and Edwards (1986). The first, 5375–5050 BP, was followed by regrowth of elm, and the second occurred 4330–4260 BP. In County Sligo, a decline of elm and pine has been recognized at around 5400 BP. Woodland was cleared and pasture developed (Dodson and Bradshaw 1987).

From *circa* 4500 BP, *Fraxinus* expanded, and *Taxus* was well represented in the later Post-glacial woodlands (Watts 1985). *Pinus* became extinct *circa* 2000–1000 BP (Watts 1984; Bennett 1984; Bradshaw and Browne 1987). Alternating phases of woodland regeneration and clearance and farming characterized the three millennia following the elm decline. Such features have been clearly identified at Fallahogy, County Londonderry (Smith and Willis 1962), Littleton Bog, County Tipperary (Mitchell 1965b), Ballyscullion, County Antrim (Pilcher *et al.* 1971) and Ballynagilly, County Tyrone (Pilcher and Smith 1979). The Bronze Age was a time of major forest reduction (Edwards 1985). The decline of *Pinus* noted above was partly due to human utilization of it over this period (Watts 1985). Agricultural activity was marked during the Bronze Age, but slackened in the Iron Age, when some woodland regrowth occurred (Mitchell 1965b,c, 1976). Renewed clearance and agricultural activity occurred from *circa* 1650 BP (Mitchell 1981b).

10.2.9 Spatial and temporal variations in Flandrian vegetation

The zonation scheme of Godwin (1940a) (Table 10.1) has subsequently been used on pollen stratigraphic sequences with the aim of demonstrating similarity in vegetation composition and synchroneity of the zone boundaries across England, Wales and Scotland. Hibbert and Switsur (1976) have concluded that synchroneity exists between pollen assemblage zones at sites in Wales and Scotland, and that these are comparable with the biozones identified at Red Moss (Hibbert *et al.* 1971) (Table 10.2). However, Smith and Pilcher (1973) have radiocarbon dated the appearance of, and frequency changes in, certain pollen taxa in different areas of the British Isles, thereby demonstrating that the Godwin zones were diachronous. Huntley and Birks (1983) have produced maps of the frequency of single pollen taxa through time which point up the fact that tree appearance was time transgressive (Birks 1986). Thus, it has become evident that pollen assemblage zones possess different characteristics in the various regions of the British Isles.

Bennett (1988a) has formally defined regional pollen assemblage zones for central East Anglia, and combined these with zones from other radiocarbon-dated palynological sequences in the British Isles to produce time–space diagrams along three transects. It has been concluded that pollen assemblage zones have a limited temporal range and spatial distribution, and that forest vegetation varied almost continually throughout the Flandrian across the entire area of the British Isles. Climatic and topographic contrasts have been cited as probable causes of the woodland mosaic. Anthropic activity has been considered to be a possible reason for analogies between pollen assemblages from England, Wales and Northern Ireland subsequent to the elm decline.

Birks (1989) has used pollen data to prepare maps that portray spatial and temporal patterns of tree dissemination over the British Isles during the Flandrian. It seems that *Betula* reached the British Isles via land then exposed in the southern North Sea Basin (sections 10.1, 10.11 and 10.12). Spreading at about 250 m per year, it was present in southern Scotland and eastern Ireland prior to 10 000 BP. *Corylus* is thought to have dispersed from west to east at about 500 m per year after appearing along the western seaboard by 9500 BP. *Ulmus* has been shown to have gained a foothold in south-east England by 9500 BP, then spread at up to 600 m per year, reaching eastern Ireland by 9000 BP. *Quercus* has been demonstrated as being resident in south-western England by the middle of the tenth millennium

BP, having moved up the west seaboard of the European mainland (Huntley and Birks 1983). It then spread east and north at up to 500 m per year until *circa* 8000 BP. *Pinus* appears to have had a restricted distribution in south-eastern England prior to 9000 BP, and by 8500 BP had moved northwards, at rates up to 700 m per year, to the southern Lake District and northern Pennines. Pine expanded in north-western Scotland *circa* 8000 BP (Birks 1972a; Pennington *et al.* 1972). This expansion may have taken place either from a refugium nearby (Huntley and Birks 1983) or, perhaps more likely, after quick progress to the region through western Ireland and south-western Scotland (Bennett 1984). Its success would have been assisted by the lack of competition from other trees (Birks 1989). By 7500 BP, pine had spread east to the upper reaches of the Dee and Spey valleys, where native stands of *Pinus sylvestris* have survived (O'Sullivan 1977). *Alnus* was well established along some parts of the coast of mid-Wales between 8900 and 8500 BP (Heyworth *et al.* 1985; Chambers and Price 1985). Alder might have possessed a widespread, if patchy, distribution across the British Isles before 8000 BP (Godwin 1975; Bennett 1986, 1988b). It had reached the Thames Estuary by 8200 BP and was in southern Argyll by 7900 BP. It has been established that *Tilia* had appeared in southern England by 7500 BP, and moved into central parts of the country at up to 500 m per year to become an important member of the arboreal vegetation (section 10.2.2). After 7000 BP, the migration of lime was reduced to at most 100 m per year, and it attained a northern limit extending from the central Lake District across the northern Pennines to the coast of Northumberland by 5500 BP (Pigott and Huntley 1980). Its present limit is still in these areas (Pigott and Huntley 1978). However, it now does not often set seed here due to insufficient summer temperatures, and the populations are most likely to be maintained by vegetative regrowth. None the less, as it was able to reach this limit *circa* 6000 BP, it is probable that summer temperatures were then 2–3°C above those of today (Pigott and Huntley 1980, 1981).

The main conclusion of Birks (1989) has been of individualistic behaviour by different tree taxa in both space and time. Arrival was independent in the various parts of the British Isles. Subsequent spread was at varying speeds and in contrasting ranks. It has been considered that following the influence of the initial climatic amelioration of the Flandrian, chance was probably the major factor in determining the pattern of spreading. As Bennett (1984) has suggested, modern analogues for earlier Flandrian forest assemblages may be lacking, according to Birks. The restricted chronostratigraphic value of pollen stratigraphic changes was also emphasized.

Bennett (1989) has gathered data on pollen plus other palaeovegetational characteristics, and used this in conjunction with edaphic, climatic and topographic information to produce a map depicting the differing composition of forest across the British Isles *circa* 5000 BP (Figure 10.5). Alder woodland was frequent but restricted to localities such as wetlands and river valleys. Fertile, non-calcareous soils, as far north as northern Yorkshire and southern Cumbria, and west to the Welsh uplands, were dominated by lime woods. Oak woodland was confined to infertile soils, while pine is likely to have been absent from south-western England and southern Wales.

10.3 Non-marine Mollusca

While these fossils have not been as intensively studied as those of plants, a reasonable amount of information has been gathered on those belonging in different habitats (Figure 10.7).

10.3.1 Terrestrial woodland and grassland assemblages

Flandrian land-snail assemblages have been most intensively studied on the chalklands of southern England, along the western coast of Scotland and in Orkney, with faunas of Boreal and Atlantic age demonstrating the existence of woodland (Evans 1972, 1975). On the chalklands, the molluscan

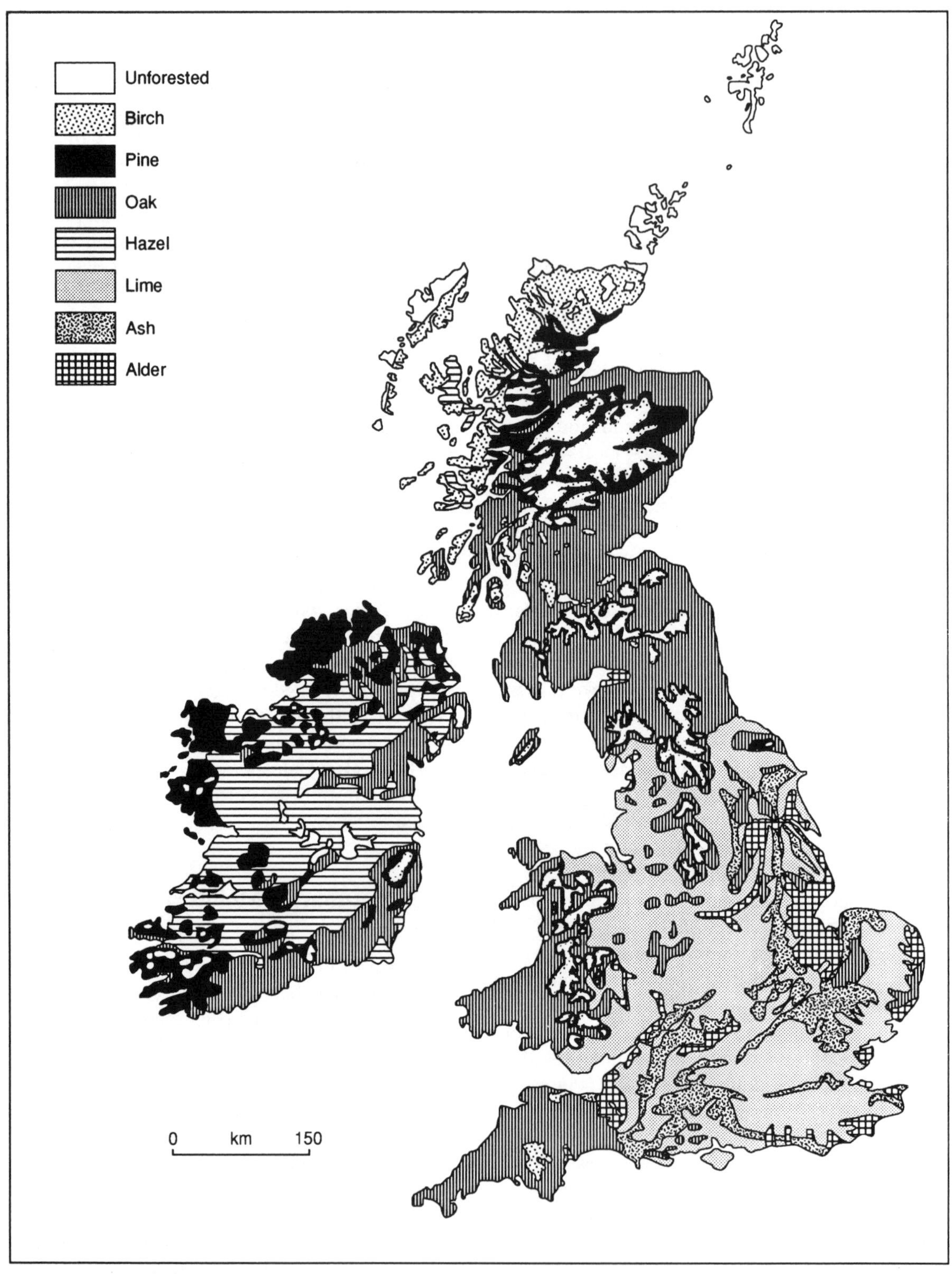

Figure 10.5 Provisional map of woodland types for the British Isles 5000 years ago. From Bennett (1989). Reproduced by permission of John Wiley and Sons Ltd. © 1989.

succession indicates forest removal by Neolithic man, together with arable and pastoral farming. In the Avebury–Windmill Hill area of north Wiltshire (section 10.1), soil buried beneath the South Street long barrow has revealed such a pattern (Evans 1971, 1972, 1975) (Figure 10.6). At Cherhill near Avebury, a buried soil with a Mesolithic occupation horizon on its surface was sealed by a tufa. A ditch was cut in this material by Neolithic inhabitants, and was subsequently infilled. The soil contains an early Postglacial woodland fauna including *Discus rotundatus*, *Carychium tridentatum* and Zonitidae. The tufa molluscan assemblage provides continuing evidence of shaded habitats, with charcoal at the base of this deposit giving a radiocarbon age of 7230 ± 140 BP. The ditch fill is characterized by Mollusca of open habitats, an admixture of shade-loving species occurring in its lower part (Evans *et al.* 1978).

Molluscan evidence has been used by Spencer (1975) to elucidate environmental changes in areas of north Cornwall and Orkney with coastal sand deposits. The succession was from *Discus* and *Carychium*, denoting woodland, to *Helicella itala*, *Cochlicella*, *Vallonia excentrica* and *Pupilla muscorum* indicative of open ground. Woodland was either removed by prehistoric man or overwhelmed by blown sand.

Kerney (1977b) has proposed a zonation scheme for Lateglacial and Postglacial deposits using terrestrial molluscs (Table 10.3). This scheme has been illustrated by Kerney *et al.* (1980) in relation to deposits at Folkestone and Wateringbury in Kent. The initial Flandrian fauna was mainly of open ground, but contained fewer species characteristic of bare soil than the underlying Lateglacial deposits (Chapter 9). Following this was a woodland assemblage with *Carychium tridentatum*, *Aegopinella* and *Discus ruderatus*; later including *Discus rotundatus*, *Oxychilus cellarius*, *Spermodea*, *Leiostyla* and *Acicula*. An open-ground fauna ensued, with *Vallonia* re-expanding and *Helix aspersa* appearing. Changes in the early Flandrian fauna have been ascribed to migration in response to climatic modifications. At certain times, an admixture of taxa with different distributional affinities resulted from this process.

Tufa began to form at Newlands Cross, just west of the City of Dublin, at 9720 ± 300 BP (Preece *et al.* 1986). The molluscan fauna then, including *Vertigo genesii* and *Columella columella*, was of open country. Colonization by *Carychium tridentatum* and *Aegopinella*, then *Leostyla anglica* and *Vitrea contracta*, followed by *Discus rotundatus* and *Spermodea lamellata*, denotes the presence of shade. Parallels could thus be drawn between this part of Ireland and Kent in respect of molluscan stratigraphy. Similar sequences had been reported from Blashenwell in the Isle of Purbeck, Dorset (Preece 1980) and from the Ancholme Valley in Lincolnshire (Preece and Robinson 1984). However, certain inter-regional variations have been observed and increased distance from Kent led to greater dissimilarities, so that at Inchrory in Banffshire (Preece *et al.* 1984), for example, the scheme is inapplicable.

The Mollusca of the Kennet and Loddon valleys in Berkshire have been analysed by Holyoak (1983c). Among land snails, *Pupilla muscorum*, *Columella columella*, *Vallonia pulchella*, *Vallonia excentrica* and *Helicella itala* continued to be present when a woodland fauna emerged in the early Flandrian, with all bar *Columella* persisting into Atlantic time. Upon woodland removal, *Cernuella virgata*, a species characteristic of unshaded herbaceous vegetation, and *Helix aspersa*, were recorded.

10.3.2 Freshwater and marsh assemblages

The freshwater molluscan succession in the Kennet Valley began with *Physa fontinalis*, *Lymnaea palustris*, *Lymnaea peregra*, *Pisidium casertanum* and *Pisidium personatum* in the Pre-Boreal. During the early Boreal, *Valvata piscinalis*, *Bithynia tentaculata*, *Lymnaea stagnalis*, *Armiger crista*, *Hippeutis complanatus*, *Sphaerium corneum*, *Pisidium milium*, *Pisidium nitidum*, *Pisidium moitessierianum* and *Acroloxus lacustris* were present. In the late Boreal, *Sphaerium lacustre*, *Pisidium amnicum*, *Pisidium*

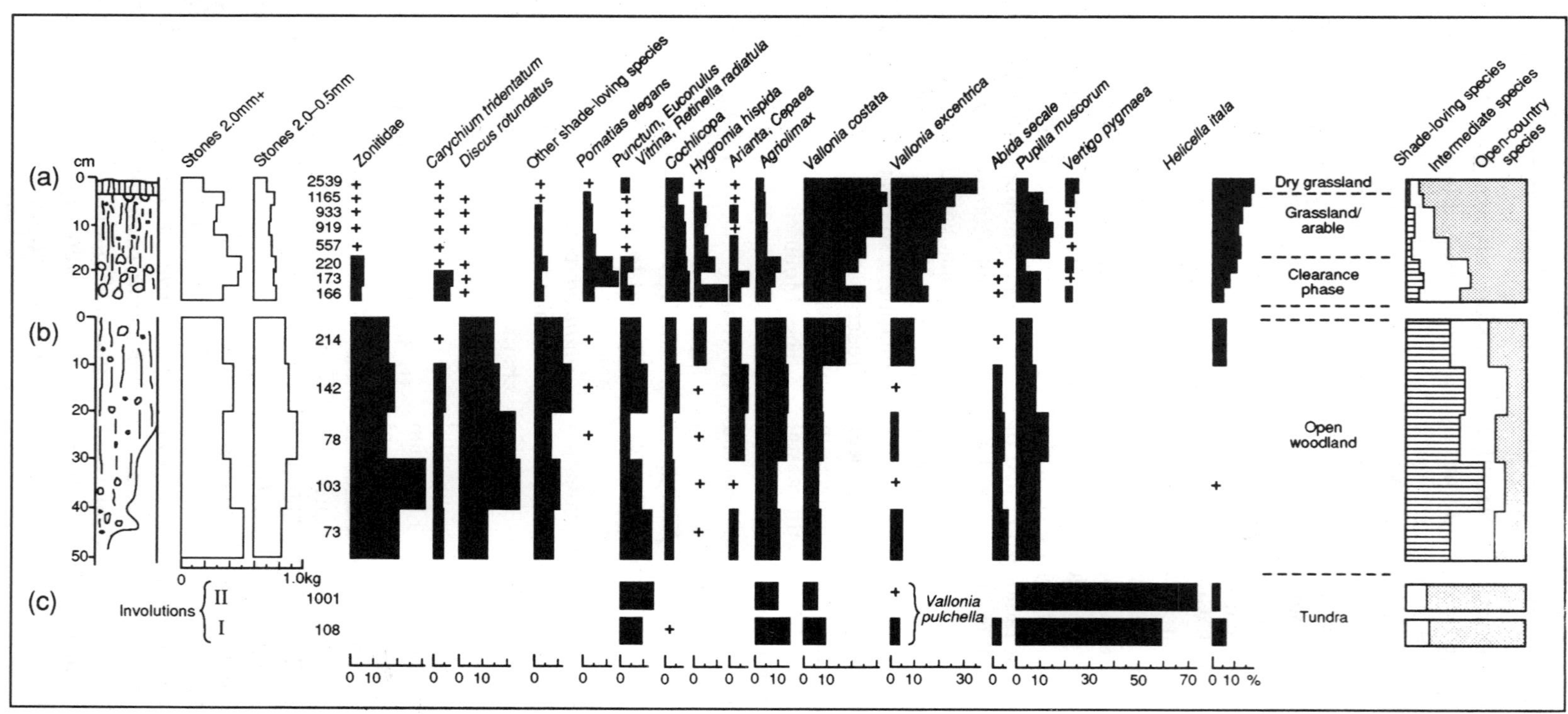

Figure 10.6 South Street molluscan diagram: (a) buried soil; (b) subsoil hollow; (c) involutions. From Evans (1972); source, Evans (1971). Reproduced from Economy and settlement in Neolithic and Early Bronze Age Britain, D. D. A. Simpson (ed.) (1971), by permission of Leicester University Press.

Table 10.3 Zonation scheme for Lateglacial and Postglacial deposits using land Mollusca. From Kerney (1977b)

^{14}*C dates* BP *for mollusc zones (Kent)*	*Mollusc zones (Kent)*	*Faunal types*	*Approximate pollen zone equivalents*	
	f	Open-ground fauna, but with *Helix aspersa*	VIIb–VIII	Postglacial
4540 ± 105 (BM-254)	e	Open-ground fauna. Decline of shade-loving species (*e.g. Discus rotundatus*). Re-expansion of *Vallonia*		
7500 ± 100 (St-3410)	d²	Closed-woodland fauna. Expansion of *Spermodea*, *Leiostyla* and *Acicula*	VIc–VIIa	
	d¹	Closed-woodland fauna. Expansion of *Oxychilus cellarius*	V–VIb	
8470 ± 190 (Q-1425)	c	Closed-woodland fauna. Expansion of *Discus rotundatus*, decline of *Discus ruderatus*		
8980 ± 100 (St-3411) 9305 ± 115 (St-3395)	b	Open-woodland fauna. Expansion of *Carychium* and *Aegopinella*. *Discus ruderatus* characteristic.		
9960 ± 170 (Q-1508)	a	Open-ground fauna, but decline of *Pupilla* and expansion of *Catholic* species	IV	
11 900 ± 160 (Q-618) 11 934 ± 210 (Q-463)	z	Open-ground fauna with *Pupilla*, *Vallonia*, *Abida* and *Trichia*	II–III	Lateglacial
13 180 ± 230 (Q-473)	y	Open-ground fauna dominated by *Pupilla* and *Vallonia*	I	

hibernicum, *Pisidium henslowanum*, *Bithynia leachi* and *Gyraulus albus* existed. The Atlantic witnessed the arrival of *Aplexa hypnorum*, *Lymnaea auricularia*, *Planorbis planorbis* and *Anisus vortex* (Holyoak 1983c).

The White Bog in County Down (Stelfox *et al.* 1972), Kildale Hall (Keen *et al.* 1984) and Bingley Bog (Keen *et al.* 1988) in Yorkshire have revealed similar early Flandrian molluscan sequences. At Kildale Hall, *Lymnaea peregra*, *Valvata piscinalis* and *Armiger crista* were among the inhabitants of a shallow lake at the start of the Postglacial. The open water was replaced by marsh where *Nesovitrea hammonis*, *Oxyloma pfeifferi* and *Punctum pygmaeum* resided. The Pre-Boreal freshwater fauna of Bingley Bog included *Physa fontinalis*, *Myxas glutinosa*, *Lymnaea stagnalis* and *Valvata piscinalis*. The speed of arrival of such a diverse assemblage of Mollusca is probably indicative of a rapid climatic amelioration.

10.3.3 Introduced species

The canalization of rivers and construction of canals to join many catchments since the late eighteenth century, together with the introduction of exotic species, has led to modifications in the molluscan fauna of the British Isles. Canal construction has afforded new migratory routes and led to the homogenization of faunas across lowlands (Boycott 1936). Freshwater introductions have been few and accidental but two have made spectacular progress. *Dreissena polymorpha*, a native of eastern Europe, introduced in the 1830s, and *Potamopyrgus jenkinsi* from New Zealand that entered *circa* 1890, have already colonized many lowland waters (Kerney 1976). A third introduction, from the southern United States of America, *Physa* cf *acuta*, has also been able to maintain a foothold in rivers and canals where waste heat pollution from industrial sources has raised the natural temperature.

10.4 Marine Mollusca

Estuarine and marine deposits associated with sea-level movements during the Flandrian (section 10.11) have yielded information on marine molluscs (Figure 10.7). Praeger (1896) has examined molluscan remains in the estuarine clays of Belfast Lough. The fauna, including species that currently reside in more southerly and westerly locations around the British Isles, has been interpreted as indicative of warmer conditions during the Atlantic Period of the Postglacial. At Fawley on Southampton Water, an assemblage consisting mainly of inter-tidal mud-dwelling species, but also containing sub-tidal and salt-marsh ones has been recovered (Hodson and West 1972). *Scrobicularia plana*, *Hydrobia ulvae*, *Littorina littorea*, *Macoma balthica* and *Mytilus edulis* are among this fauna. Other marine molluscan faunas of this age and affinity have been reported from Essex (Greensmith and Tucker 1973) and south-western Scotland (Jardine 1975).

The other major source of data concerning marine Mollusca has been coastal archaeological sites. The Mesolithic midden at Westward Ho! in north Devon contains inter-tidal shells, mainly of *Ostrea*, *Mytilus*, *Patella*, *Littorina*, *Purpurea* and *Scrobicularia* (Churchill 1965a). At Morton Tayport in Fife (section 10.1), a Later Mesolithic shell-midden is rich in species. The most frequent are *Cerastoderma edule* and *Macoma balthica* from intertidal mud and sand, while rocky shore and offshore habitats are also represented (Coles 1971). On the Hebridean island of Oronsay, *Patella* spp., *Littorina littorea* and *Nucella lapillus* are especially important in Mesolithic middens (Mellars 1978).

10.5 Vertebrates

Stuart (1982) has synthesized knowledge on Flandrian vertebrates (Table 10.4) (Figure 10.7). Large mammals such as *Mammuthus primigenius* and *Coelodonta antiquitatis*, which were characteristic of the Devensian (Chapter 9), have not been recorded from the current interglacial. Failure to adjust to the boreal forest ecosystem that succeeded that of tundra (Kowalski 1967) may have been at least a partial cause, while hunting of these slow-breeding animals by Upper Palaeolithic man during the Late Devensian could also have contributed to their demise (Stuart 1974, 1982).

Flandrian faunal history has included both wild and domesticated species. As Stuart (1982) has pointed out, non-domesticate faunal history has been influenced by three major events. One of these was the severance of land links between both the British Isles and mainland Europe, and between various parts of the British Isles, as sea-level rose (section 10.11), thereby preventing immigration of terrestrial and freshwater vertebrates. Within the British Isles, Ireland was especially affected and its faunal history is afforded separate treatment after that relating to most of the remainder of the territory. The other events were extinction, notably of large mammals, and the introduction of exotic animals, by man.

The Mesolithic sites of Thatcham and Star Carr (section 10.1) have yielded analogous faunas. At Thatcham, whose radiocarbon ages range from *circa* 10300 to 8500 BP (Wymer 1962; Churchill 1962), *Bos primigenius*, *Sus scrofa*, *Meles meles* and *Erinaceus europaeus* (hedgehog) have been recorded, together with *Oryctolagus cuniculus* (rabbit) and *Equus ferus*. The Star Carr locality, with radiocarbon dates between *circa* 9800 and 9000 BP (Clark 1954, 1972; Cloutman and Smith 1988), possesses all but the two last-named species. As Stuart (1982) has stressed, the faunas of such sites have mainly reflected animals taken during hunting, so that small vertebrates are poorly represented. Dogs (*Canis familiaris* – domesticated wolves) were used by man in both

Table 10.4 Present-day native British fauna of terrestrial vertebrates. From Stuart (1982). Reproduced by permission of Longman Group UK Ltd.

Amphibia (Smith 1969)
- *Triturus cristatus* (Laurenti), warty newt
- **Triturus vulgaris* (L.), smooth newt
- *Triturus helveticus* (Razoumoski), palmate newt
- *Bufo bufo* (L.), common toad
- **Bufo calamita* Laurenti, natterjack
- **Rana temporaria* L., common frog

Reptilia (Smith 1969)
- *Anguis fragilis* L., slow-worm
- *Lacerta agilis* L., sand lizard
- **Lacerta vivipara* Jacquin, viviparous lizard
- *Natrix natrix* (L.), grass snake
- *Coronella austriaca* Laurenti, smooth snake
- *Vipera berus* (L.), adder

Mammalia (modified from Corbet and Southern 1977) (Includes species extinct within last 2500 years, shown by †)

Insectivora
- *Sorex araneus* L., common shrew
- **Sorex minutus* L., pigmy shrew
- *Neomys fodiens* (Pennant), water shrew
- **Erinaceus europaeus* L., hedgehog? Introduced in Ireland
- *Talpa europaea* L., mole

Lagomorpha
- *Lepus capensis* L., brown hare. Introduced recently in Ireland
- **Lepus timidus* L., mountain hare

Rodentia
- **Sciurus vulgaris* L., red squirrel? Introduced in Ireland
- †*Castor fiber* L., beaver. Extinct twelfth century or later
- *Muscardinus avellanarius* (L.), dormouse
- *Clethrionomys glareolus* (Schreber), bank vole. Introduced in Ireland (probably very recently)
- *Arvicola terrestris* L., water vole
- *Microtus agrestis* (L.), field vole
- *Micromys minutus* (Pallas), harvest mouse
- **Apodemus sylvaticus* (L.), wood mouse Introduced in Ireland
- *Apodemus flavicollis* (Melchior), yellow-necked mouse

Carnivora
- †**Canis lupus* L., wolf. Extinct eighteenth century
- * *Vulpes vulpes* (L.), red fox
- † *Ursus arctos* L., brown bear. Extinct about tenth century
- * *Martes martes* (L.), pine marten
- * *Mustela erminea* L., stoat
- *Mustela nivalis* L., weasel
- *Mustela putorius* L., polecat
- * *Meles meles* (L.), badger
- * *Lutra lutra* (L.), otter
- *Felis sylvestris* Schreber, wild cat

Artiodactyla
- †**Sus scrofa* L., wild boar. Extinct probably seventeenth century
- **Cervus elaphus* L., red deer
- *Capreolus capreolus* (L.), roe deer
- †*Bos primigenius* Bojanus, aurochs. Extinct by 2500 years BP?

* Also occurs in Ireland
† Extinct

Berkshire and Yorkshire at this time (Simmons *et al.* 1981) (Table 10.5).

Small mammals were present in a Boreal fauna from Nazeing in the Lea Valley, Essex (Allison *et al.* 1952). Included in a largely woodland assemblage were *Sorex araneus* (common shrew), *Clethrionomys glareolus*, *Arvicola terrestris*, *Microtus oeconomus*, *Ochotona pusilla* (pika) and *Apodemus sylvaticus*. The pika and northern vole had been present in the Late Devensian, and must have found suitable early Flandrian habitats (Sutcliffe and Kowalski 1976). It should be noted, however, that faunas which have been dated to the outset of Flandrian time also contain species presently characteristic of temperate forests, and whose northerly distributional limits are in southern Fennoscandia. The presence of *Cervus elaphus*, *Talpa europaea* (mole), together with badger and wild boar, could be a reflection of a rapid amelioration of climate (Stuart 1982) (section 10.13).

Among the avifauna, ducks, including *Bucephala clangula* (golden-eye), grebes, divers and mergansers were present during the Boreal, when *Buteo buteo* (buzzard), *Tyto alba* (barn owl), lapwings and the stock dove also resided in the British Isles (Bramwell 1960). The rivers of south-eastern England were stocked by primary freshwater fish (those not migrating to the sea), including *Esox lucius* (pike) and *Perca fluviatilis* (perch), at this time. Colonization probably occurred from the European mainland through rivers that crossed the land connection of this area (Wheeler 1977, 1978).

Isolation from continental Europe took place *circa* 8000 BP (section 10.11). Few modifications to the resident vertebrate fauna ensued. Typical members of this in the Atlantic Period, when mixed-deciduous forest was widespread (section 10.2), were *Bos primigenius*, *Capreolus capreolus* (roe deer), *Cervus elaphus*, *Canis lupus*, *Ursus arctos* and *Lutra lutra* (otter). Closed-deciduous forest would probably not have been an ideal habitat for deer and aurochs, but the latter seems to have been frequent during the Atlantic (Grigson 1978). Elk became extinct about the conclusion of the Boreal. Succession of birch–pine woodland by that of a mixed-deciduous kind is likely to have contributed to its disappearance, as would have hunting (Simmons *et al.* 1981). Birds now included *Botaurus stellaris* (bittern), *Cygnus cygnus* (mute swan), *Haliaëtus albicilla* (white-tailed eagle), *Grus grus* (common crane) and *Alca torda* (razorbill) (Northcote 1980). *Emys orbicularis* was present in East Anglia during the Atlantic Period (Stuart 1979).

Information concerning wild fauna during the Sub-Boreal Period has been obtained mainly from Neolithic and Bronze Age archaeological sites (Smith *et al.* 1981a). The data indicate the persistence into the Early Neolithic of animals already present earlier in the Flandrian. Red deer was widely distributed, including in the Outer Hebrides and Orkney, and wild boar existed. The initial layers excavated at Windmill Hill (section 10.1) contain *Felis sylvestris* (wild cat), *Meles meles*, *Vulpes vulpes* (fox) and frog remains (Jope 1965; Grigson 1965), and probably also *Lepus capensis* (brown hare) (Smith *et al.* 1981a). The West Kennet long barrow in Wiltshire has yielded *Mustela putorius* (polecat) and *Castor fiber* among its mammalian fauna, and *Turdus merula* (blackbird) and *Corvus monedula* (jackdaw) from the Aves (Clark 1962). Dowel Cave in Derbyshire has produced remains of *Sciurus vulgaris* (red squirrel) referable to the Early Neolithic (Bramwell 1959). Bones of *Myotis bechsteinii* (Bechstein's bat), *Myotis daubentonii* (Daubenton's bat), *Myotis nattereri* (Natterer's bat) and *Myotis mystacinus* (whiskered bat), which now utilize woodland but overwinter in caves, have also been found in this locality, and at Grime's Graves (10.1). The latter site has additionally produced *Microtus agrestis*, *Clethrionomys glareolus*, *Sorex araneus*, *Apodemus sylvaticus* and *Talpa europaea*, together with remains of *Bufo bufo* (toad) and *Natrix natrix* (grass snake) (Clarke 1922).

Although it is possible that Mesolithic people herded and culled red deer (Simmons *et al.* 1981), perhaps feeding them with ivy (Simmons and Dimbleby 1974), the first evidence of domesticated animals other than the dog in the British Isles has come from Early Neolithic sites. Bones of cattle and sheep have been recovered from a barrow at Lambourn in Berkshire, which have produced a radiocarbon date of 5365 ± 180 BP (Wymer 1965). Although *Bos primigenius* and *Sus scrofa* resided wild, in view of their smaller size it seems likely that domesticated cattle and pigs were brought to the British Isles from mainland Europe, rather than *in situ* breeding having taken place. As sheep and goats lacked native wild forms, they must have been imported (Smith *et al.* 1981a). The domestic economy in southern England was dominated by cattle, according to the bone records of archaeological sites. Cattle are prominent in the faunal assemblages from Willerby Wold and Kilham barrows in eastern Yorkshire (Bramwell 1963, 1976). Pigs, of little importance in southern England at this time, may have assumed greater significance in more northerly locations (Smith *et al.* 1981a).

As Later Neolithic and Bronze Age peoples relied increasingly upon domesticates, wild animal remains from these times have been infrequently encountered, but little change from earlier Sub-Boreal time has been evident. *Microtus arvalis* (Orkney vole) has been recorded from the settlement of Skara Brae on Mainland (Clarke 1976) and from the chambered tomb of Midhowe on Rousay (Renfrew 1979; Henshall 1985) in the Orkney Islands, where it occurs today. Its only other present location in the British Isles is Guernsey, but it is found on the European mainland (Berry and Rose 1975). Its nearest genetic relation has been traced to Yugoslavia, and Berry (1985) has suggested introduction to Orkney from this area during the Neolithic.

Overkill by man probably caused the disappearance of the aurochs in the Later Bronze Age (Tinsley and Grigson 1981). Skeletal remains of this animal from Blagdon in Somerset have been radiocarbon dated to 3245 ± 37 BP (Burleigh and Clutton-Brock 1977). Wolves and brown bears ranged over considerable areas during the Bronze Age, although heightened forest clearance and more agriculture led to both decreased habitats for wild animals and a reduced requirement for hunting (Tinsley and Grigson 1981).

Late Neolithic domesticates appear to have been dominated by pigs and cattle, with the latter perhaps increasing in importance during the Early and Middle Bronze Age. The Late Bronze Age saw more sheep, and pigs declined in importance (Tinsley and Grigson 1981). At Elson's Seat, Encombe, Dorset, sheep accounted for 42% and pigs 4% of a Late Bronze Age fauna which also contained domesticated horse (Phillipson 1968). By the Middle Bronze Age, *Bos longifrons*, a smaller domesticated ox than that characteristic of the Late Neolithic (Jewell 1963), was predominant.

The wild animals of the Iron Age comprised a fauna analogous to that of the present day, although some taxa present then, including wolf and beaver, are now extinct in the British Isles (J. Turner 1981). Also as woodland clearance continued, animals such as deer assumed less significance. The reduction in tree cover also appears to have shifted the numerical balance among domesticates. Cattle and pigs, whose ancestors were woodland inhabitants, were surpassed in importance by sheep. Then, as now, the latter would have been well suited to unforested terrain, and increased in number as more pasture was created (Clark 1952).

As remarked at the start of this section, the Littletonian fauna of Ireland has been limited (van Wijngaarden-Bakker 1974, 1985). According to Devoy (1985a), a land bridge between Ireland and the remainder of the British Isles across the Malin Sea (Figure 9.23) is unlikely to have existed later than *circa* 10 200 BP. Tooley (1980) has indicated that the Irish Sea was in existence by about 9200 BP. There was thus a very restricted time for migration in the early Flandrian, so that native animals which subsequently arrived in Ireland must have done so by chance. For example, *Sus scrofa*, an early Postglacial resident, can swim strongly. Red deer appear to have immigrated during the Atlantic and badgers in the Sub-Boreal. Corbet (1961), Savage (1966) and Yalden (1982) have considered that *Apodemus sylvaticus* was introduced by humans. However, Preece *et al.* (1986) have found teeth of this species in a buried soil radiocarbon dated to younger than *circa* 7600 BP at Newlands Cross in County Dublin, and have argued for its native status. Elk, aurochs, roe deer, beaver, polecat, brown hare, mole, common shrew, weasel and voles failed to colonize. Primary freshwater fishes have never occurred naturally. However, as in the remainder of the British Isles, migratory *Salmo salar* (anadromous salmon), *Salmo trutta* (anadromous trout) and *Anguilla anguilla* (catadromous eel) penetrated rivers in Ireland. The sole native amphibian is *Triturus vulgaris* (common newt) and reptile, *Lacerta vivipara* (common lizard). As it has been able to disperse to better effect, the avifauna is reasonably diverse. Humans have introduced

Table 10.5 The anthropogenic element of the Irish terrestrial mammal fauna. From van Wijngaarden-Bakker (1985); sources, Moffat (1938), Claassens and O'Gorman (1965), Savage (1966), Deane and O'Gorman (1969), van Wijngaarden-Bakker (1974)

Species	*Type of introduction*	*Date of introduction*
Erinaceus europaeus, hedgehog	Accidental?/deliberate?	Norman times?
Canis familiaris, dog	Deliberate (domestic)	≈6500 BC?
Mustela vison, mink	Deliberate	1950 onward
Felis catus, cat	Deliberate (domestic)	First centuries AD
Cervus nippon, sika deer	Deliberate	1884
Dama dama, fallow deer	Deliberate	Norman times?
Sus domesticus, pig	Deliberate (domestic)	≈3500 BC
Bos taurus, cattle	Deliberate (domestic)	≈3500 BC
Ovis aries, sheep	Deliberate (domestic)	≈3500 BC
Capra hircus, goat	Deliberate (domestic)	≈3500 BC
Equus caballus, horse	Deliberate (domestic)	≈2000 BC
Equus asinus, donkey	Deliberate (domestic)	Eighteenth century AD
Lepus capensis, brown hare	Deliberate	1850 onward
Oryctolagus cuniculus, rabbit	Deliberate	Thirteenth century AD
Sciurus carolinensis, grey squirrel	Deliberate	1890, 1911
Glis glis, edible dormouse	Deliberate	1885 (unsuccessful)
Mus musculus, house mouse	Accidental	Roman times or earlier
Rattus rattus, black rat	Accidental	Before AD 1187
Rattus norvegicus, brown rat	Accidental	1722 onward
Clethrionomys glareolus, bank vole	Accidental	1964
Ondatra zibethica, musk rat	Deliberate	1927 (extinct 1934)

21 wild or domesticated terrestrial mammals to Ireland (Table 10.5). The earliest domesticate was *Canis familiaris*, present at Mount Sandel (section 10.1) *circa* 8500 BP (van Wijngaarden-Bakker 1985).

10.6 Insects

By comparison with the insect faunas of the Devensian, those of Flandrian age have been less studied. Coope (1979) has stated that the interpretation of Postglacial climatic changes from fossil Coleoptera has been hindered by a preponderance of sediments which have accumulated in acid conditions, and thereby contain limited and rather uniform faunas. Attention has also been drawn to the hazards involved in distinguishing natural from human-induced changes in faunas. Osborne (1976) has suggested that the lack of marked oscillations in climate during the Postglacial by comparison with the Devensian Lateglacial militates against the usefulness of insects for the detection of climatic changes in the former. Also, severance of the land link with mainland Europe placed restrictions on the coleopteran fauna of the British Isles. Moreover, human impact in later Flandrian time has led to modifications in the distribution of insects, and in some cases, to their extinction. Sites discussed below are shown in Figure 10.7.

The notion of a progressive amelioration in climate during the early Flandrian, culminating in optimum conditions *circa* 7000 BP (section 10.13), has been challenged by insect data from Lea Marston in northern Warwickshire (Osborne 1974). In deposits radiocarbon dated to *circa* 9500 BP, an insect fauna lacking a cold element, and which would have been able to inhabit woodland dominated by oak and beech in southern England today, is present. Only 10% of the fauna can be found in current arctic and alpine environments in Fennoscandia, while there are three species now

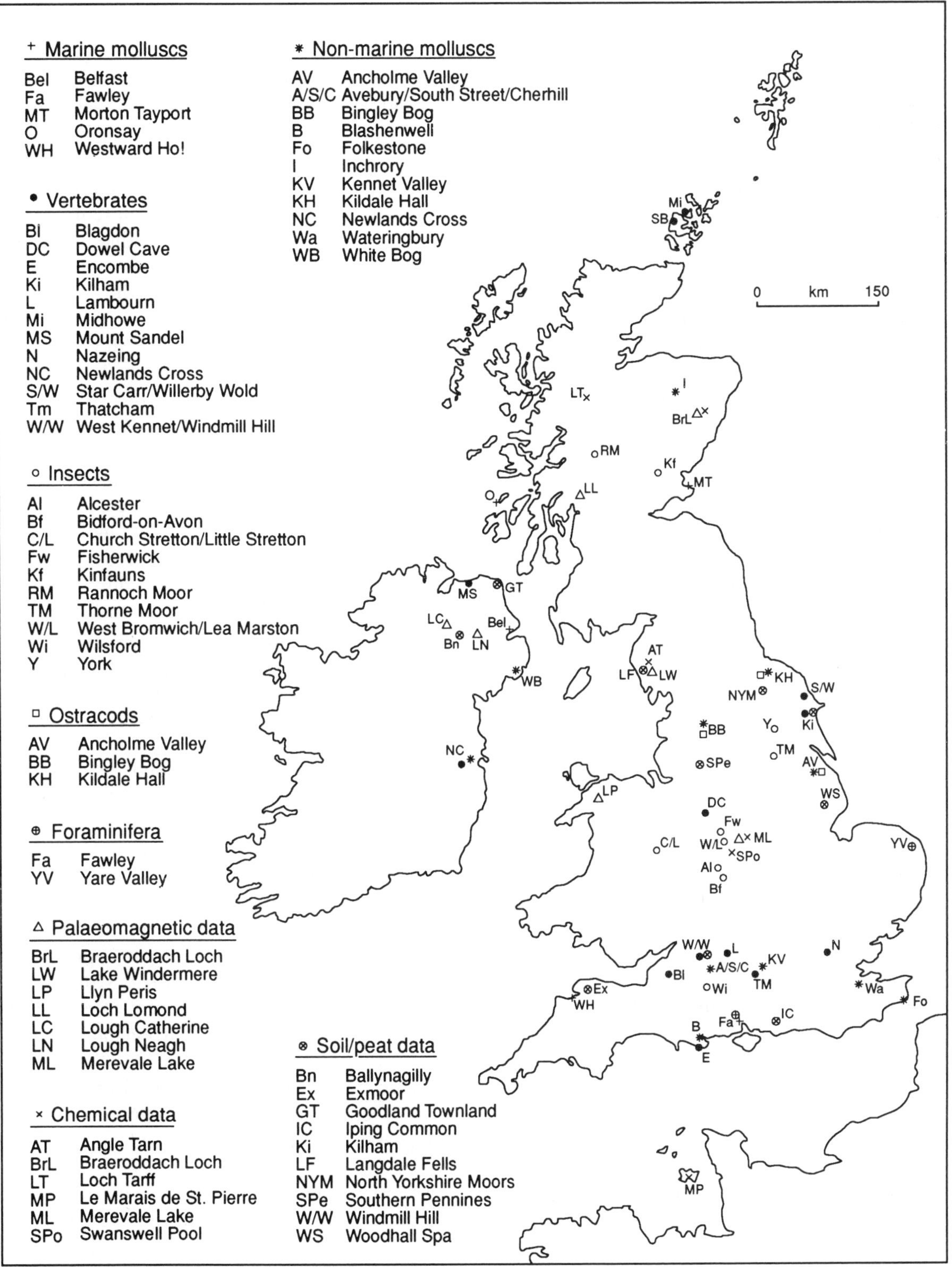

Figure 10.7 Localities referred to in sections 10.3, 10.4, 10.5, 10.6, 10.7.2, 10.7.3, 10.8 and 10.9 with types of evidence.

absent from southern England but occurring on mainland Europe south of this area. Pollen data from the locality do not reveal the incidence of *Quercus–Fagus* woodland at this time, neither has other palynological information from the English Midlands (section 10.2.2). It has been noted by Osborne, however, that the assemblage of insects which is diagnostic of the increased temperatures is not now dependent upon thermophilous broad-leaved trees for sustenance. Thus, a combination of innate mobility and suitable temperatures could have meant that insects reached northern Warwickshire before oak and beech trees, instead inhabiting the birch–willow woodland indicated by the pollen record. A deposit from West Bromwich in the Birmingham conurbation, radiocarbon dated to 9970 ± 110 BP, lacks a cold element and contains temperate insects. Above this, organic material dated to 9640 ± 100 BP possesses a temperate fauna analogous to that at Lea Marston (Osborne 1976). Higher in the sequence, sediment with a radiocarbon age of 9080 ± 455 BP reveals a continuing tendency towards a temperate woodland fauna, although a number of species associated with coniferous trees have been encountered. On Rannoch Moor in Scotland, the most abundant insect in an assemblage from the initial sediment following retreat of Loch Lomond Stadial ice (Chapter 9) is *Ochthebius foveolatus*, currently resident in the Mediterranean–central European areas (Coope 1979). After the conclusion of the Loch Lomond Stadial, only a small number of insect species tolerant of cold conditions, and which are now restricted to mountains in Scotland, remained in the British Isles (Osborne 1976). It probably took only a few centuries *circa* 10 000 BP for arctic–alpine assemblages of Coleoptera to vacate much of the British Isles and to be supplanted by aggregates of fully temperate species. Thus, steady progress towards a Postglacial climatic optimum is thought to have been unlikely (Figure 10.8) (section 10.13). The concept of a gradual move to an optimum climate is considered to have been related to the pattern of vegetation succession, which traditionally has been employed in its recognition. Compared with insects, plants respond more slowly to climatic changes and are thus less reliable as an indicator of them (Coope 1979).

Insect evidence has allowed the inference to be made that apart from a rapid rise in temperature about its opening, the tenth millennium BP lacked climatic oscillations (Osborne 1976). At Alcester in southern Warwickshire, organic deposits above river gravel have been radiocarbon dated to 8500–8300 BP (Shotton *et al.* 1977b). The insect fauna from this organic material is analogous to that of the British Isles today, save for *Cathormiocerus validiscapus*, now distributed from central France to the middle of Spain. A climate at least as temperate as present, possibly a little more so, has been suggested. Pollen data from the sediment indicate the existence of woodland dominated by pine and hazel. At Little Stretton in Shropshire, deposits dated to 8101 ± 138 BP contain the first Flandrian record of *Rhynchaenus quercus*, an obligate of oak, in the lower part of the sequence. Higher up, an increase in wood-boring insects, principally of deciduous, but some of coniferous, trees, occurs (Osborne 1972). Little Stretton, and a locality at Church Stretton, have yielded insect faunas representative of the period *circa* 7000–5000 BP (Osborne 1972). An increased number of species is evident. All of them are resident in England at present. No faunal modifications, such as range shifts or extinctions, that could have been induced by climatic changes, have been noted in this part of Shropshire over the timespan represented. Insight into woodland characteristics during the late Atlantic has been given by a deposit at Kinfauns in Perthshire, radiocarbon dated to 5180 ± 100 BP (Osborne 1978). A woodland fauna included two *Cerylon* species that live under the bark of dead logs. *Anobium* (woodworm) was also present, as was *Apoderus coryli*, a weevil that dwells on hazel.

The initial sign of a climatic oscillation later in the Flandrian comes from an organic deposit at Shustoke, about 5 km south of Lea Marston in northern Warwickshire, which has been radiocarbon dated to 4830 ± 100 BP (Kelly and

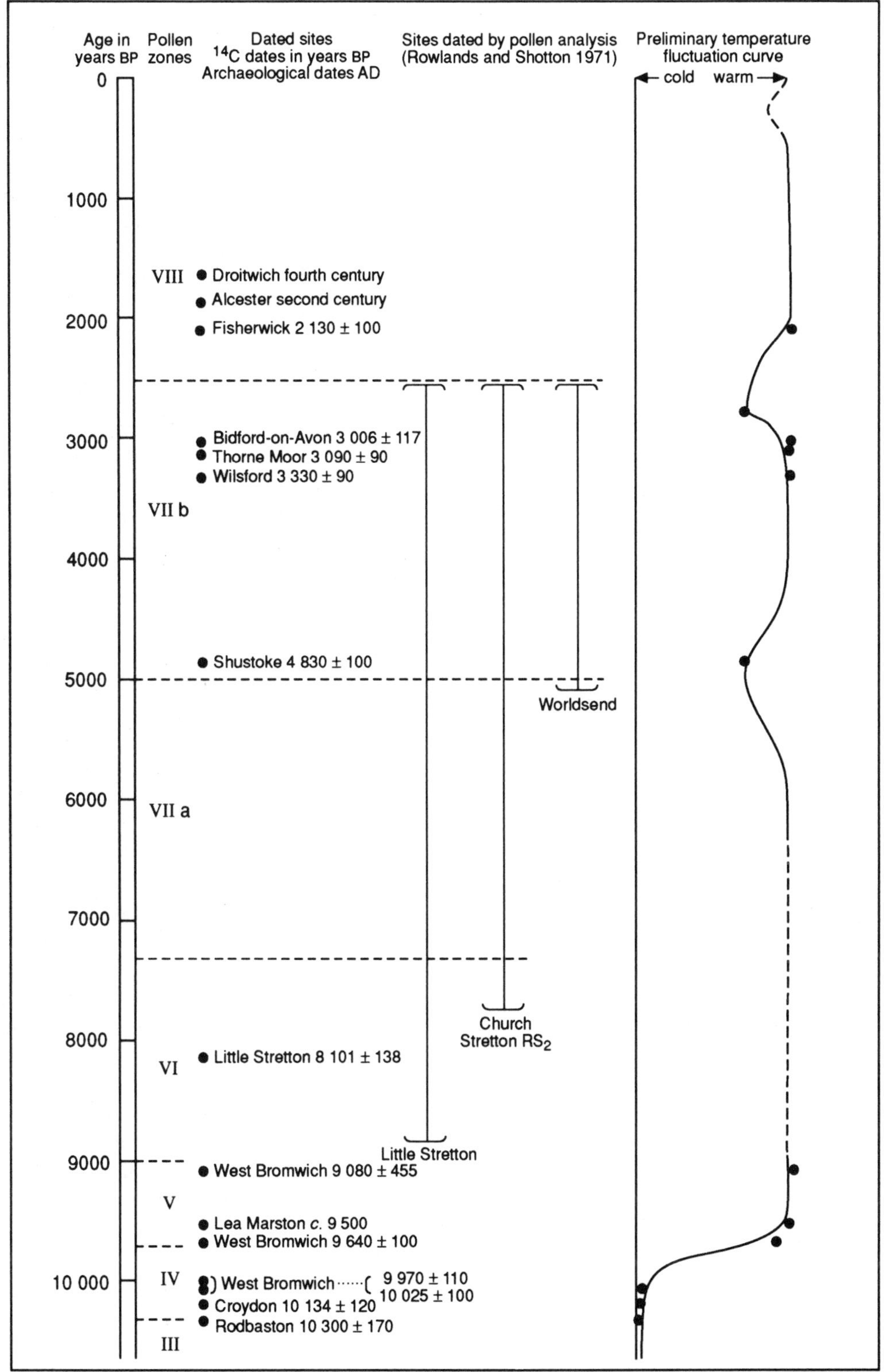

Figure 10.8 Insect evidence of climatic variation during the Flandrian Stage. From Osborne (1976).

Osborne 1964). The insects were mainly those of deciduous woodland, and most continue to inhabit the Midlands of England. Three species now with more southerly distributions have been thought to have subsequently disappeared from north Warwickshire as a result of habitat modification, and not to have been indicative of a warmer climate than today *circa* 5000 BP. Indeed, four insects, *Bembidion atroviolaceum*, *Sphaerites glabratus*, *Tropiphorus obtusus* and *Barynotus squamosus*, possess current ranges within the British Isles that are mainly in Scotland. Their presence has been interpreted as probably signifying a decline in temperatures at this time (Osborne 1976) (Figure 10.8).

The beetles from deposits in a Late Bronze Age shaft at Wilsford, near Amesbury in Wiltshire, have been reported by Osborne (1969). Material radiocarbon dated to 3330 ± 90 BP contains a thermophilous fauna. There are many individuals of two dung beetles, *Onthophagus fracticornis* and *Onthophagus nutans*, that no longer reside in the British Isles, but to their south. The same is true of *Aphodius quadriguttatus*, also a dung beetle, recovered at this site. It has been thought unlikely that dung dwellers would have disappeared subsequently in response to a reduced tree cover. Indeed, it has been hypothesized that they would have been more likely to have increased in open conditions where domestic animals were frequent. Thus, summer temperatures above those of today are suggested from this evidence, with the later demise of these insects probably due to a climatic deterioration. An insect assemblage from pasture occurs above a woodland one at Kinfauns. Apart from dung beetles, *Adelocera murina* and *Phyllopertha horticola*, grassland inhabitants, are present (Osborne 1978). A fauna analogous to that of Wilsford has been reported from a deposit aged to 3006 ± 117 BP at Pilgrim's Lock near Bidford-on-Avon in Warwickshire (Osborne 1988).

Organic material dated to *circa* 3000 BP, from Worldsend near Church Stretton, shows insects characteristic of coniferous woodland increasing in importance relative to those of deciduous trees. One of these, *Agabus wasastjernae*, is now extinct in the British Isles but is distributed to their north in Scandinavia (Osborne 1972). A deterioration in climate has been inferred which, together with human activities, may have been responsible for the extirpation of numerous woodland insects resident in the British Isles. Such insects, whose current distribution is to the south and east of the British Isles, persisted on Thorne Moor, north-east of Doncaster in Yorkshire, until at least 3090 ± 90 BP (Buckland and Kenward 1973; Buckland 1976). Examples are *Prostomis mandibularis*, an insect characteristic of mature, undisturbed forest where it dwells on rotten wood, and *Rhysodes sulcatus*, which feeds in the same habitat. The fauna however, also contains *Zimioma grossa* and *Scolytus ratzburge*. Both are associates of tree bark and now possess boreo-montane ranges, the former outside the British Isles. Osborne (1976) has suggested that their existence supports the notion of climatic cooling at this time. Buckland (1976), while stating that temperatures may have reduced if these insects were present, inclined towards a more complex cause, involving both climatic and anthropic factors, in order to explain differences between distributions revealed at Thorne Moor around the time of the transition from the Sub-Boreal to the Sub-Atlantic and those of today.

Sediments from a late Iron Age site at Fisherwick in Staffordshire, radiocarbon dated to 2130 ± 100 BP, have produced an assemblage of insects that is devoid of indicators of cold conditions. One of its members, *Heptaulacus testudinarius*, was living some considerable distance north of its current range limit in the British Isles, indicating a climate at least as warm as that which now prevails in the area (Osborne 1976). Coleopteran faunas from archaeological sites of Roman and Anglo-Saxon age have revealed species which were distributed similarly to today (Osborne 1976).

Instances of insects introduced into the British Isles by man have come from Roman, Anglo-Saxon and later historical periods. The borer *Hesperophanes fasciculatus*, found at Alcester, and likely to have emanated from wood brought by Romans in the second century AD (Osborne 1971) has since disappeared from the British Isles (Osborne 1978). A sewer in York, built during the same century by its Roman inhabitants, has yielded insects associated with stored products, including *Sitophilus granarius* (grain weevil) – a Roman or earlier introduction. Flies now found on sewage farms are also well represented in this sewer (Buckland 1976). Anglo-Danish and later medieval localities in York have revealed coleopteran faunas denoting dirty conditions. *Musca domestica* (house flies) and *Stomoxys calcitrans* (biting stable-fly) have been recovered in abundance within a leather-worker's shop and tannery, where they originated from grubs nourished on waste flesh and fat (Buckland *et al.* 1974; Buckland 1976).

10.7 Other fossils

In this category, diatoms (Figure 10.3), foraminifera and ostracods (Figure 10.7), and tree rings have been fruitful sources of palaeoenvironmental information.

10.7.1 Diatoms

Compared with those of pollen and spores, analyses of diatom assemblages have been limited and have come mainly from lacustrine deposits. The diatom stratigraphy of Blea Tarn (section 10.2.6) has been described by Haworth (1969). The inception of organic matter deposition in the earliest Flandrian was accompanied by an abrupt expansion in the diatom flora, which was dominated by planktonic species. This has been interpreted as indicative of a rapid increase in temperature at the time. Several alkaliphilous taxa, including *Asterionella formosa*, *Gyrosigma acuminatum* and *Navicula dicephala*, were present throughout the Boreal Period, although the pH then was probably more or less neutral. Alkaliphilous species underwent declines or disappeared about the Boreal–Atlantic transition, when *Achnanthes minutissima* and *Synedra nana* came to dominance. At this time, *Pinnularia* spp. appeared while the frequency of *Eunotia* spp. rose, the overall assemblage being more tolerant of acid conditions. It has been thought that the gradual disappearance of some taxa may have meant the inwash of organic material which had been building up constantly in the lake catchment, rather than a sudden change in climate, was the cause of these floristic modifications. The Atlantic witnessed minor shifts in the flora, so that basiphiles disappeared. An increase in epiphytic species probably reflected the onset of *Sphagnum* bog formation adjacent to the lake. Adjustments to the flora during the Sub-Boreal have been interpreted as probably due to natural succession. Sub-Atlantic acidification consequent upon forest removal and soil erosion, as at Devoke Water (Evans 1961), has not been detected at Blea Tarn, nor has recent eutrophication, encountered in Windermere by Pennington (1943) and in Esthwaite Water by Round (1961).

Diatom sequences in the Blelham Basin in the Lake District have been reported by Evans (1970). Differences between the flora of the tarn and kettle hole have been recognized. The latter shows no increase in the ratio of planktonic to non-planktonic forms during the early Flandrian, while in the former, non-planktonic taxa also dominated at this time. The open water of the tarn was calcareous and included *Cocconeis diminuta* and *Achnanthes suchlandtii* during the early Flandrian. A water depth of at least 12 m has been estimated for the later Boreal. An increase in *Eunotia* in the kettle hole has been linked to the emergence of an increasingly acidic environment by the end of the Boreal. By the middle of the Atlantic Period, the pelagic zone of the tarn included over 70% of planktonic forms. Their increase has been regarded as most likely the result of heightened amounts of organic compounds and of nitrogen, probably due to woodland growth, with *Alnus* being responsible for enhanced levels of nitrogen. Alkalinity was reduced in the tarn during the Atlantic, when *Eunotia* and *Frustulia* spp. either appeared or expanded in occurrence. While such a trend also characterized the late Flandrian, the tarn water remained alkaline. A phase of eutrophication can be identified close to the present day. The frequencies of *Asterionella formosa* and *Cyclotella glomerata* are diagnostic of this, with inputs to the lake from fertilizers and sewage its most likely cause. The formation of *Sphagnum* peat in the kettle hole during the late Flandrian is indicated by the presence of a suite of acidophilic diatoms including *Eunotia*, *Frustulia* and *Pinnularia*.

Evans and Walker (1977) have described the diatom floras of Llyn Clyd and Llyn Glas in Snowdonia. The Boreal and much of the Atlantic at Llyn Clyd saw *Fragilaria brevistriata* dominance, indicative of alkaline water. In late Atlantic and most of Sub-Boreal time, *Cyclotella comensis*, evidence of more neutral water, was the most important member of the flora. In the latter part of the Sub-Boreal and in the Sub-Atlantic, when *Eunotia* spp. and *Frustulia* spp. were prominent, the water became increasingly acidic. This change took place later than at Esthwaite or Blelham. The record from Llyn Glas is less precise, but an early Flandrian alkaline status can be recognized, with a change to more acidic conditions occurring at about the same time as at Llyn Clyd. Both lakes are now acidic and oligotrophic. It has been suggested that during their alkaline phase, oligotrophy most likely existed, although slightly more nutrient-rich conditions may have existed.

Loch Sionascaig in Sutherland has produced an analogous diatom record to that of Blea Tarn. However, a greater degree of acidification is evident, with lowered base status of the soils in its environs thought responsible for initiating acidic conditions within the first millennium of the Flandrian (Pennington *et al.* 1972).

Flower and Battarbee (1983) have reported diatom evidence of recent acidification of Round Loch of Glenhead in Galloway, whose surface water currently has a pH of 4.5–5.0 (Figure 10.9(a)(b)). Prior to *circa* AD 1600, when the flora included *Cyclotella kutzingiana* and *Achnanthes microcephala*, the pH ranged from 5.5 to 6.0. *Circa* AD 1850, *Achnanthes microcephala*, *Fragilaria virescens* and *Anomoeoneis vitrea* declined and were substituted by *Eunotia* spp. and *Tabellaria* spp. This change has been interpreted as the result of acid deposition. There are heavy metals and particles of soot in the lake sediments of this age, while evidence of a change in land use that could have promoted acidification (section 10.9) is lacking (Battarbee *et al.* 1985; Battarbee and Charles 1987). Additional information on pH changes at Round Loch of Glenhead has been provided by Flower *et al.* (1987). No significant shift in pH prior to the decade AD 1870–1880 has been deduced. From a value of 5.9, a swift fall ensued, with 4.9 attained by 1945 and a minimum of 4.8 by *circa* AD 1960. This level persisted until AD 1980 (Figure 10.9(b)). Battarbee *et al.* (1988) have demonstrated, from the diatom sequence, that the pH of Round Loch of Glenhead had increased slightly since about the latter time (Figure 10.9(a)). No further acidification of this still highly acidified lake could be detected after the middle of the previous decade. Diatom assemblages presented by Jones *et al.* (1989) chart environmental changes throughout the Flandrian at Round Loch of Glenhead. Six biozones have been delimited. The early Flandrian was dominated by circumneutral and acidophilous forms, a situation that continued into the later part of the stage (Figure 10.9(c)). A pH value of 5.0–5.5 until *circa* AD 1870 is indicated, with subsequent changes supporting those described above.

Diatom evidence has been used by Battarbee (1986b) to illustrate the eutrophication of Lough Erne, which extends north-west and south-east of Enniskillen in Ulster. Its recent sediments have been dated by ^{137}Cs (Pennington *et al.* 1973; Wise 1980) and ^{210}Pb (Eakins and Morrison 1978; Wise 1980) analysis (Oldfield *et al.* 1983). A rise in the ratio of planktonic to non-planktonic forms, plus one in diatom accumulation rate, signal the beginning of this episode at about the start of the current century. A subsequent, greater enhancement by nutrients has taken place since AD 1950. Increases in frequency of *Stephanodiscus* are an especially valuable indicator of eutrophication. For at least a millennium prior to *circa* AD 1900, a stable diatom flora, in which *Melosira* (*Aulacoseira*) spp. were important, existed in Loch Erne, an alkaliphilous and mesotrophic

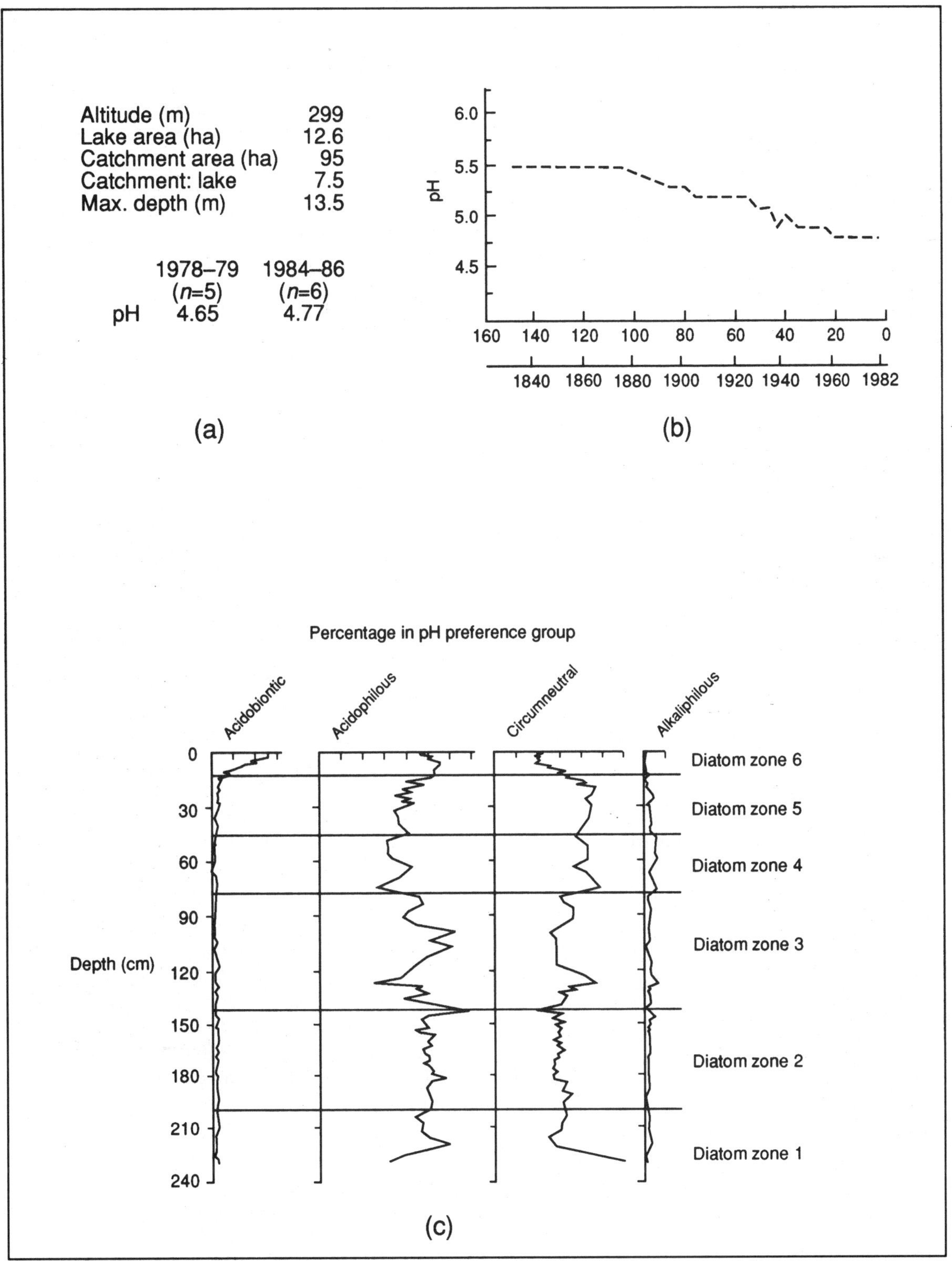

Figure 10.9 Round Loch of Glenhead. (a) Physical and pH characteristics (1978–1979 and 1984–1986). From Battarbee *et al.* (1988). Reprinted with permission from *Nature* **332**, 530–532; © 1988 Macmillan Magazines Ltd. (b) Recent pH history. From Flower *et al.* (1987). (c) Percentages of the total diatom counts grouped according to their Hustedt pH preference group in core RLGH3. From Jones *et al.* (1989).

water body. Analogous modifications early during the twentieth century AD in Lough Neagh have been deduced by Battarbee (1978).

At Loe Pool, Helston, Cornwall, laminated lake sediments deposited during approximately the last 110 years have revealed repeated sequences of diatoms (Simola *et al.* 1981). A Spring–Summer and Autumn–Winter succession has been detected in six phases from the surface to 7.2 cm depth in the sediments, by a change from *Thalassiosira pseudonana* and *Cyclotella meneghiniana* to *Asterionella formosa* and *Melosira* spp.

Jones *et al.* (1987, 1990) have described a diatom sequence from Le Marais de St. Pierre in St. Aubin's Bay, Jersey, covering approximately the last 4500 years (Figure 10.10). Freshwater, brackish and marine episodes are identifiable. During the major marine incursion, the flora was dominated by *Melosira westii* and *Paralia sulcata*.

10.7.2 Ostracoda

Early Flandrian tufa in the Ancholme Valley, Lincolnshire, possesses an ostracod fauna dominated by *Psychrodromus olivaceus* and *Eucypris pigra*, non-swimming forms typical of calcareous springs today. The ostracod assemblage lacks a clear pattern of succession but provides an insight into shifts within the local environment of deposition (Preece and Robinson 1984). An early Flandrian shell marl at Kildale Hall in north-east Yorkshire includes *Candona candona* and *Candona marchica*. A still-water lake with organic-rich litter on its bed is indicated (Keen *et al.* 1984). The early Flandrian ostracod record from Bingley Bog in Airedale contains *Candona candida*, *Herpetocypris reptans* and *Notodromas monacha*. A more or less stable lacustrine environment where aquatic plants were growing has been diagnosed (Keen *et al.* 1988).

10.7.3 Foraminifera

Estuarine deposits which began to form *circa* 3689 BP at Fawley in Hampshire have produced a foram assemblage that includes *Protelphidium anglicum*, *Ammonia beccarii* and *Elphidium articulatum*. Hyposaline, shallow water with temperatures analogous to those of today, is indicated (Hodson and West 1972). The presence of *Prototelphidium anglicum*, *Elphidium magellanicum* and *Ammonia beccarii* in marine deposits of the Yare Valley, Norfolk, is diagnostic of sedimentation either in the low inter-tidal zone or below *circa* 7500 BP. The marine environment withdrew about 5000 BP; with *Elphidium williamsonii*, *Millammina fusca*, then *Jadammina macrescens* and *Trochammina inflata* signifying progression from low to high salt-marsh in the area (Coles and Funnell 1981).

10.7.4 Tree rings

The chronological implications of tree rings are outlined in Chapter 1. Here it should be noted that information on palaeoclimates can also be gained from ring patterns, with estimates of Flandrian temperatures and rainfall having been made (Pilcher and Hughes 1982; Briffa *et al.* 1983).

Most data on tree rings in the British Isles have come from Ireland, whose dendrochronological record has been discussed by Baillie and Pilcher (1985). Remains of *Quercus robur* and *Quercus petraea*, whose rings have been considered of similar significance (Baillie 1982), have been discovered for almost the entire last 8000 year period. Sub-fossil oaks in peat bogs and constructional timber of long standing have provided the samples (Pilcher *et al.* 1977). Tree-ring chronology was developed in Ireland in response to the need for calibration of the radiocarbon timescale (Pearson *et al.* 1977, 1983; Pearson and Baillie 1983). This has resulted in the emergence of two overlapping chronologies. The first, extending from the present to *circa* 500 BC, has been derived from studies of historic and prehistoric constructional timbers. The second, ranging from the present as far as possible into the past, has utilized sub-fossil oaks. Apart from a number of hiatuses during the first millennium BP, a full dendrochronology has been established for Ireland which spans more than 7000 years

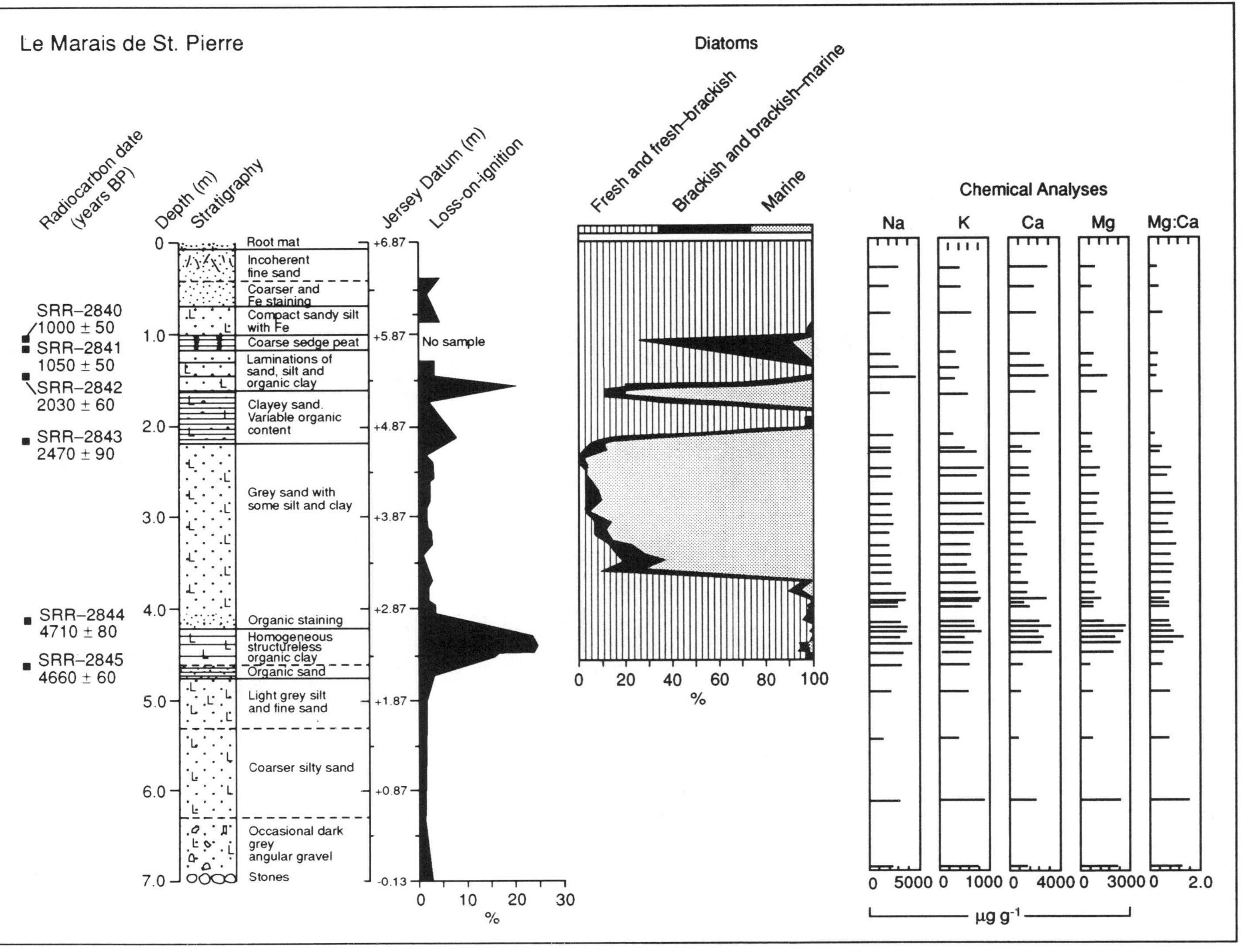

Figure 10.10 Palaeoenvironmental data from Le Marais de St. Pierre. From Jones *et al.* (1987). Reprinted with permission from *Progress in Oceanography* **18**, R. L. Jones *et al.*, Holocene sea-level changes on Jersey; © 1987 Pergamon Press Plc.

(Baillie and Pilcher 1985). In respect of basic climatic information, ring-pattern studies by Baillie (1973) in the Lough Neagh Basin and Pilcher and Baillie (1980) throughout Ireland have allowed the inference of its uniformity.

In Scotland, Bridge *et al.* (1990) have employed tree-ring evidence as part of a study of the history and climatic significance of *Pinus sylvestris* on southern Rannoch Moor, leading to the conclusion that during much of the Flandrian, pine trees grew in very low densities in stands that rarely persisted for in excess of 125 years.

10.8 Non-biological data

These have mainly been derived from studies of the magnetic and chemical characteristics of Flandrian sediments. Localities referred to during this section are shown in Figure 10.7.

10.8.1 Magnetism

Thompson (1974) and Thompson and Oldfield (1986) have described the palaeomagnetic parameters of relative declination, inclination, intensity and susceptibility, and their measurement. Superimposed upon longer-term trends such as reversals and excursions, have been secular variations in the magnetic field. The latter have been short term and local, involving oscillations in inclination or declination of 20–30° over 2000–3000 years.

Mackereth (1971) has obtained a Flandrian magnetic profile from sediments in Lake Windermere, upon which the oscillations were calibrated by radiocarbon dates. Thompson (1973) and O'Sullivan *et al.* (1973) have produced a record from the sediments of Lough Neagh that covers the period from *circa* 2000 BP to AD 1820. Hirons *et al.* (1985) have reported that the Lough Neagh sequence has been extended back to *circa* 6000 BP. The magnetostratigraphy of Lough Catherine in County Tyrone has also been studied (Thompson and Edwards 1982). A sequence from Loch Lomond (Dickson *et al.* 1978) (Figure 10.11(a)) has yielded the most information. It exhibits established traits, and has been aged to beyond 7400 BP (Figure 10.11(b)). Of particular significance is recognition of the maximum westerly declination that occurred in AD 1820. Records of declination, inclination and intensity have been kept in the British Isles since AD 1580. Over this timespan, maximum inclination was in AD 1690 (Thompson 1978). According to Thompson and Turner (1979) and Turner and Thompson (1981), 23 significant magnetic horizons can be discerned over the last 10 000 years, comprising ten secular variations in declination and 13 in inclination. These, calibrated by radiocarbon ages, have been employed in the dating and correlation of lake sediments within the British Isles.

Other properties of magnetic minerals (susceptibility and saturation isothermal remanent magnetization, for example) are linked to the mineralogy of sediments, rather than to the geomagnetic field. Such properties have assisted studies of sediment stratigraphies and budgets (Chapter 1). Ferrimagnetic minerals become embodied in sediments. Thus, knowledge of the variations in magnetic susceptibility of sediments allows quantification of differences in the erosion and deposition of these minerals (Thompson *et al.* 1975).

Edwards and Rowntree (1980) have reported a study of Braeroddach Loch in the foothills of the Grampians, in which magnetic susceptibility was employed to help indicate the rates of sediment influx over the past 10 000 years. Dearing *et al.* (1981) have used magnetic susceptibility measurements to correlate recent sediments, dated by ^{210}Pb and ^{137}Cs, from Llyn Peris in the Llanberis Pass, North Wales, as a foundation for the computation of its influx of sediment and chemicals. An increase in erosion has been established within the sparsely vegetated, non-forested, marginal upland catchment of the lake. From about AD 1750 to 1820, *circa* 5 tons of sediment per km^2 per year were being eroded. This rose to *circa* 42 tons per km^2 per year between AD 1966 and 1976. After identifying synchronous horizons in over 50 cores of sediment from Merevale Lake in northern Warwickshire by means of magnetic susceptibility and saturation isothermal remanent

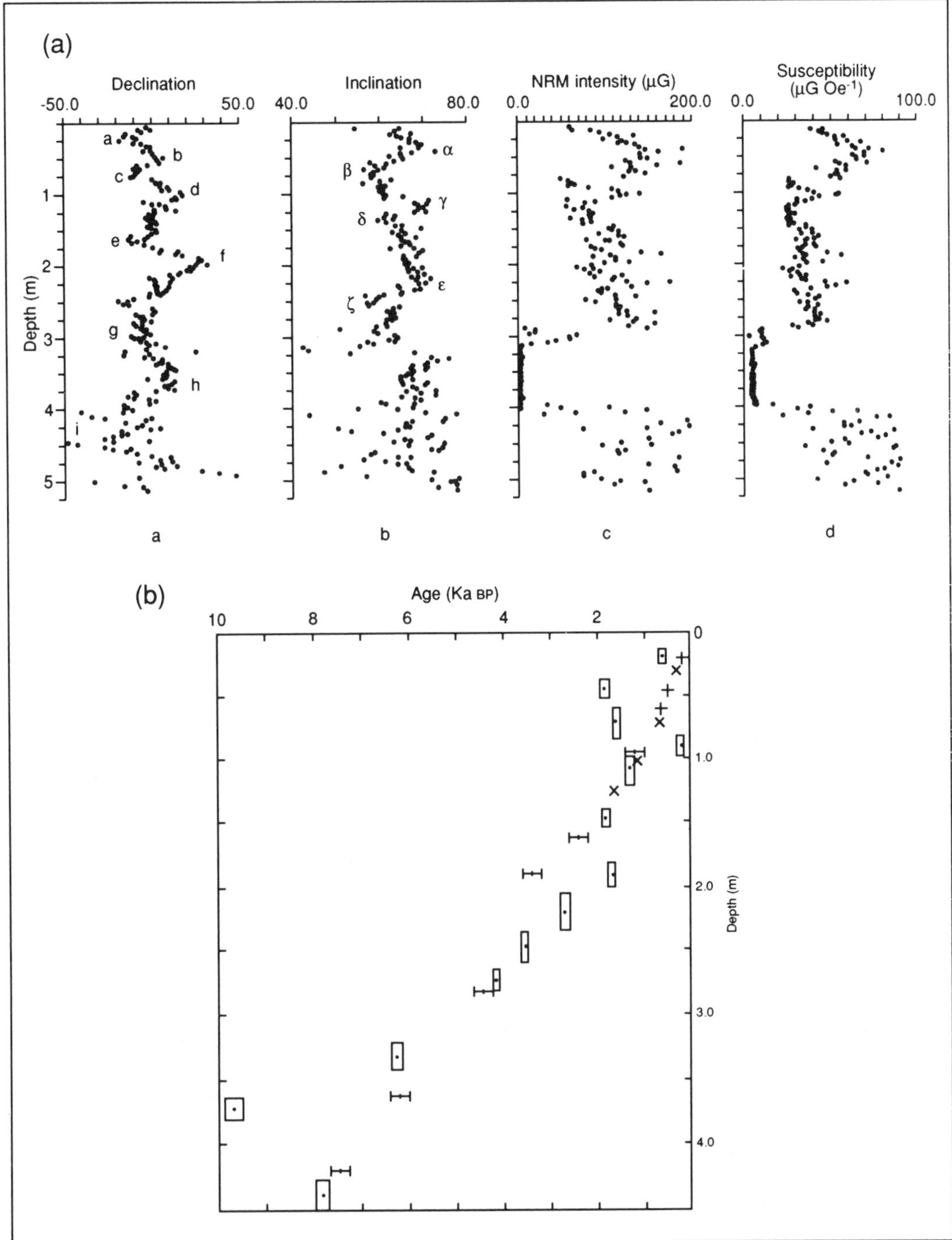

Figure 10.11 Palaeomagnetic and radiocarbon-dating evidence from Loch Lomond. (a) Palaeomagnetic measurements. The ages of turning points (a–i) in declination and (α–δ) in inclination are plotted as upright and diagonal crosses in (b). (b) Age plotted against depth for core LLRD1. Boxes represent radiocarbon dates, with widths corresponding to the ±1σ error intervals (based on statistical counting uncertainties only) and heights corresponding to depth sampled. Upright and diagonal crosses define palaeomagnetic dates from observatory and archaeomagnetic declination and inclination records, respectively. Horizontal bars represent correlations with Windermere radiocarbon dates at declination turning points and include ±1σ radiocarbon counting errors. From Dickson *et al.* (1978). Reprinted with permission from *Nature* **274**, 548–553; © 1978 Macmillan Magazines Ltd.

magnetization, and obtaining a chronology by ^{210}Pb and ^{137}Cs analysis, Foster *et al.* (1985) have calculated the total sediment volume and mass for nine periods between AD 1861 and 1982. This forested catchment has experienced low levels of sediment loss.

10.8.2 Chemical analysis

Mackereth (1966) has examined the chemistry of sediments in six Lake District lakes. Elements such as sodium, potassium and magnesium exhibit recurrent trends that have been interpreted as an indication of erosional activity within their catchments. Substantial quantities of sodium and potassium in the mineral part of the lake sediments have been interpreted as a reflection of intensive erosion of mineral soil and a low level of leaching. The latter premise is based on the fact that these elements are soluble, and thus usually susceptible to removal in downward-percolating water. Conversely, small amounts of sodium and potassium in minerogenic sediment have been linked to reduced erosion and enhanced leaching. *Circa* 10 000–9000 BP, the lake sediments possessed high values of these elements which, by 5000 BP, had reduced consequent upon soil formation and heightened leaching. An episode from the latter time until the present, when increased amounts of sodium and potassium were present, has been correlated with additional erosional, and less leaching, activity. The deposition rates of halogens have been related to oceanic air masses, from which they largely originate. Increased oceanity and a wetter climate were probably responsible for more erosion. Chemical analysis of the sediments at Angle Tarn, Bow Fell, in the Lake District, has led to the suggestion that forest removal using fire caused inwash of acid soils to the lake (Pennington 1970, 1975b). The iodine content of lake sediments in the Lake District and in Scotland has been examined by Pennington and Lishman (1971), who concluded that the level of this halide was not correlated with inter-site variations in precipitation, but more likely with soil type and temporal modifications to it.

Pennington *et al.* (1972) have suggested that heightened mineral soil erosion after forest destruction within its catchment caused a greater influx of sodium, potassium, magnesium and iron to the sediments of Loch Tarff, Inverness-shire, Scotland. From the Braeroddach locality (section 10.8.1), Edwards and Rowntree (1980) have established an analogous record to that of Mackereth in the Lake District. However, it has been shown that *circa* 5000 BP, marked rises in the amounts of sodium, potassium and magnesium did not accompany the erosion of soils which had by then undergone a substantial phase of leaching. Such increases have been demonstrated to have taken place subsequently, when sub-soil erosion began in this part of Aberdeenshire.

Jones *et al.* (1987) have measured exchangeable cation values in the sediments of a former lagoon in St. Aubin's Bay, Jersey, in order to detect changes in palaeosalinity, to which magnesium and calcium concentrations are thought to relate (Ericsson 1973; Tooley 1978a). The exchangeable cation values fluctuate broadly in harmony with the diatom stratigraphy (Figure 10.10) to confirm a complex sequence of late Flandrian freshwater, brackish and marine episodes.

Atmospheric pollutants, soot and trace metals, for example, have been recognized in recent deposits such as upland peats and lake sediments. Chambers *et al.* (1979) have suggested that older than expected radiocarbon ages on peat that formed recently on moorland in South Wales are the result of the incorporation of older carbon. It has been proposed that this is likely to have accrued from the partial combustion of fossil fuel in the industrial valleys below, and had been transported above them as particulate pollution in the atmosphere. Foster *et al.* (1991) have used chemical analysis on lake and fluvial sediments to demonstrate heavy-metal contamination in four catchments in the English Midlands over the last 150 years. It has been concluded that heavy-metal pollution at Merevale Lake in rural Warwickshire (section 10.8.1) came only from the atmosphere. At Swanswell Pool in inner-city Coventry, the heavy-metal contaminants probably have atmospheric, urban storm run-off and, possibly accidental spillage, origins. At Merevale, increases which began *circa* AD 1900, exhibited their most marked rise after 1945, and could be substantiated by documentary evidence. At Swanswell, the amount of zinc has become high during the last 132 years, and enhancement of lead values has been unequivocally demonstrated. Phosphorus and organic matter records from the four sites are diagnostic of eutrophication over the timespans represented.

Ratios of deuterium to hydrogen in radiocarbon-dated pine stumps from the Cairngorm Mountains have been established by Dubois and Ferguson (1985). Low deuterium levels *circa* 7500, 6250–5800, 4250–3870 and around 3300 BP have been correlated with periods of high rainfall.

As noted in Chapter 1, speleothem growth is most effective in temperate climates. Thus, by employing U-series dating, Atkinson *et al.* (1978) have identified an episode of speleothem growth in the British Isles from *circa* 17 000 BP to the present, and Gascoyne *et al.* (1983) have identified one from about 13 000 BP until today. The latter study has established that maximum growth in north-western England took place between 10 000 and 0 BP, with the build up of most speleothems commencing 10 000–9500 BP. Atkinson *et al.* (1986) have stated that abundant speleothem deposition has taken place in the British Isles over the last 15 000 years. In Assynt, it had restarted by 9000 ± 1000 BP, after which it was maintained.

10.9 Soil and peat formation and erosion

Although the formation of soil and peat may have been linked in certain circumstances (section 10.9.2), the nature and amount of information pertaining to both the genesis and erosion of these materials is such that each is afforded a separate treatment. Localities referred to in this section are shown in Figure 10.7.

10.9.1 Soil

A legacy of glaciation and periglaciation during the Loch Lomond Stadial was an unstable regolith and meagre soil cover within the British Isles at the beginning of the Flandrian. The dominantly physical processes operational in the landscape during the stadial would have led to a general enhancement of the base status of soil parent material. However, in this context, a contrast must have existed between uplands, where substrates generally were harder and of lower base status, and lowlands whose rocks were by and large softer and more base rich. The climatic amelioration that marked the opening of the Postglacial would have stimulated chemical and biological processes in the landscape. At first, as open vegetation achieved a more complete cover, then was replaced by woodland, soil formation would have been encouraged. An increasingly stable regolith and higher humus content in soils must have ensued. None the less, variation is likely to have occurred in both the rate and characteristics of pedogenesis at different altitudes. The combined influence of lithology and climate probably slowed soil development in the uplands, where leaching was more effective than in the lowlands. It should also be remembered (section 10.2) that considerable contrasts appear to have been present between plant communities in different parts of the British Isles during the early Flandrian. Mull humus would have been produced by stands of deciduous trees and mor by coniferous ones. However, as Simmons *et al.* (1981) have stressed, caution is required when considering plant–substrate relationships. For instance, *Pinus*, while today often found growing in an acid soil, is capable of existing in a base-rich one. In the latter circumstance, the effect of the acid pine litter would be restrained by the soil characteristics. Limbrey (1978) has suggested that succession from *Betula–Corylus* to *Pinus* forest on inherently base-poor or leached soils could have initiated podsolization. The subsequent replacement of pine by mixed-deciduous forest and the consequent accumulation of mull has been thought likely to have caused podsol formation to cease at an incipient or shallow stage. Such soils would though, thereafter have been susceptible to podsolization, according to Limbrey.

Simmons *et al.* (1981) have pointed out that modern examples of temperate deciduous forests underlain by podsols exist in Scandinavia and North America. At Woodhall Spa in Lincolnshire, an iron-humus podsol, above which is fen peat and marine clay, the former radiocarbon dated to *circa* 4100 BP, has produced palynological evidence of having supported mixed-oak forest during the mid-Flandrian (Valentine and Dalrymple 1975). According to Limbrey (1978), such a situation may have been atypical in view of evidence, produced by Dimbleby (1962), Keef *et al.* (1965) and Dimbleby and Bradley (1975), of the continued existence of mull following Mesolithic forest disturbance, and later than the Atlantic, in soils which subsequently underwent podsolization.

Certain trees, *Ulmus*, *Tilia* and *Alnus*, for example, at present appear to have a preference for mull. The emergence of mixed-deciduous forest containing such taxa would thus have been likely to cause this type of humus to accumulate and brown forest soils to develop. On parent material of lower base status, sols léssives (in which bases and sesquioxides are translocated) or podsols could have formed beneath deciduous forest (Simmons *et al.* 1981). Acid brown soils beneath Neolithic monuments in Angus and Morayshire have produced charcoal of *Quercus* (Romans and Robertson 1975).

The significance of a loessic component in soils of the British Isles has been stressed by Catt (1978). When added to siliceous or chalky soils it has provided a clay content, and more generally has added nutrients and enhanced texture. Weir *et al.* (1971) and Catt (1978) have suggested that clay migration took place in loessic soils in Kent before forest clearance. Erosion of the upper, clay-depleted soil horizons had occurred by Atlantic time.

In environments such as those of the British Isles, the amount of leaching which has occurred is related to the time that a soil has been formed (Iversen 1969). The chemical analyses of Mackereth (1966) (section 10.8.1) indicate soil acidification in the Lake District by *circa* 8000 BP, while diatom data presented by Jones *et al.* (1986, 1989) (section 10.7.1) confirm early Flandrian acidity in Galloway.

Disturbance of soil–vegetation relationships by Mesolithic man may have contributed to soil deterioration. The replacement of deep-rooting plants such as trees by shallower rooting ones such as dwarf shrubs and herbs, reduces the capacity for nutrients to be moved from depth to replenish those removed from soil by vegetation (Dimbleby 1962, 1976). Palynological data from Iping Common in Sussex indicate that hazel woodland was replaced by *Calluna* heath as a result of Mesolithic disturbance during the late Boreal (Keef *et al.* 1965). Such a vegetation modification is likely to have been accompanied by the accumulation of mor humus (Simmons *et al.* 1981). Confirmation of the existence of mull humus beneath the woodland could be provided by earthworm activity which led to Mesolithic flints being moved downwards in the soil (Evans 1975). Vegetation disturbance and subsequent soil erosion were the probable causes of blown sand accumulation over the Mesolithic land surface during the Atlantic.

Mor humus is normally not well mixed with mineral material. Thus, such a soil possesses a less stable structure than one with mull, where admixture is, as a rule, well developed. While all soils may be subject to erosion following vegetation disturbance, those with inferior structural characteristics are more at risk. Burning and trampling, which sometimes accompanied perturbation of the plant cover, would have led to more structural damage and increased the capacity of soil to erode (Simmons *et al.* 1981). Episodes of soil erosion that followed human interference with the vegetation of the North Yorkshire Moors, beginning in the Mesolithic, have been demonstrated by Simmons *et al.* (1975, 1982). Dimbleby (1976) has suggested that certain buried rendzinas on the chalklands of southern England (at Windmill Hill, for example) have been truncated, perhaps by erosion. In the Lake District (Pennington 1970, 1975b) and northern Scotland (Pennington *et al.* 1972) lithostratigraphic and chemical data from lake sediments indicate soil erosion after Neolithic vegetation clearance and farming. At Ballynagilly in Ulster, a localized episode of soil erosion has been dated to 5200 BP by means of charcoal in a mineral soil downhill of a Neolithic site (Pilcher and Smith 1979) (section 10.1). Sandy soil above the charcoal has been interpreted as the product of colluviation. Substantial amounts of eroded soil arrived in the valleys of Warwickshire and Worcestershire (Shotton 1978; Brown and Barber 1985), East Sussex (Jennings and Smyth 1987) and Jersey (Jones *et al.* 1990) during the Iron Age, when both alluviation and colluviation are considered to have been prominent (section 10.10).

Probable effects of early agriculture on soils have been discussed by Limbrey (1978). Farming practices would have led to the development of areas of low-growing, transient vegetation and expanses of bare soil. It has been concluded that lopping of vegetation and forest grazing are likely to have had a minor influence on water regimes, nutrient cycles and soils. However, more substantial clearings made for cultivation and pasture would have fostered increased rain impact and stream flow. In such circumstances, exposure of soil must have rendered its structure more susceptible to breakdown. Clay-depleted horizons of soils would have been eroded, and the removal of wood ash and topsoil by erosion would probably have been responsible for a greater loss of nutrients than leaching.

Changes in soil type between various times earlier in the Flandrian and today have been demonstrated at numerous localities within the British Isles (Evans 1971, 1972, 1975; Limbrey 1975; Keeley 1982). For example, a non-calcareous brown earth with indications of léssivage (probably resulting from woodland removal and farming practices), has been identified below a Neolithic barrow at Kilham on the Yorkshire Wolds (Manby 1971). The present soil in this locality is a rendzina that has developed in response to modern farming techniques (Evans 1972). The environs of the Neolithic site in Morayshire, associated with which was an acid brown earth (see above), have been shown currently to possess a podsol (Romans and Robertson 1975). Below Bronze Age barrows on the North Yorkshire Moors, Dimbleby (1962) has been able to identify brown soils. Various stages of leaching are evident and pollen spectra characteristic of deciduous forest vegetation are present in the soils. Over the barrows, and supporting the current heathland plant community, are podsolic soils with iron pans. As Tinsley and Grigson (1981) have pointed out, woodland clearance during the Bronze Age must have led to enhanced leaching and a deterioration in soil fertility, especially on uplands. This trend continued in the Iron Age, when widespread deforestation took place and agricultural activity intensified. Soil deterioration occurred, podsols formed and heathland extended (J. Turner 1981). Soils associated with Iron Age sites in Scotland have indicated the wide distribution of podsolic soils by that time (Romans and Robertson 1975).

Ball (1975) has noted that man also made useful modifications to soil in the highland zone of the British Isles. The creation of plaggen soils by the addition of turf, peat and seaweed to sandy material has been recognized in Ireland (Conry 1971) and Scotland.

10.9.2 Peat

Peat formation has taken place in waterlogged localities within the British Isles throughout the Flandrian. There have been a number of mire environments. Fens, such as those of East Anglia, have most frequently developed in basins under the influence of precipitation and groundwater, and have usually possessed a mesotrophic or eutrophic status. Ombrogenous mires, sustained by precipitation and of an oligotrophic nature, have been of two kinds. Raised bogs have typically formed in lowlands such as those of Lancashire, Cheshire and central Ireland, and around estuaries – Morecambe Bay and the Humber, for example. Blanket bog has been an upland formation, mantling substantial areas of gently sloping terrain on Dartmoor, the Pennines, Welsh and Scottish mountains, for instance (Godwin 1956; Birks and Birks 1980). Blanket mire began to form *circa* 9000 BP in the highlands of Scotland (Birks 1975) and between about 7500 and 2000 BP in other upland areas (Birks 1988). Evidence from the Pennines (Tallis 1964a) and Wales (Chambers 1981, 1982) has indicated that plateaux and basins at higher elevations were the sites of initial blanket peat accumulation, with lower areas subsumed later. The growth of blanket bog on mineral soil could have caused the *in situ* death of trees, remains of which have been identified in this stratigraphic context (Godwin 1956; Moore 1988). Tree regeneration may also have been restrained by soil waterlogging (Birks 1972b, 1988).

There appear to have been a number of interrelated factors involved in the initiation and spread of blanket peat. Climatic deteriorations which led to increased rainfall, lowered temperatures and hence less evapotranspiration, were almost certainly involved. The accumulation of mor humus, and the development of podsols with iron pans which impeded the downward passage of soil water were also of likely significance (Taylor and Smith 1972; Birks 1988). However, as Moore (1988) has pointed out, not all peat of this type is underlain by podsolic soil. Gleys are common, particularly in depressions, so that an iron pan was not a prerequisite for waterlogging. Soil temperatures would have been lowered by waterlogging. As a result, microbial and soil-animal activity rates would have fallen, leading to an enhanced build up of litter. Additional water would have led to decreased soil aeration and hence oxygen content. Moore has also noted that the invasion and growth of *Sphagnum* was typical in such situations. Clymo (1964) has established that *Sphagnum* removes cations from its growth medium, substituting hydrogen ions, so that acidification would have been intensified.

Human activities may also have been responsible for blanket peat growth (Moore 1975) and podsolization. The removal of trees by man and their replacement by lower growing, shallower rooted vegetation would have reduced the efficacy of transpiration and thus have led to a rise in the soil water-table. Burning of forest would have depleted nutrient supplies, thereby encouraging recolonization by plants of an acidophilous kind. Tree regeneration may have also been inhibited by grazing, which is likely to have taken place in areas that had previously been cleared and burned (Moore 1988). The start of blanket peat formation at sites in mid-Wales (Moore 1973), the North Yorkshire Moors (Simmons and Cundill 1974; Simmons *et al.* 1982) (Figure 1.9) and Exmoor (Merryfield and Moore 1974; Moore *et al.* 1984) was probably linked with Neolithic activity. Pennington (1975b) has demonstrated the inception of blanket peat growth on mineral soil and mor humus following woodland removal at two localities in the Langdale Fells, north-west England, during the Neolithic. A Neolithic site at Goodland Townland in County Antrim (Case *et al.* 1969) shows a series of environmental changes thought to have taken about 500 years (Mitchell 1972b). Firstly, forest was removed from a brown podsolic soil, which was subsequently cultivated. The site was then used as pasture and the soil became more acid. A leached soil containing a thin iron pan developed. Structural breakdown of the soil may have caused its surface to become wet. Finally, peat began to accumulate (Dimbleby 1976). Reductions in arboreal pollen frequencies and increases in those of heathland plants have accompanied radiocarbon-dated evidence of blanket peat inception during the Bronze Age in Ireland (Smith 1975). At Ballynagilly (ApSimon 1969) (section 10.1), however, peat growth was delayed until the Iron Age (Pilcher and Smith 1979).

Blanket mires have undergone recent changes in floristic composition and rates of erosion. Tallis (1964c) and Lee *et al.* (1988) have illustrated the demise of *Sphagnum* in southern Pennine peats and have correlated this with atmospheric pollution during the nineteenth and twentieth centuries. Upland peat in South Wales has revealed that base-poor grassland replaced heather moorland subsequent to industrialization in the region. Pollution by particulate matter emanating from this activity could have been involved in this modification (Chambers *et al.* 1979). Air pollution, fires and grazing have been cited by

Tallis (1985) as probable contributors to a recent heightening of peat erosion in the southern Pennines.

10.10 Rivers

A substantial quantity of the alluvium associated with the current floodplains of rivers in the British Isles has been deposited during the Post-glacial. Alluvial gravels indicative of high-energy transport are most likely to represent former river-bed sedimentation, while sand, silt and clay were probably laid down both in the channel and on the floodplain by water. Abandoned river meanders or floodplain hollows would usually have been the sites at which organic matter accumulated. The migration of a meandering river across its floodplain would have caused the sediments to be eroded and would have been responsible for breaks in the stratigraphy (Shotton 1978). In a temperate stage, the storage of groundwater and slow flow within the ground would have moderated the river regime. In addition, the development of a plant cover would have stabilized the soil and hence reduced the quantity of material liable to erosion. Rivers in the lowland zone would normally have lacked enough coarse sediment to lead to major realignments in their courses through erosive activity. The rôle of plants in the uptake and transpiration of water would also have influenced the hydrological balance (Gibbard 1988b). As river channels are adjusted in size to the flows which occupy them, flooding is likely to have been of similar frequency to that during a cold stage. Localities referred to in this section are shown in Figure 10.12.

Flandrian channel changes of the River Gipping have been described by Rose *et al.* (1980). At Sproughton near Ipswich, the channel was stable during much of the stage. The river was then restricted by cohesive material and the floodplain was built of organic and inorganic material that was deposited in backswamps. The inferences from this evidence are of an area with a well-developed cover of vegetation where infiltration rates were high, and of little variation in discharge of a river whose bed-load transport was constrained.

The climatic amelioration and rising sea-level during the early Flandrian would have influenced valley aggradation and floodplain formation. Flandrian valley-fills have been widely investigated, but as Burrin and Scaife (1984) have pointed out, many studies have not provided a comprehensive picture of the events involved. The deposition of valley sediment has been thought by Dury (1977) to be a reflection of reduced river size that resulted from hydrological change influenced by climatic modification.

Straw (1979a) has inferred that river floodplains upstream of estuaries in eastern England experienced high rates of aggradation in the early Flandrian. These were probably due to a combination of a rising sea-level (and its effect on base level) and enhanced sediment availability before a vegetation cover became well established. A decline in aggradation began *circa* 7000 BP and lasted until about 5000 BP. Although subsequent alluviation took place, the floodplains had grown to more or less their current dimensions by the latter date.

Late Flandrian valley infill has been mainly a response to land use by man (Bell 1982). Vegetation removal and farming heightened colluvial inwash and sediment input to valley floors, thereby encouraging alluviation and the development of floodplains (Burrin and Scaife 1984). Studies by Shotton (1978) in the Avon Valley in Warwickshire, and the Arrow and Severn valleys in Worcestershire, have revealed a clay or clayey silt resting upon earlier alluvium. Radiocarbon assay has established that this began to accumulate *circa* 2600 BP, as a result of Later Bronze Age–Early Iron Age vegetation disturbance and ploughing which caused soil erosion in the catchments.

The floodplain deposits and palaeohydrological characteristics of the Lower Severn Basin have been described by Brown (1983, 1987). Rising base level in the early Flandrian caused rapid aggradation in the estuary area. Movements of sea-level had negligible influence upon floodplain sedimentation above Tewkesbury. The discharge of the River Teme *circa* 7000 BP has been estimated as similar to that of today. Stable, anastomosing channels with low sinuosity probably persisted throughout the Postglacial. It has been suggested that aggradation in the Lower Severn Basin reflected sediment production in the environs, together with channel behaviour, rather than a eustatic rise of sea-level. Brown (1983) has recognized the existence of an uppermost inorganic unit in the stratigraphy of this area. This overlies semi-organic clays which accumulated in backswamps and has been established as having been produced within the last two or three millennia, and analogous to the material identified by Shotton (see above). Brown and Barber (1985) have demonstrated that a spectacular rise in sedimentation occurred in the Ripple Brook catchment near Tewkesbury during the Late Bronze Age and Early Iron Age (*circa* 2900–2300 BP), as a result of forest removal and cultivation which induced soil erosion. The past and present environmental characteristics of the Severn Basin have been dealt with by the contributors to Gregory *et al.* (1987).

Flandrian sedimentation in the Thames Valley has been of sand, silt, clay, tufa and organic material on gravels located beneath the present floodplain. Since the Neolithic, human interference with the surrounding landscape has caused the overbank deposition of silt and clay (Gibbard 1988b). A rise in the permanent water-table of the Upper Thames floodplain occurred during the Bronze Age, according to Robinson and Lambrick (1984). However, only a small quantity of alluvium was laid down, probably until late in the Iron Age (0–AD 400) when sedimentation increased. Alluviation was considerable during the Roman Period. Early Saxon time saw a drop in flooding and valley infill, a trend that was reversed between *circa* AD 800 and 1350. A further phase of reduced alluviation since the Late Medieval Period has been identified. Human activities have been linked to the deposition of fine-grained material on the River Windrush floodplain subsequent to 2660 BP by Hazelden and Jarvis (1979).

Valley sedimentation and floodplain development in the Ouse Basin, Sussex, have been reported by Burrin and Scaife (1984) and Burrin (1985), a complex sequence being suggested. *Circa* 9000 BP, when the area was well vegetated, alluviation was substantial, and may have been enhanced by Mesolithic disturbance of the plant cover. Neolithic activities, and those of Bronze Age inhabitants, led to sedimentation, and it has been postulated that most of the alluvium was emplaced by the early Iron Age. In the Combe Haven Valley, East Sussex, Iron Age forest clearance led to heightened floodplain sedimentation (Smyth 1986; Jennings and Smyth 1987). Valleys in Jersey exhibited a similar phenomenon, which began *circa* 2200 BP. Colluviation and alluviation increased when woodland was removed and farming intensified (Jones *et al.* 1990).

Alluviation in Staindale, north-east Yorkshire, has been described by Richards (1981). Wood remains, radiocarbon dated to *circa* 6270 BP are overlain by fluvial gravels. In this small catchment, the alluviation, which commenced in the Atlantic Period, was probably linked with increased slope erosion induced by a combination of wetter conditions and human disturbance of vegetation.

The development of alluvial terraces of the River Dane between Congleton and Holmes Chapel in Cheshire since the mid-Flandrian has been investigated by Hooke *et al.* (1990). Dissection of a Middle Terrace occurred prior to *circa* 4725 BP. After this (and also before 4725 BP), a lengthy episode of lateral activity by the river took place, lasting until later than *circa* AD 750. A major aggradation phase followed which, it has been inferred, was probably related to forest removal, agriculture and soil erosion in the Medieval Period. This alluvium formed a Low Terrace between about the ninth and eighteenth centuries AD. Dissection of this feature has been taking place for approximately the last 300 years, while the modern floodplain has emerged since *circa* AD 1840.

10.11 Sea-levels and crustal movements

Shennan (1983) has observed that there have been numerous investigations of Flandrian sea-level changes and crustal movements within

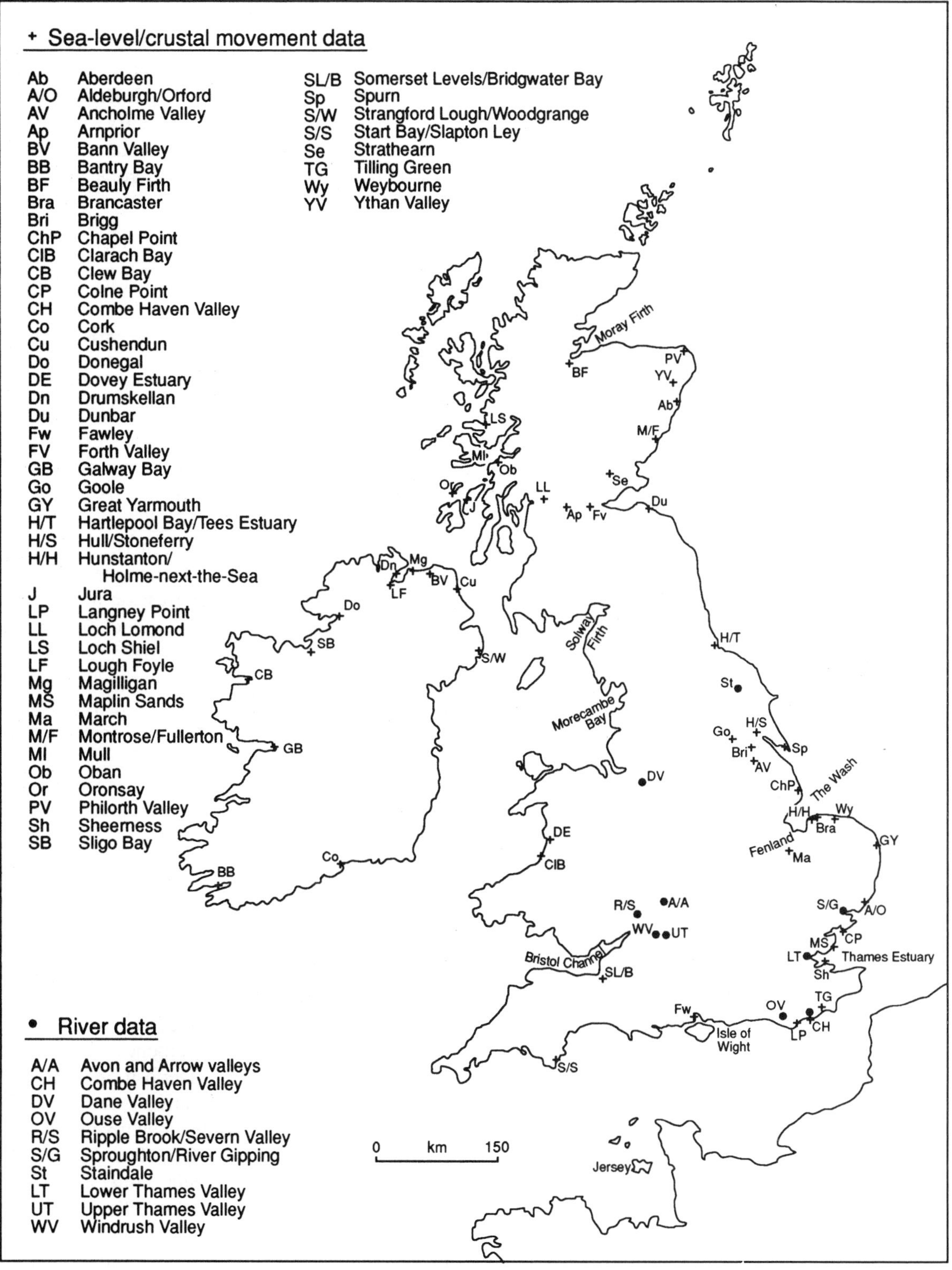

+ Sea-level/crustal movement data
Ab Aberdeen
A/O Aldeburgh/Orford
AV Ancholme Valley
Ap Arnprior
BV Bann Valley
BB Bantry Bay
BF Beauly Firth
Bra Brancaster
Bri Brigg
ChP Chapel Point
ClB Clarach Bay
CB Clew Bay
CP Colne Point
CH Combe Haven Valley
Co Cork
Cu Cushendun
Do Donegal
DE Dovey Estuary
Dn Drumskellan
Du Dunbar
Fw Fawley
FV Forth Valley
GB Galway Bay
Go Goole
GY Great Yarmouth
H/T Hartlepool Bay/Tees Estuary
H/S Hull/Stoneferry
H/H Hunstanton/ Holme-next-the-Sea
J Jura
LP Langney Point
LL Loch Lomond
LS Loch Shiel
LF Lough Foyle
Mg Magilligan
MS Maplin Sands
Ma March
M/F Montrose/Fullerton
Ml Mull
Ob Oban
Or Oronsay
PV Philorth Valley
Sh Sheerness
SB Sligo Bay
SL/B Somerset Levels/Bridgwater Bay
Sp Spurn
S/W Strangford Lough/Woodgrange
S/S Start Bay/Slapton Ley
Se Strathearn
TG Tilling Green
Wy Weybourne
YV Ythan Valley
• River data
A/A Avon and Arrow valleys
CH Combe Haven Valley
DV Dane Valley
OV Ouse Valley
R/S Ripple Brook/Severn Valley
S/G Sproughton/River Gipping
St Staindale
LT Lower Thames Valley
UT Upper Thames Valley
WV Windrush Valley
Moray Firth
Solway Firth
Morecambe Bay
The Wash
Fenland
Bristol Channel
Thames Estuary
Isle of Wight
Jersey
0 km 150

Figure 10.12 Localities from which data on Flandrian/Littletonian rivers, sea-levels and crustal movements have been obtained.

England and Wales. However, it has been usual for restricted areas to be examined by different people, so that local patterns and chronologies have emerged. Shennan has provided a synthesis of the data available for various parts of England and Wales. This, together with more recent observations, forms the basis of the account of these areas that follows. Localities referred to in this section are shown in Figure 10.12.

10.11.1 North-eastern England, East Anglia and the Thames Estuary

The coastal districts of north-eastern England, between the Tweed and Humber estuaries, have yielded only a small amount of organized information. The area of Hartlepool Bay and the Tees Estuary is situated between a region to its north and west where uplift has been taking place and one to its south where subsidence has occurred (Chapter 1). Peat beneath a marine deposit at −10.61 m OD at Thornaby in the Tees Valley has been radiocarbon dated to 9680 ± 100 BP (Gaunt and Tooley 1974; Tooley 1978b). A lithostratigraphic transgressive overlap (consisting of the progressive replacement of terrestrial deposits by marine deposits as the sea advanced over a former land surface and the water deepened (Tooley 1982)), at −2.39 m OD at West Hartlepool, has given an age of 6050 ± 90 BP, and a regressive overlap (comprising a retreat of the sea, shallowing of the water and the substitution of marine sediments by terrestrial sediments (Tooley 1982)), at −0.34 m OD, one of 5240 ± 70 BP (Gaunt and Tooley 1974; Tooley 1978c). From Hull on the Humber Estuary, the basal age of peat at −11.55 m OD is 6970 ± 100 BP; a transgressive overlap at −9.73 m OD has been dated to 6890 ± 100 BP. At Spurn, regressive overlap tendencies, *sensu* Tooley (1982), identified at −2.35 m OD, have been radiocarbon dated to 6170 ± 180 BP. At Stoneferry, a marine environment has been established between *circa* 5240 and 3775 BP (Gaunt and Tooley 1974). The relationship of deposits at Goole in Yorkshire and in the Ancholme Valley, Lincolnshire, to late Flandrian sea-levels has been considered by Smith (1958b). The passing of marine influence in the latter locality, at Island Carr, Brigg, about 14 km south of the Humber Estuary, has been dated to 2625 ± 65 BP by Smith *et al.* (1981b).

Various parts of the coastal zone between Chapel Point in Lincolnshire and Hunstanton in Norfolk, comprising mainly the Lincolnshire Marshes and Fenland around the Wash, have received considerable attention (for example, Swinnerton 1931; Godwin and Clifford 1938; Godwin 1940b; Smith 1958b; Willis 1961; Godwin and Vishnu Mittre 1975; Valentine and Dalrymple 1975; Simmons 1980; Shennan 1982, 1986a,b). According to Jelgersma (1979), the level of the southern North Sea was about 36 m below that of today *circa* 8500 BP, with its coastline considerably north of the present one. About 8300 BP, when sea-level was some 30 m lower than now, the North Sea and Southern Bight became linked (Jansen *et al.* 1979; Jelgersma 1979).

The first indication of transgression in Fenland has come from Adventurers Land, north of March in Cambridgeshire, at 6575 ± 95 BP (Shennan 1982). During the early Flandrian, Fenland and its offshore area (section 10.12) first possessed a fluvial, then a marine environment (Shennan 1986a). Between 6500 and 2500 BP, freshwater and marine sedimentation alternated in Fenland, with brackish-marine conditions extending up to 45 km inland of the current coastline. Shennan (1986b) has identified seven episodes of positive tendency (involving rises in sea-level) and six episodes of negative tendency (involving reduced or negative rates of sea-level rise) (Shennan 1982) in the region since *circa* 6500 BP, with local differences evident. Around 6400 BP, a transgressive overlap occurred at about −8 m OD. A regressive tendency may have ensued *circa* 6300–6200 BP. Deposition of mainly silty clays accompanied a positive tendency from about 6200 to 5600 BP, while a negative tendency occurred *circa* 5600–5400 BP. The Fen Clay, deposited between *circa* 5400 and 4500 BP, resulted from widespread marine-influenced sedimentation. Peat that formed *circa* 4500–3900 BP indicates a regressive overlap and negative tendency. The periods *circa* 3900–3300 and

3300–3000 BP were characterized by transgressive and regressive overlaps, respectively. From about 3000 to 1900 BP, brackish-marine sedimentation was extensive up to 1.45 m OD, with many deposits termed Romano-British Silts forming. Later than this, the sequence lacks radiocarbon dates but the following events have been inferred tentatively: *circa* 1900–1550 BP regressive episode; 1550–1150 BP transgressive episode; 1150–950 BP regressive episode; 950 BP to present, transgressive episode. Over the entire timespan, relative sea-level has been diagnosed as having risen in the region, which has been sinking as a result of consolidation of sediments and crustal subsidence. This downward movement has been calculated at 0.908 m per millennium (section 10.11.7). An average rate of sea-level rise of 1 m per 1000 years since 3000 BP has been proposed.

In Norfolk, Jennings (1952) and Lambert *et al.* (1960) have produced evidence of sea-level changes in the Broads. At Great Yarmouth, positive tendencies identified at −19.3 m OD and dated to 7580 ± 90 BP continued until *circa* 5000–4000 BP at −6.5 to −5.5 m OD. A second phase of positive tendencies commenced about 2000 BP, with marine – estuarine sedimentation reaching a maximum 1609 ± 50 BP, at *circa* −2.0 to 0.5 m OD (Coles and Funnell 1981). Elsewhere in the Norfolk Broads, transgressive overlaps occurred *circa* 7500–5650 and 2400–2200 BP, with regressive overlaps taking place from about 5000–4300 and 1800–550 BP (Alderton 1983). The north Norfolk coast from Holme-next-the-Sea to Weybourne has been examined by Funnell and Pearson (1989). Peat below salt-marsh sediment at Brancaster formed between 8410 ± 150 and 4010 ± 50 BP, according to radiocarbon evidence. Positive tendencies occurred at *circa* 6610, 5970, 4630 and 2790 BP, and negative ones between about 4520 and 4450 BP, and possibly around 3470 BP.

The Aldeburgh Marshes–Orford Marshes area of Suffolk has been investigated by Carr and Baker (1968) and Carr (1970). Basal peats from the former, at *circa* −13.7 and −12.7 m OD, have been radiocarbon dated to 8460 ± 145 BP and 8640 ± 145 BP, respectively. Minerogenic sediments occur at −12.5 m OD and peat at −9.25 m OD. The latter, dated to 7010 ± 130 BP, are overlain by marine deposits.

The Flandrian deposits of the estuaries and marshes between Colne Point and Maplin Sands have been examined by Greensmith and Tucker (1971, 1973, 1980). Peaty silt at −16.5 m OD at Foulness has been radiocarbon dated to 7516 ± 250 BP. A positive tendency is indicated by *Ostrea* shells, dated to 6620 ± 100 and 5650 ± 240 BP. Peat at *circa* −3.0 m OD has given an age of 4959 ± 65 BP, and at Colne Point, peat from about the same elevation dates to 4277 ± 95 BP, indicating regressive tendencies. A series of shell dates have been associated with a positive tendency until about 3580 ± 175 BP. Archaeological evidence has been cited for negative tendencies during the Romano-British Period in this area. Variations in height between different pointers of sea-level are probably due to unequal downwarping in the area (Greensmith and Tucker 1980). The adjoining area, the inner Thames Estuary, has been analysed by Devoy (1977, 1979, 1982). The deposition of inorganic sediments could be linked to five main transgressive episodes recognized from −25.5 to 0.4 m OD. These occurred *circa* 8200–6970, 6575–5410, 3850–2800, 2600 and 1700 BP, and contributed to a pattern of sea-level movement analogous to that recorded in coastal Essex. Overall, however, the sequence is complex, with tides, floods and differential crustal movement thought likely to have influenced the record of change.

10.11.2 The south coast of England, and Jersey

Devoy (1982) has synthesized evidence for sea-level movements along the south coast of England to the Poole area. Boreholes along the line of the Channel Tunnel have encountered wood peat at −37.0 m OD and −36.5 m OD, radiocarbon dated to 10 350 ± 120 BP and 9920 ± 120 BP, respectively. Similar material from Tilling Green in the Rother Valley, at −22.5 m OD, has given an age of 9565 ± 120 BP, and from Langney Point near Eastbourne, at −24.9 m OD, 8760 ± 75 BP (Shephard-Thorn 1975). The timing of re-entry of the Flandrian Sea into the Strait of Dover has been estimated from an age of 8250 ± 300 BP for organic material below marine deposits in the region (Delibrias *et al.* 1974; Heyworth and Kidson 1982). The Eastbourne area has been examined in detail by Jennings and Smyth (1987). *Circa* 10 000 BP, mean sea-level was below −29 m OD and a millennium later, about −25 m OD. A shift in deposition from freshwater to estuarine occurs at *circa* −26 m OD, with peat at about −25 m OD indicating the removal of marine influence until 8770 ± 50 BP. A transgressive overlap and positive tendency of sea-level has been identified at this date. By about 3400 BP, mean sea-level was *circa* OD in the Eastbourne area. Since then, the efficiency of coastal barriers, together with the incidence of high tides and storms, are thought to have played a greater rôle in marine incursions and withdrawals than movements of sea-level. Jennings and Smyth (1987) have also studied the Combe Haven Valley, between Bexhill and Hastings in East Sussex. Here, a marine transgression has been recorded between 5780 ± 80 and 5170 ± 70 BP, with increased marine influence at 2170 ± 60 BP considered to be probably due to local geomorphic, tidal and climatic factors.

At Fawley in Hampshire, Churchill (1965b) has estimated a eustatic rise in sea-level of *circa* 3 m accompanying a crustal depression of over 4.5 m since about 6500 BP. Hodson and West (1972) have discerned initial signs of the Flandrian transgression into the proto-Southampton Water earlier than 6366 ± 124 BP. A basal shell bed overlain by silty clays, sands and peats has been interpreted as having formed in marsh and fen environments. The silty clays above these were probably deposited on tidal flats, then in salt-marsh. Organic material within the lower part of the latter beds has given the radiocarbon age quoted above. Fen peat above them has been dated to 3689 BP. An estuarine deposit indicative of mainly salt-marsh conditions comprises the rest of the sequence, and probably accumulated in response to another transgression which took place in the Romano-British Period.

In Start Bay, Devon, rapid onshore movement of a shingle barrier between about 8000 and 5000 BP has been inferred. After the latter time, this migration slowed, barrier beaches were constructed and coastal lagoons impounded as the rate of sea-level rise was reduced. The Lower Ley at Slapton was a tidal lagoon until *circa* 2900 BP, when it was sealed by the barrier. Inundation of a freshwater lagoon in this locality by the sea about 1800 BP has been ascribed to a barrier breach during gales, when sea-level was approximately that of today. Subsequent stratigraphic changes signifying increases in marine influence have been interpreted against the same backdrop of sea-level, as having been produced by sediment washed over the barrier during storms (Morey 1976, 1983).

In Jersey (Jones *et al.* 1987, 1990), the pattern of sea-level movements prior to *circa* 6000 BP is consistent with that of other areas around the adjacent coasts of England and France (Larsonneur 1971; Morzadec-Kerfourn 1974; Heyworth and Kidson 1982). However, some localities on the island, notably Le Marais de St. Pierre in St. Aubin's Bay, indicate that while the rate of sea-level rise slowed following this time, there were no stillstands or regressions until after a peak was attained *circa* 4500 BP. At this maximum height, which has been calculated to have been 2–3 m above its present level, the sea would have been capable of breaching coastal sand barriers and invading lagoons, in which its presence was established for some 2000 years thereafter. Since *circa* 2500 BP, sea-level has been interpreted as having fallen to that of today. However, a complex sequence of oscillations between freshwater, brackish and marine environments over this timespan has been identified (Figure 10.10). These may have been a response to either minor transgressive and regressive episodes within the overall downward trend of sea-level, or to changes in coastal geomorphology and to extreme events such as storms. While there is evidence of marine conditions at levels above those of today between *circa* 4500 and 2500 BP in Jersey, the idea of a higher sea-level at this time is at variance with the pattern established in its environs, which demonstrates that sea-level did not exceed its present height earlier in the Flandrian (Ters 1973; Heyworth and Kidson 1982). It has been suggested that if such a sea-level is untenable on a regional basis, the most plausible explanation for these anomalous circumstances might involve recent crustal movements. Faults have been recognized bounding Jersey (Bishop and Bisson 1989), and the island may have undergone slight uplift along pre-existing lines of weakness during the late Flandrian. Alternatively, Jersey may have remained stable while surrounding areas were downwarped.

10.11.3 The Bristol Channel, western and northern Wales coasts

Around the Bristol Channel, the Bridgwater Bay – Somerset Levels area has received most attention (Kidson and Heyworth 1973, 1976, 1978; Kidson 1977; Heyworth and Kidson 1982). In what has been considered to have been an isostatically stable area, sea-level probably rose smoothly from about −25 m OD *circa* 9000 BP to around −10 m at *circa* 7000 BP, −4 m at 5000 BP and −2 m at 3000 BP, attaining its current height within the last millennium. Estuarine clays were deposited over peat in Bridgwater Bay *circa* 8400 BP. There was a subsequent progressive slackening in the pace of sea-level rise, until *circa* 6000 BP, a clay surface emerged at about OD, upon which freshwater peat formation began (Heyworth and Kidson 1982). Estuarine clay has been located beneath the Somerset Levels, where peat formation above it commenced after about 5500 BP (Dewar and Godwin 1963).

In Cardigan Bay, the Dovey Estuary received attention from Godwin and Newton (1938) and Godwin (1943). Haynes *et al.* (1977) have obtained a radiocarbon date of 8740 ± 110 BP from salt-marsh peat beneath inter-tidal deposits at −20.5 m OD in this area. A regressive overlap *circa* 6000–5100 BP has been detected by Wilks (1979), whose data on sea-level rise are comparable with those obtained from Bridgwater Bay covering the last eight millennia. In Clarach Bay, just north of Aberystwyth, marine influence has been recognized in a sedimentary sequence *circa* 6000 BP, and has been interpreted as indicating the increasing proximity of the coastline as sea-level rose (Heyworth *et al.* 1985). In North Wales, five transgressive episodes have been hinted at by Tooley (1974), with a regressive overlap at 2.43 m OD, radiocarbon dated to 4725 ± 65 BP (Tooley 1978c).

10.11.4 North-western England

Twelve phases of transgressive overlap and 12 of regressive overlap have been established from *circa* 9000 BP to the present along the coastline of north-west England from the Mersey Estuary to the Solway Firth (Tooley 1974, 1976, 1977, 1978b, 1982) (Figure 10.13). Transgressive overlaps at various elevations indicate a rapid rise in relative sea-level. This rapid rise had concluded by *circa* 6000 BP, since which time, oscillations in both directions, not exceeding 4 m, have taken place. While such changes could have been local and reflect modifications in coastal processes and isolated occurrences, an interpretation involving a more widespread action, such as eustatic variation, has been favoured in view of the mutual compatibility exhibited by the data.

10.11.5 Scotland

Dawson (1984) has reviewed the evidence for Postglacial sea-level changes in western Scotland. Here, the early Flandrian, Main Buried Beach and Low Buried Beach, identified in eastern Scotland (Sissons 1966, 1967a) (see below), have not been recognized. However, low relative sea-levels in south-western Scotland have been indicated by radiocarbon dates ranging from *circa* 9620 to 6800 BP on organic material below deposits of the Main Postglacial Transgression (Jardine 1975, 1982; Jardine and Morrison 1976; Bishop and Coope 1977). In the Solway Firth, relative sea-level rose *circa* 9400–8500 BP, and the transgression reached its maximum about 5000 BP according to Jardine (1975, 1982). In the northern

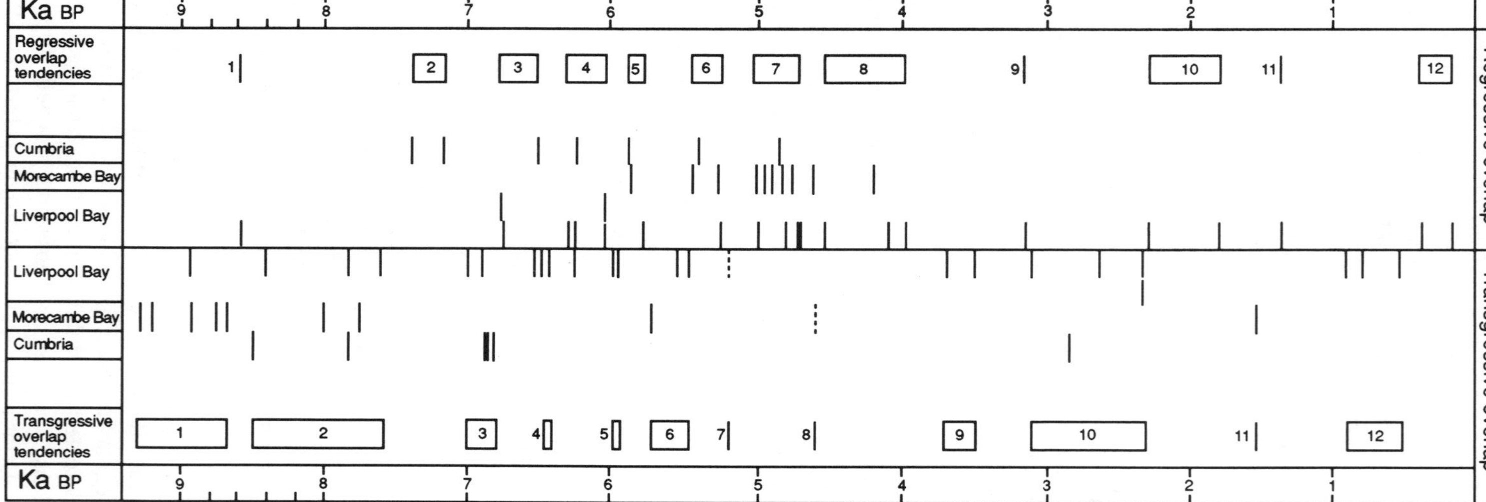

Figure 10.13 Transgressive and regressive overlap tendencies from sites in three areas – Liverpool Bay, Morecambe Bay, Cumbria (including the Isle of Man) – in north-west England. Each bar is an uncorrected radiocarbon date from transgressive and regressive overlaps or from sand dune palaeosols, and for the whole of the north-west of England these can be resolved into 12 transgressive overlap tendencies and 12 regressive overlap tendencies. Transgressive overlaps 7 and 8 have been established stratigraphically, and these are shown by dashed lines. From Tooley (1982).

Solway Firth, estuarine deposits of the Main Postglacial Transgression have been found intercalated with peats. The maximum radiocarbon age obtained from the base of the upper peat is *circa* 6650 BP, while the youngest date on the sub-estuarine deposit peat is about 6800 BP. This has prompted Sissons (1974b) to suggest that the Main Postglacial Transgression in the Solway Firth peaked between these times. However, Jardine (1975, 1982) has invoked temporal variations in this phenomenon as a result of local coastal and marine influences, with culmination *circa* 6000 BP in its western and 5000 BP in its eastern sector. Walker (1966) has intimated that in the southern part of the Solway Firth, this transgression was at its height between about 6950 and 6600 BP. Jardine (1977) has stated that the Main Postglacial Transgression had begun in northern Jura and Oronsay by *circa* 7400 BP, with regression evident in the latter area between about 6560 and 5660 BP. In the environs of Loch Lomond, marine deposition was occurring *circa* 6900 BP and had stopped by around 5450 BP (Dickson *et al.* 1978). According to Thompson and Wain-Hobson (1979), the sea withdrew from Loch Sheil, west of Glenroy, *circa* 4200 BP. Raised beaches of Flandrian age have not been identified in the Outer Hebrides. This and related evidence have led Ritchie (1966) to conclude that the transgression had not attained a height above present sea-level and took place before *circa* 5700 BP. Von Weymarn (1974) and Jardine (1977), noting the absence of the Main Postglacial Shoreline in this area, have inferred an accompanying sea-level not in excess of that of today.

Dawson (1984) has observed that landforms associated with the climax of the Main Postglacial Transgression in western Scotland were mostly beach ridges and shingle accumulations, with a relatively small number of terrace fragments. The highest elevation of the Main Postglacial Shoreline has been established at around 14 m OD in the Oban area, its altitude decreasing north-west, south-west and west from the centre of isostatic uplift (Figure 10.14) (Figure 1.12). Comparison of the isobase pattern for the Main Lateglacial Shoreline (Figure 9.22) with that for the Main Postglacial Shoreline has led Gray (1983) to propose that the centre of uplift had moved east by over 30 km during the early Flandrian. This may have been due to the build up of glaciers during the Loch Lomond Stadial which either re-lowered the crust or delayed its uplift. Subsequent to the Main Postglacial Transgression, the overall trend has been of relative regression, mainly in response to the persistence of glacio-isostatic uplift, whose pace, however, has slowed. Superimposed on this trend appear to have been minor transgressive episodes. For instance, in eastern Mull, where the Main Postglacial Shoreline has been recognized at *circa* 12 m OD, further shorelines have been discerned at about 8 and 4 m OD (Gray 1974). In the Solway Firth, the Main Postglacial Shoreline, identified at around 8 m OD, is accompanied by another shoreline 1–2 m above high-water mark (Jardine 1975). In south-western Jura, a staircase of 30 beach ridges, seaward of the Main Postglacial Shoreline, was produced during the course of the relative marine regression (Dawson 1984).

The coastline of eastern Scotland has been both extensively and intensively studied. Evidence from the western Forth Valley (Newey 1966; Sissons 1966; Sissons and Brooks 1971), lower Strathearn (Cullingford *et al.* 1980, 1989) and the inner Moray Firth (Haggart 1986; Firth and Haggart 1989) has indicated that the Main Buried Beach (a shoreline and estuarine flat) of early Flandrian age was abandoned *circa* 9600 BP due to a fall in relative sea-level. A subsequent minor increase in relative sea-level gave rise to a further shoreline and estuarine flat at lower elevations (the Low Buried Beach) in the western Forth Valley (Sissons 1966) and lower Strathearn (Cullingford *et al.* 1980, 1989) *circa* 8600 BP. This feature was also abandoned by a fall in relative sea-level that appears to have ended *circa* 8600–8200 BP in the western Forth Valley and 8300–7600 BP in lower Strathearn. Haggart (1989) has computed rates of relative sea-level fall in these areas during the early Flandrian. It was 0.3 m per 100 years in the western Forth Valley, 0.69 m per 100 years in the inner Moray Firth and 0.26 m

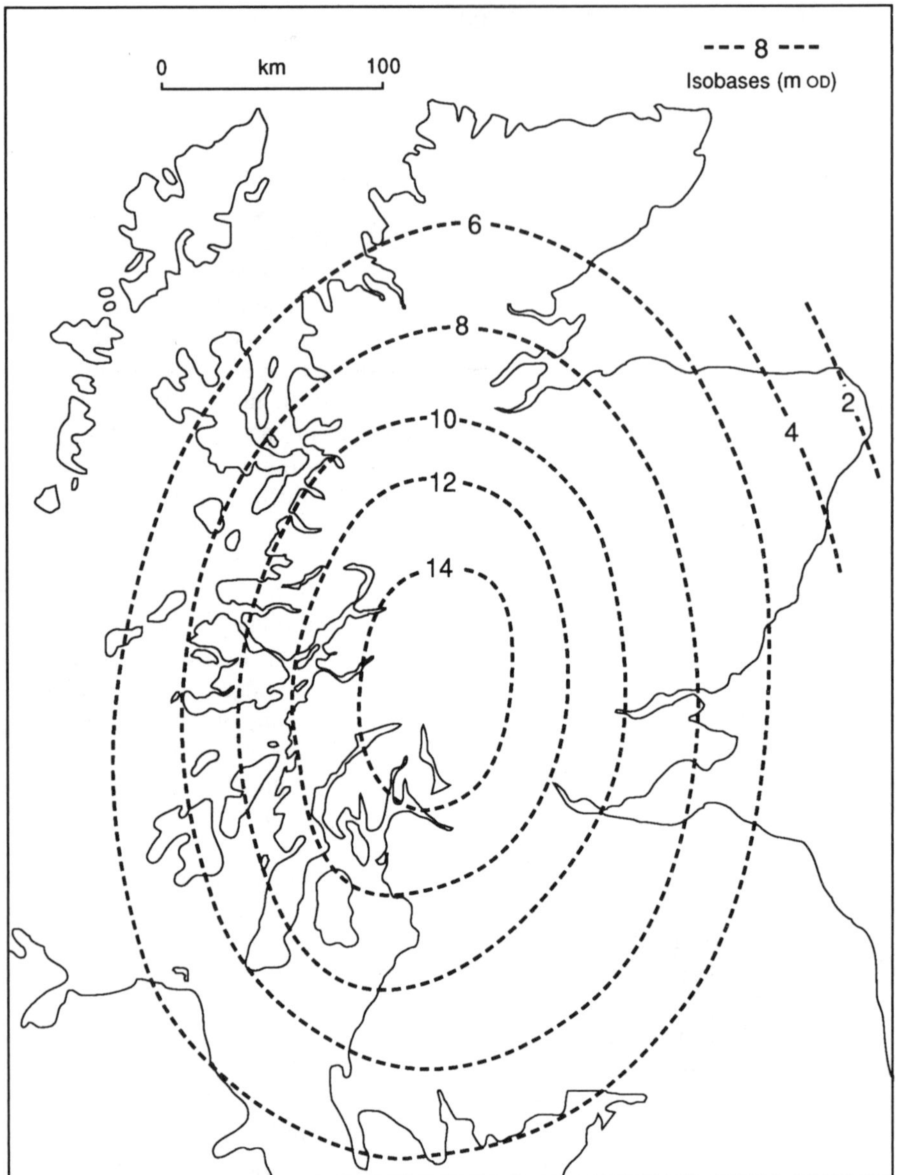

Figure 10.14 Isobases on the Main Postglacial Shoreline in Scotland. From Gray (1992).

per 100 years in lower Strathearn. It has been suggested that analogous values for the first two areas may, in view of the larger amount of isostatic uplift experienced by the former, reflect more rapid rebound in the latter over this timespan.

The main Flandrian rise in relative sea-level commenced subsequent to the dates of the last minima noted above. It has been detected at various times within the three areas of eastern Scotland from which earlier fluctuations have been described. Haggart (1989) has forwarded the idea that such differences may be explicable because of contrasting rates of isostatic uplift.

The progress of the Main Postglacial Transgression and its culmination have been examined in detail in several localities. The Main Postglacial Shoreline has been identified between Arnprior and Dunbar in south-eastern Scotland (Sissons *et al.* 1966). In the Montrose Basin, estuarine flats developed inland with sand and shingle ridges seaward, during the Main Postglacial Transgression (Smith and Cullingford 1985). The culmination of the latter is represented by a carse (estuarine clay) surface *circa* 6–7 m OD, and by sand and shingle ridges at up to about 10 m OD. A second surface *circa* 4–5 m OD in the carselands, and sand and shingle accumulation up to around 6 m OD, were produced subsequently when relative sea-level was in overall decline. At Fullerton, south of Old Montrose, in the south-west of the basin, the Main Postglacial Transgression culminated between 6983 ± 60 and 6704 ± 55 BP (Smith *et al.* 1980). It has not been possible to ascertain the date of the onset of the transgression, but comparison with eastern Fife (Chisholm 1971) has indicated that this may have been *circa* 7600 BP. Morrison *et al.* (1981) have also presented data from eastern Fife, and from the carseland margin east of Perth. Smith *et al.* (1985a) have contributed further information on the latter locality, where the Main Postglacial Transgression, in progress since about 7600 BP, peaked *circa* 6100 BP, with the sea withdrawing from the mouth of a gulley by 5735 ± 75 BP.

The Philorth Valley near Fraserburgh, located marginal to the area which was influenced by glacio-isostacy, has been examined by Smith *et al.* (1982). Peat growth continued here until at least 8465 ± 95 BP. An episode of fluvial erosion was followed by the deposition of sand, then peat, the latter forming from *circa* 7510 to 6300 BP. Marine sedimentation took place thereafter, concluding around 5700 BP. The Main Postglacial Shoreline has been identified at up to 2 m OD in this area. A further period of peat growth ensued until about 4760 BP, with estuarine deposition beginning after this time. In the lower Ythan Valley, Buchan, early Flandrian peat growth was arrested by the deposition of a marine clay of carse type (Smith *et al.* 1983). The marine transgression which was responsible for this sediment culminated between *circa* 6189 ± 95 and 4000 ± 80 BP. A carse surface traced at around 4.5 m OD, the most elevated in the area, has been equated with the Main Postglacial Shoreline. Estuarine deposits and a shoreline were associated with the Main Postglacial Transgression in the inner Moray Firth (Firth and Haggart 1989). The culmination of the transgression took place *circa* 7100–5500 BP. At the head of Beauly Firth, the Main Postglacial Shoreline has been mapped at about 9 m OD. Four lower shorelines have also been identified, and interpreted as probable representatives of either stillstands or minor rises in relative sea-level.

Within the carse of the inner Moray Firth, a thin layer of fine marine sand has been recognized, and its deposition has been radiocarbon dated to between *circa* 7400 and 7200 BP, where it wedges out into peat (Firth and Haggart 1989). Analogous material of comparable age (about 7000 BP) has been identified over a wide area of the coastline of eastern Scotland from the Dornoch Firth to the Forth Valley (Chisholm 1971; Smith *et al.* 1980, 1981, 1983; Morrison *et al.* 1981; Long *et al.* 1989). The layer has been interpreted as the product of one brief episode, which Smith *et al.* (1985b) have suggested was a storm surge. However, Dawson *et al.* (1988) and Long *et al.* (1989) have postulated that it was a tsunami deposit. The probable source of the tsunami was an earthquake-induced submarine landslide, the Second Storegga Slide. This took place on the Norwegian continental slope *circa* 750 km to the north-east of eastern Scotland between 8000 and 5000 BP, probably nearer to the former date (Jansen *et al.* 1987).

Anomalies between the sea-level curves for south-eastern and south-western Scotland have been noted by Sissons (1976), who has suggested that the greater distance of the latter area from the centre of glacio-isostatic uplift may have led to the transgression there being accomplished at greater speed. Jardine (1975, 1982), however, has disagreed, pointing out that the transgression did not terminate until 5000 BP in the eastern Solway Firth, and that the sea did not enter Loch Lomond until about 6900 BP, and remained there for some 1500 years (see above). Jardine

(1982) has also drawn attention to evidence of a relative sea-level rise *circa* 9400–8500 BP in the Solway Firth, while in the Forth Valley, relative sea-level fell between about 10 400 and 8500 BP, with the transgression concluding around 6500 BP. The distance between these areas may have been sufficient to account for such variations. As part of a comparison of Flandrian sea-level curves for Scotland, Haggart (1989) has examined the evidence of Jardine (1975) from the eastern Solway area which points to a rise in relative sea-level from 9390 ± 130 to 8135 ± 150 BP. After a re-interpretation of these data, a curve bearing a marked resemblance to those established for lower Strathearn and the inner Moray Firth over similar timespans has been forthcoming. Haggart has indicated that the Main Postglacial Transgression culminated *circa* 6600 BP in the Solway Firth. In the western Forth Valley the peak occurred prior to 6490 BP, in eastern Fife 6000–5800 BP, on Tayside *circa* 6200 BP, in lower Strathearn between 6679 and 6083 BP, and between 7100 and 5510 BP in the inner Moray Firth. The current difference in elevation between the Main Postglacial Shoreline in the western Forth Valley (*circa* 14.5 m OD) and the eastern Solway Firth (*circa* 10 m) is probably a response to a fall in the rate of isostatic uplift in the latter relative to the former since the middle Flandrian.

10.11.6 Ireland

Carter (1982) has emphasized the interaction of isostasy and eustacy in Northern Ireland during the Littletonian. In the early part of the stage, swift eustatic transgression and declining isostatic uplift allowed sea-level to attain a height of about 3 m below that of today. Variations in the elevation and age of the maximum transgression have been identified between the eastern and northern coastlines. Isostatic rebound was negligible in the former area after 7500 BP and is likely to have ceased by 5000 BP. Sea-level appears to have risen rapidly along this coast until about 7000 BP and fallen after 5000 BP to the present.

Sites around Strangford Lough (Stephens and Collins 1960; Morrison and Stephens 1960), especially Woodgrange near Downpatrick (Singh and Smith 1966, 1973), have given an important insight into Littletonian land- and sea-level changes on the eastern coast. Mean sea-level may have been 1–1.5 m OD (Belfast Ordnance Datum – approximately 0.4 m above that of Newlyn) between 5000 and 3500 BP (Carter 1982). Gravel beach ridges have been cited by Mitchell and Stephens (1974) as evidence of a middle Littletonian sea-level above that of today, while Synge (1980) has identified three isostatically tilted shorelines (*circa* 6500, 4000 and 2500 BP) on the eastern coast. These authors have suggested that Littletonian sea-levels may have attained 3.5 m OD. However, Carter (1982) has questioned the use of such features as sea-level indicators because they would have been within the reach of tides and storms, a reservation accepted by Synge (1985).

Along the northern coast, isostatic uplift after 7500 BP was in excess of eustatic rise, and sea-level reached its maximum 6500–5500 BP. A fall to just below OD ensued and thereafter a rise, but whether this led to a sea-level above that of today is unclear (Carter 1982). Sites at Cushendun, County Antrim (Movius 1940; Mitchell 1971b, 1976), Drumskellan, County Donegal (Colhoun *et al.* 1973) and Lough Foyle (McMillan 1957) (see below) have contributed relevant data. Relative shoreline chronologies for the north-east of Ulster have been established by Prior (1966) and Synge and Stephens (1966).

In a discussion of Postglacial shorelines in Ireland, Synge (1985) has described three beach levels containing marine Mollusca (6–7, 5–6 and 3 m OD) around Sligo Bay, and noted raised-beach material in Clew Bay, Galway Bay and Bantry Bay, but its absence elsewhere on the west coast. The fact that the coastline of southern Ireland has been one where submergence rather than uplift has occurred during the Littletonian was emphasized.

Carter *et al.* (1989) have discussed late Littletonian sea-levels. In the south-west of Ireland, sea-level has continued to rise, but the rate of this has probably diminished during the last 2500 years, with landform modifications frequently involved (Devoy 1983; Shaw *et al.* 1986). An analogous situation has pertained along the south coast. In Cork Harbour, land has been submerged by a rise in relative sea-level since 2100 BP (Devoy 1984, 1985b). In the south-eastern sector of the south coast, the increasing water level has initiated significant coastal-barrier migration and erosion over approximately the same timescale (Carter and Orford 1984). A general rate of relative sea-level rise of between 0.6 and 1.1 mm per year during the last 5000 years along the south-western and southern coast has been computed by Carter *et al.* (1989).

The overall pattern of Littletonian sea-level movement described by Carter (1982) (see above) for the north of Ireland has been amplified by Carter *et al.* (1989). Along the north coast, isostatic influence varied after *circa* 7500 BP, and relative sea-level reached its maximum in northern County Antrim and northern County Londonderry about 6500 BP, attaining an elevation of 3–4 m in the Bann and Foyle valleys. Relative sea-level peaked 5000–4000 BP along the north-western sector of this coast, in western Donegal, apparently at a height below present sea-level (Shaw 1985). In Donegal, sea-level may have fallen to 3 m below that of today between 4500 and 3500 BP (Shaw 1985). The north-east coast seems to have witnessed a maximum sea-level *circa* 6500 BP. Between 5500 and 2500 BP, relative sea-level fell slowly to a lesser height than is current, thereafter rising. Such a sequence has been identified from Magilligan at the mouth of Lough Foyle (Carter 1982, 1986; Wilson and Bateman 1986, 1987; Wilson and Farrington 1989; Carter and Wilson 1990), where beach ridges, dunes and palaeosols have been radiocarbon dated.

10.11.7 Overview of crustal movements

Flandrian crustal movements and sea-level changes in England, Wales and Scotland have been discussed by Shennan (1989), who has pointed out that geological, geodetic and tide-gauge information has disclosed analogous patterns reflecting relative subsidence in southern England and relative uplift in Scotland. An investigation of tide-gauge data by Woodworth (1987) has revealed a fall in relative sea-level of 1.29 ± 0.22 mm per year at Aberdeen and a rise of 0.62 ± 0.20 mm per year at Sheerness between AD 1919 and 1982. Flemming (1982) has calculated crustal movement rates that vary from 2.5 m per millennium in the highland and western coastal area of Scotland to −0.5 m per millennium in the far south-west of England. Peltier (1987) has estimated rates of present relative sea-level change in the British Isles using geophysical and glaciological data. A rise in sea-level of between zero and 0.4 mm per year in south-eastern England is indicative of relative subsidence. A −0.4 mm per year isoline joins south-western Scotland, north-western Wales and eastern Ireland and delimits an area of relative uplift. The zero contour passes through the Highlands of Scotland, the eastern coast of England to the north of the Wash, and the coast of southern England close to the Isle of Wight. Shennan (1989) has employed sea-level index points in an investigation of past and present (Figure 10.15) rates of crustal movement, concluding that linear subsidence took place in southern England and Wales, while northern England and Scotland underwent non-linear uplift. Calculations have shown that, for example, Fenland subsided 0.86 ± 0.10 mm per year, 0–6500 BP, south-eastern England 0.85 ± 0.18 mm per year, 0–8750 BP, and the Bristol Channel 0.24 ± 0.19 mm per year, 0–2500 BP. Morecambe Bay was uplifted 0.35 ± 0.12 mm per year, 0–6000 BP, the inner Firth of Forth 2.0 mm per year, 0–4000 BP, and the Moray Firth 1.6 mm per year, 0–4750 BP. Estimated rates of uplift for Scotland are in good agreement with isobase data for the Main Postglacial Shoreline (Smith *et al.* 1969; Sissons 1983; Firth and Haggart 1989).

Carter *et al.* (1989) have considered that variations between sea-level curves within Ireland are mainly the result of differential glacial loading. However, they have also suggested that geoidal movement (Pirazzoli and Grant 1987) could have been involved. Newman and Baeteman (1987) have postulated the existence of a substantial geoidal bulge adjacent to north-western Ireland during the early Postglacial and have charted an apparent easterly excursion of this 7500–3500 BP.

10.12 The offshore sequence

Balson and Cameron (1985) have indicated that peat may have formed widely on the floor of the southern North Sea early in the Flandrian (sections 10.1 and 10.11.1). However, most of this organic material is likely to have been eroded subsequently, and peat has been located only to the north-east of East Anglia and on the northern side of the Dogger Bank (Cameron *et al.* 1987). The bulk of the southern North Sea Basin became submerged *circa* 10 000–7000 BP (Eisma *et al.* 1981). About the former time, ingress of brackish-marine water began, and muddy sands and clays were laid down extensively to the east of 2°E and between 53° and 55°N during the first phase of the transgression (Cameron *et al.* 1987). Deposition on tidal flats occupied an increased area between about 9000 and 8000 BP. By the early part of the seventh millennium BP, the majority of the Southern Bight possessed fully marine conditions (Eisma *et al.* 1981). On average, between a 5 and 15 m thickness of Flandrian sediments has been recorded in the Southern Bight and German Bight (Long *et al.* 1988). In addition to peat, river sands, freshwater clays and marine sands have been identified in the southern North Sea (Robinson 1968; Gallois 1979; Eisma *et al.* 1979; Jansen *et al.* 1979; Jelgersma *et al.* 1979; Oele and Schüttenhelm 1979). The present sea-bed deposits are fine-medium sands which become gravelly when lying above till and muddy in deeper water (Cameron *et al.* 1987).

The early Flandrian deposits in the central North Sea have demonstrated a gradation from estuarine and nearshore sands to marine silts and muds (Cameron *et al.* 1987). These sediments formed prior to 7000 BP (Holmes 1977; Jansen *et al.* 1979). Over much of this region, a thin layer of mostly muddy sand is all that has been recognized as representative of subsequent sedimentation.

In the northern North Sea, depressions to the east of the expanse of Devensian till (Chapter 9) have been found to contain sands and muds, some of which have been considered to be Postglacial in age by Cameron *et al.* (1987). Close to the median line, ridges composed of shallow-marine sand have been identified above valley-fill and its adjacent sediments. These features, *circa* 5 m high, about 5 km wide and up to 20 km long, may be of early Flandrian age (Rise and Rokoengen 1984). Present sea-bed sediments here are fine-grained sands that become gravelly when till and bedrock lie beneath. The fullest records of future sedimentation in the Flandrian will probably come from open depressions on the floor of the North Sea (Cameron *et al.* 1987).

On the continental shelf west of Scotland, marine muds and silts belonging to the Jura Formation (Chapter 9) were laid down during the early Flandrian. Evidence of tidal scouring has been found within them. The modern sediments, sandy and in thin waves or small fans, comprise the Lorne Formation, whose type area has been designated in the Firth of Lorne, north of Colonsay (Davies *et al.* 1984) (Figure 9.8).

In the south-western and central Celtic Sea, part of the Melville Formation may be early Flandrian in age. At this time, mainly sandy tidal deposits were fashioned into ridges which accreted landwards during the marine transgression. These ridges were subsequently eroded and then stabilized by the emplacement of a lag deposit. Above the latter, mobile sandy material containing bedforms has been recognized, and interpreted as the contemporary response to the characteristics of the ocean current in the area (Pantin and Evans 1984).

10.13 Climate

Some of the data sources discussed earlier in this chapter have provided indirect estimates of Flandrian palaeoclimatic characteristics. It will have been apparent that such indications have not always been harmonious. Direct information on Flandrian climates in the British Isles began with statements concerning weather conditions during

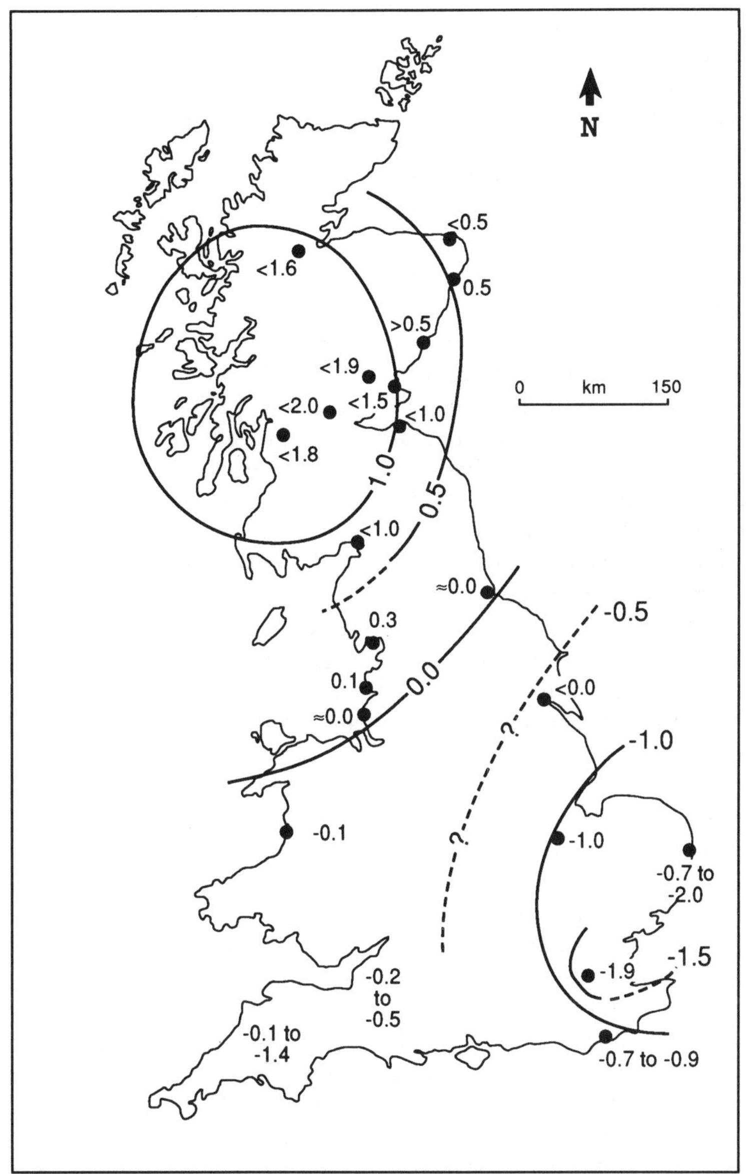

Figure 10.15 Map of estimated current rates (mm per year) of crustal movement in Great Britain. Isolines cannot be drawn for much of southern England, point estimates are shown for guidance. From Shennan (1989). Reproduced by permission of John Wiley and Sons Ltd. © 1989.

military campaigns in the first century BC. Descriptive accounts of weather commenced in the twelfth century AD as chronicles, and in the fourteenth century AD as diaries, while standard meteorological instrument readings have been taken since *circa* AD 1650 (Lamb 1982).

Predictions of past climatic parameters applicable to the British Isles have also been provided by numerical models of atmospheric circulation. These models have been based on the constituents of Croll – Milankovitch cycles, astronomically controlled variations in the amount of solar radiation received by the Earth (Roberts 1989). For example, Kutzbach (1981) has indicated that during the early Flandrian, temperature, precipitation and insolation values were more extreme than today. Kutzbach and Guetter (1986) have calculated that about 10 000–8000 BP, solar radiation in summer was *circa* 8% above its present level. Indeed, it appears likely that maximum summer temperatures during the Flandrian were attained between 9000 and 8000 BP, rather than *circa* 7000 BP, traditionally thought to have been the time of the Postglacial Climatic Optimum (see below). Kutzbach and Gallimore (1988) have predicted substantial variations in early and middle Flandrian climates.

The climatic amelioration which marked the end of the last cold stage and the beginning of the current temperate one was rapid (Lamb 1977). Seasonal and mean annual temperatures in the British Isles, reconstructed using radiocarbon-dated insect fossils (Atkinson *et al.* 1987) (Figure 10.16), demonstrate dramatic warming in the centuries around 10 000 BP. From minima *circa* 10 500 BP, a rise in summer and winter temperatures (Figure 10.16(a)) meant that by about 9800 BP, the climate was comparable with that at the optimum of the Windermere Interstadial (Chapter 9) (Figure 10.16(b)). It has been estimated that a rise in mean annual temperature of 1.7°C per century occurred around 10 000 BP. Subsequent values support both the notion of optimal mean annual temperatures before *circa* 7000 BP, and the major climatic trends *circa* 5000 BP and just prior to 3000 BP (Figure 10.16(b)), that have been postulated by Osborne (1976) (section 10.6).

Lamb *et al.* (1966) have employed physical laws to derive atmospheric circulation patterns for the British Isles *circa* 8500 BP. Against the backdrop of warmth in the Atlantic Ocean and Europe, and a substantial amount of ice in North America, a thermal gradient was set up. This caused cyclonic pressure systems to be directed north-eastwards, so that Europe would have been dominated by anticyclones. The British Isles would then have been reached by warm air from more southerly latitudes, and the climate would have been drier than today. Later shifts in major atmospheric circulation patterns during the Postglacial have been examined by Lamb (1977). The existing thermal gradient was reduced and west–east circulation bolstered by the melting of ice in North America after 8000 BP. The sub-polar low-pressure belt and sub-tropical high-pressure belt would then have been marginally to the south of their Boreal location. Summer temperature values *circa* 2°C above those of today have been suggested for the British Isles between 8000 and 6000 BP, when the optimum climate occurred. Over this period, winters would have been mild, while prevailing westerly winds and warm ocean would have led to precipitation values in excess of those of the present. Higher temperatures probably prevailed until *circa* 5000 BP, since which time there appears to have been a slow but oscillating trend to a colder climate. Cooling in arctic latitudes was probably responsible for the emergence of more powerful atmospheric circulation around 4000 BP. Between about 3000 and 2300 BP, a drop in average temperatures of up to 2°C has been postulated for the British Isles, together with an increase in precipitation. The climate 350 BC–AD 1300 has been interpreted as having been of an overall warmer and drier nature than that during the earlier part of the third millennium BP. However, wetter and colder episodes have been identified.

Lamb (1982) has charted the climatic sequence through the first and second millennia AD. Summers were warm and dry prior to AD 400, while colder and less-stable conditions characterized the sixth, eighth and ninth centuries. A warm oscillation took place from late in the tenth century until early in the fourteenth century. From the latter time until the end of the eighteenth century, there was a general lowering of temperatures accompanied by higher precipitation and more storms. A marked alteration appears to have been initiated in the middle of the sixteenth century, and until at least AD 1700 the British Isles experienced its coldest conditions since the Devensian Lateglacial, during the Little Ice Age, the characteristics of which have been discussed by Grove (1988). Indeed, the period between the

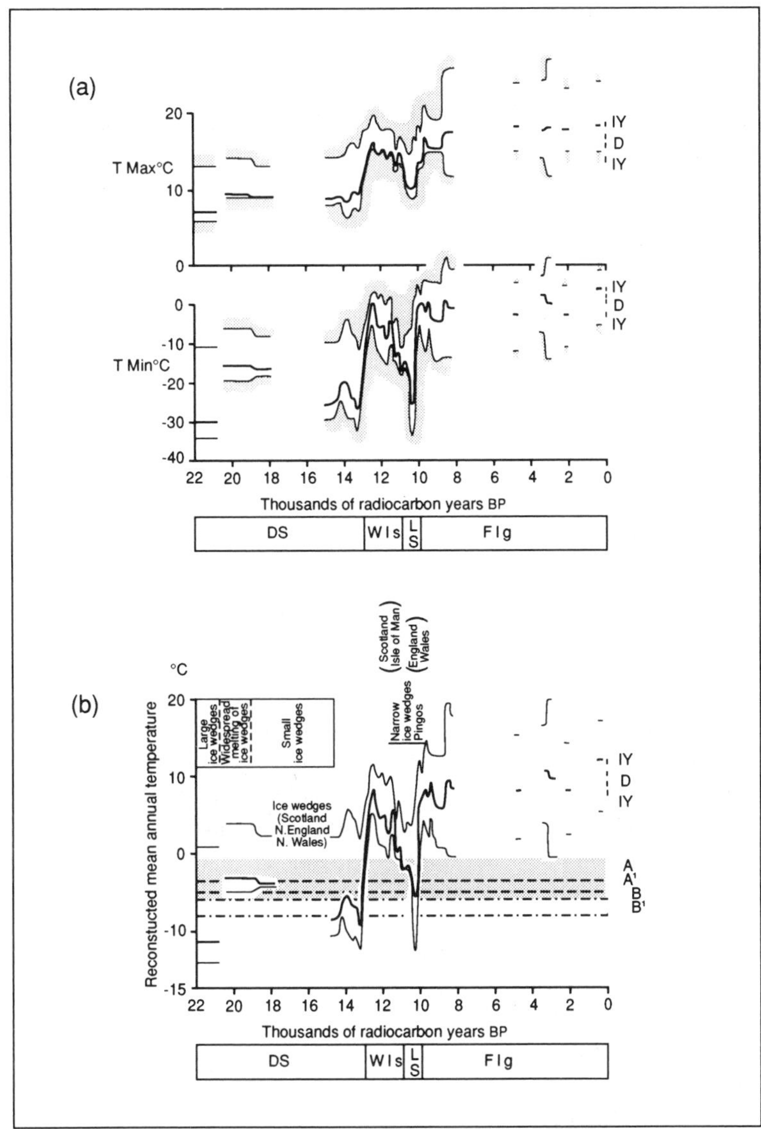

Figure 10.16 Seasonal and mean annual temperatures in the British Isles during the past 22 000 years, reconstructed using beetle remains. (a) Plots of reconstructed mean temperatures for the warmest (*T* MAX) and coldest (*T* MIN) months of the year based on time-averaged data. The shaded upper and lower bounds define the limits of the Mutual Climatic Range. The palaeotemperatures lay within these limits. The bold line shows the most probable value of palaeotemperature determined from two equations. Gaps represent periods without data. Arrows at right show the ranges of *T* MAX and *T* MIN in central England between 1659 and 1980 (Manley 1974). Inner arrows (D): decadal means. Outer arrows (IY): individual years. (b) Reconstructed values of mean annual temperature in England, Wales and southern Scotland. The continuous fine lines show upper and lower bounds of the Mutual Climatic Range. The bold line shows the most probable value of palaeotemperatures. Arrows at right show central England temperatures 1659–1980 as for (a). Shaded band shows temperature conditions within which open-system pingos can develop (Washburn 1979). Broken lines, A and A^1, show upper temperature limit for ice-wedge development according to Washburn (1979) and Hamilton *et al.* (1983). Chained lines, B and B^1, show same limit according to Péwé (1966). Box shows occurrences of ice wedges in north-west mainland Europe (Kolstrup 1980). Other legend shows approximate dates of periglacial features in Britain (Seddon and Holyoak 1985; Sissons 1974a, 1976, 1977; Watson 1977; Williams 1975). DS, Dimlington Stadial; WIs, Windermere Interstadial; LS Loch Lomond Stadial; FIg, Flandrian Interglacial. From Atkinson *et al.* (1987). Reprinted with permission from *Nature* **325**, 587–592; © 1987 Macmillan Magazines Ltd.

middle of the fifteenth and the middle of the nineteenth centuries experienced climatic traits which would be consonant with this episode. In the late seventeenth and early eighteenth centuries, sea temperatures a few hundred kilometres north of the British Isles were up to 5°C lower than those of today (Lamb 1982). Snow lay permanently on some Scottish mountains at this time, and Sugden (1977) has postulated that glaciers may have developed in certain corries in the Cairngorms between the seventeenth and nineteenth centuries. Rivers froze in winter and the frequency of storms and floods increased during the Little Ice Age. In the latter part of the seventeenth century, temperatures averaged about 1°C less than their twentieth century mean in the lowlands of England and are likely to have been as much as 2°C below it in Scotland (Lamb 1982). Climatic differences between the highland and lowland zones of the British Isles during the Flandrian have been estimated by Lamb *et al.* (1966) and Taylor (1975). Signs of a climatic recovery not long after AD 1700 have been detected, but its progress was irregular and a more consistent warming trend did not occur until the late nineteenth–early twentieth century (Manley 1971, 1974). More prevalent westerly winds and an increase in rainfall but decrease in snow and frost were characteristic of a warmer episode from about the second to the end of the fifth decade of the current century. From *circa* AD 1950 to 1980, changes in the distribution of atmospheric pressure and in temperature and rainfall values were heightened, and overall a cooling took place (Lamb 1982).

Over the last decade, a warming trend has developed but variability has remained. Reasons for the warming seem to have included a rise in the amount of carbon dioxide and other gases which assist in the retention of the earth's surface heat – the greenhouse effect. A reduction in ozone concentration, perhaps a consequence of the emission of chlorofluorocarbons into the atmosphere, has also been detected. Ozone depletion will have allowed the transmission of more ultra-violet radiation, and hence contributed to global warming (Simmons 1989). Estimates of temperature rises associated with the latter phenomenon have varied. An increase in average temperatures world-wide of up to 3°C by late in the next century has been postulated on the basis of a doubling of nineteenth-century levels of carbon dioxide. Within such a trend, regional rises of differing magnitude would occur. However, any assessment of future climatic conditions in the British Isles must set against recent human-induced modifications, knowledge of longer term natural changes in climate gained from the various lines of evidence discussed earlier in this chapter. We appear to be in the Late-temperate substage of an interglacial period *sensu* Turner and West (1968). An analogy with previous interglacials indicates that the current climatic trend is one of overall cooling towards the next cold stage. To what extent the latter phenomenon will counteract the former is uncertain.

11 Epilogue

This book has examined changes. Not only changes of environment during the most recent geological epoch, but also changes in the interpretation of the sequence of events over this timespan.

The last few decades have witnessed a substantial increase in the amount of information concerning Pleistocene environments in the British Isles. This can be ascribed to a number of factors. Firstly, to the accumulation of a larger body of data as a direct function of the passage of time and input by a greater number of investigators. Secondly, to the emergence of new methods of age determination, among which amino-acid epimerization, U-series and thermoluminescence dating have made important contributions. Thirdly, to either the initial or additional application of established techniques, such as palaeomagnetism and soil micromorphology, to Pleistocene sediments. Fourthly, to growth in the study of certain groups of fossils, insects and diatoms, for example, which are able to delimit specific environmental variables.

Beyond the range of the longest standing and most widely used method of geochronometric age determination, radiocarbon dating, the sequence of Pleistocene events in the British Isles traditionally has been ordered with reference to the geological, biological and archaeological characteristics, and to the geographical relationships of deposits and landforms. Biological evidence, especially that from plant fossils, has been paramount in the identification of temperate episodes. Orthodox sources have allowed 15 cold and temperate stages to be recognized. Four cold stages and four interglacial periods have been attributed to the Middle and Late Pleistocene.

However, data from deep-ocean sediments, and other sources such as ice and loess stratigraphy, have pointed to considerably more global cold and temperate episodes during the Pleistocene than those formulated for the British Isles. Up to 22 appear to have taken place since *circa* 800 000 BP, over a timescale which is thought to extend back about a further 1 million years.

Following these revelations, it has been realized that some of these additional events are likely to be registered in the Pleistocene deposits of the British Isles. Thus far, no long sequence of sediments of this age has been discovered. Moreover, in view of the nature of the successive cold episodes at this latitude, where erosion was effective, such sequences are unlikely to have been preserved. There have been hints of records covering a number of stages from areas such as Buchan, where erosive activity may have been reduced at certain times. Offshore, the sequence established on the continental shelf appears to contain evidence of more events in superposition than on land, but hiatuses have been identified, and the Middle and Late Pleistocene record is of three glaciations only.

Controversy concerning the status, and particularly the age of Pleistocene deposits and landforms, has long been part and parcel of their study. In a lithological context, the debate as to whether the Chalky Till of East Anglia and the Midlands of England is of Middle or Late Pleistocene age has provided an example. From a biological standpoint, the alternative interpretations of botanical and mammalian evidence in respect of the number of Late Pleistocene temperate episodes have been cited.

It seems that there have been good grounds for regarding the Chalky Till as the product of one glaciation during the Anglian. None the less, against the backdrop of the deep-ocean sequence, which has called for more rather than less stages, it has been permissible to query the resultant virtually complete absence of evidence for a major glaciation, the Wolstonian, whose presence has been widely attested elsewhere in northern and eastern Europe.

As no single locality has yielded a complete palynological record of its four substages, a composite vegetational history of the Ipswichian Interglacial has emerged. In addition to elements considered complementary to those identified at its type locality, others assumed analogous to parts of the succession have been referred to the stage. However, the mammalian faunas of some of these sites, at Aveley and Ilford for instance, exhibit significant differences from those regarded as typically Ipswichian. They could be placed earlier in time, and between those of the latter and the Hoxnian Interglacial. Over such a reduced time interval, botanical differences, such as those used to characterize the Hoxnian and Ipswichian interglacials, would probably be undetectable. Thus, episodes of different ages could have erroneously been included in the Ipswichian Stage. The deep-ocean record has indicated a temperate phase between those correlated with the Hoxnian and Ipswichian interglacials. Moreover, new localities such as Stanton Harcourt and Marsworth have produced a variety of evidence, including amino-acid ratios from the former and radiometric ages from the latter, that support this event, which was most probably an interglacial.

Other significant advances have been less controversial. Faunal data from late Lower Pleistocene and early Middle Pleistocene circumstances have permitted separation of the type Cromerian locality at West Runton and its correlatives, from sites such as Westbury-sub-Mendip and Boxgrove. The latter localities appear to possess material of a later date than the Cromerian *sensu stricto*, and representative of additional temperate intervals (with intervening cold ones) prior to the Anglian Stage. Faunal remains have also assisted in confirming a series of cold and temperate oscillations over time traditionally assigned to the Early Devensian, that have been identified in both the deep-ocean record, and in terrestrial sequences within Europe. At Stump Cross Cave, a cold-climate fauna dated to *circa* 83 000 BP has been considered a possible representative of the Oxygen Isotope Substage 5b environment, while a temperate assemblage of animals from Bacon Hole Cave, aged to about 81 000 BP, has been thought likely to represent the interstadial or interglacial conditions associated with Substage 5a.

Thus, it appears that a more flexible approach to the Pleistocene succession of the British Isles has been adopted by many workers over the last two decades. Certain of the established correlations and parts of the chronosequence have been fairly challenged. Quite sound foundations have been laid, both for the corroboration of those additional episodes already recognized and for the identification of others. As stated at its outset, this book is intended as a very provisional framework for this task.

References

Aalto, M. M., G. R. Coope and P. L. Gibbard (1984) Late Devensian river deposits beneath the Floodplain Terrace of the River Thames at Abingdon, Berkshire, England. *Proceedings of the Geologists' Association* **95**, 65–79.

Adam, K. D. (1954) Die mittelpleistozänen Faunen von Steinheim an der Murr (Wurtemberg). *Quaternaria* **1**, 131–144.

Adam, K. D. (1975) Die mittelpleistozänen Säugetier – Fauna aus dem Heppenloch bei Gutenberg (Würtemberg). *Stuttgarter Beiträge sur Naturkunde aus dem Straatlichen Museum fur Naturkunde in Stuttgart* **B3**, 1–247.

Aguirre, E. and G. Pasini (1985) The Pliocene – Pleistocene boundary. *Episodes* **8**, 116–120.

Alabaster, C. and A. Straw (1976) The Pleistocene context of faunal remains and artefacts discovered at Welton-le-Wold, Lincolnshire. *Proceedings of the Yorkshire Geological Society* **41**, 75–94.

Alderton, A. M. (1983) *Flandrian vegetational history and sea-level change of the Waveney Valley*. Ph.D. Thesis, University of Cambridge.

Allen, I. M., D. Britton and M. H. Coghlan (1970) *Metallurgical reports on British and Irish Bronze Age implements and weapons in the Pitt Rivers Museum*. Pitt Rivers Museum Occasional Papers on Technology Number 10. Oxford: Pitt Rivers Museum.

Allen, P. (1982) Great Blakenham and Creeting St. Mary. In *Field meeting guide. Suffolk*, P. Allen (ed.), Section 2, 1–18. London: Quaternary Research Association.

Allen, P. (1988) Great Blakenham. In *The Pliocene – Middle Pleistocene of East Anglia. Field guide*, P. L. Gibbard and J. A. Zalasiewicz (eds.), 87–99. Cambridge: Quaternary Research Association.

Allison, J., H. Godwin and S. H. Warren (1952) Late-glacial deposits at Nazeing in the Lea Valley, north London. *Philosophical Transactions of the Royal Society of London* **B236**, 169–240.

Andersen, S. T. (1957) New investigations of interglacial fresh-water deposits in Jutland. A preliminary report. *Eiszeitalter und Gegenwart* **8**, 181–186.

Andersen, S. T. (1961) Vegetation and its environment in Denmark in the Early Weichselian Glacial (Last Glacial). *Danmarks Geologiske Undersøgelse* Series 2, **75**, 1–175.

Anderton, R., P. H. Bridges, M. R. Leeder and B. W. Sellwood (1979) *A dynamic stratigraphy of the British Isles: a study in crustal evolution*. London: Allen and Unwin.

Andrew, R. and R. G. West (1977a) Pollen spectra from the Pliocene Crag at Orford, Suffolk. *New Phytologist* **78**, 709–714.

Andrew, R. and R. G. West (1977b) Pollen analysis from Four Ashes, Worcs. In: Early and Middle Devensian flora and vegetation. R. G. West, *Philosophical Transactions of the Royal Society of London* **B280**, 242–246.

Andrews, C. W. (1919) Description of the bones of *Elephas*. In: On a deposit of interglacial loess, and some transported preglacial freshwater clays on the Durham coast. C. T. Trechmann, *Quarterly Journal of the Geological Society of London* **75**, 201.

Andrews, J. T. (1971) *Techniques of till fabric analysis*. British Geomorphological Research Group Technical Bulletin Number 6. Norwich: Geo Abstracts.

Andrews, J. T., D. Q. Bowen and C. Kidson (1979) Amino acid ratios and the correlation of raised beach deposits in south west England and Wales. *Nature* **281**, 556–558.

Andrews, J. T., D. D. Gilbertson and A. B. Hawkins (1984) The Pleistocene succession of the Severn Estuary: a revised model based upon amino acid racemization studies. *Journal of the Geological Society of London* **141**, 967–974.

Annable, F. K. and D. D. A. Simpson (1964) *Guide catalogue of the Neolithic and Bronze Age collections in Devizes Museum*. Devizes: Wiltshire Archaeological and Natural History Society.

ApSimon, A. M. (1969) An early Neolithic house in County Tyrone. *Journal of the Royal Society of Antiquaries of Ireland* **99**, 165–168.

ApSimon, A. M. (1976) Ballynagilly and the beginning and the end of the Irish Neolithic. In *Acculturation and continuity in Atlantic Europe*, S. J. de Laet (ed.), 15–30. Bruges: de Tempel.

ApSimon, A. M. (1977) Brean Down sand cliff. In *Field handbook. Bristol*, K. Crabtree (ed.), 33–38. Bristol: Quaternary Research Association.

ApSimon, A. M., D. T. Donovan and H. Taylor (1961) The stratigraphy and archaeology of the late-glacial and post-glacial deposits at Brean Down, Somerset. *Proceedings of the University of Bristol Spelaeological Society* **9**, 67–136.

ApSimon, A. M., C. S. Gamble and M. L. Shackley (1977) Pleistocene raised beaches on Ports Down, Hampshire.

Proceedings of the Hampshire Field Club and Archaeological Society **33**, 17–32.

Archaeology Correspondent (1968) High Lodge Palaeolithic industry. *Nature* **220**, 1065–1066.

Arkell, W. J. (1943) The Pleistocene rocks at Trebetherick Point, north Cornwall: their interpretation and correlation. *Proceedings of the Geologists' Association* **54**, 141–170.

Arkell, W. J. (1947) *The geology of the country around Weymouth, Swanage, Corfe and Lulworth*. Memoirs of the Geological Survey of Great Britain. London: His Majesty's Stationery Office.

Arnold, E. N. and J. A. Burton (1978) *A field guide to the reptiles and amphibians of Britain and Europe*. London: Collins.

Ashworth, A. C. (1969) *The Late Quaternary coleopterous faunas from Rodbaston Hall, Staffordshire and Red Moss, Lancashire*. Ph.D. Thesis, University of Birmingham.

Ashworth, A. C. (1972) A Late-glacial insect fauna from Red Moss, Lancashire, England. *Entomologica Scandinavica* **3**, 211–224.

Atherden, M. A. (1976) The impact of late prehistoric cultures on the vegetation of the North York Moors. *Transactions of the Institute of British Geographers* New Series **1**, 284–300.

Atkinson, R. J. C. (1960) *Stonehenge*. London: Penguin.

Atkinson, T. C., K. R. Briffa and G. R. Coope (1987) Seasonal temperatures in Britain during the past 22 000 years, reconstructed using beetle remains. *Nature* **325**, 587–592.

Atkinson, T. C., R. S. Harmon, P. L. Smart and A. C. Waltham (1978) Palaeoclimatic and geomorphic implications of $^{230}Th/^{234}U$ dates on speleothems from Britain. *Nature* **272**, 24–28.

Atkinson, T. C., T. J. Lawson, P. L. Smart, R. S. Harmon and J. W. Hess (1986) New data on speleothem deposition and palaeoclimate in Britain over the last forty thousand years. *Journal of Quaternary Science* **1**, 67–72.

Averdieck, F. R. (1967) Die vegetationsentwicklung des Em-Interglazials und der Frühwürm-Interstadiale von Odderade/Schleswig-Holstein. *Frühe Menscheit und Umwelt* **2**, 101–125.

Avery, B. W. (1985) Argillic horizons and their significance in England and Wales. In *Soils and Quaternary landscape evolution*, J. Boardman (ed.), 69–86. Chichester: Wiley.

Avery, B. W., P. Bullock, J. A. Catt, J. H. Rayner and A. H. Weir (1982) Composition and origin of some brickearths on the Chiltern Hills, England. *Catena* **9**, 153–174.

Baden-Powell, D. F. W. (1930) Notes on raised beach Mollusca from the Isle of Portland. *Proceedings of the Malacological Society of London* **19**, 67–76.

Baden-Powell, D. F. W. (1934) On the marine gravels at March, Cambridgeshire. *Geological Magazine* **71**, 193–219.

Baden-Powell, D. F. W. (1948) The chalky boulder clays of Norfolk and Suffolk. *Geological Magazine* **85**, 279–296.

Baden-Powell, D. F. W. (1951) The interglacial beds at Hoxne, Suffolk. *Nature* **168**, 701–702.

Baden-Powell, D. F. W. (1955) Report on the marine fauna of the Clacton channels. In: The Clacton (Essex) channel deposits. S. H. Warren, *Quarterly Journal of the Geological Society of London* **111**, 301–304.

Baillie, M. G. L. (1973) A recently developed Irish tree-ring chronology. *Tree-Ring Bulletin* **33**, 15–28.

Baillie, M. G. L. (1982) *Tree-ring dating and archaeology*. London: Croom Helm.

Baillie, M. G. L. and J. R. Pilcher (1985) Dendrochronology. In *The Quaternary history of Ireland*, K. J. Edwards and W. P. Warren (eds.), 294–302. London: Academic Press.

Baker, C. A. (1977) *Quaternary stratigraphy and environments in the upper Cam Valley*. Ph.D. Thesis, University of London.

Baker, C. A. (1983) Glaciation and Thames diversion in the mid-Essex depression. In *Field guide. Diversion of the Thames*, J. Rose (ed.), 39–49. Hoddesdon: Quaternary Research Association.

Baker, C. A. and D. K. C. Jones (1980) Glaciation of the London Basin and its influence on the drainage pattern: a review and appraisal. In *The shaping of southern England*, D. K. C. Jones (ed.), 131–176. Institute of British Geographers Special Publication Number 11. London: Academic Press.

Balescu, S. and P. Haesaerts (1984) The Sangatte raised beach and the age of the opening of the Strait of Dover. *Geologie en Mijnbouw* **63**, 355–362.

Balescu, S., S. C. Packman and A. G. Wintle (1991) Chronological separation of interglacial raised beaches from northwestern Europe using thermoluminescence. *Quaternary Research* **35**, 91–102.

Ball, D. F. (1960) Relic-soil on limestone in South Wales. *Nature* **187**, 497–498.

Ball, D. F. (1975) Processes of soil degradation: a pedological point of view. In *The effect of man on the landscape: the Highland Zone*, J. G. Evans, S. Limbrey and H. Cleere (eds.), 20–27. CBA Research Report Number 11. London: Council for British Archaeology.

Ballantyne, C. K. (1987) The present-day periglaciation of upland Britain. In *Periglacial process and landforms in Britain and Ireland*, J. Boardman (ed.), 113–126. Cambridge: Cambridge University Press.

Balson, P. S. (1981) *The sedimentology and palaeoecology of the Coralline Crag (Pliocene) Suffolk*. Ph.D. Thesis, University of London.

Balson, P. S. and T. D. J. Cameron (1985) Quaternary mapping offshore East Anglia. *Modern Geology* **9**, 221–239.

Bandy, O. L. and J. A. Wilcoxon (1970) The Pliocene – Pleistocene boundary, Italy and California. *Geological Society of America Bulletin* **81**, 2939–2947.

Banham, P. H. (1968) A preliminary note on the Pleistocene stratigraphy of north-east Norfolk. *Proceedings of the Geologists' Association* **79**, 507–512.

Banham, P. H. (1971) Pleistocene beds at Corton, Suffolk. *Geological Magazine* **108**, 281–285.

Banham, P. H. (1975) The Contorted Drift of north Norfolk. *Bulletin of the Geological Society of Norfolk* **27**, 55–60.

Banham, P. H. (1977) Glacitectonites in till stratigraphy. *Boreas* **6**, 101–105.

Banham, P. H., H. Davies and R. M. S. Perrin (1975) Short field meeting in north Norfolk. *Proceedings of the Geologists' Association* **86**, 251–258.

Barber, K. E. and A. G. Brown (1987) Late Pleistocene organic deposits beneath the floodplain of the River Avon at Ibsley, Hampshire. In *Wessex and the Isle of Wight. Field guide*, K. E. Barber (ed.), 65–74. Cambridge: Quaternary Research Association.

Barber, K. E. and S. N. Twigger (1987) Late Quaternary palaeoecology of the Severn Basin. In *Palaeohydrology in practice*, K. J. Gregory, J. Lewin and J. B. Thornes (eds.), 217–250. Chichester: Wiley.

Bartley, D. D. (1962) The stratigraphy and pollen analysis of lake deposits near Tadcaster, Yorkshire. *New Phytologist* **61**, 277–287.

Bartley, D. D. (1966) Pollen analysis of some lake deposits near Bamburgh in Northumberland. *New Phytologist* **65**, 141–156.

Bartley, D. D. (1975) Pollen analytical evidence for prehistoric forest clearance in the upland area west of Rishworth, West Yorkshire. *New Phytologist* **74**, 375–381.

Bartley, D. D. and A. V. Morgan (1990) The palynological record of the King's Pool, Stafford, England. *New Phytologist* **116**, 177–194.

Bartley, D. D., C. Chambers and B. Hart-Jones (1976) The vegetational history of parts of south and east Durham. *New Phytologist* **77**, 437–468.

Bartley, D. D., I. P. Jones and R. T. Smith (1990) Studies in the Flandrian vegetational history of the Craven district of Yorkshire: the lowlands. *Journal of Ecology* **78**, 611–632.

Bassett, M. G. (1985) Towards a 'common language' in stratigraphy. *Episodes* **8**, 87–92.

Bates, M. R. (1986a) Discussion (of geology and geomorphology). In: Excavation of the Lower Palaeolithic site at Amey's Eartham Pit, Boxgrove, West Sussex: a preliminary report. M. B. Roberts, *Proceedings of the Prehistoric Society* **52**, 222–224.

Bates, M. R. (1986b) Molluscan analysis. In: Excavation of the Lower Palaeolithic site at Amey's Eartham Pit, Boxgrove, West Sussex: a preliminary report. M. B. Roberts, *Proceedings of the Prehistoric Society* **52**, 231–232.

Bates, M. R. and M. B. Roberts (1986) Geology and geomorphology. In: Excavation of the Lower Palaeolithic site at Amey's Eartham Pit, Boxgrove, West Sussex: a preliminary report. M. B. Roberts, *Proceedings of the Prehistoric Society* **52**, 216–222.

Battarbee, R. W. (1978) Observations on the recent history of Lough Neagh and its drainage basin. *Philosophical Transactions of the Royal Society of London* **B281**, 303–345.

Battarbee, R. W. (1986a) Diatom analysis. In *Handbook of Holocene palaeoecology and palaeohydrology*, B. E. Berglund (ed.), 527–570. Chichester: Wiley.

Battarbee, R. W. (1986b) The eutrophication of Lough Erne inferred from changes in the diatom assemblages of ^{210}Pb and ^{137}Cs-dated sediment cores. *Proceedings of the Royal Irish Academy* **86B**, 141–168.

Battarbee, R. W. and D. F. Charles (1987) The use of diatom assemblages in lake sediments as a means of assessing the timing, trends and causes of lake acidification. *Progress in Physical Geography* **11**, 552–580.

Battarbee, R. W., R. J. Flower, A. C. Stevenson and B. Rippey (1985) Lake acidification in Galloway: a palaeoecological test of competing hypotheses. *Nature* **314**, 350–352.

Battarbee, R. W., R. J. Flower, A. C. Stevenson, V. J. Jones, R. Harriman and P. G. Appleby (1988) Diatom and chemical evidence for reversibility of acidification of Scottish lochs. *Nature* **332**, 530–532.

Beales, P. W. (1977) Cladocera analysis from Fishers Green. In: A Hoxnian interglacial site at Fishers Green, Stevenage, Hertfordshire. P. L. Gibbard and M. M. Aalto, *New Phytologist* **78**, 520–523.

Beales, P. W. (1980) The Late Devensian and Flandrian vegetational history of Crose Mere, Shropshire. *New Phytologist* **85**, 133–161.

Beard, G. R. (1984) *Soils in Warwickshire V: Sheet SP27/37 Coventry (South)*. Soil Survey Record Number 81. Harpenden: Soil Survey of England and Wales.

Beaumont, P. (1971) Stone orientation and stone count data from the Lower Till sheet, eastern Durham. *Proceedings of the Yorkshire Geological Society* **38**, 343–360.

Beaumont, P., J. Turner and P. F. Ward (1969) An Ipswichian peat raft in glacial till at Hutton Henry, County Durham. *New Phytologist* **68**, 797–805.

Beck, R. B. (1985) Plant remains in the Stanton Harcourt Channel. In *The chronology and environmental framework of early man in the Upper Thames Valley: a new model*, D. J. Briggs, G. R. Coope and D. D. Gilbertson, 114. BAR British Series 137. Oxford: British Archaeological Reports.

Beck, R. B., B. M. Funnell and A. R. Lord (1972) Correlation of Lower Pleistocene Crag at depth in Suffolk. *Geological Magazine* **109**, 137–139.

Beckett, S. C. (1981) Pollen diagrams from Holderness, North Humberside. *Journal of Biogeography* **8**, 177–198.

Beckett, S. C. and F. A. Hibbert (1979) Vegetational change and the influence of prehistoric man in the Somerset Levels. *New Phytologist* **83**, 577–600.

Beckinsale, R. P. and L. Richardson (1964) Recent findings on the physical development of the lower Severn Valley. *Geographical Journal* **130**, 87–105.

Behre, K.-E. (1989) Biostratigraphy of the last glacial period in Europe. *Quaternary Science Reviews* **8**, 25–44.

Bell, A. M. (1904) Implementiferous sections at Wolvercote (Oxfordshire). *Quarterly Journal of the Geological Society of London* **60**, 120–132.

Bell, F. G. (1969) The occurrence of southern, steppe and halophyte elements in Weichselian (last-glacial) floras from southern Britain. *New Phytologist* **68**, 913–922.

Bell, F. G. (1970) Late Pleistocene floras from Earith, Huntingdonshire. *Philosophical Transactions of the Royal Society of London* **B258**, 347–378.

Bell, F. G., G. R. Coope, R. J. Rice and T. H. Riley (1972) Mid-Weichselian fossil-bearing deposits at Syston, Leicestershire. *Proceedings of the Geologists' Association* **83**, 197–211.

Bell, M. (1982) The effects of land-use and climate on valley sedimentation. In *Climate change in later prehistory*, A. F. Harding (ed.), 127–142. Edinburgh: Edinburgh University Press.

Bennett, K. D. (1983) Devensian Late-glacial and Flandrian vegetational history at Hockham Mere, Norfolk, England. *New Phytologist* **95**, 457–487.

Bennett, K. D. (1984) The post-glacial history of *Pinus sylvestris* in the British Isles. *Quaternary Science Reviews* **3**, 133–155.

Bennett, K. D. (1986) The rate of spread and population increase of forest trees during the postglacial. *Philosophical Transactions of the Royal Society of London* **B314**, 523–531.

Bennett, K. D. (1988a) Holocene pollen stratigraphy of central East Anglia, England, and comparison of pollen zones across the British Isles. *New Phytologist* **109**, 237–253.

Bennett, K. D. (1988b) Post-glacial vegetational history: ecological considerations. In *Handbook of vegetation science. VII Vegetation history*, B. Huntley and T. Webb III (eds.), 699–724. Dordrecht: Kluwer.

Bennett, K. D. (1989) A provisional map of forest types for the British Isles 5000 years ago. *Journal of Quaternary Science* **4**, 141–144.

Bennett, K. D., J. A. Fossitt, M. J. Sharp and V. R. Switsur (1990) Holocene vegetational and environmental history at Loch Lang, South Uist, Western Isles, Scotland. *New Phytologist* **114**, 281–298.

Beresford, M. W. and J. G. Hurst (1971) *Deserted medieval villages*. London: Lutterworth Press.

Berggren W. A., L. H. Burckle, M. B. Cita, H. B. S. Cooke, B. M. Funnell, S. Gartner, J. D. Hays, J. P. Kennett, N. D. Opdyke, L. Pastouret, N. J. Shackleton and Y. Takayanagi (1980) Towards a Quaternary time scale. *Quaternary Research* **13**, 277–302.

Berglund, B. E. (1976) *The deglaciation of southern Sweden. Presentation of a research project and a tentative radiocarbon chronology*. University of Lund, Department of Quaternary Geology Report Number 10. Lund: University of Lund.

Berglund, B. E. (1979) The deglaciation of southern Sweden 13500–10000 BP. *Boreas* **8**, 89–117.

Berglund B. E. (ed.) (1986) *Handbook of Holocene palaeoecology and palaeohydrology*. Chichester: Wiley.

Berglund B. E., S. Hakansson and E. Lagerlund (1976) Radiocarbon-dated mammoth (*Mammuthus primigenius* Blumenbach) finds in South Sweden. *Boreas* **5**, 177–191.

Berry, R. J. (1985) *The natural history of Orkney*. London: Collins.

Berry, R. J. and F. E. N. Rose (1975) Islands and the evolution of *Microtus arvalis* (Microtinae). *Journal of Zoology* **177**, 395–409.

Bersu, G. (1940) Excavations at Little Woodbury, Wiltshire. Part 1. The settlement as revealed by excavation. *Proceedings of the Prehistoric Society* **6**, 30–111.

Beschel, R. E. (1961) Dating rock surfaces by lichen growth and its application to glaciology and physiography (lichenometry). In *Geology of the Arctic*, G. O. Raasch (ed.), 1044–1062. Toronto: University of Toronto Press.

Binns, P. E., R. Harland and M. J. Hughes (1974) Glacial and postglacial sedimentation in the sea of the Hebrides. *Nature* **248**, 751–754.

Birks, H. H. (1972a) Studies in the vegetational history of Scotland. II. Two pollen diagrams from the Galloway Hills, Kirkcudbrightshire. *Journal of Ecology* **60**, 183–217.

Birks, H. H. (1972b) Studies in the vegetational history of Scotland. III. A radiocarbon-dated pollen diagram from Loch Maree, Ross and Cromarty. *New Phytologist* **71**, 731–754.

Birks, H. H. (1975) Studies in the vegetational history of Scotland. IV. Pine stumps in Scottish blanket peats. *Philosophical Transactions of the Royal Society of London* **B270**, 181–226.

Birks, H. H. and R. W. Mathewes (1978) Studies in the vegetational history of Scotland. V. Late Devensian and early Flandrian pollen and macrofossil stratigraphy at Abernethy Forest, Inverness-shire. *New Phytologist* **80**, 455–484.

Birks, H. J. B. (1973) *Past and present vegetation of the Isle of Skye: a palaeoecological study*. Cambridge: Cambridge University Press.

Birks, H. J. B. (1977) The Flandrian forest history of Scotland: a preliminary synthesis. In *British Quaternary studies: recent advances*, F. W. Shotton (ed.), 119–135. Oxford: Clarendon Press.

Birks, H. J. B. (1986) Late-Quaternary biotic changes in terrestrial and lacustrine environments, with particular reference to north-west Europe. In *Handbook of Holocene palaeoecology and palaeohydrology*, B. E. Berglund (ed.), 3–65. Chichester: Wiley.

Birks, H. J. B. (1988) Long-term ecological change in the British uplands. In *Ecological change in the uplands*, M. B. Usher and D. B. A. Thompson (eds.), 37–56. Special Publication Number 7 of the British Ecological Society. Oxford: Blackwell Scientific.

Birks, H. J. B. (1989) Holocene isochrone maps and patterns of tree-spreading in the British Isles. *Journal of Biogeography* **16**, 503–540.

Birks, H. J. B. and H. H. Birks (1980) *Quaternary palaeoecology*. London: Edward Arnold.

Birks, H. J. B. and B. J. Madsen (1979) Flandrian vegetational history of Little Loch Roag, Isle of Lewis, Scotland. *Journal of Ecology* **67**, 825–842.

Birks, H. J. B. and S. M. Peglar (1979) Interglacial pollen spectra from Sel Ayre, Shetland. *New Phytologist* **83**, 559–575.

Birks, H. J. B. and M. E. Ransom (1969) An interglacial peat at Fugla Ness, Shetland. *New Phytologist* **68**, 777–796.

Birks, H. J. B. and W. Williams (1983) The Late Quaternary vegetational history of the Inner Hebrides. *Proceedings of the Royal Society of Edinburgh* **83B**, 269–292.

Birnie, J. F. (1981) *Environmental changes in Shetland since the end of the last glaciation*. Ph.D. Thesis, University of Aberdeen.

Birnie, J. (1983) Tolsta Head: further investigations of the interstadial deposit. *Quaternary Newsletter* **41**, 18–25.

Birnie, J. (1984) Trees and shrubs in the Shetland Islands: evidence for a Postglacial climatic optimum? In *Climatic changes on a yearly to millennial basis*, N. A. Morner and W. Karlén (eds.), 155–161. Dordrecht: Reidel.

Bisat, W. S. (1939) The relationship of the 'Basement Clays' of Dimlington, Bridlington and Filey bays. *Naturalist* 133–135, 161–168.

Bisat, W. S. (1940) Older and newer drift in east Yorkshire. *Proceedings of the Yorkshire Geological Society* **24**, 137–151.

Bisat, W. S. (1946) Fluvio-glacial gravels at Stanley Ferry, near Wakefield. *Transactions of the Leeds Geological Association* **6**, 31–36.

Bishop, A. C. and G. Bisson (1989) *Classical areas of British geology: Jersey. Description of 1:25 000 Channel Islands Sheet 2*. London: Her Majesty's Stationery Office for British Geological Survey.

Bishop, M. J. (1975) Earliest record of man's presence in Britain. *Nature* **253**, 95–97.

Bishop, M. J. (1982) *The mammal fauna of the early Middle Pleistocene cavern infill site of Westbury-sub-Mendip, Somerset*. Special Papers in Palaeontology Number 28. London: Palaeontological Association.

Bishop, W. W. (1958) The Pleistocene geology and geomorphology of three gaps in the Midland Jurassic escarpment. *Philosophical Transactions of the Royal Society of London* **B241**, 255–306.

Bishop, W. W. and G. R. Coope (1977) Stratigraphical and faunal evidence for Lateglacial and early Flandrian environments in south-west Scotland. In *Studies in the Scottish Lateglacial environment*, J. M. Gray and J. J. Lowe (eds.), 61–88. Oxford: Pergamon Press.

Black, R. F. (1976) Periglacial features indicative of permafrost: ice and soil wedges. *Quaternary Research* **6**, 3–26.

Blackburn, K. B. (1952) The dating of a deposit containing an elk skeleton found at Neasham near Darlington, County Durham. *New Phytologist* **51**, 364–377.

Blackham, A. and J. R. Flenley (1984) A pollen analytical study of the Flandrian vegetational history at Skipsea Withow. In *Late Quaternary environments and man in Holderness*, D. D. Gilbertson (ed.), 159–164. BAR British Series 134. Oxford: British Archaeological Reports.

Blair, K. G. (1923) Some coleopterous remains from the peat-bed at Wolvercote, Oxfordshire. *Transactions of the Royal Entomological Society of London* **71**, 558–563.

Blytt, A. (1876) *Essay on the immigration of the Norwegian flora during alternating rainy and dry periods*. Kristiana: Cammermeyer.

Boardman, J. (1978) Grèze litées near Keswick, Cumbria. *Biuletyn Peryglacjalny* **27**, 23–34.

Boardman, J. (1985) The Troutbeck Paleosol, Cumbria, England. In *Soils and Quaternary landscape evolution*, J. Boardman (ed.), 231–260. Chichester: Wiley.

Boardman, J., J. J. Lowe, D. T. Holyoak and P. Wilson (1981) Northeastern Lake District: the valleys of Mosedale and Thornsgill. In *Field guide to eastern Cumbria*, J. Boardman (ed.), 5–39. Brighton: Quaternary Research Association.

Bohncke, S. J. P. (1988) Vegetation and habitation history of the Callanish area, Isle of Lewis, Scotland. In *The cultural landscape: past, present and future*, H. H. Birks, H. J. B. Birks, P. E. Kaland and D. Moe (eds.), 445–461. Cambridge: Cambridge University Press.

Boswell, P. G. H. (1952) The Pliocene – Pleistocene boundary in the east of England. *Proceedings of the Geologists' Association* **63**, 301–312.

Boulter, M. and W. I. Mitchell (1977) Middle Pleistocene (Gortian) deposits from Benburb, Northern Ireland. *Irish Naturalists' Journal* **19**, 2–3.

Boulter, M. C. (1971a) A palynological study of two of the Neogene plant beds in Derbyshire. *Bulletin of the British Museum Natural History (Geology)* **19**, 360–411.

Boulter, M. C. (1971b) A survey of the Neogene flora from two Derbyshire pocket deposits. *Mercian Geologist* **4**, 45–62.

Boulter, M. C. (1980) Irish Tertiary plant fossils in a European context. *Journal of Earth Sciences (Dublin)* **3**, 1–11.

Boulter M. C., T. D. Ford, M. Ijtaba and P. T. Walsh (1971) Brassington Formation: a newly recognized Tertiary formation in the southern Pennines. *Nature* **231**, 134–136.

Boulton, G. S. (1972) Modern arctic glaciers as depositional models for former ice sheets. *Quarterly Journal of the Geological Society of London* **128**, 361–393.

Boulton, G. S. and P. Worsley (1965) Late Weichselian glaciation of the Cheshire – Shropshire Basin. *Nature* **207**, 704–706.

Boulton, G. S., A. S. Jones, K. M. Clayton and M. J. Kenning (1977) A British ice-sheet model and patterns of glacial erosion and deposition in Britain. In *British Quaternary studies: recent advances*, F. W. Shotton (ed.), 231–246. Oxford: Clarendon Press.

Bowen, D. Q. (1969a) A new interpretation of the Pleistocene succession in the Bristol Channel. (Abstract). *Proceedings of the Ussher Society* **2**, 86.

Bowen, D. Q. (1969b) Views on the extent of the Weichselian Glaciation in Wales. In *Coastal Pleistocene deposits in Wales*, D. Q. Bowen (ed.), 2. Aberystwyth: Department of Geography, University College of Wales, Aberystwyth for Quaternary Research Association.

Bowen, D. Q. (1970) South-east and central South Wales. In *The glaciations of Wales and adjoining regions*, C. A. Lewis (ed.), 197–227. London: Longman.

Bowen, D. Q. (1973a) The Pleistocene history of Wales and the borderland. *Geological Journal* **8**, 207–224.

Bowen, D. Q. (1973b) The Pleistocene succession of the Irish Sea. *Proceedings of the Geologists' Association* **83**, 249–272.

Bowen, D. Q. (1973c) The excavation at Minchin Hole 1973. *Gower* **24**, 12–18.

Bowen, D. Q. (1974) The Quaternary of Wales. In *The Upper Palaeozoic and post-Palaeozic rocks of Wales*, T. R. Owen (ed.), 373–426. Cardiff: University of Wales Press.

Bowen, D. Q. (1978) *Quaternary geology: a stratigraphic framework for multidisciplinary work*. Oxford: Pergamon Press.

Bowen, D. Q. (1981) The 'South Wales End-Moraine': fifty years after. In *The Quaternary in Britain: essays, reviews and original work on the Quaternary published in honour of Lewis Penny on his retirement*, J. Neale and J. Flenley (eds.), 60–67. Oxford: Pergamon Press.

Bowen, D. Q. (1984) Introduction. In *Field guide. Wales: Gower Preseli, Fforest Fawr*, D. Q. Bowen and A. Henry (eds.), 1–17. Cambridge: Quaternary Research Association.

Bowen, D. Q. and G. A. Sykes (1988) Correlation of marine events and glaciations on the northeast Atlantic margin. *Philosophical Transactions of the Royal Society of London* **B318**, 619–635.

Bowen, D. Q., S. Hughes, G. A. Sykes and G. H. Miller (1989) Land – sea correlations in the Pleistocene based on isoleucine epimerization in non-marine molluscs. *Nature* **340**, 49–51.

Bowen, D. Q., J. Rose, A. M. McCabe and D. G. Sutherland (1986) Correlation of Quaternary glaciations in England, Ireland, Scotland and Wales. *Quaternary Science Reviews* **5**, 299–340.

Bowen, D. Q., G. A. Sykes, A. Reeves, G. H. Miller, J. T. Andrews, J. S. Brew and P. E. Hare (1985) Amino acid geochronology of raised beaches in south west Britain. *Quaternary Science Reviews* **4**, 279–318.

Bowen, H. C. and P. J. Fowler (1966) Romano-British rural settlements in Dorset and Wiltshire. In *Rural settlement in Roman Britain*, A. C. Thomas (ed.), 43–67. CBA Research Report Number 7. London: Council for British Archaeology.

Bowen, R. (1966) *Palaeotemperature analysis*. Amsterdam: Elsevier.

Boycott, A. E. (1936) The habitats of freshwater Mollusca in Britain. *Journal of Animal Ecology* **5**, 116–186.

Boyd, W. E. (1988) Early Flandrian vegetational development on the coastal plain of north Ayrshire, Scotland: evidence from multiple pollen profiles. *Journal of Biogeography* **15**, 325–337.

Boyd, W. E. and J. H. Dickson (1987) A Post-glacial pollen sequence from Loch A'Mhuilinn, north Arran: a record of vegetational history with special reference to the history of endemic *Sorbus* species. *New Phytologist* **107**, 221–244.

Boylan, P. J. (1966) The Pleistocene deposits of Kirmington, Lincolnshire. *Mercian Geologist* **1**, 339–350.

Boylan, P. J. (1967) The Pleistocene Mammalia of the Sewerby – Hessle buried cliff. *Proceedings of the Yorkshire Geological Society* **36**, 115–125.

Boylan, P. J. (1972) The scientific significance of the Kirkdale Cave hyaenas. *Yorkshire Philosophical Society Annual Report for 1971*, 38–47.

Boylan, P. J. (1977a) Kirkdale Cave. In *Guidebook for Excursion C7. Yorkshire and Lincolnshire*, J. A. Catt, 26–27. INQUA X Congress, United Kingdom. Norwich: Geo Abstracts for International Union for Quaternary Research.

Boylan, P. J. (1977b) Victoria Cave. In *Guidebook for Excursion C7. Yorkshire and Lincolnshire*, J. A. Catt, 52–53. INQUA X Congress, United Kingdom. Norwich: Geo Abstracts for International Union for Quaternary Research.

Bradley, R. J. (1978) Colonization and land use in the Late Neolithic and Early Bronze Age. In *The effect of man on the landscape: the Lowland Zone*, S. Limbrey and J. G. Evans (eds.), 95–103. CBA Research Report Number 21. London: Council for British Archaeology.

Bradley, R. (1984) *The social foundations of prehistoric Britain: themes and variations in the archaeology of power*. London: Longman.

Bradshaw, R. H. W. and P. Browne (1987) Changing patterns in the post-glacial distribution of *Pinus sylvestris* in Ireland. *Journal of Biogeography* **14**, 237–248.

Bramwell, D. (1959) The excavation of Dowel Cave, Earl Sterndale, 1958–9. *Derbyshire Archaeological Journal* **79**, 97–109.

Bramwell, D. (1960) Some research into bird distribution in Britain during the late-glacial and post-glacial periods. *Bird Report 1959–60 of the Merseyside Naturalists Association*, 51–58.

Bramwell, D. (1963) Animal bones. In: The excavation of Willerby Wold long barrow, East Riding of Yorkshire. T. G. Manby, *Proceedings of the Prehistoric Society* **29**, 204.

Bramwell, D. (1964) The excavations at Elder Bush Cave, Wetton, Staffordshire. *North Staffordshire Journal of Field Studies* **4**, 46–59.

Bramwell, D. (1976) Animal bones. In: The excavation of Kilham long barrow, East Riding of Yorkshire. T. G. Manby, *Proceedings of the Prehistoric Society* **42**, 157–158.

Bramwell, D. and F. W. Shotton (1982) Rodent remains from

the caddis-bearing tufa of Elder Bush Cave. *Quaternary Newsletter* **38**, 7–13.

Brandon, A. and M. G. Sumbler (1988) An Ipswichian fluvial deposit at Fulbeck, Lincolnshire and the chronology of the Trent terraces. *Journal of Quaternary Science* **3**, 127–133.

Brasier, M. D. (1980) *Microfossils*. London: Allen and Unwin.

Brebion, P., E. Buge, G. Fily, A. Lauriat, J.-P. Margerel and C. Pareyn (1974) Le Quaternaire ancien de Pierrepont et de St. Sauveur de Pierrepont (Manche). *Bulletin de la Société Linnéenne de Normandie* **104**, 70–106.

Bremner, A. (1931) Further problems in the glacial geology of north-eastern Scotland. *Transactions of the Edinburgh Geological Society* **12**, 147–164.

Bridge, D. M. C. (1988) Corton Cliffs. In *The Pliocene – Middle Pleistocene of East Anglia. Field guide*, P. L. Gibbard and J. A. Zalasiewicz (eds.), 119–125. Cambridge: Quaternary Research Association.

Bridge, D. M. C. and P. M. Hopson (1985) Fine gravel, heavy mineral and grain-size analyses of mid-Pleistocene glacial deposits in the lower Waveney Valley, East Anglia. *Modern Geology* **9**, 129–144.

Bridge, M. C., B. A. Haggart and J. J. Lowe (1990) The history and palaeoclimatic significance of subfossil remains of *Pinus sylvestris* in blanket peats from Scotland. *Journal of Ecology* **78**, 77–99.

Bridger, J. F. D. (1975) The Pleistocene succession in the southern part of Charnwood Forest. *Mercian Geologist* **5**, 189–203.

Bridger, J. F. D. (1981) The glaciation of Charnwood Forest, Leicestershire and its geomorphological significance. In *The Quaternary in Britain: essays, reviews and original work on the Quaternary published in honour of Lewis Penny on his retirement*, J. Neale and J. Flenley (eds.), 68–81. Oxford: Pergamon Press.

Bridgland, D. (1980) A reappraisal of Pleistocene stratigraphy in north Kent and eastern Essex, and new evidence concerning former courses of the Thames and Medway. *Quaternary Newsletter* **32**, 15–24.

Bridgland, D. R. (1988) The Pleistocene fluvial stratigraphy and palaeogeography of Essex. *Proceedings of the Geologists' Association* **99**, 291–314.

Bridgland, D. R., P. L. Gibbard and R. C. Preece (1990) The geology and significance of the interglacial sediments at Little Oakley, Essex. *Philosophical Transactions of the Royal Society of London* **B328**, 307–339.

Bridgland, D. R., D. H. Keen and D. Maddy (1986) A reinvestigation of the Bushley Green Terrace typesite, Hereford and Worcester. *Quaternary Newsletter* **50**, 1–6.

Bridgland, D. R., D. H. Keen and D. Maddy (1989) The Avon terraces: Cropthorne, Ailstone and Eckington. In *The Pleistocene of the West Midlands. Field guide*, D. H. Keen (ed.), 51–67. Cambridge: Quaternary Research Association.

Bridgland, D. R., P. L. Gibbard, P. Harding, R. A. Kemp and G. Southgate (1985) Report of Geologists' Association field meeting in north-east Essex, May 22nd–24th 1987. *Proceedings of the Geologists' Association* **99**, 315–333.

Bridgland, D. R., P. Allen, A. P. Currant, P. L. Gibbard, A. M. Lister, R. C. Preece, J. E. Robinson, A. J. Stuart and A. J. Sutcliffe (1988) New information and results from recent excavations at Barnfield Pit, Swanscombe. *Quaternary Newsletter* **46**, 25–39.

Briffa, K. R., P. D. Jones, T. M. L. Wigley, J. R. Pilcher and M. G. L. Baillie (1983) Climate reconstruction from tree rings: Part 1, basic methodology and preliminary results for England. *Journal of Climatology* **3**, 233–242.

Briggs, D. J. (1976) River terraces of the Oxford area. In *Field guide to the Oxford region*, D. A. Roe (ed.), 8–15. Oxford: Quaternary Research Association.

Briggs, D. J. and C. V. Burek (1985) Quaternary deposits in the Peak District. In *Peak District and Northern Dukeries. Field guide*, D. J. Briggs, D. D. Gilbertson and R. D. S. Jenkinson (eds.), 17–32. Cambridge: Quaternary Research Association.

Briggs, D. J. and D. D. Gilbertson (1973) The age of the Hanborough Terrace of the River Evenlode, Oxfordshire. *Proceedings of the Geologists' Association* **84**, 155–173.

Briggs, D. J. and D. D. Gilbertson (1980) Quaternary processes and environments in the Upper Thames Valley. *Transactions of the Institute of British Geographers* New Series 5, 53–65.

Briggs, D. J., G. R. Coope and D. D. Gilbertson (1985) *The chronology and environmental framework of early man in the Upper Thames Valley: a new model*. BAR British Series 137. Oxford: British Archaeological Reports.

Briggs, D. J., G. R. Coope D. D. Gilbertson, A. S. Goudie, P. J. Osborne, H. A. Osmaston, M. E. Pettit, F. W. Shotton and A. J. Stuart (1975) New interglacial site at Sugworth. *Nature* **257**, 477–479.

Bristow, C. R. and F. C. Cox (1973) The Gipping Till: a reappraisal of East Anglian glacial stratigraphy. *Journal of the Geological Society of London* **129**, 1–37.

Broecker, W. S., M. Andree, G. Bonani, W. Wolfli, H. Oeschger and M. Klas (1988) Can the Greenland climatic jumps be identified in records from ocean and land? *Quaternary Research* **30**, 1–6.

Brown, A. G. (1983) Floodplain deposits and accelerated sedimentation in the lower Severn basin. In *Background to palaeohydrology*, K. J. Gregory (ed.), 375–397. Chichester: Wiley.

Brown, A. G. (1987) Holocene floodplain sedimentation and channel response of the lower River Severn, United Kingdom. *Zeitschrift für Geomorphologie* **31**, 293–310.

Brown, A. G. and K. E. Barber (1985) Late Holocene paleoecology and sedimentary history of a small lowland catchment in central England. *Quaternary Research* **24**, 87–102.

Brown, A. P. (1977) Late-Devensian and Flandrian vegetational history of Bodmin Moor, Cornwall. *Philosophical Transactions of the Royal Society of London* **B276**, 251–320.

Brown, M. J. F., I. D. Ellis-Gruffydd, H. D. Foster and D. J. Unwin (1967) A new radio-carbon date for Wales. *Nature* **213**, 1220–1221.

Brown, R. C., D. D. Gilbertson, C. P. Green and D. H. Keen (1975) Stratigraphy and environmental significance of Pleistocene deposits at Stone, Hampshire. *Proceedings of the Geologists' Association* **86**, 349–363.

Browne, M. A. E. (1980) Erskine Bridge. In *Field guide. Glasgow region*, W. G. Jardine (ed.), 11–13. Glasgow: Quaternary Research Association.

Browne, M. A. E., D. D. Harkness, J. D. Peacock and R. G. Ward (1977) The date of deglaciation of the Paisley – Renfrew area. *Scottish Journal of Geology* **13**, 301–303.

Bryant, I. D. (1983) Facies sequences associated with some braided river deposits of Late Pleistocene age from southern Britain. In *Modern and ancient fluvial systems*, J. D. Collinson and J. Lewin (eds.), 267–275. International Association of Sedimentologists Special Publication Number 6. Oxford: Blackwell Scientific.

Bryant, I. D., D. T. Holyoak and K. A. Moseley (1983a) Late Pleistocene deposits at Brimpton, Berkshire, England. *Proceedings of the Geologists' Association* **94**, 321–343.

Bryant, I. D., P. L. Gibbard, D. T. Holyoak, V. R. Switsur and A. G. Wintle (1983b) Stratigraphy and palaeontology of Pleistocene cold-stage deposits at Alton Road Quarry, Farnham, Surrey, England. *Geological Magazine* **120**, 587–606.

Buckland, P. C. (1976) The use of insect remains in the interpretation of archaeological environments. In *Geoarchaeology: earth science and the past*, D. A. Davidson and M. L. Shackley (eds.), 369–396. London: Duckworth.

Buckland, P. C. (1982) The cover sands of north Lincolnshire and the Vale of York. In *Papers in earth studies: Lovatt Lectures – Worcester*, B. H. Adlam, C. R. Fenn and L. Morris (eds.), 143–178. Norwich: Geo Books.

Buckland, P. C. (1984) North-west Lincolnshire 10 000 years ago. In *A prospect of Lincolnshire*, N. Field and A. White (eds.), 11–17. Lincoln: Privately published.

Buckland, P. C. and H. K. Kenward (1973) Thorne Moor: a palaeoecological study of a Bronze Age site. *Nature* **241**, 405–406.

Buckland, P. C., J. R. A. Greig and H. K. Kenward (1974) An Anglo-Danish site in York. *Antiquity* **48**, 25–33.

Buckland, W. (1822) An account of an assemblage of fossil teeth and bones discovered in a cave at Kirkdale. *Philosophical Transactions of the Royal Society of London* **122**, 171–236.

Buckland, W. (1823) *Reliquiae Diluvianae*. London: John Murray.

Buge, E. (1957) Les bryozoaires du Néogène de l'Ouest de la France et leur signification stratigraphique et paleobiologique. *Mémoires Museum National Historie Naturelle* **C6**, 1–435.

Bulleid, A. and J. W. Jackson (1937) The Burtle sand beds of Somerset. *Proceedings of the Somersetshire Archaeological and Natural History Society* **83**, 171–195.

Bulleid, A. and J. W. Jackson (1941) Further notes on the Burtle sand beds of Somerset. *Proceedings of the Somersetshire Archaeological and Natural History Society* **87**, 111–114.

Bullock, P. and C. P. Murphy (1979) Evolution of a paleo-argillic brown earth (Paleudalf) from Oxfordshire, England. *Geoderma* **22**, 225–252.

Bullock, P., D. M. Carroll and R. A. Jarvis (1973) Palaeosol features in northern England. *Nature* **242**, 53–54.

Burchell, J. P. T. (1933) The Northfleet 50-foot submergence later than the Coombe Rock of post-early Mousterian times. *Archaeologia* **83**, 67–92.

Burchell, J. P. T. (1954) Loessic deposits in the Fifty-foot Terrace post-dating the Main Coombe Rock of Baker's Hole, Northfleet, Kent. *Proceedings of the Geologists' Association* **65**, 256–261.

Burgess, C. (1974) The bronze age. In *British prehistory: a new outline*, C. Renfrew (ed.), 165–232. London: Duckworth.

Burl, H. A. W. (1976) *The stone circles of the British Isles*. New Haven: Yale University Press.

Burleigh, R. and J. Clutton-Brock (1977) A radiocarbon date for *Bos primigenius* from Charterhouse Warren Farm, Mendip. *Proceedings of the University of Bristol Spelaeological Society* **14**, 255–257.

Burleigh, R., E. B. Jacobi and R. M. Jacobi (1985) Early human resettlement of the British Isles following the last glacial maximum: new evidence from Gough's Cave, Cheddar. *Quaternary Newsletter* **45**, 1–6.

Burrin, P. J. (1985) Holocence alluviation in southeast England and some implications for palaeohydrological studies. *Earth Surface Processes and Landforms* **10**, 257–271.

Burrin, P. J. and R. G. Scaife (1984) Aspects of Holocene valley sedimentation and floodplain development in southern England. *Proceedings of the Geologists' Association* **95**, 81–96.

Bush, M. B. and S. Ellis (1987) The sedimentological and vegetational history of Willow Garth. In *East Yorkshire. Field guide*, S. Ellis (ed.), 42–52. Cambridge: Quaternary Research Association.

Bush, M. B. and J. R. Flenley (1987) The age of the British chalk grassland. *Nature* **329**, 434–436.

Butterworth, A. and G. D. Lewis (1968) *Prehistoric and Roman times in the Sheffield area*. Sheffield: Sheffield City Museums.

Calkin, J. B. and J. F. N. Green (1949) Palaeoliths and terraces near Bournemouth. *Proceedings of the Prehistoric Society* **15**, 21–37.

Callow, P. (1986a) The La Cotte industries and the European Lower and Middle Palaeolithic. In *La Cotte de St. Brelade 1961–1978: excavations by C. B. M. McBurney*, P. Callow and J. M. Cornford (eds.), 377–388. Norwich: Geo Books.

Callow, P. (1986b) Interpreting the La Cotte sequence. In *La Cotte de St. Brelade 1961–1978: excavations by C. B. M. McBurney*, P. Callow and J. M. Cornford (eds.), 73–82. Norwich: Geo Books.

Callow, P. (1986c) Artefacts from the Weichselian deposits. In *La Cotte de St. Brelade 1961–1978: excavations by C. B. M. McBurney*, P. Callow and J. M. Cornford (eds.), 397–408. Norwich: Geo Books.

Callow, P. (1986d) The Channel Islands in the Old Stone Age. In *The archaeology of the Channel Islands*, P. Johnston (ed.), 3–25. Chichester: Phillimore.

Callow, P. and J. M. Cornford (1986) *La Cotte de St. Brelade 1961–1978: excavations by C. B. M. McBurney*. Norwich: Geo Books.

Cambridge, P. (1961) Mollusca of the Corton Beds. In: The glacial and interglacial deposits of Norfolk. R. G. West, *Transactions of the Norfolk and Norwich Naturalists' Society* **19**, 369–370.

Cambridge, P. G. (1977) Whatever happened to the Boytonian? A review of the marine Plio-Pleistocene of the southern North Sea basin. *Bulletin of the Geological Society of Norfolk* **29**, 23–45.

Cameron, T. D. J., C. Laban and R. T. E. Schüttenhelm (1984a) *Flemish Bight Sheet. 52°N–02°E. Quaternary geology (geologie van het Kwartaire)*. British Geological Survey/Rijks Geologische Dienst 1:250 000 Map Series. Southampton: Ordnance Survey for British Geological Survey.

Cameron, T. D. J., C. Laban and R. T. E. Schüttenhelm (1989a) Upper Pliocene and Lower Pleistocene stratigraphy in the Southern Bight of the North Sea. In *The Quaternary and Tertiary geology of the Southern Bight, North Sea*, J. P. Henriet and G. de Moor (eds.), 97–110. Brussels: Ministry of Economic Affairs Belgian Geological Survey.

Cameron, T. D. J., R. T. E. Schüttenhelm and C. Laban (1989b) Middle and Upper Pleistocene and Holocene stratigraphy in the southern North Sea between 52° and 54°N and 2° to 4°E. In *Quaternary and Tertiary geology of the Southern Bight, North Sea*, J. P. Henriet and G. de Moor (eds.), 119–135. Brussels: Ministry of Economic Affairs Belgian Geological Survey.

Cameron, T. D. J., M. S. Stoker and D. Long (1987) The history of Quaternary sedimentation in the UK sector of the North Sea Basin. *Journal of the Geological Society of London* **144**, 43–58.

Cameron, T. D. J., A. P. Bonny, D. M. Gregory and R. Harland (1984b) Lower Pleistocene dinoflagellate cyst, foraminiferal and pollen assemblages in four boreholes in the southern North Sea. *Geological Magazine* **121**, 85–97.

Cameron, T. D. J., C. Laban, C. S. Mesdag and R. T. E. Schüttenhelm (1986) *Indefatigable Sheet. 53°N–02°E. Quaternary geology (geologie van het Kwartair)*. British Geological Survey/Rijks Geologische Dienst 1:250 000 Map Series. Southampton: Ordnance Survey for British Geological Survey.

Campbell, J. B. (1977) *The Upper Palaeolithic of Britain: a study of man and nature in the late Ice Age*. Two volumes. Oxford: Clarendon Press.

Campbell, J. B. and M. Pohl (1971) Pollen analysis of the loessic deposits at La Cotte de St. Brelade, Jersey, C. I. In: The Cambridge excavations at La Cotte de St. Brelade, Jersey – a preliminary report. C. B. M. McBurney and P. Callow, *Proceedings of the Prehistoric Society* **37**, 204–206.

Campbell, J. B. and C. G. Sampson (1971) *A new analysis of Kent's Cavern, Devonshire, England*. University of Oregon Anthropological Papers Number 3. Corvallis: University of Oregon.

Campbell, S. and R. A. Shakesby (1985) Wood fragments of possible Chelford Interstadial age from till at Broughton Bay, Gower, South Wales. *Quaternary Newsletter* **47**, 33–36.

Campbell, S., J. T. Andrews and R. A. Shakesby (1982) Amino acid evidence for Devensian ice, west Gower, South Wales. *Nature* **300**, 249–251.

Carpentier, G. and J.-P. Lautridou (1982) Tourville. In *The Quaternary of Normandy. Field handbook*, J.-P. Lautridou, 31–33. Caen: Quaternary Research Association and Centre de Geomorphologie du Centre National de la Recherche Scientifique.

Carr, A. P. (1970) The evolution of Orford Ness, Suffolk before 1600 AD: geomorphological evidence. *Zeitschrift für Geomorphologie* **14**, 289–300.

Carr, A. P. and R. E. Baker (1968) Orford, Suffolk: evidence for the evolution of the area during the Quaternary. *Transactions of the Institute of British Geographers* **45**, 107–123.

Carreck, J. N. (1976) Pleistocene mammalian and molluscan remains from 'Taplow' Terrace deposits at West Thurrock, near Grays, Essex. *Proceedings of the Geologists' Association* **87**, 83–92.

Carter, D. J. (1951) Indigenous and exotic foraminifera in the Coralline Crag of Suffolk. *Geological Magazine* **88**, 236–248.

Carter, P. A., G. A. L. Johnson and J. Turner (1978) An interglacial deposit at Scandal Beck, N. W. England. *New Phytologist* **81**, 785–790.

Carter, R. W. G. (1982) Sea-level changes in Northern Ireland. *Proceedings of the Geologists' Association* **93**, 7–23.

Carter, R. W. G. (1986) The morphodynamics of beach-ridge formation: Magilligan, Northern Ireland. *Marine Geology* **73**, 191–214.

Carter, R. W. G. and J. D. Orford (1984) Coarse clastic barrier beaches: a discussion of the distinctive dynamic and morphosedimentary characteristics. *Marine Geology* **60**, 377–389.

Carter, R. W. G. and P. Wilson (1990) The geomorphological, ecological and pedological development of coastal foredunes at Magilligan Point, Northern Ireland. In *Coastal dunes: form and process*, K. F. Nordstrom, N. P. Psuty and R. W. G. Carter (eds.), 129–157. Chichester: Wiley.

Carter, R. W. G., R. J. N. Devoy and J. Shaw (1989) Late Holocene sea levels in Ireland. *Journal of Quaternary Science* **4**, 7–24.

Case, H. J., G. W. Dimbleby, G. F. Mitchell, M. E. S. Morrison and V. B. Proudfoot (1969) Land use in

Goodland Townland, County Antrim, from Neolithic times until today. *Journal of the Royal Society of Antiquaries of Ireland* **99**, 39–53.

Caseldine, C. J. (1979) Early land clearance in south-east Perthshire. *Scottish Archaeological Forum* **9**, 1–15.

Caseldine, C. J. (1980) A Lateglacial site at Stormont Loch, near Blairgowrie, eastern Scotland. In *Studies in the Lateglacial of north-western Europe*, J. J. Lowe, J. M. Gray and J. E. Robinson (eds.), 69–88. Oxford: Pergamon Press.

Caseldine, C. J. and K. J. Edwards (1982) Interstadial and last interglacial deposits covered by till in Scotland: comments and new evidence. *Boreas* **11**, 119–122.

Caseldine, C. J. and D. J. Maguire (1986) Lateglacial/early Flandrian vegetation change on northern Dartmoor, south-west England. *Journal of Biogeography* **13**, 255–264.

Catt, J. A. (1977a) General introduction. In *Guidebook for Excursion C7. Yorkshire and Lincolnshire*, J. A. Catt, 6–9. INQUA X Congress, United Kingdom. Norwich: Geo Abstracts for International Union for Quaternary Research.

Catt, J. A. (1977b) Kirmington. In *Guidebook for Excursion C7. Yorkshire and Lincolnshire*, J. A. Catt, 14–16. INQUA X Congress, United Kingdom. Norwich: Geo Abstracts for International Union for Quaternary Research.

Catt, J. A. (1977c) Loess and coversands. In *British Quaternary studies: recent advances*, F. W. Shotton (ed.), 221–229. Oxford: Clarendon Press.

Catt, J. A. (1978) The contribution of loess to soils in lowland Britain. In *The effect of man on the landscape: the Lowland Zone*, S. Limbrey and J. G. Evans (eds.), 12–20. CBA Research Report Number 21. London: Council for British Archaeology.

Catt, J. A. (1979) Soils and Quaternary geology in Britain. *Journal of Soil Science* **30**, 607–642.

Catt, J. A. (1980) Till facies associated with the Devensian glacial maximium in eastern England. *Quaternary Newsletter* **30**, 4–10.

Catt, J. A. (1981) British pre-Devensian glaciations. In *The Quaternary in Britain: essays, reviews and original work on the Quaternary published in honour of Lewis Penny on his retirement*, J. Neale and J. Flenley (eds.), 9–19. Oxford: Pergamon Press.

Catt, J. A. (1982) The Quaternary deposits of the Yorkshire Wolds. *Proceedings of the North of England Soils Discussion Group* **18**, 61–67.

Catt, J. A. (1986) *Soils and Quaternary geology: a handbook for field scientists*. Oxford: Clarendon Press.

Catt, J. A. (1987a) The Quaternary of East Yorkshire and adjacent areas. In *East Yorkshire. Field guide*, S. Ellis (ed.), 1–14. Cambridge: Quaternary Research Association.

Catt, J. A. (1987b) Dimlington. In *East Yorkshire. Field guide*, S. Ellis (ed.), 82–98. Cambridge: Quaternary Research Association.

Catt, J. A. (1987c) Sewerby. In *East Yorkshire. Field guide*, S. Ellis (ed.), 53–57. Cambridge: Quaternary Research Association.

Catt, J. A. (1988) Soils of the Plio-Pleistocene: do they distinguish types of interglacial? *Philosophical Transactions of the Royal Society of London* **B318**, 539–557.

Catt, J. A. and L. F. Penny (1966) The Pleistocene deposits of Holderness, East Yorkshire. *Proceedings of the Yorkshire Geological Society* **35**, 375–420.

Catt, J. A., A. H. Weir and P. A. Madgett (1974) The loess of eastern Yorkshire and Lincolnshire. *Proceedings of the Yorkshire Geological Society* **40**, 23–39.

Catt, J. A., W. M. Corbett, C. A. H. Hodge, P. A. Madgett, W. Tatler and A. H. Weir (1971) Loess in the soils of north Norfolk. *Journal of Soil Science* **22**, 444–452.

Cepek, A. G. (1986) Quaternary stratigraphy of the German Democratic Republic. *Quaternary Science Reviews* **5**, 359–364.

Challis, A. J. and D. W. Harding (1975) *Later prehistory from the Trent to the Tyne*. Two volumes. BAR British Series 20. Oxford: British Archaeological Reports.

Chambers, F. M. (1981) Date of blanket peat initiation in upland South Wales. *Quaternary Newsletter* **35**, 24–29.

Chambers, F. M. (1982) Two radiocarbon-dated pollen diagrams from high altitude blanket peats in South Wales. *Journal of Ecology* **70**, 445–459.

Chambers, F. M. (1983) Three radiocarbon-dated pollen diagrams from upland peats north-west of Merthyr Tydfil, South Wales. *Journal of Ecology* **71**, 475–487.

Chambers, F. M. and S.-M. Price (1985) Palaeoecology of *Alnus* (alder): early Post-glacial rise in a valley mire, north-west Wales. *New Phytologist* **101**, 333–344.

Chambers, F. M., P. Q. Dresser and A. G. Smith (1979) Radiocarbon dating evidence on the impact of atmospheric pollution on upland peats. *Nature* **282**, 829–831.

Champion, T. C. (1975) Britain in the European iron age. *Archaeologia Atlantica* **1**, 127–145.

Champion, T. C. (1979) The iron age (*c*600 BC–AD 200). A. Southern Britain and Ireland. In *Introduction to British prehistory: from the arrival of* Homo sapiens *to the Claudian invasion*, J. V. S. Megaw and D. D. A. Simpson (eds.), 344–421. Leicester: Leicester University Press.

Chandler, R. H. (1914) The Pleistocene deposits of Crayford. *Proceedings of the Geologists' Association* **25**, 61–70.

Chapelhow, R. (1965) On glaciation in North Roe, Shetland. *Geographical Journal* **131**, 60–70.

Charlesworth, E. (1835) Observations on the Crag Formation and its organic remains; with a view to establishing a division of the Tertiary strata overlying the London Clay in Suffolk. *London, Edinburgh and Dublin Philosophical Magazine and Journal of Science* **7**, 81–94.

Charlesworth, J. K. (1929) The South Wales End-Moraine. *Quaterly Journal of the Geological Society of London* **85**, 335–355.

Charlesworth, J. K. (1956) The Late-Glacial history of the Highlands and Islands of Scotland. *Transactions of the*

Royal Society of Edinburgh **62**, 769–928.

Chartres, C. J. (1980) A Quaternary soil sequence in the Kennet Valley, central southern England. *Geoderma* **23**, 125–146.

Cheshire, D. A. (1983) Till lithology in Hertfordshire and west Essex. In *Field guide. Diversion of the Thames*, J. Rose (ed.), 50–59. Hoddesdon: Quaternary Research Association.

Chisholm, J. I. (1971) The stratigraphy of the post-Glacial marine transgression in N. E. Fife. *Bulletin of the Geological Survey of Great Britain* **37**, 91–107.

Churchill, D. M. (1962) The stratigraphy of the Mesolithic sites III and IV at Thatcham, Berkshire, England. *Proceedings of the Prehistoric Society* **28**, 362–370.

Churchill, D. M. (1965a) The kitchen midden site at Westward Ho!, Devon, England: ecology, age and relation to changes in land and sea level. *Proceedings of the Prehistoric Society* **31**, 74–84.

Churchill, D. M. (1965b) The displacement of deposits formed at sea level, 6500 years ago in southern Britain. *Quaternaria* **7**, 239–249.

Clapperton, C. M. (1970) The evidence for a Cheviot ice cap. *Transactions of the Institute of British Geographers* **50**, 115–127.

Clapperton, C. M. (1971) The location and origin of glacial meltwater phenomena in the eastern Cheviot hills. *Proceedings of the Yorkshire Geological Society* **38**, 361–380.

Clapperton, C. M. and D. E. Sugden (1975) The glaciation of Buchan: – a reappraisal. In *Quaternary studies in north east Scotland*, A. M. D. Gemmell (ed.), 19–22. Aberdeen: Department of Geography, University of Aberdeen for Quaternary Research Association.

Clapperton, C. M. and D. E. Sugden (1977) The Late Devensian glaciation of north-east Scotland. In *Studies in the Scottish Lateglacial environment*, J. M. Gray and J. J. Lowe (eds.), 1–13. Oxford: Pergamon Press.

Clapperton, C. M., A. R. Gunson and D. E. Sugden (1975) Loch Lomond Readvance in the eastern Cairngorms. *Nature* **253**, 710–712.

Clark, J. G. D. (1952) *Prehistoric Europe: the economic basis*. London: Methuen.

Clark, J. G. D. (1954) *Excavations at Star Carr*. Cambridge: Cambridge University Press.

Clark, J. G. D. (1966) The invasion hypothesis in British archaeology. *Antiquity* **40**, 172–189.

Clark, J. G. D. (1972) *Star Carr: a case study in bioarchaeology*. Modular Publications in Anthropology Number 10. Reading, Massachusetts: Addison-Wesley.

Clark, M. J. and K. E. Barber (1987) Mire development from the Devensian Lateglacial to present at Church Moor, Hampshire. In *Wessex and the Isle of Wight. Field guide*, K. E. Barber (ed.), 23–32. Cambridge: Quaternary Research Association.

Clark, R. M. (1975) A calibration curve for radiocarbon dates. *Antiquity* **49**, 251–266.

Clarke, A. S. (1962) Animal bones. In *The West Kennet long barrow excavations 1955–1956*, S. Piggott, 53–55. Ministry of Works Archaeological Reports Number 4. London: Her Majesty's Stationery Office.

Clarke, D. V. (1976) *The Neolithic village at Skara Brae, Orkney. Excavations 1972–3: an interim report*. Edinburgh: Her Majesty's Stationery Office.

Clarke, D. V. and N. Sharples (1985) Settlements and subsistence in the third millennium BC. In *The prehistory of Orkney: 4000 BC–1000 AD*, C. Renfrew (ed.), 54–82. Edinburgh: Edinburgh University Press.

Clarke, M. R. and C. A. Auton (1982a) The Pleistocene depositional history of the Norfolk – Suffolk borderland. In *Report of the Institute of Geological Sciences*, Number 82/1, 23–29. London: Her Majesty's Stationery Office.

Clarke, M. R. and C. A. Auton (1982b) Ingham. In *Field meeting guide. Suffolk*, P. Allen (ed.), Section 2, 28–30. London: Quaternary Research Association.

Clarke, M. R. and C. P. Green (1987) The Pleistocene terraces of the Bournemouth – Fordingbridge area. In *Wessex and the Isle of Wight. Field guide*, K. E. Barber (ed.), 58–64. Cambridge: Quaternary Research Association.

Clarke, R. H. (1973) Cainozoic subsidence in the North Sea. *Earth and Planetary Science Letters* **18**, 329–332.

Clarke, W. G. (1922) The Grime's Graves fauna. *Proceedings of the Prehistoric Society of East Anglia* **3**, 431–433.

Clayton, K. M. (1953) The glacial chronology of part of the middle Trent basin. *Proceedings of the Geologists' Association* **64**, 198–207.

Clayton, K. M. (1957a) Some aspects of the glacial deposits of Essex. *Proceedings of the Geologists' Association* **68**, 1–21.

Clayton, K. M. (1957b) The differentiation of the glacial drifts of the East Midlands. *East Midland Geographer* **7**, 31–40.

Clayton, K. M. (1960) The landforms of parts of southern Essex. *Transactions of the Institute of British Geographers* **28**, 55–74.

Clayton, K. M. (1977) River terraces. In *British Quaternary studies: recent advances*, F. W. Shotton (ed.), 153–167. Oxford: Clarendon Press.

Clayton, K. M. (1979) The Midlands and southern Pennines. In *The geomorphology of the British Isles: eastern and central England*, A. Straw and K. M. Clayton, 141–240. London: Methuen.

Clet, M. (1982) Pierrepont-en-Cotentin. II. Bosq d' Aubigny Formation. In *The Quaternary of Normandy. Field handbook*, J.-P. Lautridou, 67–69. Caen: Quaternary Research Association and Centre de Geomorphologie du Centre National de la Recherche Scientifique.

Climap Project Members (1976) The surface of the ice-age earth. *Science* **191**, 1131–1137.

Cloutman, E. W. (1988) Palaeoenvironments in the Vale of Pickering. Part 2: Environmental history at Seamer Carr. *Proceedings of the Prehistoric Society* **54**, 21–36.

Cloutman, E. W. and A. G. Smith (1988) Palaeoenvironments in the Vale of Pickering. Part 3: Environmental history at

Star Carr. *Proceedings of the Prehistoric Society* **54**, 37–58.

Clymo, R. S. (1964) The origin of acidity in *Sphagnum* bogs. *Bryologist* **67**, 427–431.

Colalongo, M. L., G. Pasini and S. Sartoni (1981) Remarks on the Neogene/Quaternary boundary and the Vrica Section (Calabria, Italy). *Bolletino della Societa Paleontologica Italiana* **20**, 99–120.

Colalongo, M. L., G. Pasini, G. Pelosio, S. Raffi, D. Rio, G. Ruggieri, S. Sartoni, R. Selli and R. Sprovieri (1982) The Neogene – Quaternary boundary definition: a review and proposal. *Geografica Fisica e Dinamica Quaternaria* **5**, 59–68.

Coles, B. P. L. and B. M. Funnell (1981) Holocene palaeoenvironments of Broadland, England. In *Holocene marine sedimentation in the North Sea Basin*, S.-D. Nio, R. T. E. Schüttenhelm and T. C. E. van Weering (eds.), 123–137. International Association of Sedimentologists Special Publication Number 5. Oxford: Blackwell Scientific.

Coles, G., C. O. Hunt and R. D. S. Jenkinson (1985) Robin Hood's Cave: palynology. In *Peak District and northern Dukeries. Field guide*, D. J. Briggs, D. D. Gilbertson and R. D. S. Jenkinson (eds.), 178–182. Cambridge: Quaternary Research Association.

Coles, J. M. (1971) The early settlement of Scotland: excavations at Morton, Fife. *Proceedings of the Prehistoric Society* **37**, 284–366.

Coles, J. M., F. A. Hibbert and B. J. Orme (1973) Prehistoric roads and tracks in Somerset, England. 3. The Sweet Track. *Proceedings of the Prehistoric Society* **39**, 256–293.

Colhoun, E. A. (1971a) The glacial stratigraphy of the Sperrin Mountains and its relation to the glacial stratigraphy of north-west Ireland. *Proceedings of the Royal Irish Academy* **71B**, 37–52.

Colhoun, E. A. (1971b) Late Weichselian periglacial phenomena of the Sperrin Mountains, Northern Ireland. *Proceedings of the Royal Irish Academy* **71B**, 53–71.

Colhoun, E. A. (1973) Two Pleistocene sections in south-western Donegal and their relation to the last glaciation of the Glengesh plateau. *Irish Geography* **6**, 594–609.

Colhoun, E. A. and G. F. Mitchell (1971) Interglacial marine formation and late glacial freshwater formation in Shortalstown townland, Co. Wexford. *Proceedings of the Royal Irish Academy* **71B**, 211–245.

Colhoun, E. A. and F. M. Synge (1980) The cirque moraines at Loch Nahanagan, County Wicklow, Ireland. *Proceedings of the Royal Irish Academy* **80B**, 25–45.

Colhoun, E. A., A. T. Ryder and N. Stephens (1973) ^{14}C age of an oak – hazel forest bed at Drumskellan, Co. Donegal, and its relation to Late Midlandian and Littletonian raised beaches. *Irish Naturalists' Journal* **17**, 321–327.

Colhoun, E. A., J. H. Dickson, A. M. McCabe and F. W. Shotton (1972) A Middle Midlandian freshwater series at Derryvree, Maguiresbridge, County Fermanagh, Northern Ireland. *Proceedings of the Royal Society of London* **B180**, 273–292.

Connell, E. R. (1984) Kirkhill Quarry. Deposits between the lower and upper palaeosols; deposits above the upper palaeosol. In *Buchan field guide*, A. M. Hall (ed.), 62–69. Cambridge: Quaternary Research Association.

Connell, E. R. and A. M. Hall (1984a) Kirkhill Quarry. Deposits beneath the lower palaeosol. In *Buchan field guide*, A. M. Hall (ed.), 60–62. Cambridge: Quaternary Research Association.

Connell, E. R. and A. M. Hall (1984b) Boyne Bay Quarry. In *Buchan field guide*, A. M. Hall (ed.), 97–101. Cambridge: Quaternary Research Association.

Connell, E. R. and J. C. C. Romans (1984) Kirkhill Quarry. Palaeosols. In *Buchan field guide*, A. M. Hall (ed.), 70–77. Cambridge: Quaternary Research Association.

Connell, E. R., K. J. Edwards and A. M. Hall (1982) Evidence for two pre-Flandrian palaeosols in Buchan, north-east Scotland. *Nature* **297**, 570–572.

Conolly, A. P., H. Godwin and E. M. Megaw (1950) Studies in the post-glacial history of British vegetation. XI. Late-glacial deposits in Cornwall. *Philosophical Transactions of the Royal Society of London* **B234**, 397–469.

Conry, H. J. (1971) Irish plaggen soils – their distribution, origin and properties. *Journal of Soil Science* **22**, 401–416.

Conway, B. W. and J. de'A. Waechter (1977) Barnfield Pit, Swanscombe. In *Guidebook for Excursion A5. South east England and the Thames Valley*, E. R. Shephard-Thorn and J. J. Wymer, 38–44. INQUA X Congress, United Kingdom. Norwich: Geo Abstracts for International Union for Quaternary Research.

Cook, J., C. B. Stringer, A. P. Currant, H. P. Schwarcz and A. G. Wintle (1982) A review of the chronology of the European Middle Pleistocene hominid record. *Yearbook of Physical Anthropology* **25**, 19–65.

Coope, G. R. (1959) A Late Pleistocene insect fauna from Chelford, Cheshire. *Proceedings of the Royal Society of London* **B151**, 70–86.

Coope, G. R. (1962) A Pleistocene coleopterous fauna with arctic affinities from Fladbury, Worcestershire. *Quarterly Journal of the Geological Society of London* **118**, 103–123.

Coope, G. R. (1968a) An insect fauna from Mid-Weichselian deposits at Brandon, Warwickshire. *Philosophical Transactions of the Royal Society of London* **B254**, 425–456.

Coope, G. R. (1968b) Coleoptera from the 'Arctic Bed' at Barnwell Station, Cambridge, *Geological Magazine* **105**, 482–486.

Coope, G. R. (1970) Climatic interpretations of Late Weichselian Coleoptera from the British Isles. *Revue de Geographie Physique et de Géologie Dynamique* **12**, 149–155.

Coope, G. R. (1974a) Interglacial Coleoptera from Bobbitshole, Ipswich, Suffolk. *Journal of the Geological Society of London* **130**, 333–340.

Coope, G. R. (1974b) Report on the Coleoptera from Wretton. In: Late Pleistocene deposits at Wretton, Norfolk. II. Devensian deposits. R. G. West, C. A.

Dickson, J. A. Catt, A. H. Weir and B. W. Sparks, *Philosophical Transactions of the Royal Society of London* **B267**, 414–418.

Coope, G. R. (1977a) Fossil coleopteran assemblages as sensitive indicators of climatic changes during the Devensian (Last) cold stage. *Philosophical Transactions of the Royal Society of London* **B280**, 313–340.

Coope, G. R. (1977b) Quaternary Coleoptera as aids in the interpretation of environmental history. In *British Quaternary studies: recent advances*, F. W. Shotton (ed.), 55–68. Oxford: Clarendon Press.

Coope, G. R. (1979) Late Cenozoic fossil Coleoptera: evolution, biogeography and ecology. *Annual Review of Ecology and Systematics* **10**, 247–267.

Coope, G. R. (1986) Coleoptera analysis. In *Handbook of Holocene palaeoecology and palaeohydrology*, B. E. Berglund (ed.), 703–713. Chichester: Wiley.

Coope, G. R. (1989) Coleoptera from the Lower Channel, Brandon, Warwickshire. In *The Pleistocene of the West Midlands. Field guide*, D. H. Keen (ed.), 27–28. Cambridge: Quaternary Research Association.

Coope, G. R. and R. B. Angus (1975) An ecological study of a temperate interlude in the middle of the last glaciation, based on fossil Coleoptera from Isleworth, Middlesex. *Journal of Animal Ecology* **44**, 365–391.

Coope, G. R. and J. A. Brophy (1972) Late Glacial environmental changes indicated by a coleopteran succession from North Wales. *Boreas* **1**, 97–142.

Coope, G. R. and A. M. Lister (1987) Late-glacial mammoth skeletons from Condover, Shropshire, England. *Nature* **330**, 472–474.

Coope, G. R. and W. Pennington (1977) The Windermere Interstadial of the Late Devensian. In: Fossil coleopteran assemblages as sensitive indicators of climatic changes during the Devensian (Last) cold stage. G. R. Coope, *Philosophical Transactions of the Royal Society of London* **B280**, 337–339.

Coope, G. R. and C. H. S. Sands (1966) Insect faunas of the last glaciation from the Tame Valley, Warwickshire. *Proceedings of the Royal Society of London* **B165**, 389–412.

Coope, G. R., R. L. Jones and D. H. Keen (1980) The palaeoecology and age of peat at Fliquet Bay, Jersey, Channel Islands. *Journal of Biogeography* **7**, 187–195.

Coope, G. R., A. Morgan and P. J. Osborne (1971) Fossil Coleoptera as indicators of climatic fluctuations during the Last Glaciation in Britain. *Palaeogeography, Palaeoclimatology, Palaeoecology* **10**, 87–101.

Coope, G. R., F. W. Shotton and I. Strachan (1961) A Late Pleistocene fauna and flora from Upton Warren, Worcestershire. *Philosophical Transactions of the Royal Society of London* **B244**, 379–421.

Coope, G. R., R. L. Jones, D. H. Keen and P. V. Waton (1985) The flora and fauna of Late Pleistocene deposits in St. Aubin's Bay, Jersey, Channel Islands. *Proceedings of the Geologists' Association* **96**, 315–321.

Cooper, J. (1972) Last Interglacial (Ipswichian) non-marine Mollusca from Aveley, Essex. *Essex Naturalist* **33**, 9–14.

Corbet, G. B. (1961) Origin of the British insular races of small mammals and of the 'Lusitanian' fauna. *Nature* **191**, 1037–1040.

Corbet, G. B. and H. N. Southern (eds.) (1977) *The handbook of British mammals*. Second edition. Oxford: Blackwell Scientific.

Cox, A. (1969) Geomagnetic reversals. *Science* **163**, 237–245.

Cox, F. C. (1981) The 'Gipping Till' revisited. In *The Quaternary in Britain: essays, reviews and original work on the Quaternary published in honour of Lewis Penny on his retirement*, J. Neale and J. Flenley (eds.), 32–42. Oxford: Pergamon Press.

Cox, F. C. and E. F. P. Nickless (1972) Some aspects of the glacial history of central Norfolk. *Bulletin of the Geological Survey of Great Britain* **42**, 79–98.

Coxon, P. (1978) The first record of a fossil naled in Britain. *Quaternary Newsletter* **24**, 9–11.

Coxon, P. (1979) *Pleistocene environmental history in central East Anglia*. Ph.D. Thesis, University of Cambridge.

Coxon, P. (1982) The Waveney Valley: interglacial deposits. In *Field meeting guide. Suffolk*, P. Allen (ed.), Section 3, 22–32. London: Quaternary Research Association.

Coxon, P. (1985) A Hoxnian Interglacial site at Athelington, Suffolk. *New Phytologist* **99**, 611–621.

Coxon, P. (1988) Remnant periglacial features on the summit of Truskmore, Counties Sligo and Leitrim, Ireland. *Zeitschrift für Geomorphologie* **71**, 81–91.

Coxon, P. and A. Flegg (1985) A Middle Pleistocene interglacial deposit from Ballyline, County Kilkenny. *Proceedings of the Royal Irish Academy* **85B**, 107–120.

Coxon, P. and A. M. Flegg (1987) A Late Pliocene/Early Pleistocene deposit at Pollnahallia, near Headford, County Galway. *Proceedings of the Royal Irish Academy* **87B**, 15–42.

Coxon, P., A. R. Hall, A. Lister and A. J. Stuart (1980) New evidence on the vertebrate fauna, stratigraphy and palaeobotany of the interglacial deposits at Swanton Morley, Norfolk. *Geological Magazine* **117**, 525–546.

Crabtree, K. (1972) Late-glacial deposits near Capel Curig, Caernarvonshire. *New Phytologist* **71**, 1233–1243.

Craig, A. J. (1978) Pollen percentage and influx analyses in south-east Ireland: a contribution to the ecology of the Late-glacial period. *Journal of Ecology* **66**, 297–324.

Crosskey, R. W. and B. J. Taylor (1986) Fossil blackflies from Pleistocene interglacial deposits in Norfolk, England. (Diptera: Simuliidae) *Systematic Entomology* **11**, 401–412.

Cullingford, R. A. (1977) Lateglacial raised shorelines and deglaciation in the Earn – Tay area. In *Studies in the Scottish Lateglacial environment*, J. M. Gray and J. J. Lowe (eds.), 15–32. Oxford: Pergamon Press.

Cullingford, R. A., C. J. Caseldine and P. E. Gotts (1980) Early Flandrian land and sea-level changes in Lower Strathearn. *Nature* **284**, 159–161.

Cullingford, R. A., C. J. Caseldine and P. E. Gotts (1989)

Evidence of early Flandrian tidal surges in Lower Strathearn, Scotland. *Journal of Quaternary Science* **4**, 51–60.

Cunliffe, B. (1974) The iron age. In *British prehistory: a new outline*, C. Renfrew (ed.), 233–262. London: Duckworth.

Cunliffe, B. W. (1978) *Iron Age communities in Britain*. Second edition. London: Routledge and Kegan Paul.

Currant, A. P. (1985a) Vertebrates of the Summertown – Radley Terrace. In *The chronology and environmental framework of early man in the Upper Thames Valley: a new model*, D. J. Briggs, G. R. Coope and D. D. Gilbertson, 74. BAR British Series 137. Oxford: British Archaeological Reports.

Currant, A. P. (1985b) Vertebrates in the Stanton Harcourt Channel. In *The chronology and environmental framework of early man in the Upper Thames Valley: a new model*, D. J. Briggs, G. R. Coope and D. D. Gilbertson, 114–118. BAR British Series 137. Oxford: British Archaeological Reports.

Currant, A. P. (1986) Interim report on the small mammal remains. In: Excavation of the Lower Palaeolithic site at Amey's Eartham Pit, Boxgrove, West Sussex. A preliminary report. M. B. Roberts, *Proceedings of the Prehistoric Society* **52**, 229–231.

Currant, A. (1989) The Quaternary origins of the modern British mammal fauna. *Biological Journal of the Linnean Society* **38**, 23–30.

Currant, A. P., C. B. Stringer and S. N. Collcutt (1984) Bacon Hole Cave. In *Field guide. Wales: Gower, Preseli, Fforest Fawr*, D. Q. Bowen and A. Henry (eds.), 38–45. Cambridge: Quaternary Research Association.

Curry, D., C. G. Adams, M. C. Boulter, F. C. Dilley, F. E. Eames, B. M. Funnell and M. K. Wells (1978) *A correlation of Tertiary rocks in the British Isles*. Geological Society of London Special Report Number 12. London: Geological Society of London.

Dackombe, R. V. and G. S. P. Thomas (1985) Introduction. In *Field guide to the Quaternary of the Isle of Man*, R. V. Dackombe and G. S. P. Thomas (eds.), 1–10. Cambridge: Quaternary Research Association.

Dalrymple, J. B. (1957) The Pleistocene deposits of Penfold's Pit, Slindon, Sussex, and their chronology. *Proceedings of the Geologists' Association* **68**, 294–303.

Dansgaard, W., S. J. Johnsen, H. B. Clausen and C. C. Langway (1971) Climatic record revealed by the Camp Century ice core. In *The Late Cenozoic glacial ages*, K. K. Turekian (ed.), 37–56. New Haven: Yale University Press.

Dansgaard, W., H. B. Clausen, N. Gunderstrup, S. J. Johnsen, and C. Rygner (1985) Dating and climatic significance of two deep Greenland ice cores. In *Greenland ice core geophysics, geochemistry, and the environment*, C. C. Langway, H. Oeschger and W. Dansgaard (eds.), 71–76. American Geophysical Union Monograph 33. Washington, DC: American Geophysical Union.

Dardis, G. F. (1985) Till facies association in drumlins and some implications for their mode of origin. *Geografisker Annaler* **67A**, 13–22.

Dardis, G. F., W. I. Mitchell and K. R. Hirons (1985) Middle Midlandian interstadial deposits at Greenagho, near Belcoo, County Fermanagh, Northern Ireland. *Irish Journal of Earth Sciences* **7**, 1–6.

Davidson, D. A., R. L. Jones and C. Renfrew (1976) Palaeoenvironmental reconstruction and evaluation: a case study from Orkney. *Transactions of the Institute of British Geographers* New Series 1, 346–361.

Davies, H. C., M. R. Dobson and R. J. Whittington (1984) A revised seismic stratigraphy for Quaternary deposits on the inner continental shelf west of Scotland between 55°30′N and 57°30′N. *Boreas* **13**, 49–66.

Davis, M. B. and K. Faegri (1967) Forest tree pollen in south Swedish peat bog deposits. (Translation of von Post 1918). *Pollen et Spores* **9**, 375–401.

Dawson, A. G. (1980) Shore erosion by frost: an example from the Scottish Lateglacial. In *Studies in the Lateglacial of north-west Europe*, J. J. Lowe, J. M. Gray and J. E. Robinson (eds.), 45–53. Oxford: Pergamon Press.

Dawson, A. G. (1984) Quaternary sea-level changes in western Scotland. *Quaternary Science Reviews* **3**, 345–368.

Dawson, A. G., D. Long and D. E. Smith (1988) The Storegga Slides: evidence from eastern Scotland for a possible tsunami. *Marine Geology* **82**, 271–276.

Dawson, M. R. (1985) Environmental reconstructions of a late Devensian terrace sequence. Some preliminary findings. *Earth Surface Processes and Landforms* **10**, 237–246.

Dawson, M. (1988) Diamict deposits of pre-Late Devensian glacial age underlying the Severn Main Terrace at Stourport, Worcestershire: their origins and stratigraphic implications. *Proceedings of the Geologists' Association* **99**, 125–132.

Dawson, M. (1989) The Severn Valley south of Bridgnorth. In *The Pleistocene of the West Midlands. Field guide*, D. H. Keen (ed.), 78–100. Cambridge: Quaternary Research Association.

Dawson, M. R. and V. Gardiner (1987) River terraces: the general model and a palaeohydrological and sedimentological interpretation of the terraces of the Lower Severn. In *Palaeohydrology in practice: a river basin analysis*, K. J. Gregory, J. Lewin and J. B. Thornes (eds.), 269–305. Chichester: Wiley.

Dearing, J. A. (1986) Core correlation and total sediment influx. In *Handbook of Holocene palaeoecology and palaeohydrology*, B. E. Berglund (ed.), 247–270. Chichester: Wiley.

Dearing, J. A., J. K. Elner and C. M. Happey-Wood (1981) Recent sediment flux and erosional processes in a Welsh upland lake-catchment based on magnetic susceptibility measurements. *Quaternary Research* **16**, 356–372.

de Beaulieu, J. L. and M. Reille (1984) A long Upper Pleistocene pollen record from Les Echets, near Lyon, France. *Boreas* **13**, 111–132.

de Boer, G., J. W. Neale and L. F. Penny (1958) A guide to the geology of the area between Market Weighton and the Humber. *Proceedings of the Yorkshire Geological Society* **31**, 157–209.

de Geer, G. (1912) A geochronology of the last 12 000 years. *Proceedings of the XIth International Geological Congress (Stockholm)* **1**, 241–253.

de Geer, G. (1935) Dating of late-glacial varves in Scotland. *Proceedings of the Royal Society of Edinburgh* **55**, 23–26.

de Jong, J. (1988) Climatic variability during the past three million years, as indicated by vegetational evolution in northwest Europe and with emphasis on data from The Netherlands. *Philosophical Transactions of the Royal Society of London* **B318**, 603–617.

Delantey, L. J. and R. J. Whittington (1977) A re-assessment of the 'Neogene' deposits of the South Irish Sea and Nymphe Bank. *Marine Geology* **24**, 23–30.

Delibrias, G., M. T. Guillier and J. Labeyrie (1974) Gif natural radiocarbon measurements VIII. *Radiocarbon* **16**, 15–94.

Delorme, L. D. (1970) Freshwater Ostracodes of Canada, Part IV, Families Ilyocyprididae, Notodromadidae, Darwinulidae, Cytherididae and Entocytherididae. *Canadian Journal of Zoology* **48**, 1251–1259.

Destombes, J.-P., E. R. Shephard-Thorn and J. H. Redding (1975) A buried valley system in the Strait of Dover. *Philosophical Transactions of the Royal Society of London* **A279**, 243–256.

Devoy, R. J. N. (1977) Flandrian sea level changes in the Thames estuary and the implications for land subsidence in England and Wales. *Nature* **270**, 712–715.

Devoy, R. J. N. (1979) Flandrian sea level changes and vegetational history of the lower Thames estuary. *Philosophical Transactions of the Royal Society of London* **B285**, 355–407.

Devoy, R. J. N. (1982) Analysis of the geological evidence for Holocene sea-level movements in southeast England. *Proceedings of the Geologists' Association* **93**, 65–90.

Devoy, R. J. N. (1983) Late Quaternary shorelines in Ireland: an assessment of their implications for isostatic land movement and relative sea level changes. In *Shorelines and isostasy*, D. E. Smith and A. G. Dawson (eds.), 227–254. Institute of British Geographers Special Publication Number 16. London: Academic Press.

Devoy, R. J. N. (1984) Quaternary environments in the area of Cork City. *Irish Quaternary Association Newsletter* **7**, 1–18.

Devoy, R. J. N. (1985a) The problems of a Late Quaternary landbridge between Britain and Ireland. *Quaternary Science Reviews* **4**, 43–58.

Devoy, R. J. N. (1985b) Holocene changes and coastal processes on the south coast of Ireland: corals and the problem of sea level methodology in temperate waters. *Proceedings of the 5th International Coral Reef Congress, Tahiti* **3**, 173–178.

Dewar, H. S. L. and H. Godwin (1963) Archaeological discoveries in the raised bogs of the Somerset Levels, England. *Proceedings of the Prehistoric Society* **29**, 17–49.

Dickson, C. A. (1970) The study of plant macrofossils in British Quaternary deposits. In *Studies in the vegetational history of the British Isles: essays in honour of Harry Godwin*, D. Walker and R. G. West (eds.), 233–254. Cambridge: Cambridge University Press.

Dickson, C. A., J. H. Dickson and G. F. Mitchell (1970) The Late Weichselian flora of the Isle of Man. *Philosophical Transactions of the Royal Society of London* **B258**, 31–79.

Dickson, J. H., W. G. Jardine and R. J. Price (1976) Three late-Devensian sites in west-central Scotland. *Nature* **262**, 43–44.

Dickson, J. H., D. A. Stewart, R. Thompson, G. Turner, M. S. Baxter, N. D. Drndarsky and J. Rose (1978) Palynology, palaeomagnetism and radiometric dating of Flandrian marine and freshwater sediments of Loch Lomond. *Nature* **274**, 548–553.

Dimbleby, G. W. (1962) *The development of British heathlands and their soils*. Oxford Forestry Memoirs Number 23. Oxford: Clarendon Press.

Dimbleby, G. W. (1976) Climate, soil and man. *Philosophical Transactions of the Royal Society of London* **B275**, 197–208.

Dimbleby, G. W. and R. J. Bradley (1975) Evidence of pedogenesis from a Neolithic site at Rackham, Sussex. *Journal of Archaeological Science* **2**, 179–186.

Dimbleby, G. W., J. R. A. Greig and R. G. Scaife (1981) Vegetational history of the Isles of Scilly. In *Environmental aspects of coasts and islands*, D. Brothwell and G. W. Dimbleby (eds.), 127–143. BAR International Series 94. Oxford: British Archaeological Reports.

Dines, H. G. and C. P. Chatwin (1930) Pliocene sandstone from Rothamsted (Hertfordshire). *Geological Survey of Great Britain Summary of Progress for 1929*, 1–7.

Dixon, P. W. (1976) Crickley Hill, 1969–1972. In *Hillforts: later prehistoric earthworks in Britain and Ireland*, D. W. Harding (ed.), 161–175. London: Academic Press.

Dixon, R. G. (1977) *Studies of the Mollusca of the Red Crag (Pleistocene), East Anglia*. Ph.D. Thesis, University of London.

Dodson, J. R. and R. H. W. Bradshaw (1987) A history of vegetation and fire, 6600 BP to present, County Sligo, western Ireland. *Boreas* **16**, 113–123.

Donald, A. P. (1981) A pollen diagram from Pitbladdo, Fife. *Transactions of the Botanical Society of Edinburgh* **43**, 281–289.

Donner, J. J. (1957) The geology and vegetation of Late-glacial retreat stages in Scotland. *Transactions of the Royal Society of Edinburgh* **63**, 211–264.

Donner, J. J. (1962) On the Post-glacial history of the Grampian Highlands. *Societas Scientiarum Fennica Commentationes Biologicae* **22**, 1–13.

Donner, J. J. (1979) The Early or Middle Devensian peat at

Burn of Benholm, Kincardineshire. *Scottish Journal of Geology* **15**, 247–250.

Douglas, T. D. (1980) The Quaternary deposits of western Leicestershire. *Philosophical Transactions of the Royal Society of London* **B288**, 259–286.

Drewett, P. C. (1978) Field systems and land allotment in Sussex, 3rd millennium BC to 4th century AD. In *Early land allotment in the British Isles: a survey of recent work*, H. C. Bowen and P. J. Fowler (eds.), 67–80. BAR British Series 48. Oxford: British Archaeological Reports.

Dubois, A. D. and D. K. Ferguson (1985) The climatic history of pine in the Cairngorms based on radiocarbon dates and stable isotope analysis, with an account of the events leading up to its colonization. *Review of Palaeobotany and Palynology* **46**, 55–80.

Ducker, A. (1969) Der Ablauf der Holsteinwarmzeit in West-Holstein. *Eiszeitalter und Gegenwart* **20**, 46–57.

Duigan, S. L. (1956a) Pollen analysis of the Nechells interglacial deposits. *Quarterly Journal of the Geological Society of London* **112**, 373–391.

Duigan, S. L. (1956b) Interglacial plant remains from the Wolvercote Channel, Oxford. *Quarterly Journal of the Geological Society of London* **112**, 363–372.

Dury, G. H. (1977) Underfit streams: retrospect, perspect and prospect. In *River channel changes*, K. J. Gregory (ed.), 281–293. Chichester: Wiley.

Eakins, J. D. and R. T. Morrison (1978) A new procedure for the determination of lead-210 in lake and marine sediments. *International Journal of Applied Radiation and Isotopes* **29**, 531–536.

Eden, M. J. and C. P. Green (1971) Some aspects of granite weathering and tor formation on Dartmoor, England. *Geografiska Annaler* **53A**, 92–99.

Eden, R. A., I. P. Stevenson and W. Edwards (1957) *Geology of the country around Sheffield*. Memoirs of the Geological Survey of Great Britain. London: Her Majesty's Stationery Office.

Edwards, C. A. (1981) The tills of Filey Bay. In *The Quaternary in Britain: essays, reviews and original work on the Quaternary published in honour of Lewis Penny on his retirement*, J. Neale and J. Flenley (eds.), 108–118. Oxford: Pergamon Press.

Edwards, C. A. (1987) The Quaternary deposits of Filey Bay. In *East Yorkshire. Field guide*, S. Ellis (ed.), 15–21. Cambridge: Quaternary Research Association.

Edwards, K. J. (1979a) Palynological and temporal inference in the context of prehistory with special reference to the evidence from lake and peat deposits. *Journal of Archaeological Science* **6**, 255–270.

Edwards, K. J. (1979b) Environmental impact in the prehistoric period. *Scottish Archaeological Forum* **9**, 27–41.

Edwards, K. J. (1985) The anthropogenic factor in vegetational history. In *The Quaternary history of Ireland*, K. J. Edwards and W. P. Warren (eds.), 187–220. London: Academic Press.

Edwards, K. J. and K. R. Hirons (1984) Cereal pollen grains in pre-elm decline deposits: implications for the earliest agriculture in Britain and Ireland. *Journal of Archaeological Science* **11**, 71–80.

Edwards, K. J. and I. B. M. Ralston (1984) Post-glacial hunter-gatherers and vegetational history in Scotland. *Proceedings of the Society of Antiquaries of Scotland* **114**, 15–34.

Edwards, K. J. and K. M. Rowntree (1980) Radiocarbon and palaeoenvironmental evidence for changing rates of erosion at a Flandrian stage site in Scotland. In *Timescales in geomorphology*, R. A. Cullingford, D. A. Davidson and J. Lewin (eds.), 207–223. Chichester: Wiley.

Edwards, K. J. and W. P. Warren (1985) Quaternary studies in Ireland. In *The Quaternary history of Ireland*, K. J. Edwards and W. P. Warren (eds.), 1–16. London: Academic Press.

Edwards, K. J., C. J. Caseldine and D. K. Chester (1976) Possible interstadial and interglacial pollen floras from Teindland, Scotland. *Nature* **264**, 742–744.

Edwards, W., G. H. Mitchell and T. H. Whitehead (1950) *Geology of the district north and east of Leeds*. Memoirs of the Geological Survey of Great Britain. London: His Majesty's Stationery Office.

Eisma, D., W. Mook and C. Laban (1981) An early Holocene tidal flat in the Southern Bight. In *Holocene marine sedimentation in the North Sea Basin*, S.-D. Nio, R. T. E. Schüttenhelm and T. C. E. van Weering (eds.), 229–237. International Association of Sedimentologists Special Publication Number 5. Oxford: Blackwell Scientific.

Eisma, D., J. H. F. Jansen and T. C. E. van Weering (1979) Sea-floor morphology and recent sediment movement in the North Sea. In *The Quaternary history of the North Sea*, E. Oele, R. T. E. Schüttenhelm and A. J. Wiggers (eds.), 217–231. Acta Universitatis Uppsaliensis Symposium Universitatis Uppsaliensis Annum Quingentesimum Celebrantis, Number 2. Uppsala: University of Uppsala.

Ellis, A. E. (1978) *British freshwater bivalve Mollusca: keys and notes for the identification of the species*. Synopses of the British fauna (New series) Number 11. London: Academic Press for Linnean Society of London.

Embleton, C. (1970) North-eastern Wales. In *The glaciations of Wales and adjoining regions*, C. A. Lewis (ed.), 59–82. London: Longman.

Embleton, C. and C. A. M. King (1975) *Glacial and periglacial geomorphology*. Second edition. Volume 1. *Glacial geomorphology*. London: Edward Arnold.

Emiliani, C. (1955) Pleistocene temperatures. *Journal of Geology* **63**, 538–578.

Emiliani, C. (1966) Paleotemperature analysis of the Caribbean cores P6304-8 and P6034-9 and a generalized temperature curve for the last 425 000 years. *Journal of Geology* **74**, 109–126.

Epstein, S., R. P. Sharp and A. J. Gow (1970) Antarctic ice sheet: stable isotope analysis of Byrd station cores and interhemispheric climatic implications. *Science* **168**, 1570–1572.

Ericsson, B. (1973) The cation content of Swedish Post-Glacial sediments as a criterion of palaeosalinity. *Geologiska Föreningens i Stockholm Förhandlingar* **95**, 181–220.

Evans, C. D. R. and M. J. Hughes (1984) The Neogene succession of the South Western Approaches, Great Britain. *Journal of the Geological Society of London* **141**, 315–326.

Evans, G. H. (1961) *A study of the diatoms in a core from the sediments of Devoke Water.* M.Sc. Thesis, University of Wales.

Evans, G. H. (1970) Pollen and diatom analyses of Late-Quaternary deposits in the Blelham basin, north Lancashire. *New Phytologist* **69**, 821–874.

Evans, G. H. and R. Walker (1977) The Late Quaternary history of the diatom flora of Llyn Clyd and Llyn Glas, two small oligotrophic high mountain tarns in Snowdonia (Wales). *New Phytologist* **78**, 221–236.

Evans, J. G. (1966) Late-glacial and Post-glacial subaerial deposits at Pitstone, Buckinghamshire. *Proceedings of the Geologists' Association* **77**, 347–364.

Evans, J. G. (1971) Habitat changes on the calcareous soils of Britain: the impact of Neolithic man. In *Economy and settlement in Neolithic and Early Bronze Age Britain and Europe*, D. D. A. Simpson (ed.), 27–73. Leicester: Leicester University Press.

Evans, J. G. (1972) *Land snails in archaeology: with special reference to the British Isles.* London: Seminar Press.

Evans, J. G. (1975) *The environment of early man in the British Isles.* London: Paul Elek.

Evans, J. G., C. French and D. Leighton (1978) Habitat change in two Late-glacial and Post-glacial sites in southern Britain: the molluscan evidence. In *The effect of man on the landscape: the Lowland Zone*, S. Limbrey and J. G. Evans (eds.), 63–75. CBA Research Report Number 21. London: Council for British Archaeology.

Evans, P. (1971) Towards a Pleistocene time-scale. In *The Phanerozoic time-scale – a supplement.* Part 2, 123–356. Geological Society of London Special Publication Number 5. London: Geological Society of London.

Evans, W. B. and R. S. Arthurton (1973) North-west England. In *A correlation of Quaternary deposits in the British Isles*, G. F. Mitchell, L. F. Penny, F. W. Shotton and R. G. West, 28–36. Geological Society of London Special Report Number 4. London: Geological Society of London.

Eyles, C. H. and N. Eyles (1984) Glaciomarine sediments of the Isle of Man as a key to Late Pleistocene stratigraphic investigations in the Irish Sea Basin. *Geology* **12**, 359–364.

Eyles, N. and A. M. McCabe (1991) Glaciomarine deposits of the Irish Sea Basin: the role of glacio-isostatic disequilibrium. In *Glacial deposits in Great Britain and Ireland*, J. Ehlers, P. L. Gibbard and J. Rose (eds.), 311–331. Rotterdam: Balkema.

Eyles, N., C. H. Eyles and A. M. McCabe (1989) Sedimentation in an ice-contact subaqueous setting: the mid-Pleistocene 'North Sea Drifts' of Norfolk, UK. *Quaternary Science Reviews* **8**, 57–74.

Eyles, N., C. H. Eyles and A. D. Miall (1983) Lithofacies types and vertical profile models: an alternative approach to the description and environmental interpretation of glacial diamict and diamictite sequences. *Sedimentology* **30**, 393–410.

Faegri, K. and J. Iversen (1975) *Textbook of pollen analysis.* Third edition. Copenhagen: Munksgaard.

Faegri, K., P. E. Kaland and K. Krzywinsi (1989) *Textbook of pollen analysis.* Fourth edition. Chichester: Wiley.

Finberg, H. P. R. (1972) Anglo-Saxon England to 1042. In *The agrarian history of England and Wales. Volume I: II, AD 43–1042*, H. P. R. Finberg (ed.), 385–525. Cambridge: Cambridge University Press.

Finch, T. F. and F. M. Synge (1966) The drifts and soils of west Clare and the adjoining parts of counties Kerry and Limerick. *Irish Geography* **5**, 161–172.

Findlay, D. C. (1965) *The soils of the Mendip District of Somerset.* Memoirs of the Soil Survey of Great Britain. Harpenden: Soil Survey of England and Wales.

Firth, C. R. and B. A. Haggart (1989) Loch Lomond Stadial and Flandrian shorelines in the inner Moray Firth area, Scotland. *Journal of Quaternary Science* **4**, 37–50.

FitzPatrick, E. A. (1965) An interglacial soil at Teindland, Morayshire. *Nature* **207**, 621–622.

Fleming, A. (1971a) Bronze Age agriculture on the marginal lands of north-east Yorkshire. *Agricultural History Review* **19**, 1–24.

Fleming, A. (1971b) Territorial patterns in Bronze Age Wessex. *Proceedings of the Prehistoric Society* **37**, 138–166.

Fleming, A. (1978) The Dartmoor reaves. In *Early land allotment in the British Isles: a survey of recent work*, H. C. Bowen and P. J. Fowler (eds.), 17–41. BAR British Series 48. Oxford: British Archaeological Reports.

Flemming, N. C. (1982) Multiple regression analysis of earth movements and eustatic sea-level changes in the United Kingdom in the past 9000 years. *Proceedings of the Geologists' Association* **93**, 113–125.

Flenley, J. R. (1984) Towards a vegetational history of the meres of Holderness. In *Late Quaternary environments and man in Holderness*, D. D. Gilbertson (ed.), 165–175. BAR British Series 134. Oxford: British Archaeological Reports.

Flenley, J. R. and M. C. Pearson (1967) Pollen analysis of a peat from the island of Canna (Inner Hebrides). *New Phytologist* **66**, 299–306.

Flinn, D. (1978) The glaciation of the Outer Hebrides. *Geological Journal* **13**, 195–199.

Flinn, D. (1980) The glaciation of the Outer Hebrides: reply. *Scottish Journal of Geology* **16**, 85–86.

Florschütz, F., J. Menéndez-Amor and T. A. Wijmstra (1971) Palynology of a thick Quaternary succession in southern Spain. *Palaeogeography, Palaeoclimatology, Palaeoecology* **10**, 233–264.

Flower, R. J. and R. W. Battarbee (1983) Diatom evidence

for recent acidification of two Scottish lochs. *Nature* **305**, 130–133.

Flower, R. J., R. W. Battarbee and P. G. Appleby (1987) The recent palaeolimnology of acid lakes in Galloway, south-west Scotland: diatom analysis, pH trends and the rôle of afforestation. *Journal of Ecology* **75**, 797–824.

Forbes, C. L. and P. G. Cambridge (1967) The Galley Hill Pits, near St. Ives, Hunts. New finds of Pleistocene bones. *Huntingdonshire Fauna and Flora Society 19th Annual Report 1966*, 19.

Ford, D. C. and H. P. Schwarcz (1990) Uranium-series disequilibrium dating methods. In *Geomorphological techniques*. Second edition, A. S. Goudie (ed.), 394–399. London: Unwin Hyman.

Ford, T. D., M. Gascoyne and J. S. Beck (1983) Speleothem dates and Pleistocene chronology in the Peak District of Derbyshire. *Transactions of the British Cave Research Association* **10**, 103–115.

Fosse, G. and G. Carpentier (1982) Prehistoric man at Tourville-la-Riviere. In *The Quaternary of Normandy. Field handbook*, J.-P. Lautridou, 30. Caen: Quaternary Research Association and Centre de Geomorphologie du Centre National de la Recherche Scientifique.

Foster, I. D. L., S. M. Charlesworth and D. H. Keen (1991) A comparative study of heavy metal contamination and pollution in four reservoirs in the English Midlands, UK. *Hydrobiologia* **214**, 155–162.

Foster, I. D. L., J. A. Dearing, A. Simpson, A. D. Carter and P. G. Appleby (1985) Lake catchment based studies of erosion and denudation in the Merevale catchment, Warwickshire, UK. *Earth Surface Processes and Landforms* **10**, 45–68.

Francis, E. A. (1970) Quaternary (geology of Durham County). *Transactions of the Natural History Society of Northumberland, Durham and Newcastle upon Tyne* **41**, 134–152.

Franks, J. W. (1960) Interglacial deposits at Trafalgar Square, London. *New Phytologist* **59**, 145–152.

Franks, J. W., A. J. Sutcliffe, M. P. Kerney and G. R. Coope (1958) Haunt of elephant and rhinoceros: the Trafalgar Square of 100 000 years ago – new discoveries. *The Illustrated London News* June 14, 1011–1013.

French, C. A. I. (1982) An analysis of the molluscs from an Ipswichian interglacial river channel deposit at Maxey, Cambridgeshire, England. *Geological Magazine* **119**, 593–598.

French, H. M. (1976) *The periglacial environment*. London: Longman.

Fritts, H. C. (1976) *Tree rings and climate*. London: Academic Press.

Funnell, B. M. (1961a) The Palaeogene and Early Pleistocene of Norfolk. *Transactions of the Norfolk and Norwich Naturalists' Society* **19**, 340–364.

Funnell, B. M. (1961b) Foraminifera of the Corton Beds. In: The glacial and interglacial deposits of Norfolk. R. G. West, *Transactions of the Norfolk and Norwich Naturalists' Society* **19**, 370.

Funnell, B. M. (1987) Late Pliocene and Early Pleistocene stages of East Anglia and the adjacent North Sea. *Quaternary Newsletter* **52**, 1–11.

Funnell, B. M. (1988) Foraminifera in the Late Tertiary and Early Quaternary Crags of East Anglia. In *The Pliocene – Middle Pleistocene of East Anglia. Field guide*, P. L. Gibard and J. A. Zalasiewicz (eds.), 50–52. Cambridge: Quaternary Research Association.

Funnell, B. M. and I. Pearson (1989) Holocene sedimentation on the north Norfolk barrier coast in relation to relative sea-level change. *Journal of Quaternary Science* **4**, 25–36.

Funnell, B. M. and R. G. West (1962) The Early Pleistocene of Easton Bavents, Suffolk. *Quarterly Journal of the Geological Society of London* **118**, 125–141.

Funnell, B. M. and R. G. West (1977) Preglacial Pleistocene deposits of East Anglia. In *British Quaternary studies: recent advances*, F. W. Shotton (ed.), 247–265. Oxford: Clarendon Press.

Funnell, B. M., P. E. P. Norton and R. G. West (1979) The Crag at Bramerton, near Norwich, Norfolk. *Philosophical Transactions of the Royal Society of London* **B287**, 489–534.

Gallois, R. W. (1978) The Pleistocene history of west Norfolk. *Bulletin of the Geological Society of Norfolk* **30**, 3–38.

Gallois, R. W. (1979) *Geological investigations for the Wash Water Storage Scheme*. Report of the Institute of Geological Sciences, Number 78/19. London: Her Majesty's Stationery Office.

Garrard, R. A. (1977) The sediments of the South Irish Sea and Nymphe Bank area of the Celtic Sea. In *The Quaternary history of the Irish Sea*, C. Kidson and M. J. Tooley (eds.), 69–92. Geological Journal Special Issue Number 7. Liverpool: Seel House Press.

Garrard, R. A. and M. R. Dobson (1974) The nature and maximum extent of glacial sediments off the west coast of Wales. *Marine Geology* **16**, 31–44.

Gascoyne, M., A. P. Currant and T. C. Lord (1981) Ipswichian fauna of Victoria Cave and the marine palaeoclimatic record. *Nature* **294**, 652–654.

Gascoyne, M., H. P. Schwarcz and D. C. Ford (1983) Uranium-series ages of speleothem from north-west England: correlation with Quaternary climate. *Philosophical Transactions of the Royal Society of London* **B301**, 143–164.

Gaunt, G. D. (1970) The occurrence of Pleistocene ventifacts at Aldborough, near Boroughbridge, West Yorkshire. *Journal of Earth Science (Leeds)* **8**, 159–161.

Gaunt, G. D. (1974) A radiocarbon date relating to Lake Humber. *Proceedings of the Yorkshire Geological Society* **40**, 195–197.

Gaunt, G. D. (1976) The Devensian maximum ice limit in the Vale of York. *Proceedings of the Yorkshire Geological Society* **40**, 631–637.

Gaunt, G. D. (1981) Quaternary history of the southern part of the Vale of York. In *The Quaternary in Britain: essays, reviews and original work on the Quaternary published in honour of Lewis Penny on his retirement*, J. Neale and J.

Flenley (eds.), 82–97. Oxford: Pergamon Press.

Gaunt, G. D. and M. J. Tooley (1974) Evidence for Flandrian sea-level changes in the Humber estuary and adjacent areas. *Bulletin of the Geological Survey of Great Britain* **48**, 25–41.

Gaunt, G. D., D. D. Bartley and R. Harland (1974) Two interglacial deposits proved in boreholes in the southern part of the Vale of York and their bearing on contemporaneous sea levels. *Bulletin of the Geological Survey of Great Britain* **48**, 1–23.

Gaunt, G. D., G. R. Coope and J. W. Franks (1970) Quaternary deposits at Oxbow opencast coal site in the Aire Valley, Yorkshire. *Proceedings of the Yorkshire Geological Society* **38**, 175–200.

Gaunt, G. D., R. A. Jarvis and B. Matthews (1971) The late Weichselian sequence in the Vale of York. *Proceedings of the Yorkshire Geological Society* **38**, 281–284.

Gaunt, G. D., G. R. Coope, P. J. Osborne and J. W. Franks (1972) *An interglacial deposit near Austerfield, southern Yorkshire*. Report of the Institute of Geological Sciences, Number 72/4. London: Her Majesty's Stationery Office.

Geikie, A. (1894) *The Great Ice Age and its relation to the antiquity of man*. Third edition. London: Stanford.

Gennard, D. E. (1984) A palaeoecological study of the interglacial deposit at Benburb, County Tyrone. *Proceedings of the Royal Irish Academy* **84B**, 43–56.

George, T. N. (1932) The Quaternary beaches of Gower. *Proceedings of the Geologists' Association* **43**, 291–324.

George, T. N. (1970) *British regional geology: South Wales*. Third edition. London: Her Majesty's Stationery Office.

Gibbard, P. L. (1977) Pleistocene history of the Vale of St. Albans. *Philosophical Transactions of the Royal Society of London* **B280**, 445–483.

Gibbard, P. L. (1979) Middle Pleistocene drainage in the Thames Valley. *Geological Magazine* **116**, 35–44.

Gibbard, P. L. (1983) The diversion of the Thames – a review. In *Field guide. Diversion of the Thames*, J. Rose (ed.), 8–23. Hoddesdon: Quaternary Research Association.

Gibbard, P. L. (1985) *Pleistocene history of the Middle Thames Valley*. Cambridge: Cambridge University Press.

Gibbard, P. L. (1988a) Palynological problems and the vegetational sequence of the Pliocene – preglacial Pleistocene of East Anglia. In *The Pliocene – Middle Pleistocene of East Anglia. Field guide*, P. L. Gibbard and J. A. Zalasiewicz (eds.), 42–49. Cambridge: Quaternary Research Association.

Gibbard, P. L. (1988b) The history of the great northwest European rivers during the past three million years. *Philosophical Transactions of the Royal Society of London* **B318**, 559–602.

Gibbard, P. L. and M. M. Aalto (1977) A Hoxnian interglacial site at Fishers Green, Stevenage, Hertfordshire. *New Phytologist* **78**, 505–523.

Gibbard, P. L. and D. A. Cheshire (1983) Hatfield Polytechnic (Roe Hyde Pit). In *Field guide. Diversion of the Thames*, J. Rose (ed.), 110–119. Hoddesdon: Quaternary Research Association.

Gibbard, P. L. and A. R. Hall (1982) Late Devensian river deposits in the Lower Colne Valley, West London, England. *Proceedings of the Geologists' Association* **93**, 291–299.

Gibbard, P. L. and S. M. Peglar (1988) Rockhall Wood, Sutton: pollen analysis. In *The Pliocene – Middle Pleistocene of East Anglia. Field guide*, P. L. Gibbard and J. A. Zalasiewicz (eds.), 71–72. Cambridge: Quaternary Research Association.

Gibbard, P. L. and S. M. Peglar (1989) Palynology of the fossiliferous deposits at Brandon, Warwickshire. In *The Pleistocene of the West Midlands. Field guide*, D. H. Keen (ed.), 23–26. Cambridge: Quaternary Research Association.

Gibbard, P. L. and S. M. Peglar (1990) Palynology of the interglacial deposits at Little Oakley, Essex, and their correlation. *Philosophical Transactions of the Royal Society of London* **B328**, 341–357.

Gibbard, P. L. and M. Pettit (1978) The palaeobotany of interglacial deposits at Sugworth, Berkshire. *New Phytologist* **81**, 465–477.

Gibbard, P. L. and A. J. Stuart (1975) Flora and vertebrate fauna of the Barrington Beds. *Geological Magazine* **112**, 493–501.

Gibbard, P. L., I. D. Bryant and A. R. Hall (1986b) A Hoxnian interglacial doline infilling at Slade Oak Lane, Denham, Buckinghamshire, England. *Geological Magazine* **123**, 27–43.

Gibbard, P. L., V. R. Switsur and A. G. Wintle (1986a) A reappraisal of the age of silts in the Wrecclesham Gravel at Alton Road, Farnham, Surrey. *Quaternary Newsletter* **50**, 6–13.

Gibbard, P. L., A. G. Wintle and J. A. Catt (1987) Age and origin of clayey silt 'brickearth' in west London, England. *Journal of Quaternary Science* **2**, 3–9.

Gibbard, P. L., G. R. Coope, A. R. Hall, R. C. Preece and J. E. Robinson (1982) Middle Devensian deposits beneath the 'Upper Floodplain' terrace of the River Thames at Kempton Park, Sunbury, England. *Proceedings of the Geologists' Association* **93**, 275–289.

Gilbertson, D. D. (1980a) Molluscs. In: Cromerian interglacial deposits at Sugworth, near Oxford, England and their relation to the Plateau Drift of the Cotswolds and the terrace sequence of the Upper and Middle Thames. F. W. Shotton, A. S. Goudie, D. J. Briggs and H. A. Osmaston, *Philosophical Transactions of the Royal Society of London* **B289**, 63.

Gilbertson, D. D. (1980b) The palaeoecology of Middle Pleistocene Mollusca from Sugworth, Oxfordshire. *Philosophical Transactions of the Royal Society of London* **B289**, 107–118.

Gilbertson, D. and R. Beck (1974) Interim report on the Burtle sand beds at Penzoy Farm, Westonzoyland, Bridgwater. In *Field handbook. Exeter*, A. Straw (ed.), (supplement). Exeter: Quaternary Research Association.

Gilbertson, D. D. and A. B. Hawkins (1977) The Quaternary deposits at Swallow Cliff, Middlehope, County of Avon.

Proceedings of the Geologists' Association **88**, 255–266.

Gilbertson, D. D. and A. B. Hawkins (1978) The Pleistocene succession at Kenn, Somerset. *Bulletin of the Geological Survey of Great Britain*, Number 66. London: Her Majesty's Stationery Office.

Girling, M. A. (1974) Evidence from Lincolnshire of the age and intensity of the mid-Devensian temperate episode. *Nature* **250**, 270.

Girling, M. A. (1977) Tattershall Castle and Kirkby-on-Bain. In *Guidebook for Excursion C7. Yorkshire and Lincolnshire*, J. A. Catt, 19–21. INQUA X Congress, United Kingdom. Norwich: Geo Abstracts for International Union for Quaternary Research.

Gladfelter, B. G. (1975) Middle Pleistocene sedimentary sequences in East Anglia (United Kingdom). In *After the australopithecines: stratigraphy, ecology and culture change in the Middle Pleistocene*, K. W. Butzer and G. L. Isaac (eds.), 225–258. The Hague: Mouton.

Gladfelter, B. G. and R. Singer (1975) Implications of East Anglian glacial stratigraphy for the British Lower Palaeolithic. In *Quaternary studies*, R. P. Suggate and M. M. Cresswell (eds.), 139–145. Wellington: Royal Society of New Zealand.

Godwin, H. (1940a) Pollen analysis and forest history of England and Wales. *New Phytologist* **39**, 370–400.

Godwin, H. (1940b) Studies of the Post-glacial history of British vegetation III. Fenland pollen diagrams. IV. Post-glacial changes of relative land- and sea-levels in the English Fenland. *Philosophical Transactions of the Royal Society of London* **B230**, 239–303.

Godwin, H. (1943) Coastal peat beds of the British Isles and North Sea. *Journal of Ecology* **31**, 199–247.

Godwin, H. (1955) Vegetational history at Cwm Idwal: a Welsh plant refuge. *Svensk Botanisk Tidskrift* **49**, 35–43.

Godwin, H. (1956) *The history of the British flora: a factual basis for phytogeography*. First edition. Cambridge: Cambridge University Press.

Godwin, H. (1962) Vegetational history of the Kentish Chalk Downs as seen at Wingham and Frogholt. *Veröffentlichungen des Geobotanischen Instuts, Eidgenössiche technische Hochschule Rübel in Zurich* **37**, 83–99.

Godwin, H. (1968) Studies of the Post-glacial history of British vegetation XV. Organic deposits at Old Buckenham Mere, Norfolk. *New Phytologist* **67**, 95–107.

Godwin, H. (1975) *The history of the British flora: a factual basis for phytogeography*. Second edition. Cambridge: Cambridge University Press.

Godwin, H. (1978) *Fenland: its ancient past and uncertain future*. Cambridge: Cambridge University Press.

Godwin, H. and M. H. Clifford (1938) Studies of the Post-glacial history of British vegetation I. Origin and stratigraphy of Fenland deposits near Woodwalton, Hunts. II. Origin and stratigraphy of deposits in southern Fenland. *Philosophical Transactions of the Royal Society of London* **B229**, 323–406.

Godwin, H. and L. Newton (1938) The submerged forest at Borth and Ynyslas, Cardiganshire. Data for the study of Post-glacial history. *New Phytologist* **37**, 333–344.

Godwin, H. and P. A. Tallentire (1951) Studies of the Post-glacial history of British vegetation. XII. Hockham Mere, Norfolk. *Journal of Ecology* **39**, 285–307.

Godwin, H. and Vishnu-Mittre (1975) Studies of the Post-glacial history of British vegetation XVI. Flandrian deposits of the Fenland margin at Holme Fen and Whittlesey Mere, Hunts. *Philosophical Transactions of the Royal Society of London* **B270**, 561–604.

Godwin, H. and E. H. Willis (1960) Cambridge University natural radiocarbon measurements II. *Radiocarbon* **2**, 62–72.

Godwin, H., D. Walker and E. H. Willis (1957) Radiocarbon dating and Post-glacial vegetational history: Scaleby Moss. *Proceedings of the Royal Society of London* **B147**, 352–366.

Gordon, D., P. L. Smart, D. C. Ford, J. N. Andrews, T. C. Atkinson, P. J. Rowe and N. S. J. Christopher (1989) Dating of Late Pleistocene interglacial and interstadial periods in the United Kingdom from speleothem growth frequency. *Quaternary Research* **31**, 14–26.

Goudie, A. S. and M. G. Hart (1975) Pleistocene events and forms in the Oxford region. In *Oxford and its region: geographical essays*, C. G. Smith and D. I. Scargill (eds.), 3–13. Oxford: Oxford University Press.

Graham, A. (1947) Some observations on the brochs. *Proceedings of the Society of Antiquaries of Scotland* **81**, 48–99.

Gray, J. M. (1974) Lateglacial and postglacial shorelines in western Scotland. *Boreas* **3**, 129–138.

Gray, J. M. (1978) Low-level shore platforms in the south-west Scottish Highlands: altitude, age and correlation. *Transactions of the Institute of British Geographers* New Series **3**, 151–164.

Gray, J. M. (1982) The last glaciers (Loch Lomond Advance) in Snowdonia, North Wales. *Geological Journal* **17**, 111–133.

Gray, J. M. (1983) The measurement of shoreline altitudes in areas affected by glacio-isostasy, with particular reference to Scotland. In *Shorelines and isostasy*, D. E. Smith and A. G. Dawson (eds.), 97–127. Institute of British Geographers Special Publication Number 16. London: Academic Press.

Gray, J. M. (1990) The moraines and small-scale erosional features of Cwm Llydaw and Cwm Dyli. In *The Quaternary of North Wales. Field guide*, K. Addison, M. J. Edge and R. Watkins (eds.), 81–84. Coventry: Quaternary Research Association.

Gray, J. M. (1992) Raised shorelines. In *The south-west Scottish Highlands. Field guide*, M. J. C. Walker, J. M. Gray and J. J. Lowe (eds.), 11–21. Cambridge: Quaternary Research Association.

Gray, J. M. and C. L. Brooks (1972) The Loch Lomond Readvance moraines of Mull and Menteith. *Scottish*

Journal of Geology **8**, 95–103.

Gray, J. M. and J. J. Lowe (1977) The Scottish Lateglacial environment: a synthesis. In *Studies in the Scottish Lateglacial environment*, J. M. Gray and J. J. Lowe (eds.), 163–181. Oxford: Pergamon Press.

Gray, J. M. and J. J. Lowe (1982) Problems in the interpretation of small-scale erosional forms on glaciated bedrock surfaces: examples from Snowdonia, North Wales. *Proceedings of the Geologists' Association* **93**, 403–414.

Green, C. P. (1973) Pleistocene river gravels and the Stonehenge problem. *Nature* **243**, 214–216.

Green, C. P. (1974) Pleistocene gravels of the River Axe in south-western England, and their bearing on the southern limit of glaciation in Britain. *Geological Magazine* **111**, 213–220.

Green, C. P. and D. H. Keen (1987) Stratigraphy and palaeoenvironments of the Stone Point deposits: the 1975 investigation. In *Wessex and the Isle of Wight. Field guide*, K. E. Barber (ed.), 17–20. Cambridge: Quaternary Research Association.

Green, C. P. and D. F. M. McGregor (1978) Pleistocene gravel trains of the River Thames. *Proceedings of the Geologists' Association* **89**, 143–156.

Green, C. P. and D. F. M. McGregor (1980) Quaternary evolution of the River Thames. In *The shaping of southern England*, D. K. C. Jones (ed.), 177–202. Institute of British Geographers Special Publication Number 11. London: Academic Press.

Green, C. P. and D. F. M. McGregor (1983) Lithology of the Thames gravels. In *Field guide. Diversion of the Thames*, J. Rose (ed.), 24–38. Hoddesdon: Quaternary Research Association.

Green, C. P., R. W. Hey and D. F. M. McGregor (1980) Volcanic pebbles in Pleistocene gravels of the Thames in Buckinghamshire and Hertfordshire. *Geological Magazine* **117**, 59–64.

Green, C. P., D. F. M. McGregor and A. H. Evans (1982) Development of the Thames drainage system in Early and Middle Pleistocene times. *Geological Magazine* **119**, 281–290.

Green, C. P., D. H. Keen, D. F. M. McGregor, J. E. Robinson and R. B. G. Williams (1983) Stratigraphy and environmental significance of Pleistocene deposits at Fisherton, near Salisbury, Wiltshire. *Proceedings of the Geologists' Association* **94**, 17–22.

Green, C. P., G. R. Coope, A. P. Currant, D. T. Holyoak, M. Ivanovich, R. L. Jones, D. H. Keen, D. F. M. McGregor and J. E. Robinson (1984) Evidence of two temperate episodes in late Pleistocene deposits at Marsworth, UK. *Nature* **309**, 778–781.

Green, C. P., G. R. Coope, D. T. Holyoak, M. Ivanovich, R. L. Jones, D. H. Keen, R. J. Rogerson, J. E. Robinson and R. C. Young (in preparation). Stratigraphy and environmental significance of Pleistocene deposits at Stoke Goldington, Buckinghamshire.

Green, H. S. (1982) Pontnewydd Cave, Wales: a new British Middle Pleistocene hominid site. In *Handbook for the field excursion to the tufas of the Wheeler Valley and to Pontnewydd and Cefn caves*, R. C. Preece, C. Turner and H. S. Green, (unpaginated). Cambridge: Quaternary Research Association.

Green, H. S. (1984) *Pontnewydd Cave: a Lower Palaeolithic hominid site in Wales. The first report*. National Museum of Wales Quaternary Studies Monographs, Number 1. Cardiff: National Museum of Wales.

Green, H. S., C. B. Stringer, S. N. Collcutt, A. P. Currant, J. Huxtable, H. P. Schwarcz, N. Debenham, C. Embleton, P. Bull, T. I. Molleson and R. E. Bevins (1981) Pontnewydd Cave in Wales – a new Middle Pleistocene hominid site. *Nature* **294**, 707–713.

Greenslade, P. J. (1968) Habitat and altitude distribution of Carabidae (Coleoptera) in Argyll, Scotland. *Transactions of the Royal Entomological Society of London* **120**, 39–54.

Greensmith, J. T. and E. V. Tucker (1971) The effects of late Pleistocene and Holocene sea level changes in the vicinity of the River Crouch, Essex. *Proceedings of the Geologists' Association* **82**, 301–322.

Greensmith, J. T. and E. V. Tucker (1973) Holocene transgressions and regressions on the Essex coast, outer Thames Estuary. *Geologie en Mijnbouw* **52**, 193–202.

Greensmith, J. T. and E. V. Tucker (1980) Evidence for differential subsidence on the Essex coast. *Proceedings of the Geologists' Association* **91**, 169–175.

Gregory, K. J. (1962) The deglaciation of eastern Eskdale, Yorkshire. *Proceedings of the Yorkshire Geological Society* **33**, 363–380.

Gregory, K. J. (1965) Proglacial Lake Eskdale after sixty years. *Transactions of the Institute of British Geographers* **34**, 149–162.

Gregory, K. J., J. Lewin and J. B. Thornes (eds.) (1987) *Palaeohydrology in practice: a river basin analysis*. Chichester: Wiley.

Greig, J. (1982) Past and present lime woods of Europe. In *Archaeological aspects of woodland ecology*, M. Bell and S. Limbrey (eds.), 23–55. BAR International Series 146. Oxford: British Archaeological Reports.

Griffin, K. (1984) Plant macrofossils from a Quaternary deposit in the North Sea. In *Quaternary stratigraphy of the North Sea*. Abstract volume, I. Aarseth and H. P. Sejrup (eds.), 33. Symposium University of Bergen, December 1984. Bergen: University of Bergen.

Grigson, C. (1965) Faunal remains: measurements of bones, horncores, antlers and teeth. In *Windmill Hill and Avebury: excavations by Alexander Keiller 1925–1939*, I. F. Smith, 145–167. Oxford: Clarendon Press.

Grigson, C. (1978) The Late-glacial and Early Flandrian ungulates of England and Wales – an interim review. In *The effect of man on the landscape: the Lowland Zone*, S. Limbrey and J. G. Evans (eds.), 46–56. CBA Research

Report Number 21. London: Council for British Archaeology.

Grove, J. M. (1988) *The Little Ice Age*. London: Methuen.

Grün, R., H. P. Schwarcz and J. Chadam (1988) ESR dating of tooth enamel: coupled correction for U-uptake and U-series disequilibrium. *Nuclear Tracks and Radiation Measurements* **14**, 237–241.

Guiot, J., A. Pons, J. L. de Beaulieu and M. Reille (1989) A 140 000-year continental climate reconstruction from two European pollen records. *Nature* **338**, 309–313.

Haesaerts, P. and C. Dupuis (1986) Contribution à la stratigraphie des nappes alluviàles de la Somme et de l'Avre dans la région d'Amiens. *Supplement au Bulletin de l'Association Français pour l'Etude du Quaternaire* **26**, 171–186.

Haggart, B. A. (1986) Relative sea-level change in the Beauly Firth, Scotland. *Boreas* **15**, 191–207.

Haggart, B. A. (1989) Variations in the pattern and rate of isostatic uplift indicated by a comparison of Holocene sea-level curves from Scotland. *Journal of Quaternary Science* **4**, 67–76.

Hall, A. M. (1982) The 'Pliocene' gravels of Buchan: a reappraisal: discussion. *Scottish Journal of Geology* **18**, 336–338.

Hall, A. M. (1983) *Weathering and landform evolution in north-east Scotland*. Ph.D. Thesis, University of St. Andrews.

Hall, A. M. (1984a) Introduction. In *Buchan field guide*, A. M. Hall (ed.), 1–26. Cambridge: Quaternary Research Association.

Hall, A. M. (1984b) Central Buchan. In *Buchan field guide*, A. M. Hall (ed.), 28–45. Cambridge: Quaternary Research Association.

Hall, A. M. and E. R. Connell (1982) Recent excavations at the Greensand locality of Moreseat, Grampian Region. *Scottish Journal of Geology* **18**, 291–296.

Hall, A. R. (1980) Late Pleistocene deposits at Wing, Rutland. *Philosophical Transactions of the Royal Society of London* **B289**, 135–164.

Hall, A. R. (1981) Wing interglacial site. In *Field guide to the East Midlands region*, T. D. Douglas (ed.), 23–27. Leicester: Quaternary Research Association.

Hall, J. and D. K. Smyth (1973) Discussion of the relation of Palaeogene ridge and basin structures of Britain to the North Atlantic. *Earth and Planetary Science Letters* **19**, 54–60.

Hallam, J. S., B. J. N. Edwards, B. Barnes and A. J. Stuart (1973) The remains of a Late glacial elk associated with barbed points from High Furlong, near Blackpool, Lancashire. *Proceedings of the Prehistoric Society* **39**, 100–128.

Hamilton, J. R. C. (1956) *Excavations at Jarlshof: Shetland*. Ministry of Works Archaeological Reports Number 1. Edinburgh: Her Majesty's Stationery Office.

Handa, S. and P. D. Moore (1976) Studies in the vegetational history of mid-Wales. IV. Pollen analysis of some pingo basins. *New Phytologist* **77**, 205–225.

Haq, B. U., W. A. Berggren and J. A. van Couvering (1977) Corrected age of the Pliocene/Pleistocene boundary. *Nature* **269**, 483–488.

Haq, B. U., W. A. Berggren and J. A. van Couvering (1978) Corrected age of the Pliocene/Pleistocene boundary. *Nature* **272**, 287–288.

Harding, D. W. (ed.) (1976) *Hillforts: later prehistoric earthworks in Britain and Ireland*. London: Academic Press.

Harland, W. B., R. L. Armstrong, A. V. Cox, L. E. Craig, A. G. Smith and D. G. Smith (1990) *A geologic time scale 1989*. Revised edition of *A geologic time scale* (1982). Cambridge: Cambridge University Press.

Harmer, F. W. (1898) The Pleistocene deposits of the east of England. The Lenham Beds and the Coralline Crag. *Quarterly Journal of the Geological Society of London* **54**, 308–356.

Harmer, F. W. (1902) A sketch of the later Tertiary history of East Anglia. *Proceedings of the Geologists' Association* **17**, 416–479.

Harrison, C. J. O. (1977) Non-passerine birds of the Ipswichian Interglacial from the Gower Caves. *Transactions of the British Cave Research Association* **4**, 441–442.

Harrison, C. J. O. (1979a) Birds of the Cromer Forest Bed Series of the East Anglian Pleistocene. *Transactions of the Norfolk and Norwich Naturalists' Society* **24**, 277–286.

Harrison, C. J. O. (1979b) Pleistocene birds from Swanscombe, Kent. *London Naturalist* **58**, 6–8.

Hart, J. K. (1987) *The genesis of the north east Norfolk drift*. Ph.D. Thesis, University of East Anglia.

Hart, J. (1988) East Runton – West Runton cliffs: the glacial sequence. In *The Pliocene – Middle Pleistocene of East Anglia. Field guide*, P. L. Gibbard and J. A. Zalasiewicz (eds.), 158–170. Cambridge: Quaternary Research Association.

Hart, J. K. and G. S. Boulton (1991) The glacial drifts of north-eastern Norfolk. In *Glacial deposits in Great Britain and Ireland*, J. Ehlers, P. L. Gibbard and J. Rose (eds.), 233–243. Rotterdam: Balkema.

Hartley, B. R. (1972) The Roman occupation of Scotland: the evidence of Samian ware. *Britannia* **3**, 1–55.

Hawkes, C. F. C. (1931) Hillforts. *Antiquity* **5**, 60–97.

Hawkes, C. F. C. (1960) *A scheme for the British Bronze Age*. (Text of an address to a CBA conference.) London: Council for British Archaeology.

Hawkins, A. B. (1977) South west Avon: introduction. In *Field handbook. Bristol*, K. Crabtree (ed.), 2–4. Bristol: Quaternary Research Association.

Hawkins, A. B. and G. A. Kellaway (1971) Field meeting at Bristol and Bath with special reference to new evidence of glaciation. *Proceedings of the Geologists' Association* **82**, 267–292.

Hawkins, A. B. and G. A. Kellaway (1973) 'Burtle Clay' of Somerset. *Nature* **243**, 216–217.

Hawksworth, D. L. (1969) Studies on the peat deposits of the island of Foula, Shetland. *Transactions of the Botanical*

Society of Edinburgh **40**, 576–591.

Haworth, E. Y. (1969) The diatoms of a sediment core from Blea Tarn, Langdale. *Journal of Ecology* **57**, 429–439.

Haworth, E. Y. (1976) Two Late-glacial (Late Devensian) diatom assemblage profiles from northern Scotland. *New Phytologist* **77**, 227–256.

Hayes, F. L. (1978) *Palynological studies in the south-eastern United States, Bermuda and south-east Ireland*. M.Sc. Thesis, Trinity College, University of Dublin.

Haynes, J. R., R. J. Kiteley, R. C. Whatley and P. J. Wilks (1977) Microfaunas, microfloras and the environmental stratigraphy of the Late Glacial and Holocene in Cardigan Bay. *Geological Journal* **12**, 129–158.

Hazelden, J. and M. G. Jarvis (1979) Age and significance of alluvium in the Windrush Valley, Oxfordshire. *Nature* **282**, 291–292.

Headon Davies, K. (1983) Amino acid analysis of Pleistocene marine molluscs from the Gower Peninsula. *Nature* **302**, 137–139.

Headon Davies, K. and D. H. Keen (1985) The age of Pleistocene marine deposits at Portland, Dorset. *Proceedings of the Geologists' Association* **96**, 217–225.

Hecht, A. (1976) The oxygen isotope record of foraminifera in deep sea sediments. *Foraminifera* **2**, 1–43.

Hedges, J. W. (1975) Excavation of two Orcadian burnt mounds at Liddle and Beaquoy. *Proceedings of the Society of Antiquaries of Scotland* **106**, 39–98.

Henry, A. (1984). *The lithostratigraphy, biostratigraphy and chronostratigraphy of coastal Pleistocene deposits in Gower, South Wales*. Ph.D. Thesis, University of Wales.

Henshall, A. (1985) The chambered cairns. In *The prehistory of Orkney:* 4000 BC–1000 AD, C. Renfrew (ed.), 83–117. Edinburgh: Edinburgh University Press.

Herries Davies, G. L. and N. Stephens (1978) *The geomorphology of the British Isles: Ireland*. London: Methuen.

Heusser, L. and W. L. Balsam (1977) Pollen distribution in the northeast Pacific Ocean. *Quaternary Research* **7**, 45–62.

Hey, R. W. (1958) High-level gravels in and near the lower Severn Valley. *Geological Magazine* **95**, 161–168.

Hey, R. W. (1967) The Westleton Beds reconsidered. *Proceedings of the Geologists' Association* **78**, 427–445.

Hey, R. W. (1976) Provenance of far-travelled pebbles in the pre-Anglian Pleistocene of East Anglia. *Proceedings of the Geologists Association* **87**, 69–82.

Hey, R. W. (1980) Equivalents of the Westland Green Gravels in Essex and East Anglia. *Proceedings of the Geologists' Association* **91**, 279–290.

Hey, R. W. (1986) A re-examination of the Northern Drift of Oxfordshire. *Proceedings of the Geologists' Association* **97**, 291–301.

Hey, R. W. and C. A. Auton (1988) Lithostratigraphic methods in the Quaternary of East Anglia. In *The Pliocene – Middle Pleistocene of East Anglia. Field guide*, P. L. Gibbard and J. A. Zalasiewicz (eds.), 32–41. Cambridge: Quaternary Research Association.

Hey, R. W. and P. J. Brenchley (1977) Volcanic pebbles from Pleistocene gravels in Norfolk and Essex. *Geological Magazine* **114**, 219–225.

Heyworth, A. and C. Kidson (1982) Sea-level changes in southwest England and Wales. *Proceedings of the Geologists' Association* **93**, 91–111.

Heyworth, A., C. Kidson and P. Wilks (1985) Late-glacial and Holocene sediments at Clarach Bay, near Aberystwyth. *Journal of Ecology* **73**, 459–480.

Hibbert, F. A. and V. R. Switsur (1976) Radiocarbon dating of Flandrian pollen zones in Wales and northern England. *New Phytologist* **77**, 793–807.

Hibbert, F. A., V. R. Switsur and R. G. West (1971) Radiocarbon dating of Flandrian pollen zones at Red Moss, Lancashire. *Proceedings of the Royal Society of London* **B177**, 161–176.

Hicks, S. P. (1971) Pollen-analytical evidence for the effect of prehistoric agriculture on the vegetation of north Derbyshire. *New Phytologist* **70**, 647–667.

Hinsch, W. (1985) Die Molluskenfauna des Eem-Interglacials von Oldenbüttel-Schnittlohe (Nord-Ostsee-Kanal, Wes-tholstein). *Geologisches Jahrbuch* **A86**, 49–62.

Hinton, M. A. C. (1919) Description of the rodent teeth. In: On a deposit of interglacial loess, and some transported preglacial freshwater clays on the Durham coast. C. T. Trechmann, *Quaterly Journal of the Geological Society of London* **75**, 201.

Hinton, M. A. C. and A. S. Kennard (1901) Contributions to the Pleistocene geology of the Thames Valley. 1. The Grays Thurrock area, part I. *Essex Naturalist* **11**, 336–370.

Hinton, M. A. C. and A. S. Kennard (1907) Contributions to the Pleistocene geology of the Thames Valley. 1. The Grays Thurrock area, part II. *Essex Naturalist* **15**, 56–88.

Hirons, K. R. and K. J. Edwards (1986) Events at and around the first and second *Ulmus* declines: palaeoecological investigations in Co. Tyrone, Northern Ireland. *New Phytologist* **104**, 131–153.

Hirons, K. R., K. J. Edwards and R. Thompson (1985) Palaeomagnetism. In *The Quaternary history of Ireland*, K. J. Edwards and W. P. Warren (eds.), 302–308. London: Academic Press.

Hodgson, G. E. and B. M. Funnell (1987) Foraminiferal biofacies of the early Pliocene Coralline Crag. In *Micropalaeontology of carbonate environments*, M. B. Hart (ed.), 44–73. Chichester: Ellis Horwood.

Hodson, F. and I. M. West (1972) Holocene deposits of Fawley, Hampshire, and the development of Southampton Water. *Proceedings of the Geologists' Association* **83**, 421–441.

Hollin, J. T. (1977) Thames interglacial sites, Ipswichian sea levels and Antarctic ice surges. *Boreas* **6**, 33–52.

Hollingworth, S. E. and J. H. Taylor (1946) An outline of the geology of the Kettering district. *Proceedings of the Geologists' Association* **57**, 204–233.

Hollingworth, S. E., J. Allison and H. Godwin (1949) Interglacial deposits from the Histon Road, Cambridge.

Quarterly Journal of the Geological Society of London **105**, 495–509.

Holmes, R. (1977) *Quaternary deposits of the central North Sea. 5. The Quaternary geology of the UK sector of the North Sea between 56°N and 58°N*. Report of the Institute of Geological Sciences, Number 77/14. London: Her Majesty's Stationery Office.

Holmes, S. C. A. (1971) The geological mapper and the employment of his results, as illustrated in some areas of southern England. *Proceedings of the Geologists' Association* **82**, 161–186.

Holyoak, D. T. (1982) Non-marine Mollusca of the Last Glacial Period (Devensian) in Britain. *Malacologia* **22**, 727–730.

Holyoak, D. T. (1983a) A Late Pleistocene interglacial flora and molluscan fauna from Thatcham, Berkshire, with notes on Mollusca from the interglacial deposits at Aveley, Essex. *Geological Magazine* **120**, 623–629.

Holyoak, D. T. (1983b) The identity and origins of *Picea abies* (L.) Karsten from the Chelford Interstadial (Late Pleistocene) of England. *New Phytologist* **95**, 153–157.

Holyoak, D. T. (1983c) The colonization of Berkshire, England, by land and freshwater Mollusca since the Late Devensian. *Journal of Biogeography* **10**, 483–498.

Holyoak, D. T. and R. C. Preece (1983) Evidence of a high Middle Pleistocene sea-level from estuarine deposits at Bembridge, Isle of Wight, England. *Proceedings of the Geologists' Association* **94**, 231–244.

Holyoak, D. T. and R. C. Preece (1985) Late Pleistocene interglacial deposits at Tattershall, Lincolnshire. *Philosophical Transactions of the Royal Society of London* **B331**, 193–236.

Holyoak, D. T., M. Ivanovich and R. C. Preece (1983) Additional fossil and isotopic evidence for the age of the interglacial tufas at Hitchin and Icklingham. *Journal of Conchology* **31**, 260–261.

Hooke, J. M., A. M. Harvey, S. Y. Miller and C. E. Redmond (1990) The chronology and stratigraphy of the alluvial terraces of the River Dane Valley, Cheshire, N. W. England. *Earth Surface Processes and Landforms* **15**, 717–737.

Horton, A. (1970) *The drift sequence and subglacial topography in parts of the Ouse and Nene basin*. Report of the Institute of Geological Sciences, Number 70/9. London: Her Majesty's Stationery Office.

Horton, A. (1974) *The sequence of Pleistocene deposits during the construction of the Birmingham motorways*. Report of the Institute of Geological Sciences, Number 74/11. London: Her Majesty's Stationery Office.

Horton, A. (1977) Nettlebed. In *Guidebook for Excursion A5. South east England and the Thames Valley*, E. R. Shephard-Thorn and J. J. Wymer, 16–18. INQUA X Congress, United Kingdom. Norwich: Geo Abstracts for International Union for Quaternary Research.

Horton, A. (1981) Woodston Beds and river terraces. In *Field guide to the East Midlands region*, T. D. Douglas (ed.), 29–32. Leicester: Quaternary Research Association.

Horton, A. (1989a) Quinton. In *The Pleistocene of the West Midlands. Field guide*, D. H. Keen (ed.), 69–76. Cambridge: Quaternary Research Association.

Horton, A. (1989b) *Geology of the Peterborough district*. Memoirs of the British Geological Survey. London: Her Majesty's Stationery Office.

Horton, A., E. R. Shephard-Thorn and R. G. Thurrell (1974) *The geology of the new town of Milton Keynes*. Report of the Institute of Geological Sciences, Number 74/16. London: Her Majesty's Stationery Office.

Horton, A., B. C. Worssam and J. B. Whittow (1981) The Wallingford Fan Gravel. *Philosophical Transactions of the Royal Society of London* **B293**, 215–255.

Horton, A., D. H. Keen, M. H. Field, J. E. Robinson, G. R. Coope, A. P. Currant, D. K. Graham, C. P. Green and L. M. Phillips (1992) The Hoxnian Interglacial deposits at Woodston, Peterborough. *Philosophical Transactions of the Royal Society of London* **B335**, (in press).

Hoskins, W. G. (1970) *The making of the English landscape*. London: Penguin.

Huddart, D. (1972) Late Devensian glacial history. In *The Cumberland Lowland. Handbook*. Two volumes, D. Huddart and M. J. Tooley (eds.), 3–47. Durham: Quaternary Research Association.

Huddart, D. (1977) Gutterby Spa – Annaside Banks moraine, St. Bees moraine. In *Guidebook for Excursion A4. The Isle of Man, Lancashire coast and Lake District*, M. J. Tooley, 38–40. INQUA Congress, United Kingdom. Norwich: Geo Abstracts for International Union for Quaternary Research.

Huddart, D., M. J. Tooley and P. A. Carter (1977) The coasts of northwest England. In *The Quaternary history of the Irish Sea*, C. Kidson and M. J. Tooley (eds.), 119–154. Geological Journal Special Issue Number 7. Liverpool: Seel House Press.

Hull, E. (1864) *The geology of the country around Oldham and Manchester*. Memoirs of the Geological Survey of Great Britain. London: Her Majesty's Stationery Office.

Hull, E. and A. H. Green (1866) *The geology of the country around Stockport, Macclesfield, Congleton and Leek*. Memoirs of the Geological Survey of Great Britain. London: Her Majesty's Stationery Office.

Hunt, C. O. (1985) Pollen from the Eynsham Gravel at Magdalen College, Oxford. In *The chronology and environmental framework of early man in the Upper Thames Valley: a new model*, D. J. Briggs, G. R. Coope and D. D. Gilbertson, 85–87. BAR British Series 137. Oxford: British Archaeological Reports.

Hunt, C. O. (1989) The palynology and correlation of the Walton Crag (Red Crag Formation, Pliocene). *Journal of the Geological Society of London* **146**, 743–745.

Hunt, C. O., A. R. Hall and D. D. Gilbertson (1984) The palaeobotany of the Late Devensian sequence at Skipsea Withow Mere. In *Late Quaterary environments and man in Holderness*, D. D. Gilbertson (ed.), 81–108. BAR British Series 134. Oxford: British Archaeological Reports.

Hunt, T. G. and H. J. B. Birks (1982) Devensian late-glacial vegetational history at Sea Mere, Norfolk. *Journal of Biogeography* **9**, 517–538.

Huntley, B. and H. J. B. Birks (1983) *An atlas of past and present pollen maps for Europe: 0–13 000 years ago*. Cambridge: Cambridge University Press.

Huxtable, J. (1984) Thermoluminescence (TL) studies on burnt flint and stones. In *Pontnewydd Cave: a Lower Palaeolithic hominid site in Wales. The first report*, H. S. Green, 106–107. National Museum of Wales Quaternary Studies Monographs, Number 1. Cardiff: National Museum of Wales.

Huxtable, J. (1986) The thermoluminescence dates. In *La Cotte de St. Brelade 1961–1978: excavations by C. B. M. McBurney*, P. Callow and J. M. Cornford (eds.), 145–149. Norwich: Geo Books.

Huxtable, J. and M. J. Aitken (1985) Thermoluminescence dating results for the Palaeolithic site Maastricht-Belvédère. In: Maastricht-Belvédère: stratigraphy, palaeoenvironment and archaeology of the Middle and Late Pleistocene deposits. T. van Kolfschoten and W. Roebroeks (eds.), *Mededelingen Rijks Geologische Dienst* **39**, 41–44.

Hyvarinen, H. (1978) Use and definition of the term Flandrian. *Boreas* **7**, 182.

Ince, J. (1983) Two Postglacial pollen profiles from the uplands of Snowdonia, Gwynedd, North Wales. *New Phytologist* **95**, 159–172.

Innes, J. B. and I. G. Simmons (1988) Disturbance and diversity: floristic changes associated with pre-elm decline woodland recession in north east Yorkshire. In *Archaeology and the flora of the British Isles*, M. Jones (ed.), 7–20. Oxford University Committee for Archaeology Monograph Number 14; Botanical Society of the British Isles Conference Report Number 19. Oxford: Oxford University Committee for Archaeology.

Innes, J. L. (1985) Lichenometry. *Progress in Physical Geography* **9**, 187–254.

Ivanovich, M. and R. S. Harmon (eds.) (1982) *Uranium series disequilibrium: applications to environmental problems*. Oxford: Clarendon Press.

Ivanovich, M., A. M. B. Rae and M. A. Wilkins (1984) Brief report on dating the *in situ* stalagmitic floor found in the East Passage in 1982. In *Pontnewydd Cave: a Lower Palaeolithic hominid site in Wales. The first report*, H. S. Green, 98–99. National Museum of Wales Quaternary Studies Monographs. Number 1. Cardiff: National Museum of Wales.

Iversen, J. (1941) Landnam i Danmarks stenalder. *Danmarks Geologiske Undersøgelse* **II**, **66**, 1–68.

Iversen, J. (1954) The late-glacial flora of Denmark and its relation to climate and soil. *Danmarks Geologiske Undersøgelse* **II**, **80**, 87–119.

Iversen, J. (1969) Retrogressive development of a forest ecosystem demonstrated by pollen diagrams from a fossil mor. *Oikos Supplement* **12**, 35–49.

Jacobi, R. M. (1973) Aspects of the 'Mesolithic Age' in Great Britain. In *The Mesolithic in Europe*, S. K. Kozlowski (ed.), 237–265. Warsaw: University Press.

Jacobi, R. M. (1976) Britain inside and outside Mesolithic Europe. *Proceedings of the Prehistoric Society* **42**, 67–84.

Jacobi, R. M., E. B. Jacobi and R. Burleigh (1986) Kent's Cavern, Devon: dating of the 'Black Band' and human resettlement of the British Isles following the last glacial maximum. *Quaternary Newsletter* **48**, 10–12.

Jacobi, R. M., J. H. Tallis and P. A. Mellars (1976) The southern Pennine Mesolithic and the ecological record. *Journal of Archaeological Science* **3**, 307–320.

Jacobi, R. M., J. A. J. Gowlett, R. E. M. Hedges and R. Gillespie (1986) Accelerator mass spectrometry dating of Upper Palaeolithic finds, with the Poulton elk as an example. In *Studies in the Upper Palaeolithic of Britain and northwest Europe*, D. A. Roe (ed.), 121–128. BAR International Series 296. Oxford: British Archaeological Reports.

Jacobson, G. L. and R. H. W. Bradshaw (1981) The selection of sites for paleovegetational studies. *Quaternary Research* **16**, 80–96.

Jäger, K.-D. and W.-D. Heinrich (1982) The travertine at Weimar – Ehringsdorf – an interglacial site of Saalian age. *International Geological Correlation Programme 73/1/24 Report* **7**, 98–114.

Jalas, J. and J. Suominen (1973) *Atlas florae Europaeae. 2. Gymnospermae*. Helsinki: Societas Biologica Fenica Vanamo.

Jamieson, T. F. (1906) The glacial period in Aberdeenshire and the southern border of the Moray Firth. *Quarterly Journal of the Geological Society of London* **62**, 13–39.

Janossy, D. (1975) Mid-Pleistocene microfauna of continental Europe and adjoining areas. In *After the australopithecines: stratigraphy, ecology and culture change in the Middle Pleistocene*, K. W. Butzer and G. L. Isaac (eds.), 375–397. The Hague: Mouton.

Janossy, D. (1986) *Pleistocene vertebrate faunas of Hungary*. Amsterdam: Elsevier.

Jansen, E., S. Befring, T. Bugge, T. Eidvin, H. Holtedahl and H. P. Sejrup (1987) Large submarine slides on the Norwegian continental margin: sediments, transport and timing. *Marine Geology* **78**, 77–107.

Jansen, J. H. F. and A. M. Hensey (1981) Interglacial and Holocene sedimentation in the northern North Sea: an example of Eemian deposits in the Tartan Field. In *Holocene marine sedimentation in the North Sea Basin*, S.-D. Nio, R. T. E. Shüttenhelm and T. C. E. van Weering (eds.), 323–334. International Association of Sedimentologists Special Publication Number 5. Oxford: Blackwell Scientific.

Jansen, J. H. F., T. C. E. van Weering and D. Eisma (1979) Late Quaternary sedimentation in the North Sea. In *The Quaternary history of the North Sea*, E. Oele, R. T. E. Schüttenhelm and A. J. Wiggers (eds.), 175–187. Universitatis Uppsaliensis Symposium Universitatis Uppsaliensis Annum Quingentesimum Celebrantis, Number 2. Uppsala: University of Uppsala.

Jardine, W. G. (1975) Chronology of Holocene marine

transgression and regression in south-western Scotland. *Boreas* **4**, 173–196.

Jardine, W. G. (1977) The Quaternary marine record in southwest Scotland and the Scottish Hebrides. In *The Quaternary history of the Irish Sea*, C. Kidson and M. J. Tooley (eds.), 99–118. Liverpool: Seel House Press.

Jardine, W. G. (1982) Sea-level changes in Scotland during the last 18 000 years. *Proceedings of the Geologists' Association* **93**, 25–41.

Jardine, W. G. and A. Morrison (1976) The archaeological significance of Holocene coastal deposits in south-western Scotland. In *Geoarchaeology: earth science and the past*, D. A. Davidson and M. L. Shackley (eds.), 175–195. London: Duckworth.

Jeffery, D. H., C. Laban, C. S. Mesdag and R. T. E. Schüttenhelm (1988) *Silver Well Sheet. 54°N–02°E. Quaternary geology (geologie van het Kwartair)*. British Geological Survey/Rijks Geologische Dienst 1:250 000 Map Series. Southampton: Ordnance Survey for British Geological Survey.

Jelgersma, S. (1979) Sea-level changes in the North Sea basin. In *The Quaternary history of the North Sea*. E. Oele, R. T. E. Schüttenhelm and A. J. Wiggers (eds.), 233–248. Acta Universitatis Uppsaliensis Symposium Universitatis Uppsaliensis Annum Quingentesimum Celebrantis. Number 2. Uppsala: University of Uppsala.

Jelgersma, S., E. Oele and A. J. Wiggers (1979) Depositional history and coastal development in the Netherlands and the adjacent North Sea since the Eemian. In *The Quaternary history of the North Sea*, E. Oele, R. T. E. Schüttenhelm and A. J. Wiggers (eds.), 115–142. Acta Universitatis Uppsaliensis Symposium Universitatis Uppsaliensis Annum Quingentesimum Celebrantis, Number 2. Uppsala: University of Uppsala.

Jenkins, D. G. (1978) Corrected age of the Pliocene/Pleistocene boundary. *Nature* **272**, 287.

Jenkins, D. G. (1987) Was the Pliocene – Pleistocene boundary placed at the wrong stratigraphic level? *Quaternary Science Reviews* **6**, 41–42.

Jenkins, D. G. and S. D. Houghton (1987) Age, correlation and paleoecology of the St. Erth Beds and the Coralline Crag of England. *Mededelingen Werkgroep voor Tertiaire en Kwartaire Geologie* **24**, 147–156.

Jenkinson, R. D. S. (1984) *Creswell Crags: Late Pleistocene sites in the East Midlands*. BAR British Series 122. Oxford: British Archaeological Reports.

Jenkinson, R. D. S., C. O. Hunt and I. Brooks (1985) Pin Hole Cave. In *Peak District and northern Dukeries. Field guide*, D. J. Briggs, D. D. Gilbertson and R. D. S. Jenkinson (eds.), 135–138. Cambridge: Quaternary Research Association.

Jennings, J. N. (1952) *The origin of the Broads*. Royal Geographical Society Research Series, Number 2. London: Royal Geographical Society.

Jennings, S. and C. Smyth (1987) Coastal sedimentation in East Sussex during the Holocene. *Progress in Oceanography* **18**, 205–241.

Jessen, K. (1949) Studies in late Quaternary deposits and flora-history of Ireland. *Proceedings of the Royal Irish Academy* **52B**, 85–290.

Jessen, K. and A. Farrington (1938) The bogs at Ballybetagh, near Dublin, with remarks on late-glacial conditions in Ireland. *Proceedings of the Royal Irish Academy* **44B**, 205–260.

Jessen, K. and V. Milthers (1928) Stratigraphical and palaeontological studies of interglacial freshwater deposits in Jutland and northwest Germany. *Danmarks Geologiske Undersøgelse* **II**, **48**, 1–379.

Jessen, K., S. T. Andersen and A. Farrington (1959) The interglacial deposit near Gort, County Galway, Ireland. *Proceedings of the Royal Irish Academy* **60B**, 1–77.

Jewell, P. A. (1963) Cattle from British archaeological sites. In *Man and cattle*, A. E. Mourant and F. E. Zeuner (eds.), 80–100. Royal Anthropological Institute Occasional Paper Number 18. London: Royal Anthropological Institute.

Joachim, M., G. F. Mitchell and G. S. P. Thomas (1985) Glen Ballyre. In *Field guide to the Quaternary of the Isle of Man*, R. V. Dackombe and G. S. P. Thomas (eds.), 107–110. Cambridge: Quaternary Research Association.

Jobey, G. (1971) Early settlements in eastern Dumfriesshire. *Transactions of the Dumfries and Galloway Natural History and Antiquarian Society* **48**, 78–105.

Johansen, J. (1975) Pollen diagrams from the Shetland and Faroe islands. *New Phytologist* **75**, 369–387.

Johansen, J. (1985) Studies in the vegetation history of the Faroe and Shetland islands. *Annales Societatis Scientiarum Faroensis Supplementum* **11**, 1–117.

John, B. S. (1965) A possible Main Würm glaciation in west Pembrokeshire. *Nature* **207**, 622–623.

John, B. S. (1967) Further evidence for a Middle Würm interstadial and a Main Würm glaciation of south-west Wales. *Geological Magazine* **104**, 630–633.

John, B. S. (1968) Age of raised beach deposits of south-western Britain. *Nature* **218**, 665–667.

John, B. S. (1969) West Angle Bay. In *Coastal Pleistocene deposits in Wales*, D. Q. Bowen (ed.), 18–22. Aberystwyth: Department of Geography, University College of Wales, Aberystwyth for Quaternary Research Association.

John, B. S. (1970) Pembrokeshire. In *The glaciations of Wales and adjoining regions*, C. A. Lewis (ed.), 229–265. London: Longman.

John, D. T. and P. F. Fisher (1984) The stratigraphical and geomorphological significance of the Red Crag fossils at Netley Heath, Surrey: a review and re-appraisal. *Proceedings of the Geologists' Association* **95**, 235–247.

Johnsen, S. J., W. Dansgaard, H. B. Clausen and C. C. Langway (1972) Oxygen isotope profiles through the Antarctic and Greenland ice sheets. *Nature* **235**, 429–434.

Johnson, R. G. (1982) Brunhes – Matuyama reversal dated at 790 000 yr BP by marine – astronomical correlations. *Quaternary Research* **17**, 135–147.

Jones, D. K. C. (1981) *The geomorphology of the British Isles: southeast and southern England*. London: Methuen.

Jones, P. F. and M. F. Stanley (1974) Ipswichian mammalian fauna from the Beeston Terrace at Boulton Moor, near Derby. *Geological Magazine* **111**, 515-520.

Jones, R. J. A. (1983) *Soils in Staffordshire III: Sheet SK 02/12 (Needwood Forest)*. Soil Survey Record Number 80. Harpenden: Soil Survey of England and Wales.

Jones, R. L. (1976a) Late Quaternary vegetational history of the North York Moors. IV. Seamer Carrs. *Journal of Biogeography* **3**, 397–406.

Jones, R. L. (1976b) The activities of Mesolithic man: further palaeobotanical evidence from north-east Yorkshire. In *Geoarchaeology: earth science and the past*, D. A. Davidson and M. L. Shackley (eds.), 355–367. London: Duckworth.

Jones, R. L. (1977a) Late Devensian deposits from Kildale, north-east Yorkshire. *Proceedings of the Yorkshire Geological Society* **41**, 185–188.

Jones, R. L. (1977b) Late Quaternary vegetational history of the North York Moors. V. The Cleveland dales. *Journal of Biogeography* **4**, 353–362.

Jones, R. L. (1978) Late Quaternary vegetational history of the North York Moors. VI. The Cleveland moors. *Journal of Biogeography* **5**, 81–92.

Jones, R. L. (1986) The flora and vegetation of La Cotte de St. Brelade and its environs. In *La Cotte de St. Brelade 1961–1978: excavations by C. B. M. McBurney*, P. Callow and J. M. Cornford (eds.), 99–106. Norwich: Geo Books.

Jones, R. L. and G. D. Gaunt (1976) A dated late Devensian organic deposit at Cawood near Selby. *Naturalist* **101**, 121–123.

Jones, R. L., P. R. Cundill and I. G. Simmons (1979) Archaeology and palaeobotany on the North York Moors and their environs. *Yorkshire Archaeological Journal* **51**, 15–22.

Jones, R. L., D. H. Keen, J. F. Birnie and D. T. Holyoak (1987) Holocene sea-level changes on Jersey. *Progress in Oceanography* **18**, 177–204.

Jones, R. L., D. H. Keen, J. F. Birnie and P. V. Waton (1990) *Past landscapes of Jersey: environmental changes during the last ten thousand years*. St. Helier: Société Jersiaise.

Jones, V. J., A. C. Stevenson and R. W. Battarbee (1986) Lake acidification and the land-use hypothesis: a mid-post-glacial analogue. *Nature* **322**, 157–158.

Jones, V. J., A. C. Stevenson and R. W. Battarbee (1989) Acidification of lakes in Galloway: a diatom and pollen study of the Post-glacial history of the Round Loch of Glenhead. *Journal of Ecology* **77**, 1–23.

Jope, M. (1965) Faunal remains: frequencies of ages and species. In *Windmill Hill and Avebury: excavations by Alexander Keiller 1925–1939*, I. F. Smith, 142–145. Oxford: Clarendon Press.

Kahlke, H. D. (1975) The macrofaunas of continental Europe during the Middle Pleistocene: stratigraphic sequence and problems of intercorrelation. In *After the australopithecines: stratigraphy, ecology and culture change in the Middle Pleistocene*, K. W. Butzer and G. L. Isaac (eds.), 309–374. The Hague: Mouton.

Keatinge, T. H. and J. H. Dickson (1979) Mid-Flandrian changes in vegetation on Mainland Orkney. *New Phytologist* **82**, 585–612.

Keef, P. A. M., J. J. Wymer and G. W. Dimbleby (1965) A Mesolithic site on Iping Common, Sussex, England. *Proceedings of the Prehistoric Society* **31**, 85–92.

Keeley, H. C. M. (1982) Pedogenesis during the later prehistoric period in Britain. In *Climate change in later prehistory*, A. F. Harding (ed.), 114–126. Edinburgh: Edinburgh University Press.

Keen, D. H. (1978) *The Pleistocene deposits of the Channel Islands*. Report of the Institute of Geological Sciences, Number 78/26. London: Her Majesty's Stationery Office.

Keen, D. H. (1982) Depositional sequence, age and palaeoenvironment of raised beaches and head in the Channel Islands and central Channel. *Bulletin de l'Association Français pour l'Etude du Quaternaire* **19**, 3–11.

Keen, D. H. (1985) Late Pleistocene deposits and Mollusca from Portland, Dorset. *Geological Magazine* **122**, 181–186.

Keen, D. H. (1986) The Quaternary deposits of the Channel Islands. In *La Cotte de St. Brelade 1961–1978: excavations by C. B. M. McBurney*, P. Callow and J. M. Cornford (eds.), 43–52. Norwich: Geo Books.

Keen, D. H. (1987) Non-marine molluscan faunas of periglacial deposits in Britain. In *Periglacial processes and landforms in Britain and Ireland*, J. Boardman (ed.), 257–263. Cambridge: Cambridge University Press.

Keen, D. H. (1990) Significance of the record provided by Pleistocene fluvial deposits and their included molluscan faunas for palaeoenvironmental reconstruction and stratigraphy: case studies from the English Midlands. *Palaeogeography, Palaeoclimatology, Palaeoecology* **80**, 25–34.

Keen, D. H. and D. R. Bridgland (1986) An interglacial fauna from Avon No. 3 Terrace at Eckington, Worcestershire. *Proceedings of the Geologists' Association* **97**, 303–307.

Keen, D. H., R. S. Harmon and J. T. Andrews (1981) U series and amino acid dates from Jersey. *Nature* **289**, 162–164.

Keen, D. H., R. L. Jones, and J. E. Robinson (1984) A Late Devensian and early Flandrian fauna and flora from Kildale, north-east Yorkshire. *Proceedings of the Yorkshire Geological Society* **44**, 385–397.

Keen, D. H., R. L. Jones, R. A. Evans and J. E. Robinson (1988) Faunal and floral assemblages from Bingley Bog, West Yorkshire, and their significance for Late Devensian and early Flandrian environmental changes. *Proceedings of the Yorkshire Geological Society* **47**, 125–138.

Keen, D. H., J. E. Robinson, R. G. West, F. Lowry, D. R. Bridgland and N. D. W. Davey (1990) The fauna and flora of the March Gravels at Northam Pit, Eye, Cambridgeshire, England. *Geological Magazine* **127**, 453–465.

Kellaway, G. A. (1971) Glaciation and the stones of Stonehenge. *Nature* **233**, 30–35.

Kellaway, G. A. (1977) Marlborough and Salisbury Plain. In *Field handbook. Bristol*, K. Crabtree (ed.), 39–48. Bristol: Quaternary Research Association.

Kellaway, G. A., J. H. Redding, E. R. Shephard-Thorn and J.-P. Destombes (1975) The Quaternary history of the English Channel. *Philosophical Transactions of the Royal Society of London* **A279**, 189–218.

Kelly, M. R. (1964) The Middle Pleistocene of North Birmingham. *Philosophical Transactions of the Royal Society of London* **B247**, 533–592.

Kelly, M. R. (1968) Floras of Middle and Upper Pleistocene age from Brandon, Warwickshire. *Philosophical Transactions of the Royal Society of London* **B254**, 401–415.

Kelly, M. R. and P. J. Osborne (1964) Two faunas and floras from the alluvium at Shustoke, Warwickshire. *Proceedings of the Linnean Society of London* **176**, 37–65.

Kemp, R. (1983) Valley Farm Palaeosol layer. In *Field guide: Diversion of the Thames*, J. Rose (ed.), 154–158. Hoddesdon: Quaternary Research Association.

Kemp, R. A. (1985) The Valley Farm Soil in southern East Anglia. In *Soils and Quaternary landscape evolution*, J. Boardman (ed.), 179–196. Chichester: Wiley.

Kemp, R. A. (1987) The interpretation and environmental significance of a buried Middle Pleistocene soil near Ipswich Airport, Suffolk, England. *Philosophical Transactions of the Royal Society of London* **B317**, 365–391.

Kendall, P. F. (1902) A system of glacier lakes in the Cleveland Hills. *Quarterly Journal of the Geological Society of London* **58**, 471–571.

Kennard, A. S. (1942) Faunas of the High Terrace at Swanscombe. *Proceedings of the Geologists' Association* **53**, 105.

Kennard, A. S. (1944) The Crayford brickearths. *Proceedings of the Geologists' Association* **55**, 121–169.

Kennard, A. S. and B. B. Woodward (1919) Description of the non-marine Mollusca. In: On a deposit of interglacial loess, and some transported preglacial freshwater clays on the Durham coast. C. T. Trechmann, *Quarterly Journal of the Geological Society of London* **75**, 200.

Kennard, A. S. and B. B. Woodward (1922) The post-Pliocene non-marine Mollusca of the east of England. *Proceedings of the Geologists' Association* **33**, 104–142.

Kennard, A. S. and B. B. Woodward (1924) The Pleistocene non-marine Mollusca. In: The river-gravels of the Oxford district. K. S. Sandford, *Quarterly Journal of the Geological Society of London* **80**, 170–175.

Kerney, M. P. (1963) Late-glacial deposits on the Chalk of south-east England. *Philosophical Transactions of the Royal Society of London* **B246**, 203–254.

Kerney, M. P. (1965) Weichselian deposits in the Isle of Thanet, east Kent. *Proceedings of the Geologists' Association* **76**, 269–274.

Kerney, M. P. (1971) Interglacial deposits in Barnfield Pit, Swanscombe, and their molluscan fauna. *Journal of the Geological Society of London* **127**, 69–93.

Kerney, M. P. (ed.) (1976) *Atlas of the non-marine Mollusca of the British Isles*. Cambridge: Institute of Terrestrial Ecology.

Kerney, M. P. (1977a) British Quaternary non-marine Mollusca: a brief review. In *British Quaternary studies: recent advances*, F. W. Shotton (ed.), 31–42. Oxford: Clarendon Press.

Kerney, M. P. (1977b) A proposed zonation scheme for Late-glacial and Postglacial deposits using land Mollusca. *Journal of Archaeological Science* **4**, 387–390.

Kerney, M. P. (1977c) Halling. In *Guidebook for Excursion A5. South east England and the Thames Valley*, E. R. Shephard-Thorn and J. J. Wymer, 50–52. INQUA X Congress, United Kingdom. Norwich: Geo Abstracts for International Union for Quaternary Research.

Kerney, M. P. and R. A. D. Cameron (1979) *A field guide to the land snails of Britain and north-west Europe*. London: Collins.

Kerney, M. P. and G. de G. Sieveking (1977) Northfleet. In *Guidebook for Excursion A5. South-east England and the Thames Valley*, E. R. Shephard-Thorn and J. J. Wymer, 44–46. INQUA X Congress, United Kingdom. Norwich: Geo Abstracts for International Union for Quaternary Research.

Kerney, M. P., E. H. Brown and T. J. Chandler (1964) The Late-glacial and Post-glacial history of the Chalk escarpment near Brook, Kent. *Philosophical Transactions of the Royal Society of London* **B248**, 135–204.

Kerney, M. P., R. C. Preece and C. Turner (1980) Molluscan and plant biostratigraphy of some Late Devensian and Flandrian deposits in Kent. *Philosophical Transactions of the Royal Society of London* **B291**, 1–43.

Kerney, M. P., P. L. Gibbard, A. R. Hall and J. E. Robinson (1982) Middle Devensian river deposits beneath the 'Upper Floodplain' terrace of the River Thames at Isleworth, West London. *Proceedings of the Geologists' Association* **93**, 385–393.

Kesel R. H. and A. M. D. Gemmell (1981) The 'Pliocene' gravels of Buchan: a reappraisal. *Scottish Journal of Geology* **17**, 185–203.

Kidson, C. (1971) The Quaternary history of the coasts of South West England with special reference to the Bristol Channel coast. In *Exeter essays in geography in honour of Arthur Davies*, K. J. Gregory and W. L. D. Ravenhill (eds.), 1–22. Exeter: University of Exeter Press.

Kidson, C. (1977) The coast of South West England. In *The Quaternary history of the Irish Sea*, C. Kidson and M. J. Tooley (eds.), 257–298. Geological Journal Special Issue Number 7. Liverpool: Seel House Press.

Kidson, C. and J. R. Haynes (1972) Glaciation in the Somerset Levels: the evidence of the Burtle Beds. *Nature* **239**, 390–392.

Kidson, C. and A. Heyworth (1973) The Flandrian sea-level rise in the Bristol Channel. *Proceedings of the Ussher Society 2*, 565–584.

Kidson, C. and A. Heyworth (1976) The Quaternary deposits

of the Somerset Levels. *Quarterly Journal of Engineering Geology* **9**, 217–235.

Kidson, C. and A. Heyworth (1978) Holocene eustatic sea level change. *Nature* **273**, 748–750.

Kidson, C. and R. Wood (1974) The Pleistocene stratigraphy of Barnstaple Bay. *Proceedings of the Geologists' Association* **85**, 223–237.

Kidson, C., J. R. Haynes and A. Heyworth (1974) The Burtle Beds of Somerset – glacial or marine? *Nature* **251**, 211–213.

Kidson, C., D. D. Gilbertson, J. R. Haynes, A. Heyworth, C. E. Hughes and R. C. Whatley (1978) Interglacial marine deposits of the Somerset Levels, South West England. *Boreas* **7**, 215–228.

King, C. A. M. (1977) The early Quaternary landscape with consideration of neotectonic matters. In *British Quaternary studies: recent advances*, F. W. Shotton (ed.), 137–152. Oxford: Clarendon Press.

King, W. B. R. and K. P. Oakley (1936) The Pleistocene succession in the lower parts of the Thames Valley. *Proceedings of the Prehistoric Society* **2**, 52–76.

Knudsen, K. L. (1985) Foraminiferal faunas in Eemian deposits of the Oldenbüttel area near Kiel Canal, Germany. *Geologisches Jahrbuch* **A86**, 27–48.

Kowalski, K. (1967) The Pleistocene extinction of mammals in Europe. In *Pleistocene extinctions: the search for a cause*, P. S. Martin and H. E. Wright (eds.), 349–364. New Haven: Yale University Press.

Krinsley, D. H. and B. M. Funnell (1965) Environmental history of quartz sand grains from the Lower and Middle Pleistocene of Norfolk, England. *Quarterly Journal of the Geological Society of London* **121**, 36–43.

Kukla, G. J. (1970) Correlations between loesses and deep-sea sediments. *Geologiska Föreningens i Stockholm Förhandlingar* **92**, 148–180.

Kukla, G. J. (1975) Loess stratigraphy of central Europe. In *After the australopithecines: stratigraphy, ecology and culture change in the Middle Pleistocene*, K. W. Butzer and G. L. Isaac (eds.), 99–188. The Hague: Mouton.

Kukla, G. J. (1977) Pleistocene land – sea correlations I. Europe. *Earth Science Reviews* **13**, 307–374.

Kukla, G. (1987) Loess stratigraphy in central China. *Quaternary Science Reviews* **6**, 191–219.

Kuntz, G. and J. P. Lautridou (1974) Contribution à l'étude du Pliocene et du passage Pliocene – Quaternaire dans les dépôts de la fôret de la Londe, pres de Rouen. Correlations possibles avec divers gisements de Haute-Normandie. *Bulletin de l'Association Français pour l'Etude du Quaternaire* **3/4**, 117–128.

Kutzbach, J. E. (1981) Monsoon climate of the early Holocene: climate experiment with the Earth's orbital parameters for 9000 years ago. *Science* **214**, 59–61.

Kutzbach, J. E. and R. G. Gallimore (1988) Sensitivity of a coupled atmosphere/mixed ocean layer model to changes in orbital forcing at 9000 years BP. *Journal of Geophysical Research* **93**, 803–821.

Kutzbach, J. E. and P. J. Guetter (1986) The influence of changing orbital parameters and surface boundary conditions on climate simulations for the past 18 000 years. *Journal of Atmospheric Sciences* **43**, 1726–1759.

Lafrenz, H. R. (1963) Foraminiferen aus dem marinen Riss-Würm Interglazial (Eem) in Schleswig-Holstein. *Meyniana* **13**, 10–45.

Lamb, H. H. (1977) The late Quaternary history of the climate of the British Isles. In *British Quaternary studies: recent advances*, F. W. Shotton (ed.), 283–298. Oxford: Clarendon Press.

Lamb, H. H. (1982) *Climate, history and the modern world*. London: Methuen.

Lamb, H. H., R. P. W. Lewis and A. Woodroffe (1966) Atmospheric circulation and the main climatic variables between 8000 and 0 BC: meteorological evidence. In *Proceedings of the International Symposium on World Climate 8000 to 0 BC*, J. S. Sawyer (ed.), 174–217. London: Royal Meteorological Society.

Lambert, C. A., R. G. Pearson and B. W. Sparks (1963) A flora and fauna from Late Pleistocene deposits at Sidgwick Avenue, Cambridge. *Proceedings of the Linnean Society of London* **174**, 13–29.

Lambert, J. M., J. N. Jennings, C. T. Smith, C. Green and J. N. Hutchinson (1960) *The making of the Broads: a reconsideration of their origin in the light of new evidence*. Royal Geographical Society Research Series, Number 3. London: Royal Geographical Society.

Lamplugh, G. W. (1891) Final report of the committee appointed for the purpose of investigating an ancient sea-beach near Bridlington Quay. *Report of the British Association for the Advancement of Science 1890*, 375–377.

Lamplugh, G. W. (1903) *The geology of the Isle of Man*. Memoirs of the Geological Survey of Great Britain. London: Her Majesty's Stationery Office.

Larsonneur, C. (1971) *Manche central et Baie de Seine: géologie du substratum et des dépôts meubles*. Thèse, Université de Caen.

Lautridou, J.-P. (1982) St. Pierre-les-Elbeuf: older loess, tufa. In *The Quaternary of Normandy: Field handbook*, J.-P. Lautridou, 41. Caen: Quaternary Research Association and Centre de Geomorphologie du Centre National de la Recherche Scientifique.

Lautridou, J.-P., H. Duroy, P. Giresse, M. N. Le Coustumer and M. Levant (1986) Sedimentology. In *La Cotte de St. Brelade 1961–1978: excavations by C. B. M. McBurney*, P. Callow and J. M. Cornford (eds.), 83–89. Norwich: Geo Books.

Lawson, T. J. (1981) First Scottish date from the last interglacial. *Scottish Journal of Geology* **17**, 301–303.

Lawson, T. J. (1984) Reindeer in the Scottish Quaternary. *Quaternary Newsletter* **42**, 1–7.

Layard, N. F. (1912) Animal remains from the Railway Cutting at Ipswich. *Proceedings of the Suffolk Institute of Archaeology and Natural History* **14**, 59–68.

Layard, N. F. (1920) The Stoke Bone-bed, Ipswich. *Proceedings of the Prehistoric Society of East Anglia* **3**, 210–219.

Lee, J. A., J. H. Tallis and S. J. Woodin (1988) Acidic deposition and British upland vegetation. In *Ecological change in the uplands*, M. B. Usher and D. B. A. Thompson (eds.), 151–162. Special Publication Number 7 of the British Ecological Society. Oxford: Blackwell Scientific.

le Gros Clark, W. E. (1938) General features of the Swanscombe skull bones. *Journal of the Royal Anthropological Institute* **68**, 58–61.

Leopold, L. B., M. G. Wolman and J. P. Miller (1964) *Fluvial processes in geomorphology*. San Francisco: Freeman.

Leroi-Gourhan, A. (1961) Analyse pollinique de niveaux acheuléens de la Cotte de Saint Brelade (Jersey). *Internationalen Kongress für Vor-und Frühgeschichte Hamburg* **5**, 501–504.

Letzer, J. (1981) The Upper Eden Valley. In *Field guide to eastern Cumbria*, J. Boardman (ed.), 43–60. Brighton: Quaternary Research Association.

Lewis, C. A. (1970a) Introduction. In *The glaciations of Wales and adjoining regions*, C. A. Lewis (ed.), 1–20. London: Longman.

Lewis, C. A. (1970b) The upper Wye and Usk regions. In *The glaciations of Wales and adjoining regions*, C. A. Lewis (ed.), 147–173. London: Longman.

Lewis, C. A. (1985) Periglacial features. In *The Quaternary history of Ireland*, K. J. Edwards and W. P. Warren (eds.), 95–113. London: Academic Press.

Limbrey, S. (1975) *Soil science and archaeology*. London: Academic Press.

Limbrey, S. (1978) Changes in the quality and distribution of the soils of lowland Britain. In *The effect of man on the landscape: the Lowland Zone*, S. Limbrey and J. G. Evans (eds.), 21–27. CBA Research Report Number 21. London: Council for British Archaeology.

Linton, D. L. (1955) The problem of tors. *Geographical Journal* **121**, 470–487.

Linton, D. L. (1964) The origin of the Pennine tors – an essay in analysis. *Zeitschrift für Geomorphologie* **8**, 5–24.

Lister, A. M. (1989a) Mammalian faunas and the Wolstonian debate. In *The Pleistocene of the West Midlands. Field guide*, D. H. Keen (ed.), 5–12. Cambridge: Quaternary Research Association.

Lister, A. M. (1989b) Rapid dwarfing of red deer on Jersey in the Last Interglacial. *Nature* **342**, 539–542.

Lister, A. M., J. M. McGlade and A. J. Stuart (1990) The early Middle Pleistocene vertebrate fauna from Little Oakley, Essex. *Philosophical Transactions of the Royal Society of London* **B328**, 359–385.

Locke, W. W., J. T. Andrews and P. J. Webber (1979) *A manual for lichenometry*. British Geomorphological Research Group Technical Bulletin Number 26. Norwich: Geo Abstracts.

Long, D. and A. C. Morton (1987) An ash fall within the Loch Lomond Stadial. *Journal of Quaternary Science* **2**, 97–101.

Long, D. and A. C. Skinner (1985) Glacial meltwater channels in the northern isles of Shetland: comment. *Scottish Journal of Geology* **21**, 222–224.

Long, D., D. E. Smith and A. G. Dawson (1989) A Holocene tsunami deposit in eastern Scotland. *Journal of Quaternary Science* **4**, 61–66.

Long, D., C. Laban, H. Streif, T. D. J. Cameron and R. T. E. Schüttenhelm (1988) The sedimentary record of climatic variation in the southern North Sea. *Philosophical Transactions of the Royal Society of London* **B318**, 523–537.

Long, D., A. Bent, R. Harland, D. M. Gregory, D. K. Graham and A. C. Morton (1986) Late Quaternary palaeontology, sedimentology and geochemistry of a vibrocore from the Witch Ground Basin, central North Sea. *Marine Geology* **73**, 109–123.

Long, P. E. (1974) Norwich Crag at Covehithe, Suffolk. *Transactions of the Suffolk Naturalists' Society* **16**, 199–208.

Long, P. E. (1988) Covehithe: Mollusca. In *The Pliocene – Middle Pleistocene of East Anglia. Field guide*, P. L. Gibbard and J. A. Zalasiewicz (eds.), 118. Cambridge: Quaternary Research Association.

Long, P. E. and P. G. Cambridge (1988) Crag Mollusca: an overview. In *The Pliocene – Middle Pleistocene of East Anglia. Field guide*, P. L. Gibbard and J. A. Zalasiewicz (eds.), 53–56. Cambridge: Quaternary Research Association.

Lord, A. R. (1980) Interpretation of the Lateglacial marine environment of N. W. Europe by means of Foraminiferida. In *Studies in the Lateglacial of north-west Europe*, J. J. Lowe, J. M. Gray and J. E. Robinson (eds.), 103–114. Oxford: Pergamon Press.

Lord, A. R. and J. E. Robinson (1978) Marine Ostracoda from the Quaternary Nar Valley Clay, west Norfolk. *Bulletin of the Geological Society of Norfolk* **30**, 113–118.

Lowe, J. J. (1978) Radiocarbon-dated Lateglacial and early Flandrian pollen profiles from the Teith Valley, Perthshire, Scotland. *Pollen et Spores* **20**, 367–397.

Lowe, J. J. (1982) Three Flandrian pollen profiles from the Teith Valley, Perthshire, Scotland. I. Vegetational history. *New Phytologist* **90**, 355–370.

Lowe, J. J. (1984) A critical evaluation of pollen-stratigraphic investigations of pre-Late Devensian sites in Scotland. *Quaternary Science Reviews* **3**, 405–432.

Lowe, J. J. and M. J. C. Walker (1976) Radiocarbon dates and the deglaciation of Rannoch Moor, Scotland. *Nature* **246**, 632–633.

Lowe, J. J. and M. J. C. Walker (1977) The reconstruction of the Lateglacial environment in the Southern and Eastern Grampian Highlands. In *Studies in the Scottish Lateglacial environment*, J. M. Gray and J. J. Lowe (eds.), 101–118. Oxford: Pergamon Press.

Lowe, J. J. and M. J. C. Walker (1980) Problems associated with radiocarbon dating the close of the Lateglacial period in the Rannoch Moor area, Scotland. In *Studies in the Lateglacial of north-west Europe*, J. J. Lowe, J. M. Gray and J. E. Robinson (eds.), 123–137. Oxford: Pergamon Press.

Lowe, J. J. and M. J. C. Walker (1984) *Reconstructing*

Quaternary environments. London: Longman.

Lowe, J. J. and M. J. C. Walker (1986) Lateglacial and early Flandrian environmental history of the Isle of Mull, Inner Hebrides, Scotland. *Transactions of the Royal Society of Edinburgh: Earth Science* **77**, 1–20.

Luckman, B. B. (1970) The Hereford basin. In *The glaciations of Wales and adjoining regions*, C. A. Lewis (ed.), 175–196. London: Longman.

Maarleveld, G. C. (1960) Wind directions and cover sands in the Netherlands. *Biuletyn Peryglacjalny* **8**, 49–58.

Mace, A. (1959) An Upper Palaeolithic open site at Hengistbury Head, Christchurch, Hants. *Proceedings of the Prehistoric Society* **25**, 233–259.

Mackereth, F. J. H. (1966) Some chemical observations on Post-glacial lake sediments. *Philosophical Transactions of the Royal Society of London* **B250**, 165–213.

Mackereth, F. J. H. (1971) On the variation in direction of the horizontal component of remanent magnetisation in lake sediments. *Earth and Planetary Science Letters* **12**, 332–338.

MacKie, E. W. (1971) English migrants and Scottish brochs. *Glasgow Archaeological Journal* **2**, 39–71.

MacKie, E. W. (1972) Some new quernstones from brochs and duns. *Proceedings of the Society of Antiquaries of Scotland* **104**, 137–146.

MacRae, R. J. (1985) Palaeolithic archaeology of the Upper Thames Basin. In *The chronology and environmental framework of early man in the Upper Thames Valley: a new model*, D. J. Briggs, G. R. Coope and D. D. Gilbertson, 8–25. BAR British Series 137. Oxford: British Archaeological Reports.

Maddy, D. (1989) Pools Farm Pit, Brandon. In *The Pleistocene of the West Midlands. Field guide*, D. H. Keen (ed.), 14–22. Cambridge: Quaternary Research Association.

Maddy, D., D. H. Keen, D. R. Bridgland and C. P. Green (1991) A revised model for the Pleistocene development of the River Avon, Warwickshire. *Journal of the Geological Society of London* **148**, 473–484.

Madgett, P. A. and J. A. Catt (1978) Petrography, stratigraphy and weathering of Late Pleistocene tills in East Yorkshire, Lincolnshire and north Norfolk. *Proceedings of the Yorkshire Geological Society* **42**, 55–108.

Maguire, D. J. and C. J. Caseldine (1985) The former distribution of forest and moorland on northern Dartmoor. *Area* **17**, 193–203.

Maizels, J. (1985) Outwash and outwash terrace. In *The encyclopaedic dictionary of physical geography*, A. Goudie (ed.), 316–317. Oxford: Basil Blackwell.

Manby, T. G. (1971) The Kilham long barrow excavations 1965 to 1969. *Antiquity* **45**, 50–53.

Mangerud, J. (1970) Late Weichselian vegetation and ice-front oscillations in the Bergen district, western Norway. *Norsk Geografisk Tidsskrift* **24**, 121–148.

Mangerud, J. and B. E. Berglund (1978) The subdivision of the Quaternary of Norden: a discussion. *Boreas* **7**, 179–181.

Mangerud, J., S. T. Andersen., B. E. Berglund and J. J. Donner (1974) Quaternary stratigraphy of Norden, a proposal for terminology and classification. *Boreas* **3**, 109–128.

Mangerud, J., S. E. Lie, H. Furnes, I. L. Kristiansen and L. Lømo (1984) A Younger Dryas ash bed in western Norway, and its possible correlations with tephra in cores from the Norwegian Sea and the North Atlantic. *Quaternary Research* **21**, 85–104.

Manley, G. (1971) *Climate and the British scene*. Fourth (revised) impression. London: Collins.

Manley, G. (1974) Central England temperatures: monthly means 1659 to 1973. *Quarterly Journal of the Royal Meteorological Society* **100**, 389–405.

Mannion, A. M. (1986a) Plant macrofossils and their significance in Quaternary palaeoecology. Part I: Introduction. *Progress in Physical Geography* **10**, 194–214.

Mannion, A. M. (1986b) Plant macrofossils and their significance in Quaternary palaeoecology. Part II: Applications: preglacial, interglacial and interstadial deposits. *Progress in Physical Geography* **10**, 364–382.

Mannion, A. M. (1986c) Plant macrofossils and their signficance in Quaternary palaeoecology. Part III: Applications: late glacial, postglacial and archaeological deposits. *Progress in Physical Geography* **10**, 517–546.

Margerel, J. P. (1970) *Aubignyna*, nouveau genre de foraminiferes du Pliocene du Bosq d'Aubigny (Manche). *Revue de Micropaléontologie* **13**, 58–64.

Mathers, S. J. and J. A. Zalasiewicz (1985) Producing a comprehensive geological map. A case study – the Aldeburgh – Orford area of East Anglia. *Modern Geology* **9**, 207–220.

Mathers, S. J. and J. A. Zalasiewicz (1988) The Red Crag and Norwich Crag formations of southern East Anglia. *Proceedings of the Geologists' Association* **99**, 261–278.

Matthews, B. (1970) Age and origin of aeolian sand in the Vale of York. *Nature* **227**, 1234–1236.

Maw, G. (1864) On a supposed deposit of boulder clay in north Devon. *Quarterly Journal of the Geological Society of London* **20**, 445–451.

Mayhew, D. F. (1977) Avian predators as accumulators of fossil mammal material. *Boreas* **6**, 25–31.

Mayhew, D. F. (1979) The vertebrate fauna of Bramerton. In: The Crag at Bramerton, near Norwich, Norfolk. B. M. Funnell, P. E. P. Norton and R. G. West, *Philosophical Transactions of the Royal Society of London* **B287**, 531–534.

Mayhew, D. F. (1985) Preliminary report of research project on small mammal remains from British Lower Pleistocene sediments. *Quaternary Newsletter* **47**, 1–4.

Mayhew, D. F. and A. J. Stuart (1986) Stratigraphic and taxonomic revision of the fossil vole remains (Rodentia, Microtinae) from the Lower Pleistocene deposits of eastern England. *Philosophical Transactions of the Royal Society of London* **B312**, 431–485.

McBurney, C. B. M. and P. Callow (1971) The Cambridge excavations at La Cotte de St. Brelade, Jersey – a preliminary report. *Proceedings of the Prehistoric Society* **37**, 167–207.

McCabe, A. M. (1985) Glacial geomorphology. In *The Quaternary history of Ireland*, K. J. Edwards and W. P. Warren (eds.), 67–93. London: Academic Press.

McCabe, A. M. (1987) Quaternary deposits and glacial stratigraphy in Ireland. *Quaternary Science Reviews* **6**, 259–299.

McCabe, A. M. and P. G. Hoare (1978) The Late Quaternary history of east-central Ireland. *Geological Magazine* **115**, 397–413.

McCabe, A. M., J. R. Haynes and N. F. McMillan (1986b) Late-Pleistocene tidewater glaciers and glaciomarine sequences from north County Mayo, Republic of Ireland. *Journal of Quaternary Science* **1**, 73–84.

McCabe, A. M., G. F. Mitchell and F. W. Shotton (1978) An inter-till freshwater deposit at Hollymount, Maguiresbridge, Co. Fermanagh. *Proceedings of the Royal Irish Academy* **78B**, 77–89.

McCabe, A. M., G. R. Coope, D. E. Gennard and P. Doughty (1987) Freshwater organic deposits and stratified sediments between Early and Late Midlandian (Devensian) till sheets at Aghnadarragh, County Antrim, Northern Ireland. *Journal of Quaternary Science* **2**, 11–33.

McCabe, A. M., D. E. Gennard, G. R. Coope and P. Doughty (1986a) Aghnadarragh. In *Field guide to the Quaternary of South-East Ulster*, A. M. McCabe and K. R. Hirons (eds.), 142–168. Cambridge: Quaternary Research Association.

McCann, S. B. (1968) Raised shore platforms in the western isles of Scotland. In *Geography at Aberystwyth*, E. G. Bowen, H. Carter and J. A. Taylor (eds.), 22–34. Cardiff: University of Wales Press.

McCave, I. N., V. N. D. Caston and N. G. T. Fannin (1977) The Quaternary of the North Sea. In *British Quaternary studies: recent advances*, F. W. Shotton (ed.), 187–204. Oxford: Clarendon Press.

McIntyre, A. and W. F. Ruddiman (1972) Northeast Atlantic post-Eemian paleooceanography: a predictive analog of the future. *Quaternary Research* **2**, 350–354.

McIntyre, A., W. F. Ruddiman and R. Jantzen (1972) Southward penetrations of the North Atlantic Polar Front: faunal and floral evidence of large-scale surface water mass movements over the last 225 000 years. *Deep-Sea Research* **19**, 61–77.

McMillan, A. A. and J. W. Merritt (1980) *A reappraisal of the 'Tertiary' deposits of Buchan, Grampian Region*. Report of the Institute of Geological Sciences, Number 80/1. London: Her Majesty's Stationery Office.

McMillan, N. F. (1957) Quaternary deposits around Lough Foyle, North Ireland. *Proceedings of the Royal Irish Academy* **58B**, 185–205.

Megaw, J. V. S. (1979) The later bronze age (1400 BC–500 BC). In *Introduction to British prehistory: from the arrival of* Homo sapiens *to the Claudian invasion*, J. V. S. Megaw and D. D. A. Simpson (eds.), 242–343. Leicester: Leicester University Press.

Megaw, J. V. S. and D. D. A. Simpson (eds.) (1979) *Introduction to British prehistory: from the arrival of* Homo sapiens *to the Claudian invasion*. Leicester: Leicester University Press.

Meijer, T. (1985) The pre-Weichselian non-marine molluscan fauna from Maastricht – Belvédère (Southern Limburg, the Netherlands). In: Maastricht – Belvedérè: stratigraphy, palaeoenvironment and archaeology of the Middle and Late Pleistocene deposits. T. van Kolfschoten and W. Roebroeks (eds.), *Mededelingen Rijks Geologische Dienst* **39**, 75–104.

Mellars, P. A. (1969) Radiocarbon dates for a new Creswellian site. *Antiquity* **43**, 308–310.

Mellars, P. A. (1974) The palaeolithic and mesolithic. In *British prehistory: a new outline*, C. Renfrew (ed.), 41–99. London: Duckworth.

Mellars, P. (1978) Excavation and economic analysis of Mesolithic shell middens on the island of Oronsay (Inner Hebrides). In *The early Postglacial settlement of northern Europe*, P. Mellars (ed.), 371–396. London: Duckworth.

Mellars, P. (1984) Palaeolithic and Mesolithic finds from Skipsea Withow Mere deposits. In *Late Quaternary environments and man in Holderness*, D. D. Gilbertson (ed.), 177–185. BAR British Series 134. Oxford: British Archaeological Reports.

Menke, B. (1968) Beiträge zur Biostratigraphie des Mittelpleistozäns in Norddeutschland. *Meyniana* **18**, 35–42.

Menke, B. (1985) Palynologische Untersuchungen zur Transgression des Eem-Meeres in Raum Offenbüttel/Nord-Ostsee-Kanal. *Geologisches Jahrbuch* **A86**, 19–26.

Merritt, J. W. and A. A. McMillan (1982) The 'Pliocene' gravels of Buchan: a reappraisal. *Scottish Journal of Geology* **18**, 329–332.

Merryfield, D. L. and P. D. Moore (1974) Prehistoric human activity and blanket peat initiation on Exmoor. *Nature* **250**, 439–441.

Miller, G. H. and P. E. Hare (1980) Amino acid geochronology: integrity of the carbonate matrix and potential of molluscan fossils. In *Biogeochemistry of amino acids*, P. E. Hare, T. C. Hoering and K. King (eds.), 415–444. New York: Wiley.

Miller, G. H., J. T. Hollin and J. T. Andrews (1979) Aminostratigraphy of UK Pleistocene deposits. *Nature* **281**, 539–543.

Milling, M. E. (1975) Geological appraisal of foundation conditions, northern North Sea. *Proceedings of Oceanology International, Brighton 1975*, 310–319.

Mitchell, G. F. (1960) The Pleistocene history of the Irish Sea. *Advancement of Science* **17**, 313–325.

Mitchell, G. F. (1965a) The Quaternary deposits of the Ballaugh and Kirkmichael districts, Isle of Man. *Quarterly Journal of the Geological Society of London* **121**, 359–381.

Mitchell, G. F. (1965b) Littleton Bog, Tipperary: an Irish vegetational record. In *International studies on the Quaternary*, H. E. Wright and D. G. Frey (eds.), 1–16.

Geological Society of America Special Paper 84. Boulder: Geological Society of America.

Mitchell, G. F. (1965c) Littleton Bog, Tipperary: an Irish agricultural record. *Journal of the Royal Society of Antiquaries of Ireland* **95**, 121–132.

Mitchell, G. F. (1968) Glacial gravel on Lundy Island. *Transactions of the Royal Geological Society of Cornwall* **20**, 65–68.

Mitchell, G. F. (1970) Some chronological implications of the Irish Mesolithic. *Ulster Journal of Archaeology* **33**, 3–14.

Mitchell, G. F. (1971a) Coastal sections between Glen Mooar and Orrisdale. In *Isle of Man. Field guide*, G. S. P. Thomas (ed.), 3–12. Liverpool: Department of Geography, University of Liverpool for Quaternary Research Association.

Mitchell, G. F. (1971b) The Larnian culture: a minimal view. *Proceedings of the Prehistoric Society* **38**, 274–283.

Mitchell, G. F. (1972a) The Pleistocene history of the Irish Sea: second approximation. *Scientific Proceedings of the Royal Dublin Society* **A4**, 181–199.

Mitchell, G. F. (1972b) Soil deterioration associated with prehistoric agriculture in Ireland. *Proceedings of the 24th International Geological Congress Symposium 1*, 59–68.

Mitchell, G. F. (1976) *The Irish landscape*. London: Collins.

Mitchell, G. F. (1977) Raised beaches and sea-levels. In *British Quaternary studies: recent advances*, F. W. Shotton (ed.), 169–186. Oxford: Clarendon Press.

Mitchell, G. F. (1980) The search for Tertiary Ireland. *Journal of Earth Sciences (Dublin)* **3**, 13–33.

Mitchell, G. F. (1981a) The Quaternary – until 10 000 BP. In *A geology of Ireland*, C. H. Holland (ed.), 235–258. Edinburgh: Scottish Academic Press.

Mitchell, G. F. (1981b) The Littletonian Warm Stage – post 10 000 BP. In *A geology of Ireland*, C. H. Holland (ed.), 259–271. Edinburgh: Scottish Academic Press.

Mitchell, G. F. and A. R. Orme (1967) The Pleistocene deposits of the Isles of Scilly. *Quarterly Journal of the Geological Society of London* **123**, 59–92.

Mitchell, G. F. and N. Stephens (1974) Is there evidence for a Holocene sea-level higher than that of today on the coasts of Ireland? *Colloque Internationale de Centre National de la Recherche Scientifique* **219**, 115–125.

Mitchell, G. F., L. F. Penny, F. W. Shotton and R. G. West (1973a) *A correlation of Quaternary deposits in the British Isles*. Geological Society of London Special Report Number 4. London: Geological Society of London.

Mitchell, G. F., J. A. Catt, A. H. Weir, N. F. McMillan, J.-P. Margerel and R. C. Whatley (1973b) The Late Pliocene marine formation at St. Erth, Cornwall. *Philosophical Transactions of the Royal Society of London* **B266**, 1–37.

Mitchell, J. B. (1954) *Historical geography*. London: English Universities Press.

Moar, N. T. (1969a) Late Weichselian and Flandrian pollen diagrams from south-west Scotland. *New Phytologist* **68**, 433–467.

Moar, N. T. (1969b) Two pollen diagrams from the Mainland, Orkney Islands. *New Phytologist* **68**, 201–208.

Moffat, A. J., J. A. Catt, R. Webster and E. H. Brown (1986) A re-examination of the evidence for a Plio-Pleistocene marine transgression on the Chiltern Hills. 1. Structures and surfaces. *Earth Surface Processes and Landforms* **11**, 95–106.

Moir, J. R. and A. T. Hopwood (1939) Excavations at Brundon, Suffolk (1935–37). 1. Stratigraphy and archaeology. 2. Fossil mammals. *Proceedings of the Prehistoric Society* **5**, 1–32.

Molleson, T. (1977) Skeletal remains of man in the British Quaternary. In *British Quaternary studies: recent advances*, F. W. Shotton (ed.), 83–92. Oxford: Clarendon Press.

Molleson, T. and R. Burleigh (1978) A new date for Goat's Hole Cave. *Antiquity* **52**, 143–145.

Moore, P. D. (1970) Studies in the vegetational history of mid-Wales. II. The Late-glacial period in Cardiganshire. *New Phytologist* **69**, 363–375.

Moore, P. D. (1972) Studies in the vegetational history of mid-Wales. III. Early Flandrian pollen data from west Cardiganshire. *New Phytologist* **71**, 947–959.

Moore, P. D. (1973) The influence of prehistoric cultures upon the initiation and spread of blanket bog in upland Wales. *Nature* **241**, 350–353.

Moore, P. D. (1975) Origin of blanket mires. *Nature* **256**, 267–269.

Moore, P. D. (1977) Vegetational history. *Cambria* **4**, 73–83.

Moore, P. D. (1978) Studies in the vegetational history of mid-Wales. V. Stratigraphy and pollen analysis of Llyn Mire in the Wye Valley. *New Phytologist* **80**, 281–302.

Moore, P. D. (1986) Site history. In *Methods in plant ecology*. Second edition, P. D. Moore and S. B. Chapman (eds.), 525–556. Oxford: Blackwell Scientific.

Moore, P. D. (1988) The development of moorlands and upland mires. In *Archaeology and the flora of the British Isles*, M. Jones (ed.), 116–122. Oxford University Committee for Archaeology Monograph Number 14. Botanical Society of the British Isles Conference Report Number 19. Oxford: Oxford Committee for Archaeology.

Moore, P. D. and E. H. Chater (1969) The changing vegetation of west-central Wales in the light of human history. *Journal of Ecology* **57**, 361–379.

Moore, P. D., D. L. Merryfield and M. D. R. Price (1984) The vegetation and development of blanket mires. In *European mires*, P. D. Moore (ed.), 203–235. London: Academic Press.

Moore, J. W. (1954) Excavations at Flixton, Site 2. In *Excavations at Star Carr*, J. G. D. Clark, 192–194. Cambridge: Cambridge University Press.

Morey, C. R. (1976) The natural history of Slapton Ley Nature Reserve. IX. The morphology and history of the lake basins. *Field Studies* **4**, 353–368.

Morey, C. R. (1983) Barrier stability. *Quaternary Newsletter* **40**, 23–27.

Morgan, A. (1969) A Pleistocene fauna and flora from Great Billing, Northamptonshire, England. *Opuscula Entomologica* **34**, 109–129.

Morgan, A. (1970) *Weichselian insect faunas of the English*

Midlands. Ph.D. Thesis, University of Birmingham.

Morgan, A. (1973) Late Pleistocene environmental changes indicated by fossil insect faunas of the English Midlands. *Boreas* **2**, 173–212.

Morgan, A. V. (1973) The Pleistocene geology of the area north and west of Wolverhampton, Staffordshire, England. *Philosophical Transactions of the Royal society of London* **B265**, 233–297.

Morgan, A. V. and R. G. West (1988) A pollen diagram from an interglacial deposit at Trysull, Staffordshire, England. *New Phytologist* **109**, 393–397.

Mörner, N.-A. (1969) Climatic and eustatic changes during the last 15 000 years. *Geologie en Mijnbouw* **48**, 389–399.

Morrison, J., D. E. Smith, R. A. Cullingford and R. L. Jones (1981) The culmination of the Main Postglacial Transgression in the Firth of Tay area, Scotland. *Proceedings of the Geologists' Association* **92**, 197–209.

Morrison, M. E. S. and N. Stephens (1960) Stratigraphy and pollen analysis of the raised beach deposits at Ballyhalbert, Co. Down, Northern Ireland. *New Phytologist* **59**, 153–162.

Morrison, M. E. S. and N. Stephens (1965) A submerged Late Quaternary deposit at Roddans Port on the north-east coast of Ireland. *Philosophical Transactions of the Royal Society of London* **B249**, 221–255.

Morzadec-Kerfourn, M. T. (1974) *Variations de la ligne de rivage Armoricaine au Quaternaire: analyses polliniques de dépôts organiques littoraux*. Memoires de la Société Geologique et Minéralogique de Bretagne 17. Rennes: La Société Geologique et Mineralogique de Bretagne.

Morzadec-Kerfourn, M. T. (1975) Palynology of the Quaternary sediments in borehole V050. In: A buried valley system in the Strait of Dover. J.-P. Destombes, E. R. Shephard-Thorn and J. H. Redding, *Philosophical Transactions of the Royal Society of London* **A279**, 253–255.

Moseley, K. A. (1978) A preliminary report on Quaternary fossil caddis larvae (Trichoptera). *Quaternary Newsletter* **26**, 2–12.

Mottershead, D. N. (1977) The Quaternary evolution of the south coast of England. In *The Quaternary history of the Irish Sea*, C. Kidson and M. J. Tooley (eds.), 299–320. Geological Journal Special Issue Number 7. Liverpool: Seel House Press.

Mottershead, D. N., D. D. Gilbertson and D. H. Keen (1987) The raised beaches and shore platforms of Tor Bay: a re-evaluation. *Proceedings of the Geologists' Association* **98**, 241–257.

Mountain, M.-J. (1979) Exploitation and adaptation in pre-agricultural communities. In *Introduction to British prehistory: from the arrival of* Homo sapiens *to the Claudian invasion*, J. V. S. Megaw and D. D. A. Simpson (eds.), 24–76. Leicester: Leicester University Press.

Mountford, H. M. (1970) The terrestrial environment during Upper Cretaceous and Tertiary times. *Proceedings of the Geologists' Association* **81**, 181–204.

Movius, H. L. (1940) An early Post-glacial archaeological site at Cushendun, Co. Antrim. *Proceedings of the Royal Irish Academy* **46C**, 1–84.

Movius, H. L. (1953) Curran Point, Larne, Co. Antrim, the type site of the Irish Mesolithic. *Proceedings of the Royal Irish Academy* **56C**, 1–95.

Müller, H. (1974) Pollen-analytische Untersuchungen und Jahresschichten-Zählungen an der holstein-zeitlichen Kieselgur von Münster-Breloh. *Geologisches Jahrbuch* **A21**, 107–140.

Mykura, W. (1976) *British regional geology: Orkney and Shetland*. Edinburgh: Her Majesty's Stationery Office.

Mykura, W. and J. Phemister (1976) *The geology of western Shetland*. Memoirs of the Geological Survey of Great Britain. Edinburgh: Her Majesty's Stationery Office.

Newey, W. W. (1966) Pollen analysis of sub-carse peats of the Forth Valley. *Transactions of the Institute of British Geographers* **39**, 53–59.

Newey, W. W. (1968) Pollen analyses from south-east Scotland. *Transactions of the Botanical Society of Edinburgh* **40**, 424–434.

Newey, W. W. (1970) Pollen analysis of Late-Weichselian deposits at Corstorphine, Edinburgh. *New Phytologist* **69**, 1167–1177.

Newman, W. S. and C. Baeteman (1987) Holocene excursions of the northwest European geoid. *Progress in Oceanography* **18**, 287–322.

Northcote, E. M. (1980) Some Cambridgeshire Neolithic to Bronze Age birds and their presence or absence in England in the Late Glacial and early Flandrian. *Journal of Archaeological Science* **7**, 379–383.

Norton, P. E. P. (1967) Marine molluscan assemblages in the Early Pleistocene of Sidestrand, Bramerton and the Royal Society borehole at Ludham, Norfolk. *Philosophical Transactions of the Royal Society of London* **B253**, 161–200.

Norton, P. E. P. (1970) The Crag Mollusca: a conspectus. *Bulletin de la Société Belge de Géologie, de Palaeontologie et d'Hydrologie* **79**, 157–166.

Norton, P. E. P. (1977) Marine Mollusca in the East Anglian pre-glacial Pleistocene. In *British Quaternary studies: recent advances*, F. W. Shotton (ed.), 43–53. Oxford: Clarendon Press.

Norton, P. E. P. (1980) Marine mollusc faunas. In *The pre-glacial Pleistocene of the Norfolk and Suffolk coasts*, R. G. West, 125–127. Cambridge: Cambridge University Press.

Norton, P. E. P. and R. B. Beck (1972) Lower Pleistocene molluscan assemblages and pollen from the Crag of Aldeby (Norfolk) and Easton Bavents (Suffolk). *Bulletin of the Geological Society of Norfolk* **22**, 11–31.

Oakley, K. P. (1952) Swanscombe Man. *Proceedings of the Geologists' Association* **63**, 271–300.

Oakley, K. P. (1957) Stratigraphical age of the Swanscombe Skull. *American Journal of Physical Anthropology* New Series **15**, 253–260.

Oakley, K. P. (1968) The date of the 'Red Lady', of Paviland. *Antiquity* **42**, 306–307.

Oele, E. and R. T. E. Schüttenhelm (1979) Development of the North Sea after the Saalian glaciation. In *The Quaternary history of the North Sea*, E. Oele, R. T. E. Schüttenhelm and A. J. Wiggers (eds.), 191–216. Acta Universitatis Uppsaliensis Symposium Universitatis Uppsaliensis Annum Quingentesmium Celebrantis, Number 2. Uppsala: University of Uppsala.

O'Kelly, M. J. (1973) Current excavations at Newgrange, Ireland. In *Megalithic graves and ritual*, G. E. Daniel and P. Kjaerum (eds.), 137–146. Jutland Archaeological Society Publication Number 11. Copenhagen: Jutland Archaeological Society.

Oldfield, F. (1960) Studies in the Post-glacial history of British vegetation: Lowland Lonsdale. *New Phytologist* **59**, 192–217.

Oldfield, F. (1963) Pollen analysis and man's role in the ecological history of the south-east Lake District. *Geografiska Annaler* **45**, 23–40.

Oldfield, F. (1965) Problems of mid-Post-glacial pollen zonation in part of north-west England. *Journal of Ecology* **53**, 247–260.

Oldfield, F., R. W. Battarbee and J. A. Dearing (1983) New approaches to recent environmental change. *Geographical Journal* **149**, 169–181.

Osborne, P. J. (1969) An insect fauna of Late Bronze Age date from Wilsford, Wiltshire. *Journal of Animal Ecology* **38**, 555–566.

Osborne, P. J. (1971) An insect fauna from the Roman site at Alcester, Warwickshire. *Brittania* **2**, 156–165.

Osborne, P. J. (1972) Insect faunas of Late Devensian and Flandrian age from Church Stretton, Shropshire. *Philosophical Transactions of the Royal Society of London* **B263**, 327–367.

Osborne, P. J. (1974) An insect assemblage of early Flandrian age from Lea Marston, Warwickshire, and its bearing on the contemporary climate and ecology. *Quaternary Research* **4**, 471–486.

Osborne, P. J. (1976) Evidence from the insects of climatic variation during the Flandrian period: a preliminary note. *World Archaeology* **8**, 150–158.

Osborne, P. J. (1978) Insect evidence for the effect of man on the lowland landscape. In *The effect of man on the landscape: the Lowland Zone*, S. Limbrey and J. G. Evans (eds.), 32–34. CBA. Research Report Number 21. London: Council for British Archaeology.

Osborne, P. J. (1980) Coleoptera. In: Cromerian interglacial deposits at Sugworth, near Oxford, England, and their relation to the Plateau Drift of the Cotswolds and the terrace sequence of the Upper and Middle Thames. F. W. Shotton, A. S. Goudie, D. J. Briggs and H. A. Osmaston, *Philosophical Transactions of the Royal Society of London* **B289**, 63.

Osborne, P. J. (1988) A Late Bronze Age fauna from the River Avon, Warwickshire, England: its implications for the terrestrial and fluvial environment and for climate. *Journal of Archaeological Science* **15**, 715–727.

Osborne, P. J. and F. W. Shotton (1968) The fauna of the channel deposit of early Saalian age at Brandon, Warwickshire. *Philosophical Transactions of the Royal Society of London* **B254**, 417–424.

O'Sullivan, P. E. (1974) Two Flandrian pollen diagrams from the east-central Highlands of Scotland. *Pollen et Spores* **16**, 33–57.

O'Sullivan, P. E. (1976) Pollen analysis and radiocarbon dating of a core from Loch Pityoulish, eastern Highlands, Scotland. *Journal of Biogeography* **3**, 293–302.

O'Sullivan, P. E. (1977) Vegetation history and the native pinewoods. In *Native pinewoods of Scotland: Proceedings of the Aviemore Symposium 1975,* R. G. H. Bunce and J. N. R. Jeffers (eds.), 60–69. Cambridge: Institute of Terrestrial Ecology.

O'Sullivan, P. E., F. Oldfield and R. W. Battarbee (1973) Preliminary studies of Lough Neagh sediments. I. Stratigraphy, chronology and pollen analysis. In *Quaternary plant ecology: the 14th Symposium of the British Ecological Society, University of Cambridge, 28–30 March 1972*, H. J. B. Birks and R. G. West (eds.), 267–278. Oxford: Blackwell Scientific.

Owen, T. R. (1976) *The geological evolution of the British Isles*. Oxford: Pergamon Press.

Paepe, R., J. Somme, N. Cunat and C. Baeteman (1976) Flandrian, a formation or just a name? *Newsletters on Stratigraphy* **5**, 18–30.

Page, N. R. (1972) On the age of the Hoxnian interglacial. *Geological Journal* **8**, 129–142.

Palmer, S. (1977) *Mesolithic cultures of Britain*. Poole: Dolphin Press.

Pantin, H. M. and C. D. R. Evans (1984) The Quaternary history of the central and southwestern Celtic Sea. *Marine Geology* **57**, 259–293.

Pareyn, C. (1982) Pierrepont-en-Cotentin. I. Pierrepont-en-Cotentin Formation. In *The Quaternary of Normandy. Field handbook*, J.-P. Lautridou, 67. Caen: Quaternary Research Association and Centre de Geomorphologie du Centre National de la Recherche Scientifique.

Paterson, T. T. and C. F. Tebbutt (1947) Studies in the Palaeolithic succession in England. III. Palaeoliths from St. Neots, Huntingdonshire. *Proceedings of the Prehistoric Society* **13**, 1–46.

Peacock, J. D. (1966) Note on the drift sequence near Portsoy, Banffshire. *Scottish Journal of Geology* **2**, 35–37.

Peacock, J. D. (1971a) A re-interpretation of the coastal deposits of Banffshire and their place in the late-glacial history of north-east Scotland. *Bulletin of the Geological Survey of Great Britain* **37**, 81–89.

Peacock, J. D. (1971b) Marine shell radiocarbon dates and the chronology of deglaciation in western Scotland. *Nature* **230**, 43–45.

Peacock, J. D. (1975) Scottish late- and post-glacial marine deposits. In *Quaternary studies in north east Scotland*, A.

M. D. Gemmell (ed.), 45–48. Aberdeen: Department of Geography, University of Aberdeen for Quaternary Research Association.

Peacock, J. D. (1981) Scottish late-glacial marine deposits and their environmental significance. In *The Quaternary in Britain: essays, reviews and original work on the Quaternary published in honour of Lewis Penny on his retirement*, J. Neale and J. Flenley (eds.), 222–236. Oxford: Pergamon Press.

Peake, D. S. (1961) Glacial changes in the Alyn river system and their significance in the glaciology of the North Wales border. *Quarterly Journal of the Geological Society of London* **117**, 335–366.

Peake, D. S. (1981) The Devensian Glaciation on the north Welsh border. In *The Quaternary in Britain: essays, reviews and original work on the Quaternary published in honour of Lewis Penny on his retirement*, J. Neale and J. Flenley (eds.), 49–59. Oxford: Pergamon Press.

Pearson, G. W. and M. G. L. Baillie (1983) High-precision ^{14}C measurement of Irish oaks to show the natural atmospheric ^{14}C variations of the AD time period. *Radiocarbon* **25**, 187–196.

Pearson, G. W., J. R. Pilcher and M. G. L. Baillie (1983) High-precision ^{14}C measurement of Irish oaks to show the natural ^{14}C variations from 200 BC to 4000 BC. *Radiocarbon* **25**, 179–186.

Pearson, G. W., J. R. Pilcher M. G. L. Baillie and J. Hillam (1977) Absolute radiocarbon dating using a low altitude European tree-ring calibration. *Nature* **270**, 25–28.

Peglar, S. (1979) A radiocarbon dated pollen diagram from Loch of Winless, Caithness, north-east Scotland. *New Phytologist* **82**, 245–263.

Peglar, S. M. and P. L. Gibbard (1989) Pollen diagram from Waverley Wood Farm Pit. In *The Pleistocene of the West Midlands. Field guide*, D. H. Keen (ed.), (supplement). Cambridge: Quaternary Research Association.

Peglar, S. M., S. C. Fritz and H. J. B. Birks (1989) Vegetation and land-use history at Diss, Norfolk, UK. *Journal of Ecology* **77**, 203–222.

Peltier, W. R. (1987) Mechanisms of relative sea-level change and the geophysical response to ice-water loadings. In *Sea surface studies: a global view*, R. J. N. Devoy (ed.), 57–94. London: Croom Helm.

Pennington, W. (1943) Lake sediments: the bottom deposits of the north basin of Windermere with special reference to the diatom succession. *New Phytologist* **42**, 1–27.

Pennington, W. (1965) The interpretation of some Post-glacial vegetation diversities at different Lake District sites. *Proceedings of the Royal Society of London* **B161**, 310–323.

Pennington, W. (1969) *The history of British vegetation*. First edition. London: English Universities Press.

Pennington, W. (1970) Vegetational history in the north-west of England: a regional synthesis. In *Studies in the vegetational history of the British Isles: essays in honour of Harry Godwin*, D. Walker and R. G. West (eds.), 41–79. Cambridge: Cambridge University Press.

Pennington, W. (1974) *The history of British vegetation*. Second edition. London: English Universities Press.

Pennington, W. (1975a) A chronostratigraphic comparison of Late-Weichselian and Late-Devensian subdivisions, illustrated by two radiocarbon-dated profiles from western Britain. *Boreas* **4**, 157–171.

Pennington, W. (1975b) The effect of Neolithic man on the environment in north-west England: the use of absolute pollen diagrams. In *The effect of man on the landscape: the Highland Zone*, J. G. Evans, S. Limbrey and H. Cleere (eds.), 74–86. CBA Research Report Number 11. London: Council for British Archaeology.

Pennington, W. (1977a) The Late Devensian flora and vegetation of Britain. *Philosophical Transactions of the Royal Society of London* **B280**, 247–271.

Pennington, W. (1977b) Lake sediments and the Lateglacial environment in northern Scotland. In *Studies in the Scottish Lateglacial environment*, J. M. Gray and J. J. Lowe (eds.), 119–141. Oxford: Pergamon Press.

Pennington, W. (1978) Quaternary geology. In *The geology of the Lake District*, F. Moseley (ed.), 207–225. Yorkshire Geological Society Occasional Publication Number 3. Leeds: Yorkshire Geological Society.

Pennington, W. (1980) Modern pollen samples from west Greenland and the interpretation of pollen data from the British Late-glacial (Late Devensian). *New Phytologist* **84**, 171–201.

Pennington, W. and A. P. Bonny (1970) Absolute pollen diagram from the British late-glacial. *Nature* **226**, 871–873.

Pennington, W. and J. P. Lishman (1971) Iodine in lake sediments in northern England and Scotland. *Biological Reviews* **46**, 279–313.

Pennington, W., R. S. Cambray and E. M. R. Fisher (1973) Observations on lake sediments using fall-out ^{137}Cs as a tracer. *Nature* **242**, 324–326.

Pennington, W., E. Y. Haworth, A. P. Bonny and J. P. Lishman (1972) Lake sediments in northern Scotland. *Philosophical Transactions of the Royal Society of London* **B264**, 191–294.

Penny, L. F. and J. A. Catt (1972) The Speeton Shell Bed. In *Field guide. East Yorkshire and Lincolnshire*, L. F. Penny, A. Straw, J. A. Catt, J. R. Flenley, J. F. D. Bridger, P. A. Madgett and S. C. Beckett, 18. Hull: Quaternary Research Association.

Penny, L. F., G. R. Coope and J. A. Catt (1969) Age and insect fauna of the Dimlington Silts, East Yorkshire. *Nature* **224**, 65–67.

Penny, L. F., A. Straw and J. A. Catt (1972) Kirmington. In *Field guide. East Yorkshire and Lincolnshire*, L. F. Penny, A. Straw, J. A. Catt, J. R. Flenley, J. F. D. Bridger, P. A. Madgett and S. C. Beckett, 30–34. Hull: Quaternary Research Association.

Perrin, R. M. S., H. Davies and M. D. Fysh (1973) Lithology of the Chalky Boulder Clay. *Nature* **245**, 101–104.

Perrin, R. M. S., H. Davies and M. D. Fysh (1974) Distribution of late Pleistocene aeolian deposits in eastern and southern England. *Nature* **248**, 320–324.

Perrin, R. M. S., J. Rose and H. Davies (1979) The distribution, variation and origins of pre-Devensian tills in eastern England. *Philosophical Transactions of the Royal Society of London* **B287**, 535–570.

Pettit, M. and P. L. Gibbard (1980) Palaeobotany. In: Cromerian interglacial deposits at Sugworth, near Oxford, England, and their relation to the Plateau Drift of the Cotswolds and the terrace sequence of the Upper and Middle Thames. F. W. Shotton, A. S. Goudie, D. J. Briggs and H. A. Osmaston, *Philosophical Transactions of the Royal Society of London* **B289**, 63.

Péwé, T. L. (1966) Ice wedges in Alaska – classification, distribution and climatic significance. *Proceedings of the First International Permafrost Conference*, National Academy of Science – National Research Council of Canada Publication **1287**, 76–81.

Phillips, L. (1974) Vegetational history of the Ipswichian/ Eemian Interglacial in Britain and Continental Europe. *New Phytologist* **73**, 589–604.

Phillips, L. (1976) Pleistocene vegetational history and geology in Norfolk. *Philosophical Transactions of the Royal Society of London* **B275**, 215–286.

Phillipson, D. W. (1968) Excavations at Eldon's Seat, Encombe, Dorset. Part III. Animal bones. *Proceedings of the Prehistoric Society* **34**, 226–229.

Piggott, C. M. (1953) Milton Loch Crannog I: a native house of the second century AD in Kirkcudbrightshire. *Proceedings of the Society of Antiquaries of Scotland* **87**, 134–152.

Piggott, S. (1938) The Early Bronze Age in Wessex. *Proceedings of the Prehistoric Society* **4**, 52–106.

Piggott, S. (1962) *The West Kennet long barrow: excavations 1955–56*. Ministry of Works Archaeological Reports, Number 4. London: Her Majesty's Stationery Office.

Piggott, S. (1966) A scheme for the Scottish Iron Age. In *The Iron Age in northern Britain*, A. L. F. Rivet (ed.), 1–15. Edinburgh: Edinburgh University Press.

Pigott, C. D. and J. P. Huntley (1978) Factors controlling the distribution of *Tilia cordata* at the northern limits of its geographical range. I. Distribution in north-west England. *New Phytologist* **81**, 429–441.

Pigott, C. D. and J. P. Huntley (1980) Factors controlling the distribution of *Tilia cordata* at the northern limits of its geographical range. II. History in north-west England. *New Phytologist* **84**, 145–164.

Pigott, C. D. and J. P. Huntley (1981) Factors controlling the distribution of *Tilia cordata* at the northern limits of its geographical range. III. Nature and causes of seed sterility. *New Phytologist* **87**, 817–839.

Pigott, C. D. and M. E. Pigott (1963) Late-glacial and Post-glacial deposits at Malham, Yorkshire. *New Phytologist* **62**, 317–334.

Pike, K. and H. Godwin (1953) The interglacial at Clacton-on-Sea, Essex. *Quarterly Journal of the Geological Society of London* **108**, 261–272.

Pilcher, J. R. and M. G. L. Baillie (1980) Six modern oak chronologies from Ireland. *Tree Ring Bulletin* **40**, 23–34.

Pilcher, J. R. and M. Hughes (1982) The potential of dendrochronology for the study of climate change. In *Climate change in later prehistory*, A. F. Harding (ed.), 75–84. Edinburgh: Edinburgh University Press.

Pilcher, J. R. and A. G. Smith (1979) Palaeoecological investigations at Ballynagilly, a Neolithic and Bronze Age settlement in County Tyrone, Northern Ireland. *Philosophical Transactions of the Royal Society of London* **B286**, 345–369.

Pilcher, J. R., J. Hillam, M. G. L. Baillie and G. W. Pearson (1977) A long sub-fossil oak tree-ring chronology from the North of Ireland. *New Phytologist* **79**, 713–729.

Pilcher, J. R., A. G. Smith, G. W. Pearson and A. Crowder (1971) Land clearance in the Irish Neolithic: new evidence and interpretation. *Science* **172**, 560–562.

Pirazzoli, P. A. and D. R. Grant (1987) Lithospheric deformation deduced from ancient shorelines. In *Recent plate movements and deformation*, K. Kasahara (ed.), 67–72. Washington, DC: American Geophysical Union.

Pitcher, W. S., D. J. Shearman and D. C. Pugh (1954) The loess of Pegwell Bay and its associated frost soils. *Geological Magazine* **91**, 308–314.

Pocock, T. I. (1906) *The geology of the country around Macclesfield, Congleton, Crewe and Middlewich*. Memoirs of the Geological Survey of Great Britain. London: His Majesty's Stationery Office.

Pomerol, C. (1982) *The Cenozoic Era: Tertiary and Quaternary*. Chichester: Ellis Horwood.

Poole, E. G. and A. J. Whiteman (1961) The glacial drifts of the southern part of the Shropshire – Cheshire basin. *Quarterly Journal of the Geological Society of London* **117**, 91–130.

Posnansky, M. (1960) The Pleistocene succession in the middle Trent basin. *Proceedings of the Geologists' Association* **71**, 285–311.

Praeger, R. L. (1896) Report on the raised beaches of the north-east of Ireland with special reference to their fauna. *Proceedings of the Royal Irish Academy* **B4**, 30–54.

Preece, R. C. (1980) The biostratigraphy and dating of the tufa deposit at the Mesolithic site at Blashenwell, Dorset, England. *Journal of Archaeological Science* **7**, 345–362.

Preece, R. C. (1988) East Runton – West Runton cliffs: mollusca. In *The Pliocene – Middle Pleistocene of East Anglia. Field guide*, P. L. Gibbard and J. A. Zalasiewicz (eds.), 149–152. Cambridge: Quaternary Research Association.

Preece, R. C. (1989) Additions to the molluscan fauna of the early Middle Pleistocene deposits at Sugworth, near Oxford, including the first British Quaternary record of *Perforatella bidentata* (Gmelin). *Journal of Conchology* **33**, 179–182.

Preece, R. C. (1990) The molluscan fauna of the Middle Pleistocene interglacial deposits at Little Oakley, Essex, and its environmental and stratigraphical implications. *Philosophical Transactions of the Royal Society of London* **B328**, 387–407.

Preece, R. C. and J. E. Robinson (1984) Late Devensian and Flandrian environmental history of the Ancholme Valley, Lincolnshire: molluscan and ostracod evidence. *Journal of Biogeography* **11**, 319–352.

Preece, R. C. and J. D. Scourse (1987) Pleistocene sea-level history in the Bembridge area of the Isle of Wight. In *Wessex and the Isle of Wight. Field guide*, K. E. Barber (ed.), 136–149. Cambridge: Quaternary Research Association.

Preece, R. C. and P. A. Ventris (1983) An interglacial site at Galley Hill, near St. Ives, Cambridgeshire. *Bulletin of the Geological Society of Norfolk* **33**, 63–72.

Preece, R. C., K. D. Bennett and J. E. Robinson (1984) The biostratigraphy of an early Flandrian tufa at Inchrory, Glen Avon, Banffshire. *Scottish Journal of Geology* **20**, 143–159.

Preece, R. C., P. Coxon and J. E. Robinson (1986) New biostratigraphic evidence of the Post-glacial colonization of Ireland and for Mesolithic forest disturbance. *Journal of Biogeography* **13**, 487–509.

Preece, R. C., J. D. Scourse, S. D. Houghton, K. L. Knudson and D. N. Penney (1990) The Pleistocene sea-level and neotectonic history of the eastern Solent, southern England. *Philosophical Transactions of the Royal Society of London* **B328**, 425–477.

Prestwich, J. (1871) On the structure of the Crag-beds of Suffolk and Norfolk with some observations on their organic remains. Part I. The Coralline Crag of Suffolk. *Quarterly Journal of the Geological Society of London* **27**, 115–146.

Price, R. J. (1983) *Scotland's environment during the last 30 000 years*. Edinburgh: Scottish Academic Press.

Price, R. J., M. A. E. Browne and W. G. Jardine (1980) The Quaternary of the Glasgow region. In *Field guide. Glasgow region*, W. G. Jardine (ed.), 3–9. Glasgow: Quaternary Research Association.

Pringle, J. and T. N. George (1948) *British regional geology: South Wales*. Second edition. London: His Majesty's Stationery Office.

Prior, D. B. (1966) Late-glacial and post-glacial shorelines in north-east Antrim. *Irish Geography* **5**, 173–187.

Pujos-Lamy, A. (1977) Essai d'établissement d'une biostratigraphie du nannoplancton calcaire dans le Pleistocene de l'Atlantique Nord-oriental. *Boreas* **6**, 323–331.

Rackham, D. J. (1978) Evidence for changing vertebrate communities in the Middle Devensian. *Quaternary Newsletter* **25**, 1–3.

Raffi, S., S. M. Stanley and R. Marasti (1985) Biogeographic patterns and Plio-Pleistocene extinction of Bivalvia in the Mediterranean and southern North Sea. *Paleobiology* **11**, 368–388.

Raistrick, A. (1926) The glaciation of Wensleydale, Swaledale and adjoining parts of the Pennines. *Proceedings of the Yorkshire Geological Society* **20**, 366–410.

Raistrick, A. (1927) Periodicity in the glacial retreat in west Yorkshire. *Proceedings of the Yorkshire Geological Society* **21**, 24–29.

Raistrick, A. (1931) The glaciation of Wharfedale. *Proceedings of the Yorkshire Geological Society* **22**, 9–31.

Raistrick, A. (1932) The correlation of glacial retreat stages across the Pennines. *Proceedings of the Yorkshire Geological Society* **22**, 199–214.

Ralston, I. B. M. (1979) The iron age (*c*600 BC–AD 200). B. Northern Britain. In *Introduction to British prehistory: from the arrival of* Homo sapiens *to the Claudian invasion*, J. V. S. Megaw and D. D. A. Simpson (eds.), 446–501. Leicester: Leicester University Press.

Rayner, D. H. (1981) *The stratigraphy of the British Isles*. Second edition. Cambridge: Cambridge University Press.

Redda, A. (1985) The Vale of Edale: landforms and Quaternary history. In *Peak District and northern Dukeries. Field guide*, D. J. Briggs, D. D. Gilbertson and R. D. S. Jenkinson (eds.), 88–93. Cambridge: Quaternary Research Association.

Reid, C. (1882) *The geology of the country around Cromer*. Memoirs of the Geological Survey of Great Britain. London: Her Majesty's Stationery Office.

Reid, C. (1890) *The Pliocene deposits of Britain*. Memoirs of the Geological Survey of Great Britain. London: Her Majesty's Stationery Office.

Reid, C. (1899) *The origin of the British flora*. London: Dulau.

Reid, E. M. (1920) On two preglacial floras from Castle Eden (County Durham). *Quarterly Journal of the Geological Society of London* **76**, 104–144.

Reid, E. M. (1949) The Late-Glacial flora of the Lea Valley. *New Phytologist* **48**, 245–252.

Reid, P. C. and C. Downie (1973) The age of the Bridlington Crag. *Proceedings of the Yorkshire Geological Society* **39**, 315–318.

Renfrew, C. (1968) Wessex without Mycenae. *Annual of the British School of Archaeology at Athens* **63**, 277–285.

Renfrew, C. (1973a) *Before civilisation: the radiocarbon revolution and prehistoric Europe*. London: Jonathan Cape.

Renfrew, C. (1973b) Monuments, mobilisation and social organisation in Neolithic Wessex. In *The explanation of culture change: models in prehistory*, C. Renfrew (ed.), 539–558. London: Duckworth.

Renfrew, C. (ed.) (1974) *British prehistory: a new outline*. London: Duckworth.

Renfrew, C. (1979) *Investigations in Orkney*. Reports of the Research Committee of the Society of Antiquaries of London, Number 38. London: Society of Antiquaries of London.

Reynolds, P. J. (1987) Lepe Cliff: the evidence for a pre-Devensian brickearth. In *Wessex and the Isle of Wight. Field guide*, K. E. Barber (ed.), 21–22. Cambridge: Quaternary Research Association.

Rice, R. J. (1968) The Quaternary deposits of central Leicestershire. *Philosophical Transactions of the Royal Society of London* **A262**, 459–509.

Rice, R. J. (1981) The Pleistocene deposits of the area around Croft in south Leicestershire. *Philosophical Transactions of the Royal Society of London* **B293**, 385–418.

Richards, K. S. (1981) Evidence of Flandrian valley alluviation in Staindale, North York Moors. *Earth Surface Processes and Landforms* **6**, 183–186.

Rise, L. and K. Rokoengen (1984) Surficial sediments in the Norwegian sector of the North Sea between 60°30′ and 62°N. *Marine Geology* **58**, 287–317.

Ritchie, W. (1966) The post-glacial rise in sea-level and coastal changes in the Uists. *Transactions of the Institute of British Geographers* **39**, 79–86.

Roberts, B. K. (1979) *Rural settlement in Britain*. London: Hutchinson.

Roberts, M. B. (1986) Excavation of the Lower Palaeolithic site at Amey's Eartham Pit, Boxgrove, West Sussex: a preliminary report. *Proceedings of the Prehistoric Society* **52**, 215–245.

Roberts, N. (1989) *The Holocene: an environmental history*. Oxford: Basil Blackwell.

Robin, G. de Q. (1977) Ice cores and climatic change. *Philosophical Transactions of the Royal Society of London* **B280**, 143–168.

Robinson, A. H. W. (1968) The submerged glacial landscape off the Lincolnshire coast. *Transactions of the Institute of British Geographers* **44**, 119–132.

Robinson, J. E. (1980) Ostracods. In: Cromerian interglacial deposits at Sugworth, near Oxford, England, and their relation to the Plateau Drift of the Cotswolds and the terrace sequence of the Upper and Middle Thames. F. W. Shotton, A. S. Goudie, D. J. Briggs and H. A. Osmaston, *Philosophical Transactions of the Royal Society of London* **B289**, 63.

Robinson, J. E. (1990) The ostracod fauna of the Middle Pleistocene interglacial deposits at Little Oakley, Essex. *Philosophical Transactions of the Royal Society of London* **B328**, 409–423.

Robinson, M. and C. K. Ballantyne (1979) Evidence for a glacial readvance pre-dating the Loch Lomond Advance in Wester Ross. *Scottish Journal of Geology* **15**, 271–277.

Robinson, M. A. and G. H. Lambrick (1984) Holocene alluviation and hydrology in the upper Thames basin. *Nature* **308**, 809–814.

Roe, D. A. (1968) British Lower and Middle Palaeolithic hand-axe groups. *Proceedings of the Prehistoric Society* **34**, 1–82.

Roe, D. A. (1976) Palaeolithic industries in the Oxford region: some notes. In *Field guide to the Oxford region*, D. A. Roe (ed.), 36–43. Oxford: Quaternary Research Association.

Roe, D. A. (1981) *The Lower and Middle Palaeolithic periods in Britain*. London: Routledge and Kegan Paul.

Roebroeks, W. (1985) Archaeological research at the Maastricht Belvédère Pit: a review. In: Maastricht – Belvédère: stratigraphy, palaeoenvironment and archaeology of the Middle and Late Pleistocene deposits. T. van Kolfschoten and W. Roebroeks (eds.), *Mededelingen Rijks Geologische Dienst* **39**, 109–118.

Rokoengen, K. and L. Rise (1984) Quaternary stratigraphy in the northern North Sea (60°30′–62°N). In *Quaternary stratigraphy of the North Sea*. Abstract volume, I. Aarseth and H. P. Sejrup (eds.), 52–54. Symposium University of Bergen, December (1984) Bergen: University of Bergen.

Rolfe, W. D. I. (1966) Woolly rhinoceros from the Scottish Pleistocene. *Scottish Journal of Geology* **2**, 253–258.

Rolfe, W. D. I., M. P. Kerney and A. G. Davis (1958) A recent temporary section through Pleistocene deposits at Ilford. *Essex Naturalist* **30**, 93–102.

Romans, J. C. C. and L. Robertson (1975) Soils and archaeology in Scotland. In *The effect of man on the landscape: the Highland Zone*, J. G. Evans, S. Limbrey and H. Cleere (eds.), 37–39. CBA Research Report Number 11. London: Council for British Archaeology.

Rose, J. (1975) Raised beach gravels and ice wedge casts at Old Kilpatrick, near Glasgow. *Scottish Journal of Geology* **11**, 15–21.

Rose, J. (1980a) Landform development around Kisdon, upper Swaledale, Yorkshire. *Proceedings of the Yorkshire Geological Society* **43**, 201–219.

Rose, J. (1980b) Rhu. In *Field guide. Glasgow region*, W. G. Jardine (ed.), 31–37. Glasgow: Quaternary Research Association.

Rose, J. (1985) The Dimlington Stadial/Dimlington Chronozone: a proposal for naming the main glacial episode of the Late Devensian in Britain. *Boreas* **14**, 225–230.

Rose, J. (1987) Status of the Wolstonian Glaciation in the British Quaternary. *Quaternary Newsletter* **53**, 1–9.

Rose, J. (1989a) Stadial type sections in the British Quaternary. In *Quaternary type sections: imagination or reality?*, J. Rose and C. Schluchter (eds.), 45–67. Rotterdam: Balkema.

Rose, J. (1989b) Tracing the Baginton – Lillington Sands and Gravels from the West Midlands to East Anglia. In *The Pleistocene of the West Midlands. Field guide*, D. H. Keen (ed.), 102–110. Cambridge: Quaternary Research Association.

Rose, J. and P. Allen (1977) Middle Pleistocene stratigraphy in south-east Suffolk. *Journal of the Geological Society of London* **133**, 85–102.

Rose, J., P. Allen and R. W. Hey (1976) Middle Pleistocene stratigraphy in southern East Anglia. *Nature* **263**, 492–494.

Rose, J., J. Boardman, R. A. Kemp and C. A. Whiteman (1985b) Palaeosols and the interpretation of the British Quaternary stratigraphy. In *Geomorphology and soils*, K. S. Richards, R. R. Arnett and S. Ellis (eds.), 348–375. London: Allen and Unwin.

Rose, J., C. Turner, G. R. Coope and M. D. Bryan (1980) Channel changes in a lowland river catchment over the last 13 000 years. In *Timescales in geomorphology*, R. A. Cullingford, D. A. Davidson and J. Lewin (eds.), 159–175. Chichester: Wiley.

Rose, J., P. Allen, R. A. Kemp, C. A. Whiteman and N. Owen (1985a) The early Anglian Barham Soil of eastern England. In *Soils and Quaternary landscape evolution*, J. Boardman (ed.), 197–229. Chichester: Wiley.

Round, F. E. (1961) The diatoms of a core from Esthwaite Water. *New Phytologist* **60**, 43–59.

Rousseau, D.-D. and D. H. Keen (1989) Malacological records from the Upper Pleistocene at Portelet (Jersey, Channel Islands): comparisons with western and central Europe. *Boreas* **18**, 61–66.

Rowe, P. J. and T. C. Atkinson (1985) Uranium – thorium dating results from Robin Hood's Cave. In *Peak District and northern Dukeries. Field guide*, D. J. Briggs, D. D. Gilbertson and R. D. S. Jenkinson (eds.), 200–207. Cambridge: Quaternary Research Association.

Rowlands, B. M. (1971) Radiocarbon evidence of the age of an Irish Sea glaciation in the Vale of Clwyd. *Nature* **230**, 9–11.

Ruddiman, W. F. and A. McIntyre (1973) Time-transgressive deglacial retreat of Polar water from the North Atlantic. *Quaternary Research* **3**, 117–130.

Ruddiman, W. F. and A. McIntyre (1981) The North Atlantic during the last deglaciation. *Palaeogeography, Palaeoclimatology, Palaeoecology* **35**, 145–214.

Ruddiman, W. F., C. D. Sancetta and A. McIntyre (1977) Glacial/Interglacial response rate of subpolar North Atlantic waters to climatic change: the record in ocean sediments. *Philosophical Transactions of the Royal Society of London* **B280**, 119–142.

Ruddiman, W. F., A. McIntyre, V. Niebler-Hunt and J. T. Durazzi (1980) Oceanic evidence for the mechanism of rapid Northern Hemisphere glaciation. *Quaternary Research* **13**, 33–64.

Ruhe, R. V. (1956) Geomorphic surfaces and the nature of soils. *Soil Science* **82**, 441–455.

Rymer, L. (1977) A Late-glacial and early Post-glacial pollen diagram from Drimnagall, North Knapdale, Argyllshire. *New Phytologist* **79**, 211–221.

Sandford, K. S. (1924) The river-gravels of the Oxford district. *Quarterly Journal of the Geological Society of London* **80**, 113–179.

Saunders, G. E. (1968) A fabric analysis of the ground moraine deposits of the Lleyn Peninsula of south west Caernarvonshire. *Geological Journal* **6**, 105–118.

Savage, R. J. G. (1966) Irish Pleistocene mammals. *Irish Naturalists' Journal* **15**, 117–130.

Scaife, R. G. (1982) Late-Devensian and early Flandrian vegetational changes in southern England. In *Archaeological aspects of woodland ecology*, S. Limbrey and M. Bell (eds.), 57–74. BAR International Series 146, Oxford: British Archaeological Reports.

Scaife, R. G. (1984) A history of Flandrian vegetation in the Isles of Scilly: palynological investigations of the Higher Moors and Lower Moors peat mires. *Cornish Studies* **11**, 33–47.

Scaife, R. (1986a) Palynological investigation. In: Excavation of the Lower Palaeolithic site at Amey's Eartham Pit, Boxgrove, West Sussex: a preliminary report. M. B. Roberts, *Proceedings of the Prehistoric Society* **52**, 227–229.

Scaife, R. G. (1986b) Flandrian palaeobotany. In *The Isles of Scilly. Field guide*, J. D. Scourse (ed.), 28–30. Coventry: Quaternary Research Association.

Scaife, R. G. (1987) The Late-Devensian and Flandrian vegetation of the Isle of Wight. In *Wessex and the Isle of Wight. Field guide*, K. E. Barber (ed.), 156–180. Cambridge: Quaternary Research Association.

Schadla-Hall, R. T. (1987) Early man in the eastern Vale of Pickering. In *East Yorkshire. Field guide*, S. Ellis (ed.), 22–30. Cambridge: Quaternary Research Association.

Scharff, R. F., H. J. Seymour and E. T. Newton (1918) The exploration of Castlepook Cave, County Cork. *Proceedings of the Royal Irish Academy* **B34**, 33–72.

Schwarcz, H. P. (1980) Absolute age determinations of archaeological sites by uranium series dating of travertines. *Archaeometry* **22**, 3–24.

Schwarcz, H. P. (1984a) Uranium series determinations on flowstone samples 78MH6 and 81824 from Minchin Hole. In: Minchin Hole Cave. A. J. Sutcliffe and A. P. Currant. In *Field guide. Wales: Gower, Preseli, Fforest Fawr*, D.Q. Bowen and A. Henry (eds.), 36. Cambridge: Quaternary Research Association.

Schwarcz, H. P. (1984b) Uranium series determinations on stalagmite samples from Bacon Hole. In: Bacon Hole Cave. A. P. Currant, C. B. Stringer and S. N. Collcutt. In *Field guide. Wales: Gower Preseli, Fforest Fawr*, D. Q. Bowen and A. Henry (eds.), 43. Cambridge: Quaternary Research Association.

Sclater, J. G. and P. A. F. Christie (1980) Continental stretching: an explanation of the post mid-Cretaceous subsidence of the Central North Sea Basin. *Journal of Geophysical Research* **85**, 3711–3739.

Scott, K. (1986) The large mammal fauna. In *La Cotte de St. Brelade 1961–1978: excavations by C. B. M. McBurney*, P. Callow and J. M. Cornford (eds.), 109–137. Norwich: Geo Books.

Scourse, J. D. (1986) Pleistocene stratigraphy. In *The Isles of Scilly. Field guide*, J. D. Scourse (ed.), 12–28. Coventry: Quaternary Research Association.

Scourse, J. D. (1987) Periglacial sediments and landforms in the Isles of Scilly and West Cornwall. In *Periglacial processes and landforms in Britain and Ireland*, J. Boardman (ed.), 225–236. Cambridge: Cambridge University Press.

Seddon, B. (1957) Late-glacial cwm glaciers in Wales. *Journal of Glaciology* , 94–99.

Seddon, B. (1962) Late-glacial deposits at Llyn Dwythwch and Nant Ffrancon, Caernarvonshire. *Philosophical Transactions of the Royal Society of London* **B244**, 459–481.

Seddon, M. B. and D. T. Holyoak (1985) Evidence of sustained regional permafrost during deposition of.

fossiliferous Late Pleistocene river sediments at Stanton Harcourt (Oxfordshire, England). *Proceedings of the Geologists' Association* **96**, 53–71.

Sernander, R. (1908) On the evidence of Post-glacial changes of climate furnished by the peat mosses of northern Europe. *Geologiska Föreningens i Stockholm Förhandlingar* **30**, 365–478.

Shackleton, N. J. (1975) The stratigraphic record of deep-sea cores and its implications for the assessment of glacials, interglacials, stadials, and interstadials in the mid-Pleistocene. In *After the australopithecines: stratigraphy, ecology and culture change in the Middle Pleistocene*, K. W. Butzer and G. L. Isaac (eds.), 1–24. The Hague: Mouton.

Shackleton, N. J. (1977a) Oxygen isotope stratigraphy of the Middle Pleistocene. In *British Quaternary studies: recent advances*, F. W. Shotton (ed.), 1–16. Oxford: Clarendon Press.

Shackleton, N. J. (1977b) The oxygen isotope stratigraphic record of the Late Pleistocene. *Philosophical Transactions of the Royal Society of London* **B280**, 169–182.

Shackleton, N. J. and N. D. Opdyke (1973) Oxygen isotope and palaeomagnetic stratigraphy of Equatorial Pacific core V28–238: oxygen isotope temperatures and ice volumes on a 10^5 year and 10^6 year scale. *Quaternary Research* **3**, 39–55.

Shackleton, N. J. and N. D. Opdyke (1976) Oxygen isotope and palaeomagnetic stratigraphy of Equatorial Pacific core V28–239, Late Pliocene to Latest Pleistocene. In *Investigation of late Quaternary paleoceanography and paleoclimatology*, R. M. Cline and J. D. Hays (eds.), 449–464. Geological Society of America Memoir 145. Boulder: Geological Society of America.

Shackleton, N. J. and C. Turner (1967) Correlation between marine and terrestrial Pleistocene successions. *Nature* **216**, 1079–1082.

Shackley, M. L. (1973) A contextual study of the Mousterian industry from Great Pan Farm, Isle of Wight. *Proceedings of the Isle of Wight Natural History and Archaeological Society* **6**, 542–554.

Shakesby, R. A. (1978) Dispersal of glacial erratics from Lennoxtown, Stirlingshire. *Scottish Journal of Geology* **14**, 81–86.

Shaw, J. (1985) *Holocene coastal evolution in Co. Donegal*. D.Phil. Thesis, University of Ulster.

Shaw, J., J. D. Orford and R. W. G. Carter (1986) Fahamore and Inch. In *Corca Dhuibhe*, W. P. Warren (ed.), 34–37 and 40–43. Irish Association for Quaternary Studies Field Guide Number 9. Dublin: Irish Association for Quaternary Studies.

Shennan, I. (1982) Interpretation of Flandrian sea-level data from the Fenland, England. *Proceedings of the Geologists' Association* **93**, 53–63.

Shennan, I. (1983) Flandrian and Late Devensian sea-level changes and crustal movements in England and Wales. In *Shorelines and isostasy*, D. E. Smith and A. G. Dawson (eds.), 255–283. Institute of British Geographers Special Publication Number 16. London: Academic Press.

Shennan, I. (1986a) Flandrian sea-level changes in the Fenland. I: The geographic setting and evidence of relative sea-level changes. *Journal of Quaternary Science* **1**, 119–154.

Shennan, I. (1986b) Flandrian sea-level changes in the Fenland. II: Tendencies of sea-level movement, altitudinal changes, and local and regional factors. *Journal of Quaternary Science* **1**, 155–179.

Shennan, I. (1989) Holocene crustal movements and sea-level changes in Great Britain. *Journal of Quaternary Science* **4**, 77–89.

Shephard-Thorn, E. R. (1975) The Quaternary of the Weald – a review. *Proceedings of the Geologists' Association* **86**, 537–548.

Shephard-Thorn, E. R. (1977) Pegwell Bay. In *Guidebook for Excursion A5. South-east England and the Thames Valley*, E. R. Shephard-Thorn and J. J. Wymer, 54–58. INQUA X Congress, United Kingdom. Norwich: Geo Abstracts for International Union for Quaternary Research.

Shephard-Thorn, E. R. and J. J. Wymer (1977) Introduction. In *Guidebook for Excursion A5. South-east England and the Thames Valley*, E. R. Shephard-Thorn and J. J. Wymer, 7–11. INQUA X Congress, United Kingdom. Norwich: Geo Abstracts for International Union for Quaternary Research.

Shotton, F. W. (1953) Pleistocene deposits of the area between Coventry, Rugby and Leamington and their bearing upon the topographic development of the Midlands. *Philosophical Transactions of the Royal Society of London* **B237**, 209–260.

Shotton, F. W. (1962) The physical background of Britain in the Pleistocene. *Advancement of Science* **19**, 193–206.

Shotton, F. W. (1968) The Pleistocene succession around Brandon, Warwickshire. *Philosophical Transactions of the Royal Society of London* **B254**, 387–400.

Shotton, F. W. (1972) An example of hard-water error in radiocarbon dating of vegetable matter. *Nature* **240**, 460–461.

Shotton, F. W. (1973) English Midlands. In *A correlation of Quaternary deposits in the British Isles*, G. F. Mitchell, L. F. Penny, F. W. Shotton and R. G. West, 18–22. Geological Society of London Special Report Number 4. London: Geological Society of London.

Shotton, F. W. (1976) Amplification of the Wolstonian Stage of the British Pleistocene. *Geological Magazine* **113**, 241–250.

Shotton, F. W. (1977a) British dating work with radioactive isotopes. In *British Quaternary studies: recent advances*, F. W. Shotton (ed.), 17–29. Oxford: Clarendon Press.

Shotton, F. W. (1977b) The Devensian Stage: its development, limits and substages. *Philosophical Transactions of the Royal Society of London* **B280**, 107–118.

Shotton, F. W. (1977c) The Quaternary of the English

Midlands. In *Guidebook for Excursion A2. The English Midlands*, F. W. Shotton, 5–18. INQUA X Congress, United Kingdom. Norwich: Geo Abstracts for International Union for Quaternary Research.

Shotton, F. W. (1978) Archaeological inferences from the study of alluvium in the lower Severn – Avon valleys. In *The effect of man on the landscape: the Lowland Zone*, S. Limbrey and J. G. Evans (eds.), 27–32. CBA Research Report Number 21. London: Council for British Archaeology.

Shotton, F. W. (1981) Major contributions of north-east England to the advancement of Quaternary studies. In *The Quaternary in Britain: essays, reviews and original work on the Quaternary published in honour of Lewis Penny on his retirement*, J. Neale and J. Flenley (eds.), 137–145. Oxford: Pergamon Press.

Shotton, F. W. (1983a) Observations on the type Wolstonian glacial sequence. *Quaternary Newsletter* **40**, 28–36.

Shotton, F. W. (1983b) The Wolstonian Stage of the British Pleistocene in and around its type area of the English Midlands. *Quaternary Science Reviews* **2**, 261–280.

Shotton, F. W. (1985) IGCP 24: Quaternary glaciations in the Northern Hemisphere – final report by the British correspondent. *Quaternary Newsletter* **45**, 28–36.

Shotton, F. W. (1986) Glaciations in the United Kingdom. *Quaternary Science Reviews* **5**, 293–297.

Shotton, F. W. (1989a) The exposures at Waverley Wood Farm (SP 3262 7135) north of Leamington Spa. In *The Pleistocene of the West Midlands. Field guide*, D. H. Keen (ed.), 30–33. Cambridge: Quaternary Research Association.

Shotton, F. W. (1989b) The Wolston sequence and its position within the Pleistocene. In *The Pleistocene of the West Midlands. Field guide*, D. H. Keen (ed.), 1–4. Cambridge: Quaternary Research Association.

Shotton, F. W. and G. R. Coope (1983) Exposures in the Power House Terrace of the River Stour at Wilden, Worcestershire, England. *Proceedings of the Geologists' Association* **94**, 33–44.

Shotton, F. W. and P. J. Osborne (1965) The fauna of the Hoxnian interglacial deposits of Nechells, Birmingham. *Philosophical Transactions of the Royal Society of London* **B248**, 353–378.

Shotton, F. W. and R. E. G. Williams (1971) Birmingham University radiocarbon dates V. *Radiocarbon* **13**, 141–156.

Shotton, F. W. and R. E. G. Williams (1973) Birmingham University radiocarbon dates VI. *Radiocarbon* **15**, 1–12.

Shotton, F. W., P. H. Banham and W. W. Bishop (1977a) Glacial – interglacial stratigraphy of the Quaternary in Midland and eastern England. In *British Quaternary studies: recent advances*, F. W. Shotton (ed.), 267–282. Oxford: Clarendon Press.

Shotton, F. W., D. J. Blundell and R. E. G. Williams (1970) Birmingham University radiocarbon dates IV. *Radiocarbon* **12**, 385–399.

Shotton, F. W., P. J. Osborne and J. R. A. Greig (1977b) The fossil content of a Flandrian deposit at Alcester, Warwickshire. *Proceedings of the Coventry and District Natural History and Scientific Society* **5**, 19–32.

Shotton, F. W., A. J. Sutcliffe and R. G. West (1962) The fauna and flora from the Brick Pit at Lexden, Essex. *Essex Naturalist* **31**, 15–22.

Shotton, F. W., A. S. Goudie, D. J. Briggs and H. A. Osmaston (1980) Cromerian interglacial deposits at Sugworth, near Oxford, England, and their relation to the Plateau Drift of the Cotswolds and the terrace sequence of the Upper and Middle Thames. *Philosophical Transactions of the Royal Society of London* **B289**, 55–86.

Shotton, F. W., A. J. Sutcliffe, D. Q. Bowen, A. P. Currant, G. R. Coope, R. Harmon, N. J. Shackleton, C. B. Stringer, C. Turner, R. G. West and J. Wymer (1983) Interglacials after the Hoxnian in Britain. *Quaternary Newsletter* **39**, 19–25.

Sieveking, G. de G., I. H. Longworth, M. J. Hughes and A. J. Clark (1973) A new survey of Grime's Graves – first report. *Proceedings of the Prehistoric Society* **39**, 182–218.

Simmons, B. B. (1980) Iron Age and Roman coasts around the Wash. In *Archaeology and coastal change*, F. A. Thompson (ed.), 56–73. Society of Antiquaries of London Occasional Paper, New Series 1. London: Society of Antiquaries of London.

Simmons, I. G. (1964) Pollen diagrams from Dartmoor. *New Phytologist* **65**, 165–180.

Simmons, I. G. (1969) Pollen diagrams from the North York Moors. *New Phytologist* **68**, 807–827.

Simmons, I. G. (1989) *Changing the face of the earth: culture, environment, history*. Oxford: Basil Blackwell.

Simmons, I. G. and P. R. Cundill (1974) Late Quaternary vegetational history of the North York Moors. I. Pollen analyses of blanket peats. *Journal of Biogeography* **1**, 159–169.

Simmons, I. G. and J. B. Innes (1981) Tree remains in a North York Moors peat profile. *Nature* **294**, 76–78.

Simmons, I. G. and J. B. Innes (1988a) Late Quaternary vegetational history of the North York Moors. VIII. Correlation of Flandrian II litho- and pollen stratigraphy at North Gill, Glaisdale Moor. *Journal of Biogeography* **15**, 249–272.

Simmons, I. G. and J. B. Innes (1988b) Late Quaternary vegetational history of the North York Moors. X. Investigations on East Bilsdale Moor. *Journal of Biogeography* **15**, 299–324.

Simmons, I. G., G. W. Dimbleby and C. Grigson (1981) The Mesolithic. In *The environment in British prehistory*, I. G. Simmons and M. J. Tooley (eds.), 82–124. London: Duckworth.

Simmons, I. G., J. I. Rand and K. Crabtree (1983) A further pollen analytical study of the Blacklane peat section on Dartmoor, England. *New Phytologist* **94**, 655–667.

Simmons, I. G., M. A. Atherden, P. R. Cundill and R. L. Jones (1975) Inorganic layers in soligenous mires of the

North Yorkshire Moors. *Journal of Biogeography* **2**, 49–56.

Simmons, I. G., M. A. Atherden, P. R. Cundill, J. B. Innes and R. L. Jones (1982) Prehistoric environments. In *Prehistoric and Roman archaeology of north-east Yorkshire*, D. A. Spratt (ed.), 33–99. BAR British Series 104. Oxford: British Archaeological Reports.

Simola, H. L. K., M. A. Coard and P. E. O'Sullivan (1981) Annual laminations in the sediments of Loe Pool, Cornwall. *Nature* **290**, 238–241.

Simpkins, K. (1974) The late-glacial deposits at Glanllynnau, Caernarvonshire. *New Phytologist* **73**, 605–618.

Simpson, D. D. A. (1968) Food vessels: associations and chronology. In *Studies in ancient Europe: essays presented to Stuart Piggott*, J. M. Coles and D. D. A Simpson (eds.), 197–212. Leicester: Leicester University Press.

Simpson, D. D. A. (1979) The first agricultural communities (*c*.3500–2500 BC). The later neolithic (*c*.2500–1700 BC). The early bronze age (*c*.2000–1300 BC). In *Introduction to British prehistory: from the arrival of* Homo sapiens *to the Claudian invasion*, J. V. S. Megaw and D. D. A. Simpson (eds.), 78–129, 130–177, 178–241. Leicester: Leicester University Press.

Simpson, I. M. and R. G. West (1958) On the stratigraphy and palaeobotany of a Late-Pleistocene organic deposit at Chelford, Cheshire. *New Phytologist* **57**, 239–250.

Simpson, J. B. (1961) The Tertiary flora of Mull and Ardnamurchan. *Transactions of the Royal Society of Edinburgh* **64**, 421–468.

Singer, R., J. J. Wymer, B. G. Gladfelter and R. Wolff (1973) Excavation of the Clactonian industry at the Golf Course, Clacton-on-Sea, Essex. *Proceedings of the Prehistoric Society* **39**, 6–74.

Singh, G. (1970) Late-glacial vegetational history of Lecale, Co. Down. *Proceedings of the Royal Irish Academy* **69B**, 189–216.

Singh, G. and A. G. Smith (1966) The Post-glacial marine transgression in Northern Ireland – conclusions from estuarine and 'raised beach' deposits: a contrast. *Palaeobotanist* **15**, 230–234.

Singh, G. and A. G. Smith (1973) Post-glacial vegetational history and relative land and sea-level changes in Lecale, Co. Down. *Proceedings of the Royal Irish Academy* **73B**, 1–51.

Sissons, J. B. (1963) The Perth Readvance in central Scotland. Part I. *Scottish Geographical Magazine* **79**, 151–163.

Sissons, J. B. (1964) The Perth Readvance in central Scotland. Part II. *Scottish Geographical Magazine* **80**, 28–36.

Sissons, J. B. (1966) Relative sea-level changes between 10 300 and 8300 BP in part of the Carse of Stirling. *Transactions of the Institute of British Geographers* **39**, 19–29.

Sissons, J. B. (1967a) *The evolution of Scotland's scenery*. Edinburgh: Oliver and Boyd.

Sissons, J. B. (1967b) Glacial stages and radiocarbon dates in Scotland. *Scottish Journal of Geology* **3**, 375–381.

Sissons, J. B. (1969) Drift stratigraphy and buried morphological features in the Grangemouth – Falkirk – Airth area, central Scotland. *Transactions of the Institute of British Geographers* **48**, 19–50.

Sissons, J. B. (1972) The last glaciers in part of the south-east Grampians. *Scottish Geographical Magazine* **88**, 168–181.

Sissons, J. B. (1974a) The Quaternary in Scotland: a review. *Scottish Journal of Geology* **10**, 311–337.

Sissons, J. B. (1974b) Late-glacial marine erosion in Scotland. *Boreas* **3**, 41–48.

Sissons, J. B. (1976) *The geomorphology of the British Isles: Scotland*. London: Methuen.

Sissons, J. B. (1979a) Palaeoclimatic inferences from former glaciers in Scotland and the Lake District. *Nature* **278**, 518–521.

Sissons, J. B. (1979b) The Loch Lomond Stadial in the British Isles. *Nature* **280**, 199–203.

Sissons, J. B. (1980a) The Loch Lomond Advance in the Lake District, northern England. *Transactions of the Royal Society of Edinburgh: Earth Sciences* **71**, 13–27.

Sissons, J. B. (1980b) Palaeoclimatic inferences from Loch Lomond Advance glaciers. In *Studies in the Lateglacial of north-west Europe*, J. J. Lowe, J. M. Gray and J. E. Robinson (eds.), 31–43. Oxford: Pergamon Press.

Sissons, J. B. (1981a) British shore platforms and ice-sheets. *Nature* **291**, 473–475.

Sissons, J. B. (1981b) The last Scottish ice-sheet: facts and speculative discussion. *Boreas* **10**, 1–17.

Sissons, J. B. (1982) The so-called high 'interglacial' rock shoreline of western Scotland. *Transactions of the Institute of British Geographers* New Series **7**, 205–216.

Sissons, J. B. (1983) Shorelines and isostasy in Scotland. In *Shorelines and isostasy*, D. E. Smith and A. G. Dawson (eds.), 209–225. Institute of British Geographers Special Publication Number 16. London: Academic Press.

Sissons, J. B. and C. L. Brooks (1971) Dating of early postglacial land and sea-level changes in the western Forth Valley. *Nature* **234**, 124–127.

Sissons, J. B. and A. G. Dawson (1981) Former sea-levels and ice limits in part of Wester Ross, northwest Scotland. *Proceedings of the Geologists' Association* **92**, 115–124.

Sissons, J. B. and D. E. Smith (1965) Raised shorelines associated with the Perth Readvance in the Forth Valley and their relation to glacial isostasy. *Transactions of the Royal Society of Edinburgh* **66**, 143–168.

Sissons, J. B. and M. J. C. Walker (1974) Late-glacial site in the central Grampian Highlands. *Nature* **249**, 822–824.

Sissons, J. B., D. E. Smith and R. A. Cullingford (1966) Late-glacial and Post-glacial shorelines in south-east Scotland. *Transactions of the Institute of British Geographers* **39**, 9–18.

Skinner, A. C. and D. M. Gregory (1983) Quaternary stratigraphy in the northern North Sea. *Boreas* **12**, 145–152.

Small, R. J., M. J. Clark and J. Lewin (1970) The periglacial rock-stream at Clatford Bottom, Marlborough Downs,

Wiltshire. *Proceedings of the Geologists' Association* **81**, 87–98.

Smalley, I. J. (1967) The subsidence of the North Sea Basin and the geomorphology of Britain. *Mercian Geologist* **2**, 267–278.

Smith, A. G. (1958a) Two lacustrine deposits in the south of the English Lake District. *New Phytologist* **57**, 363–386.

Smith, A. G. (1958b) Post-glacial deposits in south Yorkshire and north Lincolnshire. *New Phytologist* **57**, 19–49.

Smith, A. G. (1959) The mires of south-western Westmorland: stratigraphy and pollen analysis. *New Phytologist* **58**, 105–127.

Smith, A. G. (1975) Neolithic and Bronze Age landscape changes in northern Ireland. In *The effect of man on the landscape: the Highland Zone*, J. G. Evans, S. Limbrey and H. Cleere (eds.), 64–74. CBA Research Report Number 11. London: Council for British Archaeology.

Smith, A. G. (1981) Palynology of a Mesolithic – Neolithic site in County Antrim, Northern Ireland. *Proceedings of the Fourth International Palynological Congress, Lucknow* **3**, 248–257.

Smith, A. G. (1984) Newferry and the Boreal – Atlantic transition. *New Phytologist* **98**, 35–55.

Smith, A. G. and E. W. Cloutman (1988) Reconstruction of Holocene vegetation history in three dimensions at Waun – Fignen – Felen, an upland site in South Wales. *Philosophical Transactions of the Royal Society of London* **B322**, 159–219.

Smith, A. G. and J. R. Pilcher (1973) Radiocarbon dates and vegetational history of the British Isles. *New Phytologist* **72**, 903–914.

Smith, A. G. and E. H. Willis (1962) Radiocarbon dating of the Fallahogy Landnam phase. *Ulster Journal of Archaeology* **24/25**, 16–24.

Smith, A. G., E. A. Brown and C. A. Green (1981b) Environmental evidence from pollen analysis and stratigraphy. In *The Brigg 'raft' and her prehistoric environment*, S. McGrail (ed.), 134–145. National Maritime Museum, Greenwich, Archaeological Series, Number 6. BAR British Series 89. Oxford: British Archaeological Reports.

Smith, A. G., C. Grigson, G. Hillman and M. J. Tooley (1981a) The Neolithic. In *The environment in British prehistory*, I. G. Simmons and M. J. Tooley (eds.), 125–209. London: Duckworth.

Smith, B. (1927) Borings through the glacial drifts of the northern plain of the Isle of Man. *Summary of Progress of the Geological Survey of Great Britain* **3**, 14–23.

Smith, D. B. (1981) The Quaternary geology of the Sunderland district, north east England. In *The Quaternary in Britain: essays, reviews and original work on the Quaternary published in honour of Lewis Penny on his retirement*, J. Neale and J. Flenley (eds.), 146–167. Oxford: Pergamon Press.

Smith, D. B. and E. A. Francis (1967) *Geology of the country between Durham and West Hartlepool*. Memoirs of the Geological Survey of Great Britain. London: Her Majesty's Stationery Office.

Smith, D. E. and R. A. Cullingford (1985) Flandrian relative sea-level changes in the Montrose Basin area. *Scottish Geographical Magazine* **101**, 91–105.

Smith, D. E., R. A. Cullingford and C. L. Brooks (1983) Flandrian relative sea level changes in the Ythan Valley, northeast Scotland. *Earth Surface Processes and Landforms* **8**, 423–438.

Smith, D. E., R. A. Cullingford and B. A. Haggart (1985b) A major coastal flood during the Holocene in eastern Scotland. *Eiszeitalter und Gegenwart* **35**, 109–118.

Smith, D.E., R. A. Cullingford and W. P. Seymour (1982) Flandrian relative sea-level changes in the Philorth valley, north-east Scotland. *Transactions of the Institute of British Geographers* New Series 7, 321–336.

Smith, D. E., J. B. Sissons and R. A. Cullingford (1969) Isobases for the Main Perth Raised Shoreline in south-east Scotland as determined by trend-surface analysis. *Transactions of the Institute of British Geographers* **46**, 45–52.

Smith, D. E., A. G. Dawson, R. A. Cullingford and D. D. Harkness (1985a) The stratigraphy of Flandrian relative sea-level changes at a site in Tayside, Scotland. *Earth Surface Processes and Landforms* **10**, 17–25.

Smith, D. E., J. Morrison, R. L. Jones and R. A. Cullingford (1980) Dating the Main Postglacial Shoreline in the Montrose area, Scotland. In *Timescales in geomorphology*, R. A. Cullingford, D. A. Davidson and J. Lewin (eds.), 225–245. Chichester: Wiley.

Smith, E. G., G. H. Rhys and R. A. Eden (1967) *Geology of the country around Chesterfield, Matlock and Mansfield*. Memoirs of the Geological Survey of Great Britain. London: Her Majesty's Stationery Office.

Smith, E. G., G. H. Rhys and R. F. Goosens (1973) *Geology of the country around East Retford, Worksop and Gainsborough*. Memoirs of the Geological Survey of Great Britain. London. Her Majesty's Stationery Office.

Smith, I. F. (1965) *Windmill Hill and Avebury: excavations by Alexander Keiller 1925–1939*. Oxford: Clarendon Press.

Smith, I. F. (1974) The neolithic. In *British prehistory: a new outline*, C. Renfrew (ed.), 100–136. London: Duckworth.

Smith, I. F. and D. D. A. Simpson (1966) Excavation of a round barrow on Overton Hill, N. Wilts. *Proceedings of the Prehistoric Society* **32**, 122–155.

Smith, R. A. (1911) A Palaeolithic industry at Northfleet, Kent. *Archaeologia* **62**, 515–532.

Smith, R. T. (1977) Sherburn-in-Elmet. In *Guidebook for Excursion C7. Yorkshire and Lincolnshire*, J. A. Catt, 17–18. INQUA X Congress, United Kingdom. Norwich: Geo Abstracts for International Union for Quaternary Research.

Smyth, C. (1986) A palaeoenvironmental investigation of Flandrian valley deposits from the Combe Haven Valley, East Sussex. *Quaternary Studies* **2**, 22–33.

Southgate, G. A. (1984) Thermoluminescence dating of the

Kempton Park Silts. *Quaternary Newsletter* **43**, 1–9.

Southgate, G. A. (1985) Thermoluminescence dating of beach and dune sands: potential of single-grain measurements. *Nuclear Tracks and Radiation Measurements* **10**, 743–747.

Sparks, B. W. (1956) The non-marine Mollusca of the Hoxne interglacial. In: The Quaternary deposits at Hoxne, Suffolk. R. G. West, *Philosophical Transactions of the Royal Society of London* **B239**, 351–354.

Sparks, B. W. (1957) The non-marine Mollusca of the interglacial deposits at Bobbitshole, Ipswich. *Philosophical Transactions of the Royal Society of London* **B241**, 33–44.

Sparks, B. W. (1961) The ecological interpretation of Quaternary non-marine Mollusca. *Proceedings of the Linnean Society of London* **172**, 71–80.

Sparks, B. W. (1964) The distribution of non-marine Mollusca in the Last Interglacial in south-east England. *Proceedings of the Malacological Society of London* **36**, 7–25.

Sparks, B. W. (1976) The non-marine Mollusca from Swanton Morley Sample D. In: Pleistocene vegetational history and geology in Norfolk. L. Phillips, *Philosophical Transactions of the Royal Society of London* **B275**, 278–281.

Sparks, B. W. (1980) Land and freshwater Mollusca of the West Runton Freshwater Bed. In *The pre-glacial Pleistocene of the Norfolk and Suffolk coasts*, R. G. West, 25–27. Cambridge: Cambridge University Press.

Sparks, B. W. and R. G. West (1959) The palaeoecology of the interglacial deposits at Histon Road, Cambridge. *Eiszeitalter und Gegenwart* **10**, 123–143.

Sparks, B. W. and R. G. West (1963) The interglacial deposits at Stutton, Suffolk. *Proceedings of the Geologists' Association* **74**, 419–432.

Sparks, B. W. and R. G. West (1968) Interglacial deposits at Wortwell, Norfolk. *Geological Magazine* **105**, 471–481.

Sparks, B. W. and R. G. West (1970) Late Pleistocene deposits at Wretton, Norfolk. 1. Ipswichian interglacial deposits. *Philosophical Transactions of the Royal Society of London* **B258**, 1–30.

Sparks, B. W. and R. G. West (1972) *The Ice Age in Britain*. London: Methuen.

Sparks, B. W., R. B. G. Williams and F. G. Bell (1972) Presumed ground-ice depressions in East Anglia. *Proceedings of the Royal Society of London* **A327**, 329–343.

Sparks, B. W., R. G. West, R. B. G. Williams and M. Ransom (1969) Hoxnian interglacial deposits near Hatfield, Herts. *Proceedings of the Geologists' Association* **80**, 243–267.

Spencer, H. E. P. (1956) The Hoxne mammalian remains. In: The Quaternary deposits at Hoxne, Suffolk. R. G. West, *Philosophical Transactions of the Royal Society of London* **B239**, 354.

Spencer, H. E. P. (1970) A contribution to the geological history of Suffolk: 4. The interglacial epochs. *Transactions of the Suffolk Naturalists' Society* **15**, 148–196.

Spencer, P. J. (1975) Habitat change in coastal sand-dune areas: the molluscan evidence. In *The effect of man on the landscape: the Highland Zone*, J. G. Evans, S. Limbrey and H. Cleere (eds.), 96–103. CBA Research Report Number 11. London: Council for British Archaeology.

Spratt, D. A. and I. G. Simmons (1976) Prehistoric activity and environment on the North York Moors. *Journal of Archaeological Science* **3**, 193–210.

Stather, J. W. (1905) Investigation of the fossiliferous drift deposits at Kirmington, Lincolnshire, and at various localities in the East Riding of Yorkshire. *Report of the British Association for 1904*, 272–274.

Stather, J. W. (1910) The Bielsbeck fossiliferous beds. *Transactions of the Hull Geological Society* **6**, 103–109.

Stelfox, A. W., J. G. J. Kuiper, N. F. McMillan and G. F. Mitchell (1972) The Late-glacial and Post-glacial Mollusca of the White Bog, Co. Down. *Proceedings of the Royal Irish Academy* **72B**, 185–207.

Stephens, N. (1961) Pleistocene events in north Devon. *Proceedings of the Geologists' Association* **72**, 469–472.

Stephens, N. (1966) Some Pleistocene deposits in north Devon. *Biuletyn Peryglacjalny* **15**, 103–114.

Stephens, N. (1970a) The lower Severn valley. In *The glaciations of Wales and adjoining regions*, C. A. Lewis (ed.), 107–124. London: Longman.

Stephens, N. (1970b) The West Country and Southern Ireland. In *The glaciations of Wales and adjoining regions*, C. A. Lewis (ed.), 267–314. London: Longman.

Stephens, N. (1974a) Some aspects of the Quaternary of South-West England. In *Field handbook. Exeter*, A. Straw (ed.), 5–7. Exeter: Quaternary Research Association.

Stephens, N. (1974b) The Chard area and the Axe Valley sections. In *Field handbook. Exeter*, A. Straw (ed.), 46–51. Exeter: Quaternary Research Association.

Stephens, N. (1980) Introduction. In *Field handbook. West Cornwall Meeting*, P. C. Sims (ed.), 1–7. Plymouth: Quaternary Research Association.

Stephens, N. and A. E. P. Collins (1960) The Quaternary deposits at Ringneill Quay and Ardmillan, Co. Down. *Proceedings of the Royal Irish Academy* **61C**, 41–77.

Stephens, N. and A. M. McCabe (1977) Late-Pleistocene ice movements and patterns of Late- and Post-Glacial shorelines on the coast of Ulster, Ireland. In *The Quaternary history of the Irish Sea*, C. Kidson and M. J. Tooley (eds.), 179–198. Geological Journal Special Issue Number 7. Liverpool: Seel House Press.

Stevens, L. A. (1960) The interglacial of the Nar Valley, Norfolk. *Quarterly Journal of the Geological Society of London* **115**, 291–315.

Stevenson, A. C. and P. D. Moore (1982) Pollen analysis of an interglacial deposit at West Angle, Dyfed, Wales. *New Phytologist* **90**, 327–337.

Stevenson, I. P. and G. D. Gaunt (1971) *Geology of the country around Chapel-en-le-Frith*. Memoirs of the Geological Survey of Great Britain. London: Her Majesty's Stationery Office.

Stinton, F. (1985) Quaternary fish otoliths. *Proceedings of the Geologists' Association* **96**, 199–215.

Stoker, M. S. and A. Bent (1985) Middle Pleistocene glacial and glaciomarine sediments in the west central North Sea. *Boreas* **14**, 325–332.

Stoker, M. S. and D. Long (1984) A relict ice-scoured erosion surface in the central North Sea. *Marine Geology* **61**, 85–93.

Stoker, M. S., D. Long and J. A. Fyfe (1985a) The Quaternary succession in the central North Sea. *Newsletters on Stratigraphy* **14**, 119–128.

Stoker, M. S., D. Long and J. A. Fyfe (1985b) *A revised Quaternary stratigraphy for the central North Sea.* Report of the British Geological Survey, Number 17/2. London: Her Majesty's Stationery Office.

Stoker, M. S., D. Long, A. C. Skinner and D. Evans (1985c) The Quaternary succession on the northern United Kingdom continental shelf and slope: implications for regional geotechnical investigations. In *Offshore site investigation*, D. A. Ardus (ed.), 45–61. Society for Underwater Technology Conference Proceedings. London: Graham and Trotman.

Stoker, M. S., A. C. Skinner, J. A. Fyfe and D. Long (1983) Palaeomagnetic evidence for early Pleistocene in the central and northern North Sea. *Nature* **304**, 332–334.

Straw, A. (1957) Some glacial features in east Lincolnshire. *East Midland Geographer* **1**, 41–48.

Straw, A. (1958) The glacial sequence in Lincolnshire. *East Midland Geographer* **2**, 29–40.

Straw, A. (1960) The limit of the 'Last' Glaciation in north Norfolk. *Proceedings of the Geologists' Association* **71**, 379–390.

Straw, A. (1961) Drifts, meltwater channels and ice-margins in the Lincolnshire Wolds. *Transactions of the Institute of British Geographers* **29**, 115–128.

Straw, A. (1965) A reassessment of the Chalky Boulder Clay or Marly Drift of north Norfolk. *Zeitschrift für Geomorphologie* **9**, 209–221.

Straw, A. (1966) The development of the middle and lower Bain valley, east Lincolnshire. *Transactions of the Institute of British Geographers* **40**, 145–154.

Straw, A. (1967) The Penultimate or Gipping Glaciation in north Norfolk. *Transactions of the Norfolk and Norwich Naturalists' Society* **21**, 21–24.

Straw, A. (1969) Pleistocene events in Lincolnshire: a survey and revised nomenclature. *Transactions of the Lincolnshire Naturalists' Union* **17**, 85–98.

Straw, A. (1979a) Eastern England. In *The geomorphology of the British Isles: eastern and central England*, A. Straw and K. M. Clayton, 3–139. London: Methuen.

Straw, A. (1979b) The geomorphological significance of the Wolstonian glaciation of eastern England. *Transactions of the Institute of British Geographers* New Series 4, 540–549.

Straw, A. (1979c) An Early Devensian glaciation in eastern England. *Quaternary Newsletter* **28**, 18–24.

Straw, A. (1982) The Wolstonian Glaciation in East Anglia. In *Field meeting guide. Suffolk*, P. Allen (ed.), Section 1, 36–39. London: Quaternary Research Association.

Straw, A. (1983) Pre-Devensian glaciation of Lincolnshire (eastern England) and adjacent areas. *Quaternary Science Reviews* **2**, 239–260.

Strickland, H. E. (1835) An account of land and freshwater shells found associated with the bones of land quadrupeds beneath diluvial gravels at Cropthorne in Worcestershire. *Proceedings of the Geological Society of London* **2**, 111–112.

Stringer, C. B. (1974) Population relationships of later Pleistocene hominids: a multivariate study of available crania. *Journal of Archaeological Science* **1**, 317–342.

Stringer, C. B. (1975) A preliminary report on new excavations at Bacon Hole Cave. *Gower* **26**, 32–37.

Stringer, C. B. (1977) Evidence of climatic change and human occupation during the Last Interglacial at Bacon Hole Cave, Gower. *Gower* **28**, 36–44.

Stringer, C. B. and A. P. Currant (1986) Hominid specimens from La Cotte de St. Brelade. In *La Cotte de St. Brelade 1961–1978: excavations by C. B. M. McBurney*, P. Callow and J. M. Cornford (eds.), 155–158. Norwich: Geo Books.

Stringer, C. B., A. P. Currant, H. P. Schwarcz and S. N. Collcutt (1986) Age of Pleistocene faunas from Bacon Hole, Wales. *Nature* **320**, 59–62.

Stuart, A. J. (1974) Pleistocene history of the British vertebrate fauna. *Biological Reviews* **49**, 225–266.

Stuart, A. J. (1975) The vertebrate fauna of the type Cromerian. *Boreas* **4**, 63–76.

Stuart, A. J. (1976) The history of the mammal fauna during the Ipswichian/Last interglacial in England. *Philosophical Transactions of the Royal Society of London* **B276**, 221–250.

Stuart, A. J. (1977) The vertebrates of the Last Cold Stage in Britain and Ireland. *Philosophical Transactions of the Royal Society of London* **B280**, 295–312.

Stuart, A. J. (1979) Pleistocene occurrences of the European pond tortoise (*Emys orbicularis* L.) in Britain. *Boreas* **8**, 359–371.

Stuart, A. J. (1980) Vertebrates. In: Cromerian interglacial deposits at Sugworth, near Oxford, England, and their relation to the Plateau Drift of the Cotswolds and the terrace sequence of the Upper and Middle Thames. F. W. Shotton, A. S. Goudie, D. J. Briggs and H. A. Osmaston, *Philosophical Transactions of the Royal Society of London* **B289**, 62.

Stuart, A. J. (1982) *Pleistocene vertebrates in the British Isles*. London: Longman.

Stuart, A. J. (1985) Midlandian faunas. In *The Quaternary history of Ireland*, K. J. Edwards and W. P. Warren (eds.), 222–233. London: Academic Press.

Stuart, A. J. (1988a) Preglacial Pleistocene vertebrate faunas of East Anglia. In *The Pliocene – Middle Pleistocene of East Anglia. Field guide*, P. L. Gibbard and J. A. Zalasiewicz (eds.), 57–64. Cambridge: Quaternary Research Association.

Stuart, A. J. (1988b) East Runton – West Runton cliffs: vertebrate fauna. In *The Pliocene – Middle Pleistocene of East Anglia. Field guide*, P. L. Gibbard and J. A. Zalasiewicz (eds.), 152–157. Cambridge: Quaternary Research Association.

Stuart, A. J. and R. G. West (1976) Late Cromerian fauna and flora at Ostend, Norfolk. *Geological Magazine* **113**, 469–473.

Sturdy, R. G., R. H. Allen, P. Bullock, J. A. Catt and S. Greenfield (1979) Paleosols developed on chalky boulder clay in Essex. *Journal of Soil Science* **30**, 117–137.

Suess, H. E. (1970) Bristlecone pine calibration of the radiocarbon time-scale 5200 BC to the present. In *Radiocarbon variations and absolute chronology*, I. U. Olsson (ed.), 303–311. Stockholm: Almquist and Wiksell.

Suess, H. E. and R. M. Clark (1976) A calibration curve for radiocarbon dates. *Antiquity* **50**, 61–63.

Sugden, D. E. (1970) Landforms of deglaciation in the Cairngorm Mountains, Scotland. *Transactions of the Institute of British Geographers* **51**, 201–219.

Sugden, D. E. (1977) Did glaciers form in the Cairngorms in the seventeenth to nineteenth centuries? *Cairngorm Club Journal* **18**, 189–201.

Sugden, D. E. and C. M. Clapperton (1975) The deglaciation of Upper Deeside and the Cairngorm mountains. In *Quaternary studies in north east Scotland*, A. M. D. Gemmell (ed.), 30–38. Aberdeen: Department of Geography, University of Aberdeen for Quaternary Research Association.

Suggate, R. P. and R. G. West (1959) On the extent of the Last Glaciation in eastern England. *Proceedings of the Royal Society of London* **B150**, 263–283.

Sumbler, M. G. (1983a) A new look at the type Wolstonian glacial deposits of Central England. *Proceedings of the Geologists' Association* **94**, 23–31.

Sumbler, M. G. (1983b) The type Wolstonian sequence – some further comments. *Quaternary Newsletter* **40**, 36–39.

Sutcliffe, A. J. (1960) Joint Mitnor Cave, Buckfastleigh. *Transactions and Proceedings of the Torquay Natural History Society* **13**, 1–26.

Sutcliffe, A. J. (1964) The mammalian fauna. In *The Swanscombe skull*, C. D. Ovey (ed.), 85–111. Royal Anthropological Institute Occasional Paper Number 20. London: Royal Anthropological Institute.

Sutcliffe, A. J. (1974a) The Torbryan Caves, including Tornewton Cave. In *Field handbook. Exeter*, A. Straw (ed.), 20–22. Exeter: Quaternary Research Association.

Sutcliffe, A. J. (1974b) The William Pengelly Cave Studies Centre and Joint Mitnor Cave. In *Field handbook. Exeter*, A. Straw (ed.), 16–19. Exeter: Quaternary Research Association.

Sutcliffe, A. J. (1975) A hazard in the interpretation of glacial – interglacial sequences. *Quaternary Newsletter* **17**, 1–3.

Sutcliffe, A. J. (1976) The British glacial – interglacial sequence: reply to Professor R. G. West. *Quaternary Newsletter* **18**, 1–7.

Sutcliffe, A. J. (1981) Progress report on excavations in Minchin Hole, Gower. *Quaternary Newsletter* **33**, 1–17.

Sutcliffe, A. J. and D. Q. Bowen (1973) Preliminary report on excavations in Minchin Hole, April–May 1973. *Newsletter of the William Pengelly Cave Studies Trust* **21**, 12–25.

Sutcliffe, A. J. and A. P. Currant (1984) Minchin Hole Cave. In *Field guide. Wales: Gower, Preseli, Fforest Fawr*, D. Q. Bowen and A. Henry (eds.), 33–37. Cambridge: Quaternary Research Association.

Sutcliffe, A. J. and K. Kowalski (1976) Pleistocene rodents of the British Isles. *Bulletin of the British Museum Natural History (Geology)* **27**, 33–147.

Sutcliffe, A. J. and F. E. Zeuner (1962) Excavations in the Torbryan caves, Devonshire. I. Tornewton Cave. *Proceedings of the Devon Archaeological Exploration Society* **5**, 127–145.

Sutcliffe, A. J., A. P. Currant and C. B. Stringer (1987) Evidence of sea-level change from coastal caves with raised beach deposits, terrestrial faunas and dated stalagmites. *Progress in Oceanography* **18**, 243–271.

Sutcliffe, A. J., T. C. Lord, R. S. Harmon, M. Ivanovich, A. Rae and J. W. Hess (1984) A mammalian fauna of northern character in Yorkshire at *ca* 83 000 years BP. *Quaternary Newsletter* **43**, 9–12.

Sutcliffe, A. J., T. C. Lord, R. S. Harmon, M. Ivanovich, A. Rae and J. W. Hess (1985) Wolverine in northern England at about 83 000 yr BP: faunal evidence for climatic change during Isotope Stage 5. *Quaternary Research* **24**, 73–86.

Sutherland, D. G. (1980) Problems of radiocarbon dating deposits from newly deglaciated terrain: examples from the Scottish Lateglacial. In *Studies in the Lateglacial of north-west Europe*, J. J. Lowe, J. M. Gray and J. E. Robinson (eds.), 139–149. Oxford: Pergamon Press.

Sutherland, D. G. (1981a) The high-level marine shell beds of Scotland and the build up of the last Scottish ice sheet. *Boreas* **10**, 247–254.

Sutherland, D. G. (1981b) *The raised shorelines and deglaciation of the Loch Long/Loch Fyne area, western Scotland*. Ph.D. Thesis, University of Edinburgh.

Sutherland, D. G. (1984) The Quaternary deposits and landforms of Scotland and the neighbouring shelves: a review. *Quaternary Science Reviews* **3**, 157–254.

Sutherland, D. G. and M. J. C. Walker (1984) A late Devensian ice-free area and possible interglacial site on the Isle of Lewis, Scotland. *Nature* **309**, 701–703.

Sutherland, D. G., C. K. Ballantyne and M. J. C. Walker (1984) Late Quaternary glaciation and environmental change on St. Kilda, Scotland, and their palaeoclimatic significance. *Boreas* **13**, 261–272.

Swinnerton, H. H. (1931) The post-glacial deposits of the Lincolnshire coasts. *Quarterly Journal of the Geological Society of London* **87**, 360–375.

Sylvester-Bradley, P. C. (1965) On *Cytherissa lacustris* (Sars) and other ostracods from Nechells. In: The fauna of the Hoxnian interglacial deposits of Nechells, Birmingham. F.

W. Shotton and P. J. Osborne, *Philosophical Transactions of the Royal Society of London* **B248**, 375–377.

Synge, F. M. (1956) The glaciation of north-east Scotland. *Scottish Geographical Magazine* **72**, 129–143.

Synge, F. M. (1963) A correlation between the drifts of south-east Ireland and those of west Wales. *Irish Geography* **4**, 360–366.

Synge, F. M. (1969) The Würm ice limit in the west of Ireland. In *Quaternary geology and climate*, 89–92. National Academy of Sciences Publication 1701. Washington, DC: National Academy of Sciences.

Synge, F. M. (1970) The Pleistocene period in Wales. In *The glaciations of Wales and adjoining regions*, C. A. Lewis (ed.), 315–350. London: Longman.

Synge, F. M. (1973) The glaciation of south Wicklow and the adjoining parts of the neighbouring counties. *Irish Geography* **6**, 561–569.

Synge, F. M. (1977) The coasts of Leinster (Ireland). In *The Quaternary history of the Irish Sea*, C. Kidson and M. J. Tooley (eds.), 199–222. Geological Journal Special Issue Number 7. Liverpool: Seel House Press.

Synge, F. M. (1979) Quaternary glaciation in Ireland. *Quaternary Newsletter* **28**, 1–18.

Synge, F. M. (1980) Raised beaches in Ireland. *Bulletin de l'Association Français pour l'Etude Quaternaire* **17**, 77–79.

Synge, F. M. (1981) Quaternary glaciation and changes of sea level in the south of Ireland. *Geologie en Mijnbouw* **60**, 305–315.

Synge, F. M. (1985) Coastal evolution. In *The Quaternary history of Ireland*, K. J. Edwards and W. P. Warren (eds.), 115–131. London: Academic Press.

Synge, F. M. and N. Stephens (1960) The Quaternary Period in Ireland: an assessment. *Irish Geography* **4**, 121–130.

Synge, F. M. and N. Stephens (1966) Late- and post-glacial shorelines and ice limits in Argyll and north-east Ulster. *Transactions of the Institute of British Geographers* **39**, 101–125.

Szabo, B. J. and D. Collins (1975) Ages of fossil bones from British interglacial sites. *Nature* **254**, 680–682.

Tallentire, P. A. (1953) Studies in the Post-glacial history of British vegetation. XIII. Lopham Little Fen, a Late-glacial site in central East Anglia. *Journal of Ecology* **41**, 361–373.

Tallis, J. H. (1964a) The pre-peat vegetation of the southern Pennines. *New Phytologist* **63**, 363–373.

Tallis, J. H. (1964b) Studies on southern Pennine peats. II. The pattern of erosion. *Journal of Ecology* **52**, 333–344.

Tallis, J. H. (1964c) Studies on southern Pennine peats. III. The behaviour of *Sphagnum*. *Journal of Ecology* **52**, 345–353.

Tallis, J. H. (1985) Mass movement and erosion of a southern Pennine blanket peat. *Journal of Ecology* **73**, 283–315.

Tallis, J. H. and J. McGuire (1972) Central Rossendale: the evolution of an upland vegetation. I. The clearance of woodland. *Journal of Ecology* **60**, 721–737.

Tauber, H. (1965) Differential pollen dispersion and the interpretation of pollen diagrams. *Danmarks Geologiske Undersøgelse* **II**, **89**, 1–69.

Tauxe, L., N. D. Opdyke, G. Pasini and C. Elmi (1983) Age of the Plio-Pleistocene boundary in the Vrica Section, southern Italy. *Nature* **304**, 125–129.

Taylor, C. C. (1978) Aspects of village mobility in medieval and later times. In *The effect of man on the landscape: the Lowland Zone*, S. Limbrey and J. G. Evans (eds.), 126–134. CBA Research Report Number 21. London: Council for British Archaeology.

Taylor, C. W. (1956) Erratics of the Saunton and Fremington areas. *Report and Transactions of the Devonshire Association for the Advancement of Science* **88**, 52–64.

Taylor, J. A. (1975) The role of climatic factors in environmental and cultural changes in prehistoric times. In *The effect of man on the landscape: the Highland Zone*, J. G. Evans, S. Limbrey and H. Cleere (eds.), 6–19. CBA Research Report Number 11. London: Council for British Archaeology.

Taylor, J. A. and R. T. Smith (1982) Climatic peat – a misnomer? *Proceedings of the Fourth International Peat Congress, Helsinki*, 471–484.

Tebble, N. (1976) *British bivalve seashells: a handbook for identification*. Second edition. Edinburgh: Her Majesty's Stationery Office for Royal Scottish Museum.

Ters, M. (1973) Les variations du niveau marin depuis 10 000 ans, le long de littoral Atlantique Français. In *Le Quaternaire: geodynamique stratigraphie et environment*, 114–135. Centre National de la Recherche Scientifique. Paris: Comité National Français de l'INQUA.

Ter Wee, M. W. (1983) The Elsterian Glaciation in the Netherlands. In *Glacial deposits in north-west Europe*, J. Ehlers (ed.), 413–415. Balkema: Rotterdam.

Thew, N. M. (1985) The Eynsham Gravel at Magdalen College, Oxford: non-marine Mollusca. In *The chronology and environmental framework of early man in the Upper Thames Valley: a new model*, D. J. Briggs, G. R. Coope and D. D. Gilbertson, 87–89. BAR British Series 137. Oxford: British Archaeological Reports.

Thomas, G. S. P. (1971) Cliff sections south of Glen Mooar. In *Isle of Man. Field guide*, G. S. P. Thomas (ed.), 16–20. Liverpool: Department of Geography, University of Liverpool, for Quaternary Research Association.

Thomas, G. S. P. (1976) The Quaternary stratigraphy of the Isle of Man. *Proceedings of the Geologists' Association* **87**, 307–323.

Thomas, G. S. P. (1977) The Quaternary of the Isle of Man. In *The Quaternary history of the Irish Sea*, C. Kidson and M. J. Tooley (eds.), 155–178. Geological Journal Special Issue Number 7. Liverpool: Seel House Press.

Thomas, G. S. P. (1985a) Glen Mooar. In *Field guide to the Quaternary of the Isle of Man*, R. V. Dackombe and G. S. P. Thomas (eds.), 26–29. Cambridge: Quaternary Research Association.

Thomas, G. S. P. (1985b) Sub-surface succession. In *Field guide to the Quaternary of the Isle of Man*, R. V.

Dackombe and G. S. P. Thomas (eds.), 81–83. Cambridge: Quaternary Research Association.

Thomas, G. S. P. (1985c) The Late Devensian glaciation along the border of north-east Wales. *Geological Journal* **20**, 319–340.

Thomas, G. S. P. (1989) The Late Devensian glaciation along the western margin of the Cheshire – Shropshire lowland. *Journal of Quaternary Science* **4**, 167–181.

Thomas, G. S. P. and R. V. Dackombe (1985) Comment and reply on 'Glaciomarine sediments of the Isle of Man as a key to late Pleistocene stratigraphic investigations in the Irish Sea Basin'. *Geology* **13**, 445–447.

Thomas, G. S. P. and G. Hardy (1985) Ballure. In *Field guide to the Quaternary of the Isle of Man*, R. V. Dackombe and G. S. P. Thomas (eds.), 19–24. Cambridge: Quaternary Research Association.

Thompson, R. (1973) Palaeolimnology and palaeomagnetism. *Nature* **242**, 182–184.

Thompson, R. (1974) Palaeomagnetism. *Science Progress* **62**, 349–373.

Thompson, R. (1978) European palaeomagnetic secular variation 13 000–0 BP. *Polish Archives of Hydrobiology* **25**, 413–418.

Thompson, R. and K. J. Edwards (1982) A Holocene palaeomagnetic record and a geomagnetic master curve from Ireland. *Boreas* **11**, 335–349.

Thompson, R. and F. Oldfield (1986) *Environmental magnetism*. London: Allen and Unwin.

Thompson, R. and G. M. Turner (1979) British geomagnetic master curve 10 000–0 yr BP from dating European sediments. *Geophysics Research Letters* **6**, 249–252.

Thompson, R. and T. Wain-Hobson (1979) Palaeomagnetic and stratigraphic study of the Loch Shiel marine regression and overlying gyttja. *Journal of the Geological Society of London* **136**, 383–388.

Thompson, R., R. W. Battarbee, P. E. O'Sullivan and F. Oldfield (1975) Magnetic susceptibility of lake sediments. *Limnology and Oceanography* **20**, 687–698.

Thomson, M. E. and R. A. Eden (1977) *Quaternary deposits of the central North Sea. 3. The Quaternary sequence in the west central North Sea.* Report of the Institute of Geological Sciences, Number 77/12. London: Her Majesty's Stationery Office.

Thorley, A. (1971) Vegetational history of the Vale of Brooks. In *Guide to Sussex Excursions: Institute of British Geographers*, R. B. G. Williams (ed.), 47–50. Brighton: University of Sussex for Institute of British Geographers.

Thorley, A. (1981) Pollen analytical evidence relating to the vegetation history of the Chalk. *Journal of Biogeography* **8**, 93–106.

Tinsley, H. M. and C. Grigson (1981) The Bronze Age. In *The environment in British prehistory*, I. G. Simmons and M. J. Tooley (eds.), 210–249. London: Duckworth.

Tipping, R. (1990) Biostratigraphic dating of the Cwm Idwal moraines. In *The Quaternary of North Wales. Field guide*, K. Addison, M. J. Edge and R. Watkins (eds.), 96–98. Coventry: Quaternary Research Association.

Tomlinson, M. E. (1925) River terraces of the lower valley of the Warwickshire Avon. *Quarterly Journal of the Geological Society of London* **81**, 137–163.

Tomlinson, M. E. (1935) The superficial deposits of the country north of Stratford-on-Avon. *Quarterly Journal of the Geological Society of London* **91**, 423–460.

Tomlinson, M. E. (1963) The Pleistocene chronology of the Midlands. *Proceedings of the Geologists' Association* **74**, 187–202.

Tooley, M. J. (1974) Sea-level changes during the last 9000 years in north-west England. *Geographical Journal* **140**, 18–42.

Tooley, M. J. (1976) Flandrian sea-level changes in west Lancashire and their implications for the 'Hillhouse coastline'. *Geological Journal* **11**, 37–52.

Tooley, M. J. (1977) The Quaternary history of north-west England and the Isle of Man. In *Guidebook for Excursion A4. The Isle of Man, Lancashire coast and Lake District*, M. J. Tooley, 5–7. INQUA X Congress, United Kingdom. Norwich: Geo Abstracts for International Union for Quaternary Research.

Tooley, M. J. (1978a) Holocene sea-level changes: problems of interpretation. *Geologiska Föreningens i Stockholm Forhandlingar* **100**, 203–212.

Tooley, M. J. (1978b) *Sea-level changes: north-west England during the Flandrian Stage*. Oxford: Clarendon Press.

Tooley, M. J. (1978c) The history of Hartlepool Bay. *International Journal of Nautical Archaeology and Underwater Exploration* **7**, 71–75.

Tooley, M. J. (1980) Theories of coastal change in north-west England. In *Archaeology and coastal change*, F. H. Thompson (ed.), 74–86. Society of Antiquaries of London Occasional Paper, New Series 1. London: Society of Antiquaries of London.

Tooley, M. J. (1982) Sea-level changes in northern England. *Proceedings of the Geologists' Association* **93**, 43–51.

Tooley, M. J. and B. Kear (1977) Mere Sands Wood (Shirdley Hill Sand). In *Guidebook for Excursion A4. The Isle of Man, Lancashire coast and Lake District*, M. J. Tooley, 9–10. INQUA X Congress United Kingdom. Norwich: Geo Abstracts for International Union for Quaternary Research.

Tratman, E. K. (1964) Picken's Hole, Crook Peak, Somerset; a Pleistocene site: preliminary note. *Proceedings of the University of Bristol Spelaeological Society* **10**, 112–115.

Tratman, E. K., D. T. Donovan and J. B. Campbell (1971) The Hyaena Den (Wookey Hole), Mendip Hills, Somerset. *Proceedings of the University of Bristol Spelaeologoical Society* **12**, 245–279.

Traverse, A. and R. N. Ginsburg (1966) Palynology of the surface sediments of Great Bahama Bank, as related to water movement and sedimentation. *Marine Geology* **4**, 417–459.

Trechmann, C. T. (1915) The Scandinavian Drift of the Durham coast and the general glaciology of south-east

Durham. *Quarterly Journal of the Geological Society of London* **71**, 53–82.

Trechmann, C. T. (1919) On a deposit of interglacial loess and some transported preglacial freshwater clays on the Durham coast. *Quarterly Journal of the Geological Society of London* **75**, 173–203.

Trechmann, C. T. (1931) The Scandinavian Drift or Basement Clay on the Durham coast. *Proceedings of the Geologists' Association* **42**, 292–294.

Trotter, F. M. and S. E. Hollingworth (1932) The glacial sequence in the North of England. *Geological Magazine* **69**, 374–380.

Tuffreau, A. (1982) The transition Lower/Middle Palaeolithic in northern France. In *The transition from Lower to Middle Palaeolithic and the origin of modern man*, A. Ronen (ed.), 137–151. BAR International Series 151. Oxford: British Archaeological Reports.

Tufnell, L. (1985) Periglacial landforms in the Cross Fell – Knock Fell area of the north Pennines. In *Field guide to the periglacial landforms of northern England*, J. Boardman (ed.), 4–14. Cambridge: Quaternary Research Association.

Turner, A. (1981) Ipswichian mammal faunas, cave deposits and hyaena activity. *Quaternary Newsletter* **33**, 17–22.

Turner, C. (1970) The Middle Pleistocene deposits at Marks Tey, Essex. *Philosophical Transactions of the Royal Society of London* **B257**, 373–440.

Turner, C. (1973a) High Lodge, Mildenhall. In *Field handbook. Clacton*, J. Rose and C. Turner (eds.), (unpaginated). Clacton: Quaternary Research Association.

Turner, C. (1973b) Pleistocene stratigraphy of south-east Essex. In *Field handbook. Clacton*, J. Rose and C. Turner (eds.), (unpaginated). Clacton: Quaternary Research Association.

Turner, C. (1975) The correlation and duration of Middle Pleistocene interglacial periods in northwest Europe. In *After the australopithecines: stratigraphy, ecology and culture change in the Middle Pleistocene*, K. W. Butzer and G. L. Isaac (eds.), 259–308. The Hague: Mouton.

Turner, C. (1977) Stoke Tunnel and Stoke Lane bone-bed, Ipswich. In *Guidebook for Excursions A1 and C1. East Anglia*, R. G. West, 50–53. INQUA X Congress, United Kingdom. Norwich: Geo Abstracts for International Union for Quaternary Research.

Turner, C. (1983) Nettlebed. In *Field guide. The diversion of the Thames*, J. Rose (ed.), 66–68. Hoddesdon: Quaternary Research Association.

Turner, C. and M. P. Kerney (1971) A note on the age of the freshwater beds of the Clacton Channel. In: Interglacial deposits in Barnfield Pit, Swanscombe, and their molluscan fauna. M. P. Kerney, *Journal of the Geological Society of London* **127**, 87–93.

Turner, C. and R. G. West (1968) The subdivision and zonation of interglacial periods. *Eiszeitalter und Gegenwart* **19**, 93–101.

Turner, G. M. and R. Thompson (1981) Lake sediment record of the geomagnetic secular variation in Britain during Holocene times. *Geophysical Journal of the Royal Astronomical Society* **65**, 703–725.

Turner, J. (1964a) Surface sample analysis from Ayrshire, Scotland. *Pollen et Spores* **6**, 583–592.

Turner, J. (1964b) The anthropogenic factor in vegetational history. I. Tregaron and Whixall Mosses. *New Phytologist* **63**, 73–90.

Turner, J. (1970) Post-Neolithic disturbance of British vegetation. In *Studies in the vegetational history of the British Isles: essays in honour of Harry Godwin*, D. Walker and R. G. West (eds.), 97–116. Cambridge: Cambridge University Press.

Turner, J. (1981) The Iron Age. In *The environment in British prehistory*, I. G. Simmons and M. J. Tooley (eds.), 250–281. London: Duckworth.

Turner, J. and J. Hodgson (1979) Studies in the vegetational history of the northern Pennines. I. Variations in the composition of the early Flandrian forests. *Journal of Ecology* **67**, 629–646.

Turner, J. and J. Hodgson (1981) Studies in the vegetational history of the northern Pennines. II. An atypical diagram from Pow Hill, Co. Durham. *Journal of Ecology* **69**, 171–188.

Turner, J. and J. Hodgson (1983) Studies in the vegetational history of the northern Pennines. III. Variations in the composition of the mid-Flandrian forests. *Journal of Ecology* **71**, 95–118.

Turner, J., V. P. Hewetson, F. A. Hibbert, K. H. Lowry and C. Chambers (1973) The history of the vegetation and flora of Widdybank Fell and the Cow Green reservoir basin, Upper Teesdale. *Philosophical Transactions of the Royal Society of London* **B265**, 327–408.

Usinger, H. (1975) Pollenanalytische und stratigraphische Untersuchungen an zwei Spätglazial-Vorkommen in Schleswig-Holstein. *Mitteilungen der Arbeitsgemeinschaft Geobotanik in Schleswig-Holstein und Hamburg* **25**, 1–183.

Vail, P. R., R. M. Mitchum and S. Thompson (1977) Global cycles of relative changes of sea level. In *Seismic stratigraphy: applications to hydrocarbon exploration*, C. E. Payton (ed.), 83–89. American Association of Petroleum Geologists Memoirs 26. Tulsa: American Association of Petroleum Geologists.

Valentine, K. W. G. and J. B. Dalrymple (1975) The identification, lateral variation and chronology of two buried paleocatenas at Woodhall Spa and West Runton, England. *Quaternary Research* **5**, 551–590.

Vandenberghe, J. (1985) Paleoenvironment and stratigraphy during the Last Glacial in the Belgian – Dutch border region. *Quaternary Research* **24**, 23–38.

Vandermeersh, B. (1978) Etude préliminaire du crâne humain du gisement paleolithique de Biache-Saint-Vaast. *Bulletin de l'Association Française pour l'Etude du Quaternaire* **1–3**, 65–67.

van der Hammen, T. (1951) Late Glacial flora and periglacial phenomena in the Netherlands. *Leidsche Geologische Mededelingen* **17**, 71–183.

van der Hammen, T., T. A. Wijmstra and W. H. Zagwijn (1971) The floral record of the Late Cenozoic of Europe. In *The Late Cenozoic glacial ages*, K. K. Turekian (ed.), 391–424. New Haven: Yale University Press.

van der Hammen, T., G. C. Maarleveld, J. C. Vogel and W. H. Zagwijn (1967) Stratigraphy, climatic succession and radiocarbon dating of the Last Glacial in the Netherlands. *Geologie en Mijnbouw* **46**, 79–95.

van der Meulen, A. J. and W. H. Zagwijn (1974) *Microtus (Allophaiomys) pliocaenicus* from the Lower Pleistocene near Brielle, the Netherlands. *Scripta Geologica* **21**, 1–12.

van Montfrans, H. M. (1971) Palaeomagnetic dating in the North Sea Basin. *Earth and Planetary Science Letters* **11**, 226–235.

van Staalduinen, C. J., H. A. van Adrichem Boogaert, M. J. M. Bless, J. W. C. Doppert, H. M. Harsveldt, H. M. van Montfrans, E. Oele, R. A. Wermuth and W. H. Zagwijn (1979) The geology of the Netherlands. *Mededelingen Rijks Geologische Dienst* **31**, 9–49.

van Vliet-Lanoë, B. (1986) Micromorphology. In *La Cotte de St. Brelade 1961–1978: excavations by C. B. M. McBurney*, P. Callow and J. M. Cornford (eds.), 91–96. Norwich: Geo Books.

van Vliet-Lanoë, B. (1988) *Le rôle de la glace de segregation dans les formations superficielles de l'Europe de l'Ouest: processus et heritages*. Two volumes. Thèse de doctorat d'etat, Université de Paris I - Sorbonne; Centre de Géomorphologie du Centre National de la Recherche Scientifique, Caen.

van Wijngaarden-Bakker, L. H. (1974) The animal remains from the Beaker settlement at Newgrange, Co. Meath: first report. *Proceedings of the Royal Irish Academy* **74C**, 313–383.

van Wijngaarden-Bakker, L. H. (1985) Littletonian faunas. In *The Quaternary history of Ireland*, K. J. Edwards and W. P. Warren (eds.), 233–249. London: Academic Press.

Vasari, Y. (1977) Radiocarbon dating of the Lateglacial and early Flandrian vegetational succession in the Scottish Highlands and the Isle of Skye. In *Studies in the Scottish Lateglacial environment*, J. M. Gray and J. J. Lowe (eds.), 143–162. Oxford: Pergamon Press.

Vasari Y. and A. Vasari (1968) Late- and Post-glacial macrophytic vegetation in the lochs of northern Scotland. *Acta Botanica Fennica* **8**, 1–120.

Ventris, P. A. (1985) *Pleistocene environmental history of the Nar Valley, Norfolk*. Ph.D. Thesis, University of Cambridge.

Ventris, P. A. (1986) The Nar Valley. In *The Nar Valley and north Norfolk. Field guide*, R. G. West and C. A. Whiteman (eds.), 6–55. Coventry: Quaternary Research Association.

Vita-Finzi, C. (1986) *Recent earth movements: an introduction to neotectonics*. London: Academic Press.

Vogel, J. C. and H. T. Waterbolk (1963) Groningen radiocarbon dates, IV. *Radiocarbon* **5**, 163–202.

Vogel, J. C. and W. H. Zagwijn (1967) Groningen radiocarbon dates, VI. *Radiocarbon* **9**, 63–106.

von Post, L. (1916) Om skogsträdpollen i sydsvenska torfmosslager-följder. *Geologiska Föreningens i Stockholm Förhandlingar* **38**, 384.

von Post, L. (1918) Skogsträdpollen i sydsvenska torfmosslager-följder. *Förhandlingar Skandinaviske Naturforskermøte 1916* **16**, 433–465.

von Weymarn, J. A. (1974) *Coastline development in Lewis and Harris, Outer Hebrides, with particular reference to the effects of glaciation*. Ph.D. Thesis, University of Aberdeen.

von Weymarn, J. A. (1979) A new concept of glaciation in Lewis and Harris, Outer Hebrides. *Proceedings of the Royal Society of Edinburgh* **77B**, 97–105.

von Weymarn, J. and K. J. Edwards (1973) Interstadial site on the island of Lewis, Scotland. *Nature* **246**, 473–474.

Waechter, J. d'A. (1971) Swanscombe 1970. *Proceedings of the Royal Anthropological Institute* **1970**, 43–49.

Wainwright, G. J. and I. H. Longworth (1971) *Durrington Walls: excavations 1966–68*. Reports of the Research Committee of the Society of Antiquaries of London, Number 29. London: Society of Antiquaries of London.

Walcott, R. I. (1970) Isostatic response to loading of the crust in Canada. *Canadian Journal of Earth Sciences* **7**, 716–727.

Walcott, R. I. (1973) Structure of the earth from glacio-isostatic rebound. *Annual Review of Earth and Planetary Sciences* **1**, 15–37.

Walker, D. (1953) The interglacial deposits at Histon Road, Cambridge. *Quarterly Journal of the Geological Society of London* **108**, 273–282.

Walker, D. (1955) Late-glacial deposits at Lunds, Yorkshire. *New Phytologist* **54**, 343–349.

Walker, D. (1966) The Late Quaternary history of the Cumberland Lowland. *Philosophical Transactions of the Royal Society of London* **B251**, 1–210.

Walker, D. and H. Godwin (1954) Lake stratigraphy, pollen analysis and vegetational history. In *Excavations at Star Carr*, J. G. D. Clark, 25–68. Cambridge: Cambridge University Press.

Walker, M. J. C. (1975a) Late Glacial and Early Postglacial environmental history of the central Grampian Highlands, Scotland. *Journal of Biogeography* **2**, 265–284.

Walker, M. J. C. (1975b) Two Lateglacial pollen diagrams from the eastern Grampian Highlands of Scotland. *Pollen et Spores* **17**, 67–92.

Walker, M. J. C. (1975c) A pollen diagram from the Pass of Drumochter, central Grampian Highlands, Scotland. *Transactions of the Botanical Society of Edinburgh* **42**, 335–343.

Walker, M. J. C. (1977) Corrydon: a Lateglacial profile from Glenshee, south east Grampian Highlands, Scotland. *Pollen et Spores* **19**, 391–406.

Walker, M. J. C. (1980) Late-Glacial history of the Brecon Beacons, South Wales. *Nature* **287**, 133–135.

Walker, M. J. C. (1982) The Late-glacial and early Flandrian deposits at Traeth Mawr, Brecon Beacons, South Wales. *New Phytologist* **90**, 177–194.

Walker, M. J. C. (1984a) Pollen analysis and Quaternary research in Scotland. *Quaternary Science Reviews* **3**, 369–404.

Walker, M. J. C. (1984b) A pollen diagram from St. Kilda, Outer Hebrides, Scotland. *New Phytologist* **97**, 99–113.

Walker, M. J. C. and J. J. Lowe (1979) Postglacial environmental history of Rannoch Moor, Scotland. II. Pollen diagrams and radiocarbon dates from the Rannoch Station and Corrour areas. *Journal of Biogeography* **6**, 349–362.

Walker, M. J. C. and J. J. Lowe (1981) Postglacial environmental history of Rannoch Moor, Scotland. III. Early and mid-Flandrian pollen-stratigraphic data from sites on western Rannoch Moor and near Fort William. *Journal of Biogeography* **8**, 475–491.

Walker, M. J. C. and J. J. Lowe (1982) Lateglacial and early Flandrian chronology of the Isle of Mull, Scotland. *Nature* **296**, 558–561.

Walker, M. J. C., J. M. Gray and J. J. Lowe (1985) Introduction. In *Field guide. Isle of Mull, Inner Hebrides, Scotland*, M. J. C. Walker, J. M. Gray and J. J. Lowe (eds.), 1–11. Cambridge: Quaternary Research Association.

Walker, M. J. C., C. K. Ballantyne, J. J. Lowe and D. G. Sutherland (1988) A reinterpretation of the Lateglacial environmental history of the Isle of Skye, Inner Hebrides, Scotland. *Journal of Quaternary Science* **3**, 135–146.

Walsh, P. T. and E. H. Brown (1971) Solution subsidence outliers containing probable Tertiary sediment in north-east Wales. *Geological Journal* **7**, 299–320.

Walsh, P. T., M. C. Boulter, M. Ijtaba and D. M. Urbani (1972) The preservation of the Neogene Brassington Formation of the southern Pennines and its bearing on the evolution of Upland Britain. *Journal of the Geological Society of London* **128**, 519–559.

Warren, S. H. (1912) On a Late Glacial stage in the valley of the River Lea. *Quarterly Journal of the Geological Society of London* **68**, 213–229.

Warren, S. H. (1951) The Clacton flint industry: a new interpretation. *Proceedings of the Geologists' Association* **62**, 107–135.

Warren, S. H. (1955) The Clacton (Essex) channel deposits. *Quarterly Journal of the Geological Society of London* **111**, 283–307.

Warren, W. P. (1979) The stratigraphic position and age of the Gortian Interglacial deposits. *Bulletin of the Geological Survey of Ireland* **2**, 315–332.

Warren, W. P. (1985) Stratigraphy. In *The Quaternary history of Ireland*, K. J. Edwards and W. P. Warren (eds.), 39–65. London: Academic Press.

Washburn, A. L. (1979) *Geocryology: a survey of periglacial processes and environments*. London: Edward Arnold.

Waters, R. S. (1961) Involutions and ice-wedges in Devon. *Nature* **189**, 389–390.

Waters, R. S. (1964) The Pleistocene legacy to the geomorphology of Dartmoor. In *Dartmoor essays*, I. G. Simmons (ed.), 73–96. Exeter: Devonshire Association for the Advancement of Science.

Waters, R. S. (1974) Dartmoor: tors and periglaciation. In *Field handbook. Exeter*, A. Straw (ed.), 14–15. Exeter: Quaternary Research Association.

Waters, R. S. and R. H. Johnson (1958) The terraces of the Derbyshire Derwent. *East Midland Geographer* **2**, 3–15.

Waton, P. V. (1982) Man's impact on the chalklands: some new pollen evidence. In *Archaeological aspects of woodland ecology*, S. Limbrey and M. Bell (eds.), 75–91. BAR International Series 146. Oxford: British Archaeological Reports.

Waton, P. V. (1986) Palynological evidence for early and permanent woodland on the chalk of central Hampshire. In *The scientific study of flint and chert: Proceedings of the Fourth International Flint Symposium*, G. de G. Sieveking and M. B. Hart (eds.), 169–174. Cambridge: Cambridge University Press.

Waton, P. V. and K. E. Barber (1987) Rimsmoor, Dorset: biostratigraphy and chronology of an infilled doline. In *Wessex and the Isle of Wight. Field guide*, K. E. Barber (ed.), 75–80. Cambridge: Quaternary Research Association.

Watson, E. (1970) The Cardigan Bay area. In *The glaciations of Wales and adjoining regions*, C. A. Lewis (ed.), 125–145. London: Longman.

Watson, E. (1977) The periglacial environment of Great Britain during the Devensian. *Philosophical Transactions of the Royal Society of London* **B280**, 183–198.

Watson, E. and S. Watson (1967) The periglacial origin of the drifts at Morfa-bychan, near Aberystwyth. *Geological Journal* **5**, 419–440.

Watts, W. A. (1959a) Pollen spectra from the interglacial deposits at Kirmington, Lincolnshire. *Proceedings of the Yorkshire Geological Society* **32**, 145–152.

Watts, W. A. (1959b) Interglacial deposits at Kilbeg and Newtown, Co. Waterford. *Proceedings of the Royal Irish Academy* **60B**, 79–134.

Watts, W. A. (1964) Interglacial deposits at Baggotstown, near Bruff, Co. Limerick. *Proceedings of the Royal Irish Academy* **63B**, 167–189.

Watts, W. A. (1967) Interglacial deposits in Kildromin townland, near Herbertstown, Co. Limerick. *Proceedings of the Royal Irish Academy* **65B**, 339–348.

Watts, W. A. (1977) The Late Devensian vegetation of Ireland. *Philosophical Transactions of the Royal Society of London* **B280**, 273–293.

Watts, W. A. (1980) Regional variation in the response of vegetation to Lateglacial climatic events in Europe. In

Studies in the Lateglacial of north-west Europe, J. J. Lowe, J. M. Gray and J. E. Robinson (eds.), 1–21. Oxford: Pergamon Press.

Watts, W. A. (1984) The Holocene vegetation of the Burren, western Ireland. In *Lake sediments and environmental history: studies in palaeolimnology and palaeoecology in honour of Winifred Tutin*, E. Y. Haworth and J. W. G. Lund (eds.), 359–376. Leicester: Leicester University Press.

Watts, W. A. (1985) Quaternary vegetation cycles. In *The Quaternary history of Ireland*, K. J. Edwards and W. P. Warren (eds.), 155–185. London: Academic Press.

Webb, J. A. and P. D. Moore (1982) The Late Devensian vegetational history of the Whitlaw Mosses, south-east Scotland. *New Phytologist* **91**, 341–398.

Weir, A. H., J. A. Catt and P. A. Madgett (1971) Postglacial soil formation in the loess of Pegwell Bay, Kent (England). *Geoderma* **5**, 131–149.

West, R. G. (1955) The glaciations and interglacials of East Anglia: a summary and discussion of recent research. *Quaternaria* **2**, 45–52.

West, R. G. (1956) The Quaternary deposits at Hoxne, Suffolk. *Philosophical Transactions of the Royal Society of London* **B239**, 265–356.

West, R. G. (1957) Interglacial deposits at Bobbitshole, Ipswich. *Philosophical Transactions of the Royal Society of London* **B241**, 1–31.

West, R. G. (1961a) Vegetational history of the Early Pleistocene of the Royal Society borehole at Ludham, Norfolk. *Proceedings of the Royal Society of London* **B155**, 437–453.

West, R. G. (1961b) The glacial and interglacial deposits of Norfolk. *Transactions of the Norfolk and Norwich Naturalists' Society* **19**, 365–375.

West, R. G. (1963) Problems of the British Quaternary. *Proceedings of the Geologists' Association* **74**, 147–186.

West, R. G. (1968) *Pleistocene geology and biology: with especial reference to the British Isles*. First edition. London: Longman.

West, R. G. (1969a) A note on pollen analyses from the Speeton Shell Bed. *Proceedings of the Geologists' Association* **80**, 217–218.

West, R. G. (1969b) Pollen analyses from interglacial deposits at Aveley and Grays, Essex. *Proceedings of the Geologists' Association* **80**, 271–282.

West, R. G. (1970) Pollen zones in the Pleistocene of Great Britain and their correlation. *New Phytologist* **69**, 1179–1183.

West, R. G. (1972) Relative land – sea-level changes in south-eastern England during the Pleistocene. *Philosophical Transactions of the Royal Society of London* **A272**, 87–98.

West, R. G. (1977a) *Pleistocene geology and biology: with especial reference to the British Isles*. Second edition. London: Longman.

West, R. G. (1977b) Early and Middle Devensian flora and vegetation. *Philosophical Transactions of the Royal Society of London* **B280**, 229–246.

West, R. G. (1979) Further on the Flandrian. *Boreas* **8**, 426.

West, R. G. (1980a) *The pre-glacial Pleistocene of the Norfolk and Suffolk coasts*. Cambridge: Cambridge University Press.

West, R. G. (1980b) Pleistocene forest history in East Anglia. *New Phytologist* **85**, 571–622.

West, R. G. (1981a) Palaeobotany and Pleistocene stratigraphy in Britain. *New Phytologist* **87**, 127–137.

West, R. G. (1981b) A contribution to the Pleistocene of Suffolk: an interglacial site at Sicklesmere, near Bury St. Edmunds. In *The Quaternary in Britain: essays, reviews and original work on the Quaternary published in honour of Lewis Penny on his retirement*, J. Neale and J. Flenley (eds.), 43–48. Oxford: Pergamon Press.

West, R. G. (1987) A note on the March Gravels and Fenland sea levels. *Bulletin of the Geological Society of Norfolk* **37**, 27–34.

West, R. G. (1988a) The record of the cold stages. *Philosophical Transactions of the Royal Society of London* **B318**, 505–522.

West, R. G. (1988b) A pollen diagram from Norwich Crag at Outney Common, Bungay, Suffolk. *New Phytologist* **110**, 603–606.

West, R. G. and J. J. Donner (1956) The glaciations of East Anglia and the East Midlands: a differentiation based on stone-orientation measurements of the tills. *Quarterly Journal of the Geological Society of London* **112**, 69–91

West, R. G. and P. E. P. Norton (1974) The Icenian Crag of south-east Suffolk. *Philosophical Transactions of the Royal Society of London* **B268**, 1–28.

West, R. G. and B. W. Sparks (1960) Coastal interglacial deposits of the English Channel. *Philosophical Transactions of the Royal Society of London* **B243**, 95–133.

West, R. G. and D. G. Wilson (1966) Cromer Forest Bed Series. *Nature* **209**, 497–498.

West, R. G. and D. G. Wilson (1968) Plant remains from the Corton Beds, Lowestoft, Suffolk. *Geological Magazine* **105**, 116–123.

West, R. G., B. M. Funnell and P. E. P. Norton (1980) An Early Pleistocene cold marine episode in the North Sea: pollen and faunal assemblages at Covehithe, Suffolk, England. *Boreas* **9**, 1–10.

West, R. G., C. A. Lambert and B. W. Sparks (1964) Interglacial deposits at Ilford, Essex. *Philosophical Transactions of the Royal Society of London* **B247**, 185–212.

West, R. G., R. J. N. Devoy, B. M. Funnell and J. E. Robinson (1984) Pleistocene deposits at Earnley, Bracklesham Bay, Sussex. *Philosophical Transactions of the Royal Society of London* **B306**, 137–157.

West, R. G., C. A. Dickson, J. A. Catt, A. H. Weir and B. W. Sparks (1974) Late Pleistocene deposits at Wretton, Norfolk. II. Devensian deposits. *Philosophical*

Transactions of the Royal Society of London **B267**, 337–420.

Wheeler, A. (1977) The origin and distribution of the freshwater fishes of the British Isles. *Journal of Biogeography* **4**, 1–24.

Wheeler, A. (1978) Why were there no fish remains at Star Carr? *Journal of Archaeological Science* **5**, 85–89.

Wheeler, R. E. M. (1943) *Maiden Castle, Dorset*. Reports of the Research Committee of the Society of Antiquaries of London, Number 12. London: Society of Antiquaries of London.

Whitehead, P. F. (1989) The development and sequence of deposition of the Avon Valley river-terraces. In *The Pleistocene of the West Midlands. Field guide*, D. H. Keen (ed.), 37–41. Cambridge: Quaternary Research Association.

Whittle, A. (1978) Resources and population in the British Neolithic. *Antiquity* **52**, 34–42.

Whittow, J. B. (1976) The Wallingford Fan Gravels. In *Field guide to the Oxford region*, D. A. Roe (ed.), 44. Oxford: Quaternary Research Association.

Whittow, J. B. and D. F. Ball (1970) North-west Wales. In *The glaciations of Wales and adjoining regions*, C. A. Lewis (ed.), 21–58. London: Longman.

Wijmstra, T. A. (1969) Palynology of the first 30 metres of a 120 m deep section in northern Greece. *Acta Botanica Neerlandica* **18**, 511–527.

Wilkinson, I. P. (1980) Coralline Crag Ostracoda and their environmental and stratigraphical significance. *Proceedings of the Geologists' Association* **91**, 291–306.

Wilks, P. J. (1979) Mid-Holocene sea-level and sedimentation interactions in the Dovey estuary area, Wales. *Palaeogeography, Palaeoclimatology, Palaeoecology* **26**, 17–36.

Williams, R. B. G. (1964) Fossil patterned ground in eastern England. *Biuletyn Peryglacjalny* **14**, 337–349.

Williams, R. B. G. (1968) Some estimates of periglacial erosion in southern and eastern England. *Biuletyn Peryglacjalny* **17**, 311–335.

Williams, R. B. G. (1975) The British climate during the Last Glaciation: an interpretation based on periglacial phenomena. In *Ice ages: ancient and modern*, A. E. Wright and F. Moseley (eds.), 95–120. Geological Journal Special Issue Number 6. Liverpool: Seel House Press.

Williams, R. B. G. (1987) Frost weathered mantles on the Chalk. In *Periglacial processes and landforms in Britain and Ireland*, J. Boardman (ed.), 127–133. Cambridge: Cambridge University Press.

Willis, E. H. (1961) Marine transgression sequences in the English Fenlands. *Annals of the New York Academy of Science* **95**, 368–376.

Wills, L. J. (1938) The Pleistocene development of the Severn from Bridgnorth to the sea. *Quarterly Journal of the Geological Society of London* **94**, 161–242.

Wilson, A. T. (1964) Origin of ice ages: an ice shelf theory for Pleistocene glaciation. *Nature* **201**, 147–149.

Wilson, D. G. (1973) Notable plant records from the Cromer Forest Bed Series. *New Phytologist* **72**, 1207–1234.

Wilson, P. and R. Bateman (1986) Nature and palaeoenvironmental significance of a buried soil sequence from Magilligan Foreland, Northern Ireland. *Boreas* **15**, 137–153.

Wilson, P. and R. Bateman (1987) Pedogenic and geomorphic evolution of a buried dune palaeo-catena at Magilligan Foreland, Northern Ireland. *Catena* **14**, 501–517.

Wilson, P. and O. Farrington (1989) Radiocarbon dating of the Holocene evolution of Magilligan Foreland, Co. Londonderry. *Proceedings of the Royal Irish Academy* **89B**, 1–23.

Wilson, S. J. (1991) The correlation of the Speeton Shell Bed, Filey Bay, Yorkshire, to an oxygen isotope stage. *Proceedings of the Yorkshire Geological Society* **48**, 223–226.

Wintle, A. G. (1981) Thermoluminescence dating of late Devensian loesses in southern England. *Nature* **289**, 479–480.

Wintle, A. G. and J. A. Catt (1985) Thermoluminescence dating of Dimlington Stadial deposits in eastern England. *Boreas* **14**, 231–234.

Wintle, A. G. and D. J. Huntley (1982) Thermoluminescence dating of sediments. *Quaternary Science Reviews* **1**, 31–53.

Wirtz, D. (1953) Zur Stratigraphie des Pleistozäns im Westen der Britischen Inseln. *Neues Jahrbuch für Geologie und Paläontologie* **96**, 267–303.

Wise, S. M. (1980) Caesium-137 and Lead-210: a review of the techniques and some applications in geomorphology. In *Timescales in geomorphology*, R. A. Cullingford, D. A. Davidson and J. Lewin (eds.), 109–127. Chichester: Wiley.

Woillard, G. M. (1978) Grande Pile peat bog: a continuous pollen record for the last 140 000 years. *Quaternary Research* **9**, 1–21.

Woillard, G. M. and W. G. Mook (1982) Carbon-14 dates at Grand Pile: correlation of land and sea chronologies. *Science* **215**, 159–161.

Woldstedt, P. (1952) Interglaziale Meereshochstände in Nordwest-Europa als Bezugsflächen für tektonische und isostatische Bewegungen. *Eiszeitalter und Gegenwart* **2**, 5–12.

Woldstedt, P. (1966) Der Ablauf des Eiszeitalters. *Eiszeitalter und Gegenwart* **17**, 153–158.

Wood, A. and A. W. Woodland (1968) Borehole at Mochras, west of Llanbedr, Merionethshire. *Nature* **219**, 1352–1354.

Wood, S. V. (1848–1882) *A monograph of the Crag Mollusca*. Four parts; three supplements. London: Palaeontographical Society.

Wood, T. R. (1974) Quaternary deposits around Fremington. In *Field handbook. Exeter*, A. Straw (ed.), 30–34. Exeter:

Quaternary Research Association.

Woodcock, A. G. (1981) *The Lower and Middle Palaeolithic Periods in Sussex*. BAR British Series 94. Oxford: British Archaeological Reports.

Woodman, P. C. (1981a) A Mesolithic camp in Ireland. *Scientific American* **245**, 92–100.

Woodman, P. C. (1981b) Problems of the Mesolithic survival in Ireland. In *Proceedings of the Second Mesolithic in Europe Symposium, Potsdam 1978*, B. Gramsch (ed.), 201–210. Veröffentlichungen des Museums für ur-und Fruhgeschichte Band 14/15, 1980. Berlin: VEB Deutscher Verlag der Wissenschaften.

Woodman, P. C. (1985) Prehistoric settlement and environment. In *The Quaternary history of Ireland*, K. J. Edwards and W. P. Warren (eds.), 251–278. London: Academic Press.

Woodworth, P. L. (1987) Trends in UK mean sea level. *Marine Geodesy* **11**, 57–87.

Woolacott, D. (1920) On an exposure of sands and gravels containing marine shells at Easington, Co. Durham. *Geological Magazine* **57**, 307–311.

Woolacott, D. (1922) On the 60–foot raised beach at Easington, Co. Durham. *Geological Magazine* **59**, 64–74.

Wooldridge, S. W. (1927) The Pliocene history of the London Basin. *Proceedings of the Geologists' Association* **38**, 49–132.

Wooldridge, S. W. (1960) The Pleistocene succession in the London Basin. *Proceedings of the Geologists' Association* **71**, 113–129.

Wooldridge, S. W. and D. L. Linton (1955) *Structure, surface and drainage in south-east England*. Second edition. London: George Philip.

Worsley, P. (1966) Some Weichselian fossil frost wedges from east Cheshire. *Mercian Geologist* **1**, 357–365.

Worsley, P. (1967) Problems in naming the Pleistocene deposits of the north-east Cheshire Plain. *Mercian Geologist* **2**, 51–55.

Worsley, P. (1970) The Cheshire – Shropshire lowlands. In *The glaciations of Wales and adjoining regions*, C. A. Lewis (ed.), 83–106. London: Longman.

Worsley, P. (1977) Periglaciation. In *British Quaternary studies: recent advances,* F. W. Shotton (ed.), 205–219. Oxford: Clarendon Press.

Worsley, P. (1978) Chelford. In *Field handbook. Keele*, E. A. Francis, H. Davies, E. Derbyshire, M. P. Lee and P. Worsley, 29–36. Keele: Quaternary Research Association.

Worsley, P. (1980) Problems in radiocarbon dating the Chelford Interstadial of England. In *Timescales in geomorphology*, R. A. Cullingford, D. A. Davidson and J. Lewin (eds.), 289–304. Chichester: Wiley.

Worsley, P. (1985) Pleistocene history of the Cheshire – Shropshire Plain. In *The geomorphology of north-west England*, R. H. Johnson (ed.), 201–221. Manchester: Manchester University Press.

Worsley, P. (1987) Permafrost stratigraphy in Britain – a first approximation. In *Periglacial processes and landforms in Britain and Ireland*, J. Boardman (ed.), 89–99. Cambridge: Cambridge University Press.

Worsley, P. (1990a) Lichenometry. In *Geomorphological techniques*. Second edition, A. S. Goudie (ed.), 422–428. London: Unwin Hyman.

Worsley, P. (1990b) Radiocarbon dating: principles, application and sample collection. In *Geomorphological techniques*. Second edition, A. S. Goudie (ed.), 383–393. London: Unwin Hyman.

Worsley, P., G. R. Coope, T. R. Good, D. T. Holyoak and J. E. Robinson (1983) A Pleistocene succession from beneath Chelford Sands at Oakwood Quarry, Chelford, Cheshire. *Geological Journal* **18**, 307–324.

Wyatt, R. J. (1971) New evidence for drift-filled valleys in north-east Leicestershire and south Lincolnshire. *Bulletin of the Geological Survey of Great Britain* **37**, 29–55.

Wyatt, R. J., A. Horton and R. J. Kenna (1971) Drift-filled channels on the Leicestershire – Lincolnshire border. *Bulletin of the Geological Survey of Great Britain* **37**, 57–79.

Wymer, J. J. (1962) Excavations at the Maglemosian sites at Thatcham, Berkshire, England. *Proceedings of the Prehistoric Society* **28**, 329–361.

Wymer, J. J. (1965) Excavation of the Lambourn long barrow, 1964. *Berkshire Archaeological Journal* **62**, 1–16.

Wymer, J. J. (1968) *Lower Palaeolithic archaeology in Britain: as represented by the Thames Valley*. London: John Baker.

Wymer, J. J. (1974) Clactonian and Acheulian industries in Britain: their chronology and significance. *Proceedings of the Geologists' Association* **85**, 391–421.

Wymer, J. J. (1977) The archaeology of man in the British Quaternary. In *British Quaternary studies: recent advances*, F. W. Shotton (ed.), 93–106. Oxford: Clarendon Press.

Wymer, J. J. (1981) The Palaeolithic. In *The environment in British prehistory*, I. G. Simmons and M. J. Tooley (eds.), 49–81. London: Duckworth.

Wymer, J. (1985) *The Palaeolithic sites of East Anglia*. Norwich: Geo Books.

Wymer, J. (1988) Palaeolithic archaeology and the British Quaternary sequence. *Quaternary Science Reviews* **7**, 79–97.

Wymer, J. J. and J. Rose (1976) A long blade industry from Sproughton, Suffolk. *East Anglian Archaeology Report* **3**, 1–10.

Wymer, J. J. and A. Straw (1977) Hand-axes from beneath glacial till at Welton-le-Wold, Lincolnshire, and the distribution of palaeoliths in Britain. *Proceedings of the Prehistoric Society* **43**, 355–360.

Wymer, J. J., R. M. Jacobi and J. Rose (1975) Late Devensian and early Flandrian barbed points from Sproughton, Suffolk. *Proceedings of the Prehistoric Society* **41**, 235–241.

Yalden, D. W. (1982) When did the mammal fauna of the British Isles arrive? *Mammal Review* **12**, 1–57.

York, D. and R. M. Farquhar (1972) *The earth's age and geochronology*. Oxford: Pergamon Press.

Zagwijn, W. H. (1960) Aspects of the Pliocene and Early Pleistocene vegetation in the Netherlands. *Mededelingen Geologie Stichting* CIII, **5**, 1–78.

Zagwijn, W. H. (1961) Vegetation, climate and radiocarbon datings in the Late Pleistocene of the Netherlands. Part I: Eemian and Early Weichselian. *Mededelingen Geologische Stichting* Nieuwe Serie 14, 15–45.

Zagwijn, W. H. (1973) Pollen analytic studies of Holsteinian and Saalian beds in the northern Netherlands. *Mededelingen Rijks Geologische Dienst* Nieuwe Serie 24, 139–156.

Zagwijn, W. H. (1974a) The Pliocene – Pleistocene boundary in western and southern Europe. *Boreas* **3**, 75–97.

Zagwijn, W. H. (1974b) The palaeogeographic evolution of the Netherlands during the Quaternary. *Geologie en Mijnbouw* **53**, 369–385.

Zagwijn, W. H. (1974c) Vegetation, climate and radiocarbon datings in the Late Pleistocene of the Netherlands. Part II: Middle Weichselian. *Mededelingen Rijks Geologische Dienst* Nieuwe Serie 25, 101–111.

Zagwijn, W. H. (1975) Variations in climate as shown by pollen analysis, especially in the Lower Pleistocene of Europe. In *Ice ages: ancient and modern*, A. E. Wright and F. Moseley (eds.), 137–152. Geological Journal Special Issue Number 6. Liverpool: Seel House Press.

Zagwijn, W. H. (1979) Early and Middle Pleistocene coastlines in the southern North Sea Basin. In *The Quaternary history of the North Sea*, E. Oele, R. T. E. Schüttenhelm and A. J. Wiggers (eds.), 31–42. Acta Universitatis Uppsaliensis Symposium Universitatis Uppsaliensis Annum Quingentesimum Celebrantis, Number 2. Uppsala: University of Uppsala.

Zagwijn, W. H. (1985) An outline of the Quaternary stratigraphy of the Netherlands. *Geologie en Mijnbouw* **64**, 17–24.

Zagwijn, W. H. (1989) The Netherlands during the Tertiary and the Quaternary: a case history of coastal lowland evolution. *Geologie en Mijnbouw* **68**, 107–120.

Zagwijn, W. H., H. M. van Montfrans and J. G. Zandstra (1971) Subdivision of the 'Cromerian' in the Netherlands; pollen-analysis, palaeomagnetism and sedimentary petrology. *Geologie en Mijnbouw* **50**, 41–58.

Zalasiewicz, J. A. and P. L. Gibbard (1988) The Pliocene to early Middle Pleistocene of East Anglia: an overview. In *The Pliocene – Middle Pleistocene of East Anglia. Field guide*, P. L. Gibbard and J. A. Zalasiewicz (eds.), 1–31. Cambridge: Quaternary Research association.

Zalasiewicz, J. A. and S. J. Mathers (1985) Lithostratigraphy of the Red and Norwich Crags of the Aldeburgh – Orford area, south-east Suffolk. *Geological Magazine* **122**, 287–296.

Zalasiewicz, J. A., S. J. Mathers, M. J. Hughes, S. M. Peglar, R. Harland, R. A. Nicholson, G. S. Boulton, P. Cambridge and G. P. Wealthall (1988) Stratigraphy and palaeoenvironments of the Red Crag and Norwich Crag formations between Aldeburgh and Sizewell, Suffolk, England. *Philosophical Transactions of the Royal Society of London* **B322**, 221–272.

Zeuner, F. E. (1937) A comparison of the Pleistocene of East Anglia with that of Germany. *Proceedings of the Prehistoric Society* **3**, 136–157.

Zeuner, F. E. (1940) *The age of Neanderthal Man, with notes on the Cotte de St. Brelade, Jersey, C. I.* Institute of Archaeology, University of London, Occasional Papers Number 3. London: Institute of Archaeology.

Zeuner, F. E. (1946) *Cervus elaphus jerseyensis* and other fossils in the 25-foot beach of Belle Hougue Cave, Jersey, C. I. *Bulletin Annuel de la Société Jersiaise* **14**, 238–250.

Zeuner, F. E. (1959) *The Pleistocene period: its climate, chronology and faunal successions*. Second edition. London: Hutchinson.

Index

Aberdeen 269
  Ground Formation 54, 70, 72, 82, 83
  – Lammermuir Readvance 196
Abernethy 216
Aberystwyth 263
Abida secale 194
Abies 36, 43, 63, 65, 86–89, 96, 116, 117, 128, 157
Abingdon 168, 189
Abra 91
Aby Grange 185
Acer 62, 84, 115, 116, 126–129
  A. monspessulanum 128
Acheulian 69, 93, 95, 96, 103, 107, 115, 136
Achnanthes microcephala 246
  A. minutissima 245
  A. suchlandtii 245
Acicula 233
Acroloxus lacustris 233
Adelocera murina 244
Adventurers Land 260
Aegopinella 233
Africa 19, 134
Agabus wasastjernae 244
Aghnadarragh 156, 157, 200
  Interstadial 157
Agonum exaratum 156
Ailstone 114, 121
Airaphilus elongatus 135
Aire(dale) 22, 165, 186, 226, 248
Akera bullata 91
Alca torda 238
Alces alces 194, 207
  A. gallicus 57
  A. latifrons 66
Alcester 242, 244
Aldeburgh 39, 43, 45, 50, 261
Aldeby 47
Alderley Edge 145
Allerød Interstadial 178, 184, 187, 189, 194, 206
Alnus 36, 43, 45, 49, 57, 58, 62, 63, 84, 86–8, 125, 127, 128, 129, 221, 222, 225–229, 231, 245, 254
Alopex lagopus 169, 194
Alpine Orogeny 31, 33
Alvaston 134
Alyn 175
Amara alpina 147, 151, 164, 165, 192
  A. torrida 202
Amersfoort 160
  Interstadial 159, 160
Ammonia beccarii 248
Anancus avernensis 47
Ancholme 185, 233, 248, 260
Ancylus 89
  A. fluviatilis 189
Angle Tarn 252
Anguilla anguilla 239
Anisus vortex 167, 235
  A. vorticulus 131
Anobium 242
Anomoeoneis vitrea 246
Anotylus gibbulus 21, 117
Anser fabalis 153
Aonyx reevi 52
Aphodius 164, 168
  A. bonvouloiri 165
  A. quadriguttatus 244
Aplexa hypnorum 235
Apodemus sylvaticus 19, 66, 134, 237–9
Apoderus coryli 242
Archidiskodon meridionalis 46, 47, 52, 57, 67
Arctica islandica 204
Arctostaphylos 194
Ardleigh 62, 63
Ardnamurchan 33
Arenig Mountains 192
Armeria 150
Armiger crista 233, 235
Armthorpe 128, 135
Arnprior 267
Arpedium brachypterum 101, 157, 192
Arran 156, 204, 228, 229
Arrow (Warwickshire) 257
Arvicola cantiana 66, 67, 92, 133
  A. terrestris 237
Asaphidion cyanicome 202
Asia 24, 36, 116
Astarte borealis 47
  A. montagui 59
Asterionella formosa 245, 246, 248
Astralium rugosum 20, 132
Athelington 84
Atlantic Ocean 28, 38, 43, 45, 71, 76, 143, 144, 207, 272
Austerfield 128, 135
Avebury 214, 233
Aveley 109, 111, 120, 121, 126, 131–133, 135, 137, 276
Avon (Warwickshire – Worcestershire) 100, 257
  terraces 100, 113, 114, 131, 134, 136, 137, 147, 163, 164, 175
  (Hampshire)
  terraces 127, 195
Axe 107
Azeca goodalli (menkeana) 116, 118, 131
Azolla 88
  A. filiculoides 95

Bacon Hole 112, 113, 131, 133,

140–142, 152, 276
Badger Hole 169, 223
Baggotstown 88, 129
Baggy Point 140
Baginton
- Lillington Gravel 79, 98, 100, 101
Sand 79, 98, 100
Bain 127, 148, 165
Baker's Hole 107, 115, 120, 122
Balcom Canyon 38
Baltic Sea 45
Ballybetagh 202
Ballycroneen Till 200
Ballyline 88, 89
Ballynagilly 214, 230, 255, 256
Ballyscullion 230
Bamburgh 184
Banc-y-Warren 192
Bann 269
Bantega Interstadial 121
Bantry Bay 268
Barfield Tarn 225
Barford (Norfolk) 84
Barham
Soil 63, 73, 77
Coversand 77
Loess 77
Sands and Gravels 77
Bar Hill 175
Barnstaple 80
Barnwell Station 166
Barra Formation 203
Barrington 133, 136
Beds 125
Barynotus squamosus 244
Baschurch Pools 222
Basement Till 105, 106, 121, 178
Bavelian 54
Bawtry 128
Beaconsfield Gravel 52
Beaker Culture 214
Beaulieu (River) 127
Beauly Firth 267
Beeston (Norfolk) 55
Belfast
Lough 236
Ordnance Datum 268
Belgium 206
Belgrandia 131, 132
B. marginata 20, 89, 130–132, 151
Belle Lake 201
Belmont Till 105
Bembidion aeneum 164
B. atroviolaceum 244
B. elongatum 135
B. hasti 147
B. grisvardi 192
B. lapponicum 202
Benburb 88
Berula erecta 126
Berwyn Mountains 192
Betula 36, 43, 49, 57, 62, 63, 73, 84, 87, 88, 115, 116, 128, 129, 145, 147, 151, 152, 154, 155, 184, 186, 187, 190, 193, 195, 198, 199, 202, 207, 218, 220, 225–230, 253
B. nana 152, 154, 164–167, 174, 184, 194
B. pubescens 154, 187
Bewdley 175
Bidford-on-Avon 244
Bielsbeck 119
Bingley 226
Bog 186, 235, 248
Birmingham 102
Nechells 80, 87, 91–3, 102
Quinton 80, 87, 102, 106, 120
West Bromwich 242
Bishopbriggs 170
Bison 169
B. priscus 163, 194
Bithynia 89, 91, 131
B. leachi 235
B. tentaculata 89, 91, 131, 132, 233
B. troscheli 65, 66, 131
Bittium reticulatum 91
Blackhall (Castle Eden) 33, 62, 64, 66, 67
Blackhall Till 179
Blacklane Brook 223
Black
Mountain (Carmarthenshire) 222
Mountains 192
Park Gravel 79
Rock (Brighton) 139, 141
Blackwater 115
Blagdon 239
Blake Event 27, 140
Blashenwell 233
Blastophagus piniperde 147
Bleaklow 227
Blea Tarn (Langdale) 225, 245, 246
Blelham 246
Basin 245
Bog 187
Tarn 245
Blessington 200
Bloak Moss 229
Bobbitshole 111, 123, 130, 133, 135, 137, 138, 141
Bodmin Moor 193, 223
Bognor Regis 62, 70
Bølling Interstadial 178, 184, 186, 187, 189, 194, 206, 207
Boreaphilus henningianus 146, 156, 164
B. nordenskioeldi 155, 162, 164
Bos 192
B. longifrons 239
B. primigenius 115, 133, 207, 236, 238
Bosq d'Aubigny 36, 37
Bosworth Clays and Silts 102
Botaurus stellaris 238
Bothrideres contractus 135
Bournemouth 136, 195
Bourton-on-the-Water 168
Boxgrove 62–70, 96, 276
Boyn Hill Gravel 115
Boyne
Valley (Ireland) 214
Bay (Scotland) 106
Brachytemnus submuricatus 135
Bradybaena fruticum 131
Braeroddach Loch 250, 252
Bramerton 49, 58
Brancaster 261
Brandon (Warwickshire) 100, 164, 170
Brassington Formation 31, 33
Brean Down 193
Breckland 185, 221
Brecon Beacons 192, 193, 222
Bridlington Crag 59, 81, 91
Brimham Rocks 186
Brimpton 153, 159, 168, 194

Interstadial 154, 160
Bristol 80, 193
Channel 107, 263, 269
Brittany 205
Bridgnorth 175
Bridgwater Bay 263
Broads 261
Broadstairs 194
Brook 194
Brørup 160
Interstadial 159, 160, 162
Broughton Bay 152
Brown Bank Formation 159
Broxbourne 166
Brue 107
Brundon 111, 134, 136
Brunhes Epoch 27, 71
Brunssumian 37
Bubbenhall Clay 80, 100
Buccella inusitata 47
Bucephala clangula 238
Buchan 81, 88, 95, 106, 128, 156, 170, 195, 267, 275
Gravels 33, 54
Buckingham 61
Bufo bufo 238
Builth Wells 222
Bungay 49, 84
Burnham 78
Burnswark 216
Burren 229
Burtle Beds 107, 113, 132
Bury-St-Edmunds 79, 84
Bushley Green 113
Buteo buteo 238
Buxton 150
Buxus 88

Caccobius schreberi 135
Cairngorms 198, 199, 229, 253, 274
Calabria 38
Calanthus melanocephalus 163
Calcethorpe Till 103, 105
California 38
Calonectris diomeda 133
Cam 123, 125, 151, 166
Loch (Sutherland) 198
Cambridge 103, 123, 165, 166, 216
Histon Road 123, 130, 137, 141
Sidgwick Avenue 151
Canada 10, 22
Candidula crayfordensis 131
Candona angulata 135
C. candida 248
C. candona 248
C. cf *lozeki* 167
C. marchica 248
Canis familiaris 236, 240
C. lupus 116, 194, 238
Cannabis 222
Canna Formation 108, 109
Cannock Forest 217
Canterbury 216, 224
Capel Curig 190
Capreolus capreolus 66, 91, 238
Carabus arvensis 164
C. hortensis 163
Cardiff 192
Cardigan Bay 190, 192, 263
Cardium (Cerastoderma) 66, 91
C. edule 132, 236
Carinocytheresis 93
Carpinus 43, 45, 49, 58, 62, 63, 86, 88, 115, 117, 123, 125–129, 134, 135, 218, 221
Carya 37, 43
Carychium 233
C. tridentatum 233
Caryophyllaceae 163, 198
Arenaria ciliata 165, 168, 169
Cerastium alpinum 169
Viscaria alpina 163
Castle Law 216
Castlepook Cave 170
Castor fiber 66, 91, 92, 133–135, 238
Cat Hole 168
Cathormiocerus validiscapus 242
Catinella arenaria 154, 185, 194
Cawood 184
Celtic Sea 36, 37, 54, 109, 193, 200, 205, 270
Western Approaches 31
Central Rossendale 227
Cepaea (Helix) nemoralis 115, 142
Cernuella virgata 233
Cervus elaphus 133, 134, 194, 238
Cerylon 242
Chalky Till 78, 105, 275, 276
Channel
Islands 96, 139, 224
River System 76, 81
Tunnel 261
Chapel Point 260
Chard Gap 107
Charnwood Forest 102, 217
Chatteris (Block Fen) 126, 138, 141
Chelford 102, 145, 147, 160
Interstadial 63, 146, 147, 150, 152, 154, 159, 162
Sands Formation 147
Chelmsford 103
Chenopodiaceae 46, 47, 154, 198
Cherhill 233
Chesterfield 106
Cheviot Hills 181
Chillesford 49
Chiltern Drift 47, 137
Chilterns 37, 42, 45, 78
Chlamys varia 91
Christchurch (Hampshire) 136
Gravels 155
Church
Moor (Hampshire) 195
Stretton 175, 242, 244
Chydorus sphaericus 93
Cibicides lobulatus 47
Clactonian 93, 95
Clacton-on-Sea 78, 87, 91, 93, 95, 96
Channel Complex 92
Golf Course 92, 93
Clangula hyemalis 153
Clarach Bay 263
Clausilia bidentata 65, 116
C. pumila 20, 118, 131
Cledlyn 222
Clethrionomys glareolus 134, 238
Cletwr 222
Cleveland 215, 225
Clew Bay 268
Clieves Hills 189
Clifton Gorge 80
Clogher (Valley) 157
Clyde 156, 216
Beds 12, 204, 205
Firth of 195, 198, 204
Cocconeis diminuta 245
Cochlicella 233
Cochlicopa lubrica 154
Coelodonta 169

C. antiquitatis 107, 115, 163, 164, 169, 194, 236
Colchester 84, 133, 216
Coledale 187
Coleshill 164
College Farm Clay 58
Colnbrook 190
Colne
Point 261
(River) (Essex) 87
(Valley) (Buckinghamshire) 190
Colonsay 204, 270
Columella 233
C. columella 20, 151, 154, 164, 166, 169, 185, 189, 194, 195, 233
Colymbetes dolabratus 167
Combe Haven (Valley) 224, 258, 262
Compositae 57, 59, 154, 163, 170
Artemisia 150, 175, 184–186, 189, 193, 199, 202
Condover 175
Congleton 258
Connachtian 82, 108
Contorted Drift 76
Coolteen 201
Coombe Rock 107, 116, 120, 153, 177
Copford 91
Corbicula 131, 132
C. fluminalis 89, 91, 115–117, 130, 131
Cork Harbour 269
Cornus sanguinea 116
Corton 55, 75, 76
Beds 76
Interstadial 76
Corvus monedula 238
Corylus 33, 37, 43, 45, 58, 62, 63, 84, 86, 88, 115, 117, 125, 127–129, 218, 221–230, 253
C. avellana 128
Cotentin Peninsula 36
Cotswolds 79, 80, 137, 213
Court Hill Channel 80
Courtmacsherry Raised Beach 96, 108, 140, 200
Covehithe 47, 52, 55
Coventry 79, 98, 102, 137, 164, 252
Coygan Cave *152, 168*
Crag(s) 19, 22, 23, 33–4, 36–9, 40, 43, 46, 52, 59, 69
Basin 45, 55, 70
Coralline 33, 34, 36–9, 42
Norwich 33, 38, 39, 42, 46, 49, 50, 52, 55, 58
Red (Butley, Walton) 30, 33, 38, 41, 44–6
Nodule Bed 38
Weybourne 55, 58, 76
Craig - Cerrig - gleisiad 193
Craig-y-Fro 193
Cranesmoor 223
Craven District 120, 142, 150, 165
Crayford 111, 121, 131, 133, 136, 138
Creeting Formation 58
Creswell Crags 119, 150, 165, 185
Cricetus cricetus 66
Crickley Hill 216
Crocidura 111
Crocuta crocuta 111, 133, 134, 250, 294
Croll–Milankovitch cycles 271
Cromer 76
Forest-bed Series 38, 55
Formation 55, 59, 67, 126
Till Formation (North Sea Drift) 75–77, 79
Cromerian Complex 54, 61, 62, 64, 71–3, 83
Cropthorne 132
Crose Mere 175, 221
Crossbrae Farm (Buchan) 156, 170, 195
Cross Fell 186
Croyde 80
Crypticus quisquilicus 162
Cudmore Grove 87, 91–3
Cumberland Lowland 225
Cushendun 268
Cwm
Cywion 190
Idwal 190
Llydaw 190
Cybister laterali - marginalis 135
Cyclotella comensis 246
C. glomerata 246
C. kutzingiana 246
C. meneghiniana 248
Cygnus bewicki 66
C. cygnus 238
Cyperaceae 59, 88, 126, 127, 154, 163, 165, 170, 186, 187, 189, 194, 198
Carex bigelowi 164
Cyprideis torosa 93, 135
Cyprinotus salinus 163
Cytherissa lacustris 93
Cytheropteron testudo 38

Dama dama 66, 91, 109, 133, 134
Dane (River) 258
Darlington 178, 184, 226
Dartmoor 107, 193, 215, 223, 255
Dead Man's Cave 184
Dee (Aberdeenshire) 199, 231
Delamere Forest 217
Denekamp Interstadial 171
Denham 78, 87
Denmark 138, 141, 187, 206, 207
Deroceras 65
Derryvree 157, 170
Cold Phase 170
Interstadial Complex 170
Derwent (Derbyshire) 106, 134, 137, 150
(Durham-Northumberland) 227
Deverel–Rimbury pottery 215
Devil's Kneadingtrough 194
Devoke Water 245
Diacheila arctica 147, 155, 157, 164, 202
D. polita 147, 165, 167
Dicerorhinus 18, 134
D. etruscus 67
D. hemitoechus 92, 109, 111, 112, 116, 119, 133, 134, 153
D. kirchbergensis 111
Dicrostonyx 169
D. torquatus 133, 134, 164
Dierden's Pit 92
Dimlington 91, 171
Stadial 171, 178, 193
Din Moss 229
Diphasium (Lycopodium) alpinum 184
Discus 233
D. rotundatus 116, 131, 233
D. ruderatus 65, 131, 154, 233
Diss 79

Mere 185, 218
Domesday Book 217
Donacea semicuprea 116
Doncaster 81, 128, 139, 181, 244
Dornoch Firth 267
Dovey Estuary 263
Dowel Cave 238
Draba incana 163, 165
Dreissena polymorpha 235
Droitwich 163, 218
Drumskellan 268
Dryas 202
D. octopetala 170, 186
Dryophthorus corticalis 135
Dublin 218, 233
Dunbar 267
Dun Fell 186
Dungeness 59
Dunsmore Gravel 98
Durham City 179

Earith 151, 165
Earn 204
Earnley 88, 91, 93, 95
Easington 132, 139
Eastbourne 262
Eastend Green Till 78
East
Mersea 111, 134, 136
Moor (Derbyshire) 227
Retford 106
Runton 58, 59
Tilbury Marshes Gravel 168
Easton Bavents 46, 47, 52, 59
Eburonian 54, 58
Eckington 132
Edale 189
Eden 106, 186
Edgworth 227
Eemian 105, 121, 126, 138, 140, 141, 159
Ehenside Tarn 225
Ehringsdorf 122
Elan (Valley) 192, 222
Elder Bush Cave 128, 133, 150
Ellesmere 175
Readvance 177
Elphidiidae 45
Elphidium 22
E. articulatum 135, 248
E. clavatum 22
E. frigidum 43
E. haagensis 43
E. magellanicum 248
E. obiculare 47
El. subarcticum 47
E. williamsonii 248
Elphidiella hannai 23, 46, 49
Elson's Seat 239
Elsterian 76, 82, 83, 103
Emys orbicularis 19, 92, 133, 134, 238
Enborne 154, 194
Encalypta ciliata 154
English Channel 22, 43, 68, 76, 81, 96, 107, 109, 132, 135, 141, 170
Fosse Dangeard 159
Strait of Dover 42, 76, 141, 262
Eppleworth 178
Equus 66, 111, 116, 119, 134, 169
E. ferus (*caballus*) (*przewalskii*) 91, 92, 100, 107, 133, 134, 153, 185, 194, 236
E. stenonis 46, 47
Ericales 43, 45–7, 49, 57, 58, 147, 151
Ericaceae 33
Bruckenthalia 130, 151
Calluna 151, 170, 218, 221, 229, 254
Empetrum 86, 130, 151, 169, 194, 198, 199, 202, 206, 221, 223, 225, 228, 229
Erinaceus europaeus 236
Erith 131
Errol Beds 20, 205
Erskine Bridge 156
Escrick Moraine 148, 181, 184
Esox lucius 238
Esthwaite Water 245, 246
Eston Nab 215
Eucladoceros falconeri 47, 52
Eucommia 37
Eucypris pigra 248
Eunotia 245, 246
Europe 14, 19, 24, 33, 36, 38, 54, 66, 67, 89, 96, 97, 100, 105, 122, 126, 131, 133–136, 146, 159, 164, 165, 187, 190, 192, 202, 206–208, 213–217, 231, 235, 236, 238, 239, 240, 242, 272
Evenlode (Valley) 52, 79
Evesham 98, 132
Exmoor 223, 256
Eye 84, 126, 132
Eynsham Gravel 115, 127, 131, 134, 136

Fagus 63, 218, 242
Fallahogy 230
Farmoor 189, 190
Farnham 24, 154, 169
Faroe Islands 204
Fawley 236, 248, 262
Felis sylvestris 238
Fen Clay 260
Fenland 95, 125, 260, 269
Basin 103, 165
Fennoscandia 146, 147, 154, 238, 240
Fermanagh Stadial 157, 170
Feronia blandulus 165
F. macra 192
Fforest Fawr 192, 193
Filey 91
Bay 105, 128, 178
Brigg 105
Filipendula 186, 187, 193, 229
Finland 146
Firth of Lorne 270
Fisherton 154
Fisherwick 244
Fladbury 164
Flanders 42
Moss (Scotland) 229
Flixton 184
Folkestone 194, 224, 233
Fordingbridge 195
Forest of
Arden 217
Dean 218
Forêt de la Londe 37, 42
Forth 216, 265, 267–9
Estuary 196, 204, 205
Foulness 261
Four Ashes 127, 135, 143, 147, 162, 163, 171, 174
Foyle 269
Fragilaria brevistriata 246
F. virescens 246
France 42, 121, 122, 134, 141, 155,

159, 242, 262
Fraserburgh 267
Fraxinus 62, 129, 218, 221, 228, 230
Fremington 193
Till 80
Frogholt 224
Frustulia 245, 246
Fulbeck 134
Fullerton 267

Gainsborough 106
Galley Hill 126, 131, 133
Galloway Hills 229
Galway Bay 268
Gaul 217
Gauss Epoch 38
Gavel Pot 150
Gazella anglica 52
Gentiana verna 184
Georyssus crenulatus 162
Gephyrocapsa carribeanica 71
G. oceania 71
Germany 69, 72, 83, 96, 97, 122, 138, 140, 141, 167, 206, 207
Gerrards Cross Gravel 52
Gibbula 91
Gimingham Sands 76
Gipping 58, 152, 185, 257
Glaciation 103
Till 77, 84, 103, 105
Glanllynnau 190, 191, 192
Glasgow 156, 195, 198, 204
Glaux maritima 152, 163, 166
Glenavy Stadial 200
Glenroy 265
Goodland Townland 256
Goole 128, 139, 260
Gort 82, 88
Gough's Cave 194
Gower Peninsula 106, 111, 120, 139, 140, 152, 168, 190, 192
Gramineae 43, 45, 46, 49, 57, 59, 87, 126–128, 147, 154, 163, 165, 170, 187, 189, 193, 195, 198, 202, 228
Glyceria maxima 116
Grampians 195, 198, 199, 228, 250
Grand Pile 160
Gransmoor Quarry 226
Grays 111, 126
Great
Billing 164
Blakenham 58, 77
Yarmouth 76, 261
Greenagho 157
Greenland 19
Grime's Graves 214, 238
Groenlandia densa 126, 166, 168
Grus grus 238
Guernsey 239
Gulo gulo 107, 150
Gunton Stadial 75
Gymnusa variegata 154
Gyraulus albus 20, 235
G. (Planorbis) laevis 20, 66, 131, 163
Gyrosigma acuminatum 245

Haliaëtus albicilla 238
Halling 194, 206
Hallstatt Culture 215, 216, 217
Happisburgh 55
Harkstead 111, 120, 133
Harleston 125
Hartlepool 106, 128, 184
Bay 260
Harwood Dale Moor 136
Haslingden 227
Hatfield (Hertfordshire) 87, 89
Hawks Tor 193
Haynesina germanica 135
Hebrides 139, 156, 159, 195, 203, 204, 228, 236, 238, 265
Sea 159
Formation 203
Hedera 43, 63, 127
Helianthemum 150, 185, 186
Helicella itala 155, 195, 233
Helix aspersa 233
Helophorus 101
H. glacilis 190
H. jacutus 147, 151
H. sibiricus 155, 162, 192
Hengelo Interstadial 171
Hengistbury Head 210
Heptaulacus testudinarius 244
Herpetocypris reptans 248
Hesperophanes fasciculatus 244
Hessle Till 178
Heterhelus scutellaris 117
Heterocypris salina 135
Hidden Interregnum 129
Higher Gravel Train 52, 61
High Lodge (Mildenhall) 62, 63, 67, 69, 70, 105
Hippeutis complanatus 233
Hippopotamus 18, 92, 109, 111, 112, 119, 133–5
H. amphibius 19, 115, 116, 133, 134
Hippophaë 84, 88, 89, 185
Hitchin 87
Hockham Mere 185, 218
Hoddesdon 78
Holderness 91, 139, 148, 178, 182, 226
Hollymount
(County Fermanagh) 157
Cold Phase 157
(County Laois) 33
Holme-next-the-Sea 261
Holmes Chapel 258
Holsteinian 96, 105, 121
Homo sapiens neanderthalensis 93, 115, 122, 155
H. sapiens sapiens 93, 192
Hope's Nose 113, 140
Homotherium latidens 67
Hoogeveen Interstadial 122
Horwich 174, 227
Howardian Hills 181
Hoxne 84, 89, 91–3, 95, 97, 119
Hull (River) 148
Humber 29, 181
Estuary 91, 255, 260
Hungary 69
Hungry Hill Gravels 151
Hunstanton 260
Till 178
Hutton Henry 128
Hyaena Den 155, 169
Stratum 133
Hydrobia 91, 95, 132
H. ulvae 45, 91, 236
Hydrocharis morsus-ranae 126
Hydrochus flavipennis 165

Ibsley 127
Iceland 143
Idle 128
Ilex 43, 63, 117, 127
Ilford 109, 111, 120, 126, 131, 133,

137, 138, 276
Ilfordian 119, 120
Ilyocypris gibba 154
Iping Common 254
Inchrory 233
Inverbervie (Burn of Benholm) 156
Ingham Sand and Gravel 79
Ipswich 77, 103, 123, 185, 257
Bramford Road 152
Irish Sea 42, 82, 109, 140, 153, 174, 205, 213, 239
Glaciation 80, 106, 175
Island Carr (Brigg) 260
Islay 204, 205
Ile de Ré 20
Isle of
Man 108, 157, 202, 213
Ballure 140
Glen Ballyre 202
Point of Ayre 108
Wight 70, 96, 195, 224, 269
Bembridge 63, 65, 66, 70, 127, 139, 141
Great Pan Farm 136, 155
Lane End 127
Isleworth 167, 168
Isorhipis melasoides 135
Italy 38

Jadammina macrescens 248
Jaramilo Event 71
Jersey 20, 25, 30, 224, 255, 258, 262, 263
Belcroute Bay 107, 113
Belle Hougue Cave 132, 134, 139, 142
Fliquet 155
La Cotte de St. Brelade 24, 107, 115, 122, 127, 137, 155
Le Marais de St. Pierre 248, 262
Portelet Bay 169
St. Aubin's Bay 156, 248, 252, 262
St. Brelade's Bay 155
Joint Mitnor Cave 133
Juncus gerardii 166
Juniperus 43, 88, 154, 166, 169, 170, 184, 185, 187, 190, 192–194, 198, 199, 202, 221–223, 228, 229
Jura 265
Formation 204, 270
Jutland 159

Kelvin 170
Kempton Park 24, 167
Gravel 167, 168
Kenilworth 137
Kenn 80, 132, 193
Kennet 95, 127, 154, 194, 210, 233
Kent's Cavern 62, 67, 69, 70, 136, 169, 194, 210
Kesgrave(s) 59, 61, 62, 73
Formation 52, 58, 59, 61, 79
Sands and Gravels 59, 62, 69, 73, 77, 79
Keskadale 187
Keswick 187
Kettering 61
Kilbeg 88
Kildale Hall 184, 225, 235, 248
Kildromin 88
Kingston-upon-Hull (Hull) 178, 260
Kilham 238, 255
Kilmaurs 156
Kinfauns 242, 244
Kintyre 204
Kirkdale Cave 133
Kirkham Abbey Gorge 181
Kirkhill 81, 88, 95, 106, 128, 137, 156
Kirmington 81, 88, 89, 91, 95, 103
Knock Fell 186
Koenigia islandica 156, 170

Lacerta vivipara 239
Lagurus lagurus 107
Lake
District 81, 95, 174, 184, 186, 187, 189, 225, 231, 245, 252, 254
Fenland 148, 149
Harrison 79, 98, 100, 102, 120
Humber 149, 181
Maw 107
Pickering 181
Lambourn 238
Langdale Fells 256
Langham 128, 135, 136, 139
Langley Silt Complex 190
Langney Point 262
Larnian 213
La Tene Culture 216, 217
Late Weichselian 206
Lauria cylindracea 112
Lea 166, 237
Marston 240, 242
Leamington Spa 63, 98
Le Castella 38
Leeds 136, 148, 165, 186, 226
Leicester 98, 102
Leiostyla 233
L. anglica 233
Lemmus lemmus 91, 92, 133, 134
Lenham Beds 31, 37
Lepe Cliff 137
Lepidurus arcticus 147
Lepus capensis 238
L. timidus 92
les Echets 160
Levallois(ian) 115, 120, 122, 136, 152, 155, 168
Leven (Yorkshire) 184, 225
Lewis (Isle of) 82, 195
Toa Galson 128, 156
Tolsta Head 170
Lexden 133
Limax 65
Limerick 200
Lincolnshire Marsh 139
Ling Bank Formation 95
Lingulodinium machaerophorum 136
Linton-Sutton Gravels 148
Linum perenne 168
Liquidambar 33, 36, 37
Little
Ice Age 272, 274
Oakley 62, 63, 65, 66, 67, 69–71
Paxton 152
Rissington 168
Sole Formation 36, 37, 54
Stretton 242
Woodbury 216
Littleton Bog 208, 230
Littorina 132, 236
L. littorea 236
Llanwern 139
Lleyn Peninsula 190, 205
Llyn
Clyd 246

Glas 246
Peris 250
Loch
Carron 197
Lomond 198, 250, 265, 267
Lomond Stadial 171, 174, 177, 181, 184, 186, 187, 189, 190, 192–199, 201, 204–206, 222, 228, 242, 252, 265
Maree 229
Sheil 265
Sionascaig 229, 246
Tarff 252
Loddon 233
Loe Pool 23, 248
London 109, 126, 190, 218
Colney 78
Long Hole 152, 168
Lopham Little Fen 185
Lorne Formation 270
Low
Buried Beach 263, 265
Furness 106, 186
Lower
Gravel Train 52, 61
Marsh Till 148
Pleniglacial 159
Wolston Clay 98, 100
Lough
Catherine 250
Erne 246
Foyle 268, 269
Nahanagan 202
Neagh 156, 200, 248, 250
Basin 250
Lowestoft 76
Till 76–80, 84, 102, 103, 105
Till Formation 76, 77
Stadial 76, 77
Lowland
Lonsdale 225
Craven 226
Ludham (borehole) 43–47
Lunds 186
Lutra lutra 238
Lycopodium 198, 199
L. (Huperzia) selago 155, 187
Lycopus europaeus 166
Lymnaea auricularia 235
L. palustris 233
L. peregra 66, 163, 233, 235
L. stagnalis 233, 235
Lynch Hill Gravel 115

Maastricht–Belvédère 122
Macaca 92
Macoma 91
M. arctica 49
M. balthica 20, 57, 59, 66, 71, 115, 132, 236
M. calcarea 45, 47, 199
Magilligan 269
Maglemosian 210
Maiden Castle 216
Maidenhall 136
Maidstone 194, 224
Main
Buried Beach 263, 265
Head 107
Lateglacial Shoreline 204, 265
Perth Shoreline 196, 204, 205
Postglacial Shoreline 12, 213, 265, 267–9
Postglacial Transgression 263, 265–268
Wester Ross Shoreline 198
Maldon Till 77, 103
Malham 186
Malin
Formation 108, 109
Sea 159, 205, 239
Mammuthus 116, 134, 153, 155
M. primigenius 100, 107, 109, 111, 115, 116, 134, 154, 156, 163–165, 175, 207, 236
M. trogontherii 109, 111
Mam Tor 215
Manchester 227
Manea 132
Manifold 106, 128, 150
Maplin Sands 261
March 260
Gravels 126, 131, 132, 135, 138
Marks Tey 23, 84, 89, 91, 93, 97
Marlborough Downs 47
Marlow 168
Marly Drift 77, 103
marnes à *Nassa* 36, 37
Marr Bank Formation 203
Marsworth (Pitstone) 116–121, 134, 177, 178, 276
Pitstone Soil 177, 178
Matuyama
Epoch 27, 38, 71
–Brunhes boundary 70
Maxey 125, 131
Mediterranean 36, 135, 242
Medway 78
Megaceros (*Megaloceros*) 133
M. giganteus 92, 133, 194, 202
M. verticornis 66
Meles meles 134, 236, 238
Melosira (*Aulacoseira*) 246, 248
M. westii 248
Melville Formation 205, 270
Menapian 54
Mendip Hills 155, 169, 193, 210, 223
Merevale Lake 250, 252, 253
Mersey Estuary 263
Merthyr Tydfil 222
Messingham 185
Sands 185
Meuse 61, 76, 213
Microtendipes 93
Microtus 66
M. agrestis 238
M. arvalis 239
M. nivalis 107
M. oeconomus 112, 116, 237
Middle Pleniglacial 171
Midland Valley 106, 156, 198, 199
Millammina fusca 248
Mimomys 52, 57, 67
M. blanci 47
M. pliocaenicus 47, 59
M. savini 66, 67
Milton Loch 216
Minchin Hole 111–113, 132, 133, 140, 142
(D/L) Stage 113, 140, 141
Minworth 164
Mochras (borehole) 31
Modiolus 66
Moershoofd Interstadial 171
Mole (Valley) 78
Monastirian 138
Montrose Basin 267
Mooar Till 108
Moray Firth 195, 265, 267–9
Mordon Carr 226
Morecambe Bay 139, 225, 255, 269
Moreseat 106

Moreton-in-Marsh 98, 100, 102, 105, 116
Morton Tayport 213, 238
Mosedale 187
Mother Grundy's Parlour 184, 227
Mount Sandel 213, 240
Mousterian 69, 136, 150, 152, 155, 168
of Acheulian Tradition 136, 152
Much Hadham 78
Mull (Isle of) 33, 195, 198, 204, 228, 265
Mullock Bridge 192
Mundesley 126, 133
Musca domestica 245
Mustela putorius 238
Mya truncata 66
Mynydd
Bach (Cardiganshire) 222
Illtydd 193, 222
Myotis bechsteinii 238
M. daubentonii 238
M. mystacinus 238
M. nattereri 238
Myrica 33, 128
Mysella bidentata 91
Mytilus 66, 91, 236
M. edulis 91, 132, 236
Myxas glutinosa 154, 235

Nadder 154
Nahanagan Stadial 201, 202
Najas flexilis 166
N. minor 123
Nant Ffrancon 190, 222
Nar Valley 87, 95, 126, 181
Clay 91, 93
Formation 89
Natrix natrix 238
Navicula dicephala 245
Nazeing 166, 237
Neasham 184
Fen 226
Needwood Forest 137
Nene 80, 164
Neogloboquadrina atlantica 38
Neptunea contraria 66
Neritoides Beach 111, 113, 140
Nesovitrea hammonis 154, 235
Netherlands 37–39, 42, 46, 54, 58, 61, 64, 71, 76, 97, 121–123, 138, 141, 159, 171, 206
Netley Heath 42, 45
Nettlebed 63
Gravel 37, 63
Interglacial deposit 63
Newer Drift 15
Newferry 229
New Forest 195, 223, 224
Newgrange 214
Newlands Cross 233, 239
Newmarket 103
Newport (Gwent) 192
Newtondale 181
Fen Bogs 226
Newtown (County Waterford) 88
New Zealand 235
Nidderdale 186
Normandy 36, 37, 42
North
America 24, 36, 169, 254, 272
Gill 226
Downs 31, 37, 42
Lechlade 168
Sea 20, 27, 29, 31, 33, 37, 38, 40, 42, 43, 45–47, 54, 61, 70, 76, 82, 83, 91, 95, 108, 138, 140, 159, 171, 178, 203, 213, 230, 260, 270
Dogger Bank 270
Forth Approaches 70
German Bight 270
Southern Bight 260, 270
Yorkshire Moors 81, 95, 106, 178, 181, 213, 215, 225, 226, 254, 255, 256
Northern (Plateau) Drift 47, 52, 60, 62, 78, 79, 137
Northfleet 107, 115, 131, 137
Northmoor 189
Notaris acridulus 116
N. aethiops 151, 154
Norway 19, 26, 47, 76, 203, 207
Norwich 75, 84
Brickearth 76
Norwegian Trench 203
Notodromas monacha 248
Nucula 91
Nucella lapillus 236
Nyssa 36

Oadby Till 79, 102
Oakwood Till 102, 147
Oban 204, 265
Ochotona pusilla 237
Ochthebius foveolatus 242
O. viridus 164
Odderade Interstadial 159, 160, 162
Odostomia unidentata 91
Older Drift 15
Older *Dryas* 186, 189, 193
Olophorum boreale 165
Oodes gracilis 135
Onthophagus fracticornis 244
O. nutans 244
O. opacicollis 135
Orford 36, 37, 261
Orkney 195, 199, 216, 228, 229, 231, 233, 238, 239
Maes Howe 214
Midhowe 239
Ring of Brodgar 214
Rousay 239
Skara Brae 214, 239
Oronsay 213, 236, 265
Orwell 123
Oryctolagus cuniculus 236
Ostend (Norfolk) 66–8
Ostrea 91, 236, 261
O. edulis 91, 132
Otiorrhynchus 101
O. arcticus 151
O. ligneus 162
O. nodosus 151
Oulton Beds 76
Ouse
(East Anglia) 80, 102, 117, 185
terraces 102, 117, 126, 151
(Sussex) 258
Ovibos moschatus 134
Oxbow 165
Oxford 62, 79, 115, 131, 134
Magdalen College 115, 127
Oxychilus cellarius 233
Oxyloma pfeifferi 154, 164, 168, 235
Oxytelus opacus 68

Paisley 204
Pakefield 62
Palaeoloxodon 133, 134
P. antiquus 18, 91, 92, 100, 109,

112, 119, 133, 134, 153
Palmae 33
Panthera 134
P. gombazogensis 67
P. leo 115, 116, 134
Paralia sulcata 248
Pararotalia serrata 43
Paris Basin 45
Parrett 107
Pas de Calais 42
Paston 58
Patella Beach 111–13, 133, 139, 142
Patella 236
P. vulgata 132
Paviland
Cave 192
Glaciation 106, 120
Moraine 106
Peak District 31, 33, 81, 106, 128, 150, 189
Peelo Formation 76
Pegwell Bay 194
Pelochares vesicolor 68
Pelocypris atabulbosa 118
Pelophila borealis 147, 164
Penkridge 174
Pennard (D/L) Stage 113, 140
Pennines 33, 213, 227, 231, 255–7
Penrith 187
Pentney 126
Sand and Gravel 181
Perca fluviatilis 238
Perforatella bidentata 66
Perth 205, 267
Readvance 196
Peterborough 87, 95
ware 214
Peterhead 81
Peterlee 179
Phalacrocorax carbo 66
Philonthus linki 163
Philorth 267
Phyllopertha horticola 244
Physa cf *acuta* 236
P. fontinalis 233, 235
Phytia myosotis 132
Picea 36, 43, 45, 57, 58, 62, 63, 87, 88, 97, 115–117, 128, 129, 145–147, 151, 152, 157, 159
P. abies 154, 160
P. omorika 160
P. omorikoides type 160
Picken's Hole 169
Pillaton Hall 174
Pilgrim's Lock 144
Pin Hole Cave 150, 165, 185
Pinnularia 245, 246
Pinus 33, 36, 39, 43, 45, 46, 49, 57, 58, 62, 63, 73, 84, 86–88, 97, 116, 117, 123, 125, 128, 129, 145, 147, 151, 152, 154, 185, 192, 218, 221, 222, 224–231, 253
P. sylvestris 154, 231, 250
P. longaeva 23
Pisidium 65, 91
P. amnicum 233
P. casertanum 66, 164
P. henslowanum 235
P. hibernicum 233
P. milium 233
P. moitessierianum 167, 233
P. nitidum 233
P. obtusale lapponicum 163
P. personatum 233
P. supinum 154
Pitymys 66
Planorbis planorbis 235
Plantago 128, 154
P. maritima 163
Pleasure Gardens Till 76
Pliomys episcopalis 67
Pohlia wahlenbergii var. *glacialis* 178
Polar Front 143, 198
Pollnahallia 43
Polygonum viviparum 164
Pontnewydd Cave 24, 115, 122
Poole 261
Populus 62, 228
Portland 113, 139, 210
East 22, 113, 132, 135, 139, 140
West 113, 155
Portlandia arctica 19, 205
Postglacial Climatic Optimum 272
Potamogeton distinctus 63
Potentilla 228
P. crantzii 165
Potomida littoralis 131
Poulton-le-Fylde 189
Praetiglian 39, 43
Prostomis mandibularis 244
Protelphidium anglicum 248
Proto-
Maglemosian 189, 210
Soar 98, 100, 102
Pseudamnicola confusa 132
Psychrodromus olivaceus 248
Pteridium 229
Pterocarya 36, 37, 43, 45, 46
Pterostichus blandulus 162
Potamopyrgus jenkinsi 235
Punctum pygmaeum 235
Pupilla muscorum 91, 112, 115, 116, 131, 151, 154, 155, 164, 166–9, 189, 194, 195, 233
Purfleet 126, 138
Purpurea 236
Pycnoglypta lurida 101, 155, 156, 189, 192
Pycnomerus terebrans 135
Pyramidula rupestris 112

Queensford 168
Quercus 23, 36, 43, 45, 49, 58, 62, 63, 84, 86, 88, 92, 115, 116, 125–129, 218, 221–230, 242, 254
Q. petraea 248
Q. robur 248
Quinqueloculina 22

Rangifer tarandus 100, 150, 156, 163, 184, 185, 194, 202
Rannoch Moor 195, 242, 250
Ranunculus platanifolius 189
Red Moss 174, 208, 221, 227, 230
Reindeer
Cave 170
Stratum 169
Reuverian 36, 39, 43
Rhaxella chert 47, 52
Rhododendron 88
Rhine 45, 61, 76, 89, 213
Rhynchaenus 92
R. foliorum 164
R. quercus 135, 242
Rhyncolus elongatus 157
R. strangulatus 157
Rhysodes sulcatus 244
Ribble 226
Ridgacre Till 106, 121

Rievaulx 217
Rimsmoor 224
Rinyo-Clacton (Grooved) ware 214
Ripple Brook 222, 258
Robin Hood's Cave 119, 150, 165, 185, 227
Rochford Channel 115
Rockhall Wood (Suffolk) 36
Rodbaston 174
Roddans Port 201, 208
Romano-British Silts 261
Roos 182
Roosting Hill (Beetley) 103, 126, 133, 151
Rothamsted 42, 45
Rother (Sussex) 261
Round Loch of Glenhead 246
Rubiaceae 170
Rubus chamaemorus 151
Rugby 98
Rumex 154, 175, 187, 189, 190, 192, 193, 198, 199, 202, 228
R. acetosa 185
Russia 146

Saalian 100, 103, 108, 121, 122
Salisbury 154
Plain 47
Salix 43, 58, 116, 154, 164, 169, 170, 185, 198, 199, 202, 225, 227, 228
S. herbacea 152, 157, 163, 164, 167, 168, 170, 187, 202
Salmo 91
S. salar 239
S. trutta 239
Salvinia natans 123
Salwarpe 163
Sangatte 141
Satwell Gravel 52
Saunton 80, 140
Sauveterrian 210
Saxifragaceae 155
Saxifraga oppositifolia 157, 165, 184
Scaleby Moss 225
Scandal Beck 128
Scandinavia 76, 108, 135, 163, 199 203, 206, 208, 217, 244, 254
Scandinavian Drift 106
Sciadopitys 33, 36, 43
Scilly Isles 36, 81, 96, 107, 153, 169, 193, 200, 205
St. Mary's 224
Sciurus vulgaris 238
Scolytus carpini 135
S. ratzburgi 147, 244
Scorpidium turgescens 154
Scottia browniana 68
Scottish Readvance 186
Scrobicularia 236
S. plana 91, 132, 236
Seamer
Carr 184
Carrs 184
Sea Mere 186
Second Storegga Slide 267
Seine 121
Selaginella 187
S. selaginoides 157
Selby 81, 148, 181, 184
Selsey 127, 132, 133, 136, 139, 141
Sequoia 36, 43
Serripes groenlandicus 45, 47
Settle 133, 150, 186
Severn 29, 102, 163, 257
Basin 80, 163, 167, 257, 258
Estuary 139
terraces 80, 91, 100, 113, 114, 137, 147, 163, 175, 177
Sewerby 105, 132, 134, 139, 149
Shackleton Formation 54, 82
Sheerness 269
Sheffield 106, 227
Shepperton Gravel 190
Sherburn-in-Elmet 136
Shetland 129, 195, 199, 203, 228
Fugla Ness 82, 88, 106, 128, 129
Jarlshof 215
Mousa 216
Murraster 129
Sel Ayre 128, 129
Shirdley Hill Sand Formation 189
Shoalstone 132, 139, 140
Shortalstown 129
Shrewsbury 175, 222
Shustoke 242
Siberia 10, 152, 168, 169
Sicklesmere 84, 93
Sidestrand 76
Simplocaria metallica 146, 156, 164
Sitophilus granarius 245
Sizewell 39, 43, 45, 47, 50
Skerryvore Formation 54
Skiddaw Formation 187
Skipsea
(Drab) Till 178, 179
Withow Mere 184
Skye 198, 199, 228
Slapton Ley 262
Sleaford 216
Sligo Bay 268
Slindon 96
Formation 96
Sands 63, 71, 96
Snowdonia 190, 192, 222
Soar 164
Solihull 102
Solway 186, 216
Firth 263, 265, 267, 268
Somerset Levels 214, 223, 263
Somersham 125, 138
Sorbus 228
Sorex araneus 237, 238
S. runtonensis 66
S. savini 67
Southampton Water 236, 262
Southend-on-Sea 78, 115
Southern Uplands 195
South
Irish End Moraine 108, 200
Street (Wiltshire) 233
Tyne 227
Uist 228, 229
Southwold 46, 55
Spain 242
Speeton 91
Shell Bed 91, 95, 105, 115, 128
Spermophilus 134
Sperrin Mountains 200
Spermodea 233
S. lamellata 131, 233
Spey 82, 106, 199, 228, 231
Sphaerites glabratus 244
Sphaerium corneum 233
S. lacustre 233
Sphagnum 151, 245, 246, 256
Sphenia binghami 91
Spiniferites 136
Spisula solida 45
S. subtruncata 45
Springfield Till 77, 84, 103

Sproughton 185, 210, 257
Spurn 260
  Point 105
St. Cross South Elmham 84
St. Erth 34, 37
  Beds 34, 36, 37
St. Germain temperate episodes 159, 160, 162
St. Ives 126
St. Kilda 82, 106, 195
  Hirta 170
St. Pierre-les-Elbeuf 121
St. Sauveur de Pierrepont 42
Stafford 174, 221, 222
  King's Pool 221
Staindale 258
Stainmore 178, 179, 181, 227
Standlake Common 168
Stanton Formation 159, 171, 203
Stanton Harcourt (Linch Hill) 115, 117–121, 152, 168, 276
  Gravel 115, 116, 120
  Channel 115, 116
Star Carr 184, 210, 213, 226, 236
Start Bay 262
Stephanodiscus 246
  S. astraea 23
Stevenage 77
  Fishers Green 87, 93
Steyne Wood Clay 63, 66, 68, 71, 96
Stirling 196, 229
Stockport Formation 147, 174
Stoke
  Goldington 102, 117–19
  Tunnel Bone-bed (Ipswich) 19, 111, 120, 133, 134, 136
Stomodes gracilis 21
  S. gyrosicollis 117
Stomoxys calcitrans 245
Stone (Point) 127, 132, 133, 137, 139
Stoneferry 260
Stonehenge 47, 214, 215
Stour
  (Essex-Suffolk) 123, 131, 134
  (Worcestershire) 177
Stourport 102, 147, 177
Stradbroke (borehole) 39, 43, 45, 46
Strangford Lough 268
Strata Florida 217
Stratford-upon-Avon 79, 102, 114
Strathearn 265, 266, 268
Strathmore 195
Stump Cross Cave 150, 276
Sturton 185
Stutton 111, 120, 121, 123, 131–133, 136–138
Suaeda maritima 152, 166
Sub-Basement Clay 81
Succinidae 65
Sugworth 62–68, 70, 79
Sun Hole 155, 169, 194, 223
Sus scrofa 207, 238, 239
Sussex Downs 215
Sutton Courtenay 168
Swaledale 186
Swallow Cliff (Middlehope) 132, 139
Swanscombe 24, 89, 92, 95–97, 107, 119
Swansea
  Bay 192
  Valley 222
Swanswell Pool 252, 253
Swanton Morley 103, 126, 131, 133, 137
Swarte Bank Formation 82
Sweden 23, 163, 206, 207
Symplocos 33
Synedra nana 245

Tabellaria 246
Tachinus arcticus 168
  T. jacutus 169
Tadcaster 184, 226
Talpa europaea 238
Tame 164
Tanousia (*Nematurella*) *runtoniana* 20, 65
Tantytarsus 93
Taplow Gravel 115
Tattershall 103, 127, 131, 135, 165
  Castle 127, 131, 132, 135, 139, 141, 142, 165
  Thorpe 141, 142, 165
Taxodiaceae 39, 43
  Taxodium 33, 36, 37
Taxus 84, 87, 88, 115, 126–129, 220, 230
Tay (side) 204, 268
  Estuary 196, 204, 206
Tees (dale) 178, 227, 260
  Estuary 260
Teindland 106, 128, 137, 169
Teme 258
Tendring Plateau 62
Tewkesbury 175, 222, 257, 258
Thalassiosira pseudonana 248
Thalictrum 185, 193
  T. alpinum 163–5, 169
Thames 29, 45, 52, 54, 61, 62, 63, 65, 76–78, 80, 81, 83, 84, 95, 105, 126, 127, 131, 133, 137, 138, 167, 168, 190, 213, 258
  Estuary 231, 261
  terraces 60, 73, 78, 79, 89, 100, 105, 109, 114–116, 119, 120, 136, 152, 167, 168, 189
Thatcham 127, 131, 210, 236
Thatcher Rock 113, 132, 139, 140
Threlkeld Formation 187
Thelnetham 185
Theodoxus serratiliniformis (*danubialis*) 89
Thetford 185, 214, 218
Thornaby 260
Thorne Moor 244
Thorpe Bulmer 184
Thrussington Till 102
Thurrock 111
  Little 126, 131, 138
  West 126, 138
Tiglian 37, 43, 46, 52
Tilia 62, 63, 84, 88, 128, 129, 218, 220–222, 225, 226, 228, 231, 254
Tilling Green 261
Timmia norvegica 154
Tipperary 200
Torbay 113, 132, 139
Tornewton Cave 107, 111, 133, 169
Tourville 121
Traeth Mawr 193
Trafalgar Square 109, 111, 126, 133–5, 141
Trebetherick Point 140, 193
Trechus rivularis 154
Tregaron 222
Tremadoc

Bay 31
Basin 31
Tremeirchion Caves 190
Trent 81
terraces 103, 137, 164, 177
Trichia hispida 89, 154, 194
Triglochin maritima 152, 163
Trimingham 77
Triturus vulgaris 239
Trochammina inflata 248
Trochus 132
Trogontherium cuvieri 92
Tropiphorus obtusus 244
Troutbeck Palaeosol 137
Truskmore Plateau 200
Trysull 80, 87
Tsuga 33, 37, 43, 45, 46, 49, 58
Turbonilla elegantissima 91
Turdus merula 238
Tweed 181
Estuary 260
Tyne (dale) 178, 216
Type X 89
Typha latifolia 187
Tyto alba 238

Uamh an Claonaite 142
Ullapool 197
Ulmus 36, 43, 45, 49, 57, 58, 62, 84, 116, 128, 129, 218, 222–230, 254
Ulster 157, 246, 255, 268
Unio littoralis 115
United States of America 23, 236
Unnamed (D/L) Stage 113, 140
Upper
Head 193
Marsh Till 178
Pleniglacial 206
Wolston Clay 98
Upton Warren 22, 163, 164, 167
Interstadial (Interstadial Complex) 162–171
Ure 186
Ursus 92, 116, 169
U. arctos 107, 165, 169, 194, 238
U. deningeri 67
U. spelaeus 67
Usk (River) 192
Uxbridge 78

Vale of
Brooks 224
Clwyd 190
Glamorgan 192
Pickering 148, 181, 184, 210, 226
St. Albans 77, 78
York 81, 106, 148–50, 178, 181, 184, 186, 226
Valgus hemipterus 68
Valley Farm Soil 14, 63, 69, 70, 77
Vallonia 233
V. costata 194
V. excentrica 131, 195, 233
V. pulchella 154, 194, 233
V. pulchella var. *enniensis* (*V. enniensis*) 130, 132
Valvata 91
V. cristata 89, 131
V. goldfussiana 65
V. naticina 89
V. piscinalis 66, 89, 91, 118, 131, 132, 233, 235
Vedde Ash Bed 203
Vertigo genesii 185, 194, 233
V. parcedentata 151
V. substriata 154
Victoria Cave 133, 136, 142, 150, 186
Villafranchian 38
Vitis 43
Vitrea contracta 233
Vitrinobrachium breve 66
Vrica 38
Vulpes vulpes 194, 238

Waalian 54
Wallingford Fan Gravel 78
Walton
Common 181
on-the-Naze 34, 39
Wandle 78
Ware 77, 78
Till 77, 78
Warren House Till 106, 121
Wash 80, 260, 269
–Fen Basin 81
Wateringbury 233
Watford 78
Waun-Fignen-Felen 222
Waveney 59, 125, 185
Waverley Wood Farm (Warwickshire) 62, 65–7, 70, 71
Weald 107, 194, 213, 217, 218
Weardale 178
Wear Till 179
Wee Bankie Formation 203
Weichselian (Wurmian) 159
Welton-le-Wold 103
Welton Till 103, 106, 121
Wensleydale 186
Wensum 126
Wessex 214, 215, 216
Culture 215
West
Angle Bay 80, 88
Drayton 115, 122, 190
Hartlepool 260
House Moss 225
Kennet 238
Mill Gravel 77, 78
Moor (near Doncaster) 181
Overton 214
Runton 47, 55, 59, 60, 61, 62, 66, 67, 71, 76, 276
Freshwater Bed 65
Westbury-sub-Mendip 59, 63, 67–72, 83, 276
Wester Ross Readvance 197, 198
Western Highlands 195, 204
Westland Green Gravel 52, 61
Westleton Beds 52, 59
Westward Ho! 236
Wey 78, 154
Weybourne 55, 103, 261
Wharfe (dale) 186, 226
Whitchurch (Shropshire) 175, 222
White Bog 235
Whixall Moss 222
Wicklow 200, 205
Mountains 200, 202
Willerby Wold 238
Willow Garth 184
Wimblingdon 132
Windermere 187, 189, 245, 250
Interstadial (Lateglacial Interstadial) 171, 174, 177, 178, 181, 184–187, 189, 194–196, 198, 199, 201, 202, 204–206, 272
Low Wray Bay 184, 187, 189
Wilsford 215, 244

Winchester 218, 224
Windmill Hill 214, 233, 238, 254
Windrush 258
Wing 102, 127, 129, 147
Wingham 224
Winter Hill Gravel 78, 79
Winterton Shoal Formation 82
Wirksworth 218
Wissey 125, 150
Witham 134
  terraces 103, 137
Withernsea (Purple) Till 178, 179, 182, 203
Wivenhoe 62
Wolf Crags Formation 187
Wolston 98, 100, 120
Wolvercote Channel 116
Wolverhampton 127, 174, 175
Wombourn Gravels 102
Woodgrange 201, 268
  Interstadial 201, 202, 205
Woodhall Spa 254
Woore 175
Woodston Beds 87, 89, 91
Worksop 106
Worldsend 244
Wortwell 125, 131, 133
Wragby Till 81, 103, 105, 127
Wrecclesham Gravel 154
Wretton 125, 130–132, 137, 138, 150
  Interstadial 150, 154, 159
Wrexham 175
  Delta Terrace 175
Wye (Derbyshire) 106
  (Herefordshire) 192, 222

Xenocyon lycaonoides 67

Yare 248
Yarmouth Roads Formation 61, 82
Yoldia (*Leda*) *myalis* 47, 49, 59, 66
  Leda myalis Bed 66
York 181, 218, 226, 245
  Moraine 148, 181
Yorkshire Wolds 47, 81, 106, 178, 184, 215, 216, 226, 255
Younger *Dryas* Stadial 206, 207
Ythan 267
Yugoslavia 160, 239

Zimioma grossa 244
Zonitidae 233